Sky Atlas 2000.0 Companion

2nd Edition

Sky Atlas 2000.0 Companion

2nd Edition

Robert A. Strong and Roger W. Sinnott

Descriptions and data for all 2,700 star clusters, nebulae, and galaxies

shown in Sky Atlas 2000.0, 2nd Edition,

by Wil Tirion and Roger W. Sinnott

Published by Sky Publishing Corporation
49 Bay State Road, Cambridge, Massachusetts 02138 U.S.A.   http://www.skypub.com

and by the Press Syndicate of the University of Cambridge
The Pitt Building, Trumpington Street, Cambridge, United Kingdom

Cambridge University Press
The Edinburgh Building, Cambridge CB2 2RU, U.K.   http://www.cambridge.org
40 West 20th Street, New York, NY 10011-4211, U.S.A.   http://www.cup.org
10 Stamford Road, Oakleigh, VIC 3166, Australia
Ruiz de Alarcón 13, 28014 Madrid, Spain
Dock House, The Waterfront, Cape Town 8001, South Africa

© 2000 Sky Publishing Corporation

All rights reserved. Except for brief passages quoted in a review, no part
of this book may be reproduced by any mechanical, photographic, or
electronic process, nor may it be stored in an information retrieval
system, transmitted, or otherwise copied for public or private use,
without the written permission of the publisher.

Printed in Canada

Library of Congress Cataloging-in-Publication Data

Strong, Robert A., 1951-
    Sky atlas 2000.0 companion, second edition / Robert A. Strong and Roger W. Sinnott.
        p.   cm.
    "A complete listing of the locations, characteristics, and special features of the 2,700
    deep-sky objects plotted in Sky atlas 2000.0, 2nd edition by Wil Tirion and
    Roger W. Sinnott."
    Includes bibliographical references and index.
    ISBN 0-933346-95-6 (alk. paper)
    1. Stars — Atlases. 2. Astronomy — Charts, diagrams, etc. I. Sinnott, Roger W. II. Tirion, Wil.
    Sky atlas 2000.0. III. Title.

G1000.2 S8 2000
523.8'022'3 — dc21
                                                                                        00-030011

A catalog record for this book is available from the British Library.

ISBN 0-933346-95-6 (Sky Publishing Corporation)
ISBN 0-521-00882-4 (Cambridge University Press)

Contents

vii	Preface
ix	Introduction
xviii	Bibliography
1	Alphabetical List of Deep-Sky Objects
228	*Sky Atlas 2000.0* Chart Key
231	Appendix: Objects Listed by Chart Number

Preface

How many times has someone come up to you at a star party, peered into your telescope, and said, "Wow! How far away is that thing?" or, "How big is it?" For anything other than the Moon or M31, your response was probably something like, "It's really, really far." Years ago I began scribbling notes on the sizes, distances, and other interesting facts about many deep-sky objects and carrying them with me to public star parties sponsored by my local Rockland Astronomy Club. That way I had the information at my fingertips. Sometimes a person would ask to see an object whose location I wasn't sure of. And because there was no index to my favorite set of star charts, Wil Tirion's *Sky Atlas 2000.0*, I had no simple way to find it. So I added chart numbers to my notes, too. As time went on I needed cross-references to other objects on each chart, so my "little project" kept growing. Before long I had accumulated a sizable amount of data.

For my 41st birthday in October 1992, I treated myself to a week at the Star Hill Inn in the mountains near Sapello, New Mexico. Anyone who's done that knows how totally relaxing the experience can be — beautiful days and wonderfully clear nights. It was there, on one of those perfect nights, that I got the idea to write my own book. The name *Sky Atlas 2000.0 Companion* came naturally enough, but getting it published was not as easy. After a few turn-downs I got up the nerve to go the self-publishing route. *Sky Atlas 2000.0 Companion* was born in June 1994.

To say the least, my first sale was exhilarating. The buyer actually asked me to sign his copy! Even more exciting was the satisfaction I got when people came up to me at star parties and said, "Hey, I really *use* your book!" Now *that* was cool!

About two years ago Rick Fienberg of Sky Publishing contacted me and asked if I'd be interested in doing a new companion to the second edition of *Sky Atlas 2000.0*. It had just been created jointly by Wil Tirion and Roger Sinnott, and I was pleased to find out from Rick that I'd be working with Roger on this project. Roger began by generating lists of objects from the same databases that were used to create the atlas itself. I then incorporated many of the notes from my first edition, updating them from more modern references as needed. Since the new edition of the atlas contains somewhat more deep-sky objects and a different, more homogeneous selection of them around the sky, there were quite a few cases that needed researching from scratch.

Users of the earlier *Companion* will notice an entirely new format and feel. These improvements are largely attributable to the book-crafting talents of project manager Sally MacGillivray and designer Lynn Sternbergh. We think you'll find the two-column arrangement, with short data lines and descriptive paragraphs immediately under them, easier to read. These paragraphs could have various lengths without causing layout problems, so we had the flexibility to wax on about many of the sky's most famous sights and include comments from skilled observers of the past and present.

Finally, this new *Companion* owes much to the eagle-eyed attention of E. Talmadge Mentall, who had played a key role working earlier with Roger on the *Millennium Star Atlas*. Tal made several meticulous passes through the entire manuscript, drawing on his long experience as an observer and paying careful heed to factual accuracy, style, consistency in cross-references, and countless other details.

We hope you get as much enjoyment out of using this *Companion* as we had putting it together.

Clear skies,

Bob Strong
August 2000

Introduction

Our aim in this book has been to provide a handy reference guide to every deep-sky (nonstellar) object plotted on the charts of *Sky Atlas 2000.0*, 2nd edition. We have brought together a wealth of information about every galaxy, star cluster, and nebula labeled in the atlas, including any alternate name(s) by which the object is likely to be known. But you won't find any mention (except in passing) of individual stars, including variable and double stars. Other works, such as the two-volume *Sky Catalogue 2000.0* or the *Celestia 2000* CD-ROM, should be consulted for these.

The main section of the book (pages 1–227) is a complete listing of these deep-sky objects, arranged in alphabetical order. Here is where to turn if you already have an object in mind and want to know three things: (1) whether or not it is shown in the atlas; (2) if so, what chart or charts it appears on; and (3) its celestial coordinates, size, brightness, and a description of its physical characteristics and appearance in a backyard telescope.

Next, the chart key on pages 228–229 is a miniature, simplified version of that found at the end of the atlas itself — a useful reminder of the chart layout.

Finally, the appendix on pages 231–281 alphabetically lists all the deep-sky objects again, this time by chart number. Each of the atlas's large charts (1 through 26) is dealt with in turn, followed by each of the close-up charts of selected regions (A1 through A5, B1, and B2). This section of the book can be of help when you want to explore a particular region of sky. Let's say you are already viewing the galaxy M33, which appears on Chart 4 of the atlas. You can turn to page 237 for a compact "hit list" of other objects that, because they are also on Chart 4, are similarly placed for viewing at the same time of night or season of the year. The appendix includes only the bare essentials: object name, type, celestial coordinates, apparent magnitude, and constellation. The coordinates allow you to quickly find each new object on Chart 4, and they may also be used with your telescope's setting circles or computerized Go To mount to find it in the sky. The magnitude column lets you scan for objects that might be easier targets than others. And since you have the new object's name, you can (if you wish) look up more information about it in the book's main section.

Since the chart key and appendix are largely self-explanatory to those using the atlas, the rest of this introduction will focus on the book's main section.

Names and Data Lines

Entries in the main section fall in two categories: primary and alternate. The primary name or designation is always one taken from the actual charts of the atlas, and under this name we give full information as to object type, coordinates, size, magnitude, and description. If there are additional names by which an object is commonly known or listed in other observing guides, these names are considered alternates and are included with cross-references to the primary entry.

When several designations for the same object are given in the atlas itself, we used a simple rule to pick the primary one. It is always the popular or common name (if actually shown on the charts) or, failing that, the Messier (M) number. Otherwise the primary name is some other designation actually used on the charts. For example, Chart 8 shows an object bearing three labels: Ring Nebula, M57, and 6720 (for NGC 6720). By our precedence rule, Ring Nebula is the primary name under which the full information may be found. If you look up M57, NGC 6720, or a less familiar name like Donut Nebula or Smoke Ring Nebula, you are directed to "*See* Ring Nebula." Many alternate names and designations are included in this *Companion* whether or not they are found in the atlas itself.

Keep in mind that, unlike the atlas, we include the prefix NGC for all numbers from J. L. E. Dreyer's *New General Catalogue*. Similarly, we use the prefix IC (rather than I.) for those from Dreyer's first or second *Index Catalogue*. Galaxies from Peter Nilson's *Uppsala General Catalogue* are prefixed UGC (rather than U).

Immediately after the name, the first data line contains the object's type, constellation, right ascension, declination, and angular size. The types and

Introduction

their abbreviations are explained on pages xi–xvi, and a table of the standard three-letter abbreviations of the constellations is found on page xvii. Like the coordinates shown in the atlas, all right ascensions and declinations are for equinox 2000.0. Most of the angular sizes are expressed in arcminutes ($'$), but those of very large objects are given in degrees (°), and those of very small ones in arcseconds ($''$). There are 60$''$ in 1$'$, and 60$'$ in 1°. For ill-defined bright and dark nebulae, sizes are stated less precisely than for galaxies, whose long and short dimensions have usually been measured with greater care. In a few cases the diameter is preceded by "<" (less than) or ">" (greater than), if an object has proved difficult to measure accurately.

The second data line contains the chart number(s) on which the object is found, its magnitude, and in the case of a galaxy the position angle (p.a.). Our aim throughout has been to provide visual, photovisual, or photoelectric V magnitudes of objects whenever possible, because these are measured in wavelengths nearly matching those to which the dark-adapted human eye is most sensitive. All such values are flagged "m_v" in the data lines of the main section and listed as default values in the descriptions and the appendix. If only a photographic or photoelectric B value is available, it is preceded by "m_p" or "phot." in the main section and followed by "p" in the appendix. Because these values have been measured in blue light, they are less reliable for judging how easily the object may be seen in a telescope. Often a poorly determined value is rounded to the nearest whole magnitude. A colon (:) following a magnitude also denotes uncertainty. Dark nebulae have no magnitude, of course. Oddly enough, the magnitudes of bright nebulae have never been systematically measured by astronomers — a worthwhile project waiting to be undertaken.

Position angles are listed for all galaxies in the atlas except those that are nearly face on to our view, hence circular. This angle describes the orientation of the long axis, measured counterclockwise on the sky from celestial north around through east and south. For example, p.a. 45° means the galaxy extends toward the northeast (and also toward the southwest at 225°), while p.a. 90° means it runs precisely east and west (toward 90° and 270°). The value less than 180° is the one listed here.

Descriptive Paragraphs

The heart and soul of this book are the descriptive paragraphs following entries in the main section. We have done our best to organize this information in a fairly predictable sequence, as follows: (1) alternate or popular names, if any; (2) directly observed characteristics, such as shape or star count; (3) quotes from well-known observers; (4) quantities that have been indirectly deduced, such as age and distance; (5) coded subdivisions of an object's type, such as the Hubble class for a galaxy; and (6) other objects in or near the same telescopic field.

Our general characterizations come in part from the coded remarks in *NGC 2000.0* or the data columns of *Sky Catalogue 2000.0, Vol. 2*. But these clues to visibility are hardly homogeneous, made as they were by various astronomers using a wide range of equipment. What looked "bright" to an early observer like William or John Herschel with their large reflectors (the original source of many *NGC* descriptions) may be all but invisible in a modern 3-inch. Objects we now know to be galaxies were sometimes called "resolvable" in the *NGC*, but the old-time observers were surely misled by a mottled appearance due to nebulous knots, and we have tried to suppress such errors. On the other hand, if the *NGC* called a globular cluster resolvable, it probably is — if not visually in a backyard telescope, then certainly with a CCD imaging system on one.

The distance to a celestial object in light-years (ly) is often the very first question asked by a newcomer to the hobby. But the distance is the single quantity often hardest to pin down, despite all the techniques and theories in the arsenal of modern astronomers. The values quoted here are tentative and subject to revision, often by 20 percent or more. The same caution applies to an object's physical extent in light-years. When it is given, we have deduced it from the object's tabulated angular size and distance. (There is more about the distances to galaxies on pages xi–xii.)

A remark from a well-known observer is always set

apart in a sentence by itself, with that person's last name in parentheses. These are not necessarily literal quotations, for we have often abridged them or adjusted the punctuation and style to improve readability. But they are the observer's own words, as found in their published works listed in the bibliography on pages xviii–xix. After all, a thoughtful visual description of a deep-sky object from 150 years ago can be just apt as one written today, despite the fact that early observers lacked our modern understanding of the object's intrinsic nature.

To make sense of these remarks you need to have some idea of each observer's place in history and the size of the telescope used. When he counted the stars in Praesepe (the Beehive Cluster), the great Italian astronomer Galileo Galilei (1564–1642) had only a crude refractor with an effective aperture of less than 40 millimeters (1.6 inches). During his lifetime, the French comet hunter Charles Messier (1730–1817) used more than a dozen telescopes, including several 3½-inch refractors. Messier's largest telescope was a 7½-inch Gregorian reflector, but it probably had no more light grasp than a modern 5-inch, given that its mirrors were made of old-fashioned speculum metal.

During what has been called the Golden Age of British amateur astronomy, William Henry Smyth (1788–1865) made the notes for his *Bedford Catalogue* using an excellent 5.9-inch refractor. Thomas W. Webb (1807–85), who wrote about what could be seen in "common telescopes," stated that he had in mind refractors in the 3- to 5-inch range. William F. Denning (1848–1931) often told how things looked in his 10-inch silvered-glass reflector, probably the equivalent of a good 10-inch today.

The American writer Garrett P. Serviss (1851–1929) was fond of describing what could be seen with the naked eye or modest opera glasses; modern binoculars will show much more. Among the more recent observers we've quoted, Leland S. Copeland (1886–1973) and Walter Scott Houston (1912–93) wrote broadly about what could be seen in 3- to 12-inch telescopes. But while penning modern visual descriptions for their Messier books, John H. Mallas used a 4-inch Unitron refractor and Stephen James O'Meara a 4-inch Tele Vue Genesis refractor, intentionally limiting themselves to about the same light-gathering power available to Messier himself.

Types of Objects

Sometimes the popular name of a celestial object conveys its type as understood by modern astronomers. This is true, for example, of the Dumbbell Nebula, Beehive Cluster, and Sombrero Galaxy. Such names, however, are not an infallible guide. Bode's Nebulae and Coddington's Nebula are carryovers from before the time when external galaxies were recognized as such, and names like the Coal Sack, Omega Centauri, or White-Eyed Pea may seem downright mysterious.

Equally uninformative are the designations in the widest use. Messier's famous catalog (prefix M) includes such disparate celestial sights as the open cluster M45 (Pleiades), the globular cluster M13, the bright nebula M42, and the galaxy M51, along with a handful of other types. The same is true of the NGC (*New General Catalogue*) and IC (*Index Catalogue*) objects. Caldwell numbers, from the popular new listing created in 1995 by British astronomy popularizer Patrick Moore, are preceded by C and included as cross-references in this book even though they are not used in the atlas itself.

However, many other designations *do* help to identify an object's type. They come from special studies done by a particular astronomer or undertaken at a specific observatory. The meanings of all the designation prefixes used in this book are explained under the type headings below.

Galaxies (Gx). Apart from those bearing NGC or IC numbers, designations of galaxies can begin with E (European Southern Observatory), F (Fairall), M-(*Morphological Catalogue of Galaxies)*, Mrk (Markarian), UGC (*Uppsala General Catalogue)*, II Zw and III Zw (Zwicky). On the charts of the atlas, NGC is omitted as a prefix because it is so common, IC is abbreviated I., and UGC is shortened to U.

For most of the galaxies in this book, tentative distances are given in millions of light-years. But the cosmic distance scale is very much a focus of modern research, and the figures given in this book must be taken as educated guesses. For all but the nearest galaxies, the distances we cite are based on the measured

Introduction

redshifts of their spectral lines. Early in the 20th century, the American astronomer Edwin Hubble established that the farther away a galaxy is, the higher the velocity at which it is receding from our own Milky Way galaxy as part of the general expansion of the universe. Even though Hubble's law is a widely accepted feature of modern cosmology, the exact value of the proportionality constant, H_0, has been hotly debated since Hubble's day. His own estimate in 1929 was that a galaxy recedes at 500 kilometers per second for each megaparsec (3.26 million light-years) of distance from our own Milky Way. But later studies showed this value to be way too high. By the 1970s most astronomers favored a value near 55, but more recent work has pushed H_0 back up to around 70 or 75. For the distances to galaxies in this book, we have adopted $H_0 = 75$. In other words, each galaxy's recessional velocity in km/sec has been divided by 75, then multiplied by 3.26 to express the result in millions of light-years.

But Hubble's law is not a reliable guide for the very nearest galaxies, such as those in the Local Group, which are close enough to our own Milky Way that any recessional velocity due to the universe's expansion is small compared to the galaxy's idiosyncratic motion as it interacts gravitationally with its neighbors. Fortunately these galaxies are so near that large ground-based telescopes and the Hubble Space Telescope (HST) can study individual stars within them. For a few Local Group members, we cite new values inferred from the distances to similar stars in our own galaxy, as measured by European Space Agency astronomers during their highly successful Hipparcos mission in 1989–93.

The galaxies in the Virgo Galaxy Cluster, with such familiar showpieces as M87 and M49, sit squarely in a problematical "middle ground" for astronomers. The members of this cluster are a little too far for their individual stars to be studied well, but they are not so far away that Hubble's law can serve as a useful yardstick. For some members we have adopted a value of 41 million light-years; for others we retained the Hubble-law distance if it did not deviate too much from that figure. Modern estimates of the Virgo cluster's distance range from about 40 to 75 million light-years, giving some idea of the uncertainties that still apply to the extragalactic distance scale at the dawn of the 21st century.

The upshot is that our use of two significant digits for distances to these and other galaxies (as in 63 or 280 million light-years) gives a false sense of the accuracy to which they are really known. But they do hint at these objects' *relative* distances.

Galaxies differ widely in appearance, as a glance at any HST portfolio or the *Carnegie Atlas of Galaxies* makes clear. To describe this profusion of forms and shapes and call attention to evolutionary clues, astronomers have long recognized three broad categories: ellipticals, E, which are spheroidal in shape and appear to consist of mainly older stars with little intervening gas or dust; spirals, S, whose stars, dust, and nebulous knots are arranged in a distinctly spiral pattern; and irregulars, Ir, which lack the symmetry of the other two.

In 1936 Hubble published his famous "tuning fork" diagram, an updated version of which appears in Figure 1. The ellipticals form a progression from E0 (nearly round) to E7 (highly flattened) and end with the lenticular form, S0. From here the true spirals diverge in two distinct branches, S and SB, based on the appearance of their central regions. Those with a simple nuclear bulge are called S, while those that include a bright central bar are SB, or S(B) if the presence of the bar is uncertain. A lowercase letter a, b, c, d, or more rarely m, can be added to a normal or a barred spiral's type to indicate subtler details of form. The sequence begins with galaxies having a very prominent bulge and tightly coiled spiral arms, and it ends with those showing hardly any nucleus at all and very loose, open arms.

Numerous galaxy specialists since Hubble have elaborated on this scheme, sometimes in a confusing or ambiguous manner. For example, one common way to indicate that a galaxy falls between Sa and Sb is to call it Sb–, but another notation for the same thing is Sab. Some researchers have worked on objects in one hemisphere only, so an all-sky compilation like ours is apt to bring in a mixture of classification schemes. For the purposes of this book a simplified approach seemed better. If a galaxy has been classified between two subdivisions, we have arbitrarily reassigned it to one or the other. We have also done away with Roman

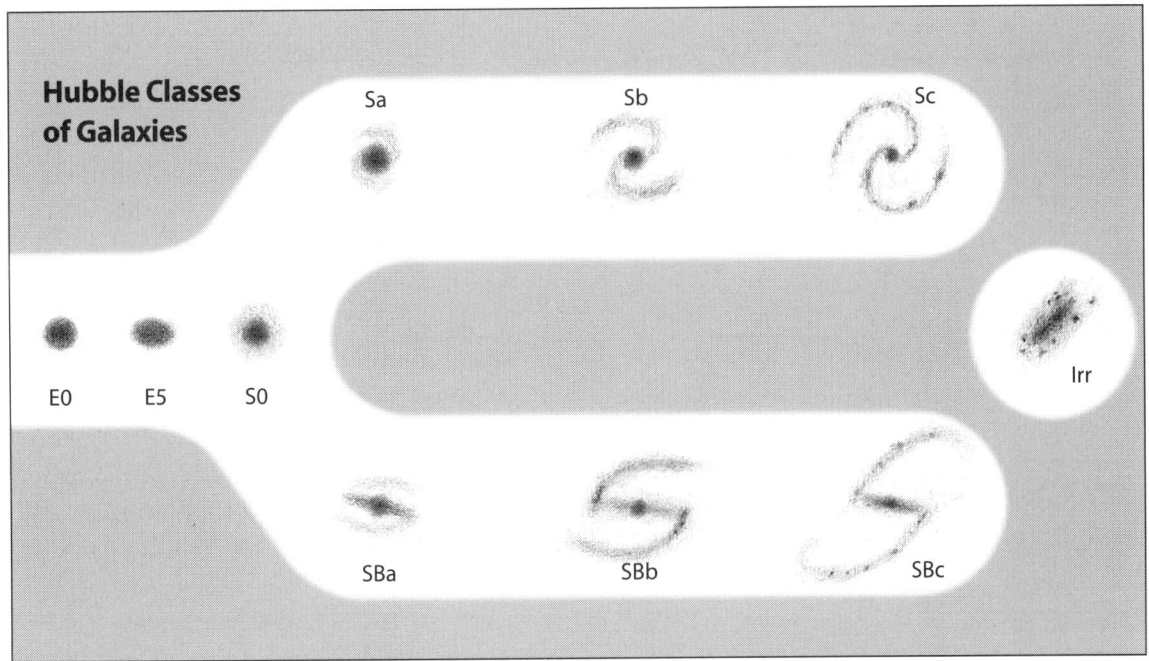

Figure 1. Galaxies of various Hubble classes, as sketched by Ronald J. Buta. This University of Alabama astronomer notes that galaxies of all types come in a wide variety of sizes. His purpose here is to show differences in shape, not size.

numerals indicating the David Dunlap Observatory luminosity class or, alternatively, the Ir type. A plus sign following a class (S+, Ir+) indicates it is resolvable into stars, a minus sign (S–, Ir–) that it is not. Peculiar galaxies are assigned to Hubble class P, or else flagged with a lowercase "p" added to some other class.

Planetary Nebulae (PN). Most planetary nebulae shown in the atlas are listed here by NGC or IC number, but some of these intriguing objects are better known by popular names: the Ring Nebula, Dumbbell, Owl, Helix, and so on. Messier listed only four planetaries, but the Caldwell catalog contains 13.

For many planetaries, we also include the PK (Perek and Kohoutek) designation, derived by concatenating the 1950.0 galactic longitude and latitude (both truncated to the nearest degree) with a decimal point and integer representing the object's sequential number within its 1°-by-1° sky area. These are numerically quite similar, but not identical, to the PN designations used in the newer compilations by Agnès Acker and colleagues. We sometimes include alternate designations that recognize the astronomer who discovered or first described the object: Abell, Baade, Cn (Cannon), Fg (Fleming), Hb (Hubble), He (Henize), Hu (Humason), J (Jonckheere), JE (Barbieri and Sulentic), M 1- through M 4- (Minkowski), Me (Merrill), Mz (Menzel), Na (Nassau, Stephenson, and Caprioli), PuWe (Purgathofer and Weinberger), Sp (Shapley), SwSt (Swings and Struve), and Vy (Vyssotsky). For most planetaries, our verbal descriptions rely in part on the Verontsov-Velyaminov (V-V) class, which is stated near the end of the paragraph and explained further in *Sky Catalogue 2000.0, Vol. 2.*

At summer star parties in the Northern Hemisphere, probably more people line up for a telescopic view of the Ring Nebula than anything else. But struggle as they might, very few ever catch sight of the central star that is so obvious on photographs. Lowell Observatory astronomer Brian A. Skiff notes that even such well-equipped 19th-century observers as John Herschel, Heinrich d'Arrest, Leopold Trouvelot, and Lord Rosse omitted the star in their drawings of the Ring. Asaph Hall, discoverer of the moons of Mars, failed to see it with the U.S. Naval Observatory 26-inch

Introduction

refractor. Paradoxically, however, observers with much smaller scopes have sometimes succeeded. As Skiff explains, "The bright background of the center of the nebula makes seeing the central star more difficult than would be the case if it were an isolated star." He adds that to pick it up in a telescope as small as a 10-inch requires not only quite high magnification (say, 500×) but subarcsecond seeing.

Don't expect to see the Ring Nebula's central star unless you can see even fainter stars nearby. For our chart (Figure 2), Skiff has provided V magnitudes of many stars in the immediate vicinity of the Ring, based on careful photoelectric measurements that he and Arne Hendon (U.S. Naval Observatory) made.

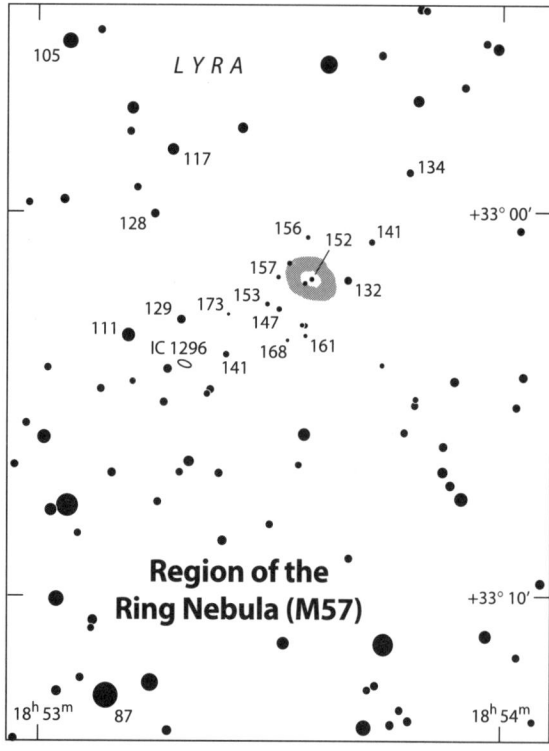

Figure 2. The region of the Ring Nebula in Lyra, with visual magnitudes for a number of faint stars in the same telescopic field. (To avoid confusion with stars, magnitude values are given to tenths with the decimal point omitted.) During summer months, this chart is handy for checking a telescope's magnitude limit, as well as for showing how close an observer is to glimpsing the nebula's central star. South is up. Data courtesy Brian A. Skiff.

In this observing guide, we have tried to include the central star's visual magnitude for as many planetaries as possible.

Open Clusters (OC, +C, C+). Some open clusters consist of more than 100 stars in a small patch of sky, forming a group so eye-catching that there can be no doubt the stars are physically associated in space. But other, poorer congregations may be chance alignments in the line of sight. Our type OC refers to objects that have been classed at one time or another as open clusters, based on their appearance in a telescope or on photographs. But some of the sparser ones, like Brocchi's Cluster in Vulpecula, are now known not to be true clusters after all (as our paragraph on this object explains).

Evidence of a physical connection is stronger if a cluster appears bathed in diffuse nebulosity — a lingering remnant of the interstellar gas and dust in which the stars were born, perhaps within the past million years or so. Sometimes this nebulosity is mentioned in the description, and if it is pronounced it may also be included in the type itself, with a notation like C+BNe or BNr+C. The C is put first if the object is more commonly thought of as a cluster than as a nebula.

By far the most famous example of a cluster-in-nebula is the Trapezium in Orion. Using a primitive 23-foot-long refractor in 1656, Christiaan Huygens was apparently the first astronomer to notice "three stars, nearly touching one another" at the center of the Great Orion Nebula. Then in 1684 he found that the group consisted of four stars, not three. As recounted in a 1960 article by Joseph Ashbrook, the Trapezium has been thoroughly studied as a multiple system ever since, the four main stars being designated A, B, C, and D from west to east, as shown in the diagram at right (Figure 3). In 1826 Wilhelm Struve added a fifth star, E, and within six years John Herschel caught sight of a sixth, F, with his 18-inch reflector. These faint attendants still challenge observers today, some of whom have glimpsed them with much smaller telescopes. *Sky & Telescope* columnist Sue French reports, "With my 4.1-inch Astro-Physics Traveler at 87×, under decent seeing, I find both E and F to be rather easy. Under mediocre conditions, I sometimes need to boost the power to 127× to hold F steadily in view. If the seeing is poor, F disappears — but it has to be really rotten before E is rendered invisible." (Both A and B, by the way, are eclipsing binary stars.)

In 1888, using the just-completed 36-inch Lick Observatory refractor, Alvan G. Clark noticed the faint star G, and Edward Emerson Barnard saw the exceedingly faint pair H and H′. Barnard even suspected a 10th star, lying about one-third of the way from C to B. While G and the fainter companions are well beyond the visual reach of amateur telescopes, some can perhaps be captured with a backyard CCD setup. The existence of Barnard's 10th star has been confirmed in images taken not only with HST but also with a laser-compensated 1.5-meter reflector at the Starfire Optical Range in New

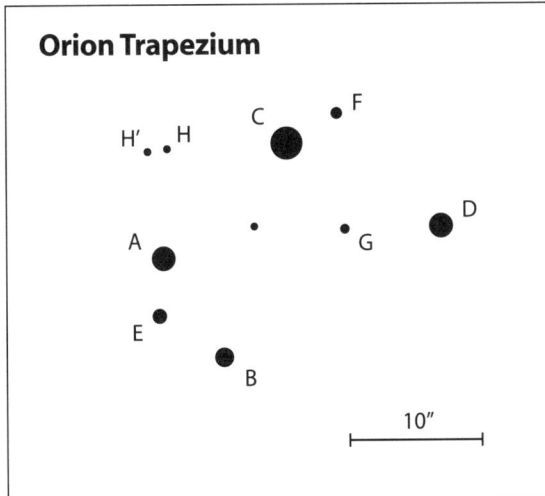

Figure 3. **A highly magnified chart of the famous Trapezium stars at the heart of the Great Orion Nebula. The brightest star, θ¹ Orionis C, has a visual magnitude of 5.1, while that of G is about 16. South is up. Generations of astronomers have scrutinized this region to test their telescopes' optics.**

Mexico. Both instruments have obtained tantalizing views of "proplyds" (young stars encircled by protoplanetary disks) in this active star-forming region.

The specialized open-cluster designations found in this book carry the prefixes Bas (Basel), Be (Berkeley), Bi (Biurakan), Blanco, Cr (Collinder), Do (Dolidze), H (Harvard), Ho (Hogg), K (King), Ly (Lyngå), Mel (Melotte), Mrk (Markarian), Pi (Pismis), Ru (Ruprecht), St (Stock), Steph (Stephenson), Tom (Tombaugh), Tr (Trumpler), and Wa (Waterloo). Not all of them are so labeled in the atlas. Our descriptions are based partly on the cluster's Trumpler class, which is also mentioned near the end of the paragraph and explained further in *Sky Catalogue 2000.0, Vol. 2*.

Globular Clusters (GC). Despite their round appearance and abundance of tightly packed stars, up to 100,000 or more, globular clusters exhibit a surprising individuality. But these aspects stand out much better in the eyepiece than they do on photographs or digital images, as the quotes from observers bring out.

Most globulars are known by their NGC, IC, or M numbers. But two special prefixes are also found in this book: Pal (Palomar) and Ru (Ruprecht). We have characterized globulars in part from the Shapley-Sawyer (S-S) class, which is also listed at the end of the paragraph and discussed in *Sky Catalogue 2000.0, Vol. 2*.

Dark Nebulae (DN). Both *Sky Atlas 2000.0* and this *Companion* use the prefix B for the dark nebulae cataloged, and for the most part discovered, by the American astronomer Edward Emerson Barnard (1857–1923). His famous photographic atlas, posthumously published in the 1920s, still turns up occasionally in the rare book market.

Apart from B for Barnard, the following prefixes are also found here: Be (Bernes), LDN (Lynds), and SL (Sandqvist and Lindroos).

Bright Nebulae (BN, BNe, BNr, BNer). The letter or letters after BN give some idea of what makes a bright nebula shine. Thus, "e" denotes an emission nebula, and "r" a reflection nebula. Sometimes both mechanisms occur in the same object.

In a pure emission nebula (BNe), luminous energy is released whenever wayward electrons are captured by ionized hydrogen atoms in space. Typically these photons have a wavelength of 6563 angstroms, giving the nebula a characteristic red hue on color images. Because the dark-adapted human eye is colorblind, this ruddy tint is almost never reported by visual observers.

A reflection nebula (BNr), however, contains dust that does not shine by itself but simply reflects the light from one or more nearby stars. Frequently these are hot, young stars of spectral type *O* or *B*, which have a bluish tint already, and the scattering process

Introduction

makes the light of the nebula appear bluer still. A rare exception is IC 4592 in Scorpius, whose outer fringe gets its red-orange color from the bright *M*-type supergiant, Antares.

The special prefixes for bright nebulae found in this book are Ced (Cederblad), Gum, LBN (Lynds), LH (Lucke and Hodge), M1 (Minkowski), R (Rodgers, Campbell, and Whiteoak), Sh2 (Sharpless), Simeis, vdB (van den Bergh), and vdBH (van den Bergh and Herbst).

Supernova Remnant (SNR). The only object listed in this book with the type SNR is the Crab Nebula, which was actually seen to explode as a supernova in the year A.D. 1054 and was first observed as a telescopic nebula seven centuries later, by John Bevis in England. But a number of other large, nebulous structures are very likely remnants of similar catastrophes in the remote past, including the Vela SNR (as its name implies), Simeis 147, the Veil Nebula, and even Barnard's Loop. They are simply classed with other bright nebulae.

Abbreviations

Within the text descriptions, our aim has been to avoid the cryptic codes found in most other works, especially the *NGC* and *IC*. Still, some shorthand notations help to improve readability. Since the time of Johann Bayer (1572–1625), almost every bright star has been assigned a lowercase Greek letter followed by the genitive of the constellation within which it lies. These Greek letters are listed in Table 1. Astronomers use standard three-letter abbreviations for the 88 constellations, and these are listed in Table 2.

The abbreviations used for the designations and types of objects are explained under the type headings discussed earlier.

Directions on the celestial sphere are specified in a manner much like compass headings on Earth. The main ones are N (north), E (east), S (south), and W (west), and these can be used in combinations such as NW (northwest) or SSE (south-southeast). A few additional abbreviations are found throughout: approx. (approximately), km/sec (kilometers per second), ly (light-years), mag. (magnitude), and phot. mag. (photographic magnitude). This last term refers to a value measured in blue light, rather than the yellow-green to which the eye is most sensitive.

Table 1. The Greek Alphabet

α	alpha	ι	iota	ρ	rho
β	beta	κ	kappa	σ	sigma
γ	gamma	λ	lambda	τ	tau
δ	delta	μ	mu	υ	upsilon
ε	epsilon	ν	nu	φ	phi
ζ	zeta	ξ	xi	χ	chi
η	eta	ο	omicron	ψ	psi
θ	theta	π	pi	ω	omega

Table 2. Constellation Names and Abbreviations

After each abbreviation is the full constellation name, followed by the genitive or possessive form used with star designations.

And	Andromeda, Andromedae	Crv	Corvus, Corvi	Oph	Ophiuchus, Ophiuchi
Ant	Antlia, Antliae	CVn	Canes Venatici, Canum Venaticorum	Ori	Orion, Orionis
Aps	Apus, Apodis			Pav	Pavo, Pavonis
Aql	Aquila, Aquilae	Cyg	Cygnus, Cygni	Peg	Pegasus, Pegasi
Aqr	Aquarius, Aquarii	Del	Delphinus, Delphini	Per	Perseus, Persei
Ara	Ara, Arae	Dor	Dorado, Doradus	Phe	Phoenix, Phoenicis
Ari	Aries, Arietis	Dra	Draco, Draconis	Pic	Pictor, Pictoris
Aur	Auriga, Aurigae	Equ	Equuleus, Equulei	PsA	Piscis Austrinus, Piscis Austrini
Boo	Boötes, Boötis	Eri	Eridanus, Eridani		
Cae	Caelum, Caeli	For	Fornax, Fornacis	Psc	Pisces, Piscium
Cam	Camelopardalis, Camelopardalis	Gem	Gemini, Geminorum	Pup	Puppis, Puppis
		Gru	Grus, Gruis	Pyx	Pyxis, Pyxidis
Cap	Capricornus, Capricorni	Her	Hercules, Herculis	Ret	Reticulum, Reticuli
Car	Carina, Carinae	Hor	Horologium, Horologii	Scl	Sculptor, Sculptoris
Cas	Cassiopeia, Cassiopeiae	Hya	Hydra, Hydrae	Sco	Scorpius, Scorpii
Cen	Centaurus, Centauri	Hyi	Hydrus, Hydri	Sct	Scutum, Scuti
Cep	Cepheus, Cephei	Ind	Indus, Indi	Ser	Serpens, Serpentis
Cet	Cetus, Ceti	Lac	Lacerta, Lacertae	Sex	Sextans, Sextantis
Cha	Chamaeleon, Chamaeleontis	Leo	Leo, Leonis	Sge	Sagitta, Sagittae
Cir	Circinus, Circini	Lep	Lepus, Leporis	Sgr	Sagittarius, Sagittarii
CMa	Canis Major, Canis Majoris	Lib	Libra, Librae	Tau	Taurus, Tauri
CMi	Canis Minor, Canis Minoris	LMi	Leo Minor, Leonis Minoris	Tel	Telescopium, Telescopii
Cnc	Cancer, Cancri	Lup	Lupus, Lupi	TrA	Triangulum Australe, Trianguli Australis
Col	Columba, Columbae	Lyn	Lynx, Lyncis		
Com	Coma Berenices, Comae Berenices	Lyr	Lyra, Lyrae	Tri	Triangulum, Trianguli
		Men	Mensa, Mensae	Tuc	Tucana, Tucanae
CrA	Corona Australis, Coronae Australis	Mic	Microscopium, Microscopii	UMa	Ursa Major, Ursae Majoris
		Mon	Monoceros, Monocerotis	UMi	Ursa Minor, Ursae Minoris
CrB	Corona Borealis, Coronae Borealis			Vel	Vela, Velorum
		Mus	Musca, Muscae	Vir	Virgo, Virginis
Crt	Crater, Crateris	Nor	Norma, Normae	Vol	Volans, Volantis
Cru	Crux, Crucis	Oct	Octans, Octantis	Vul	Vulpecula, Vulpeculae

Bibliography

Ackers, Agnès, et al., *The Strasbourg-ESO Catalogue of Galactic Planetary Nebulae*, Garching bei München, 1992: European Southern Observatory.

Ashbrook, Joseph, "The Trapezium in Orion," *Sky & Telescope*, **19**, 213, February 1960.

Barnard, Edward Emerson, *A Photographic Atlas of Selected Regions of the Milky Way*, Washington, DC, 1923 (Part I) and 1927 (Part II): Carnegie Institution.

Burnham, Robert, Jr., *Burnham's Celestial Handbook*, New York, 1978: Dover Publications.

Clark, Roger N., *Visual Astronomy of the Deep Sky*, Cambridge, MA, 1990: Sky Publishing Corp. and Cambridge University Press.

Copeland, Leland S., "An Amateur's Tour of Planetary Nebulae," *Sky & Telescope*, **19**, 214, February 1960.

Copeland, Leland S., "The Buried Treasure of Monoceros," *Sky & Telescope*, **16**, 248, March 1957.

Cragin, Murray, James Lucyk, and Barry Rappaport, *The Deep Sky Field Guide to Uranometria 2000.0*, Richmond, VA, 1993: Willmann-Bell.

Denning, William F., *Telescopic Work for Starlight Evenings*, London, 1891: Taylor and Francis.

Harrington, Philip S., *The Deep Sky: An Introduction*, Cambridge, MA, 1997: Sky Publishing Corp.

Hirshfeld, Alan, and Roger W. Sinnott, eds., *Sky Catalogue 2000.0, Volume 2: Double Stars, Variable Stars and Nonstellar Objects*, Cambridge, MA, 1985: Sky Publishing Corp. and Cambridge University Press.

Hirshfeld, Alan, Roger W. Sinnott, and François Ochsenbein, *Sky Catalogue 2000.0, Volume 1: Stars to Magnitude 8.0*, 2nd edition, Cambridge, MA, 1991: Sky Publishing Corp. and Cambridge University Press.

Houston, Walter Scott, *Deep-Sky Wonders*, Cambridge, MA, 1999: Sky Publishing Corp.

Hynes, S. J., *Planetary Nebulae*, Richmond, VA, 1991: Willmann-Bell.

Jones, Kenneth Glyn, *Messier's Nebulae and Star Clusters*, 2nd edition, Cambridge, 1991: Cambridge University Press.

Luginbuhl, Christian B., and Brian A. Skiff, *Observing Handbook and Catalogue of Deep-Sky Objects*, Cambridge, 1989: Cambridge University Press.

Mallas, John H., and Evered Kreimer, *The Messier Album*, Cambridge, MA, 1978: Sky Publishing Corp.

Moore, Patrick, and M. Barlow Pepin, "Beyond Messier: The Caldwell Catalog," *Sky & Telescope*, **90**, 6:38, December 1995.

Mullaney, James, *Celestial Harvest: 300-Plus Showpieces of the Heavens for Telescope Viewing & Contemplation*, Exton, PA, 1998: James Mullaney.

Nilson, Peter, *Uppsala General Catalogue of Galaxies*, Uppsala, 1973: Uppsala Astronomical Observatory.

O'Meara, Stephen James, *Deep-Sky Companions: The Messier Objects*, Cambridge, MA, 1998: Sky Publishing Corp. and Cambridge University Press.

Paturel, Georges, *Lyon-Meudon Extragalactic Database*, Lyon, 1999: CRAL-Observatoire de Lyon (http://www-obs.univ-lyon1.fr/).

Sandage, Allan, and John Bedke, *The Carnegie Atlas of Galaxies*, two volumes, Washington, DC, 1994: Carnegie Institution.

Serviss, Garrett P., *Astronomy with an Opera-Glass*, New York, 1888: D. Appleton and Co.

Serviss, Garrett P., *Astronomy with the Naked Eye*, New York, 1908: Harper & Brothers Publishers.

Serviss, Garrett P., *Curiosities of the Sky*, New York, 1909: Harper & Brothers Publishers.

Serviss, Garrett P., *Round the Year with the Stars*, New York, 1910: Harper & Brothers Publishers.

Sinnott, Roger W., ed., *NGC 2000.0: The Complete New*

General Catalogue and Index Catalogues of Nebulae and Star Clusters by J. L. E. Dreyer, Cambridge, MA, 1988: Sky Publishing Corp. and Cambridge University Press.

Sinnott, Roger W., and Michael A. C. Perryman, *Millennium Star Atlas,* three volumes, Cambridge, MA, 1997: Sky Publishing Corp. and European Space Agency.

Skiff, Brian A., "Exploring the Hubble Sequence by Eye," *Sky & Telescope,* **99**, 5:120, May 2000.

Smyth, William Henry, *A Cycle of Celestial Objects, Volume the Second: The Bedford Catalogue,* Richmond, VA, 1986: Willmann-Bell. Originally published in London, 1844, by John W. Parker.

Tirion, Wil, and Roger W. Sinnott, *Sky Atlas 2000.0,* 2nd edition, Cambridge, MA, 1998: Sky Publishing Corp. and Cambridge University Press.

Turon, Catherine, et al., *Celestia 2000: The Hipparcos and Tycho Catalogues,* CD-ROM, Noordwijk, 1998: European Space Agency.

Webb, T. W., *Celestial Objects for Common Telescopes,* London, 1881: Longmans, Green, and Co.

Alphabetical List of Deep-Sky Objects

30 Doradus

See Tarantula Nebula.

47 Tucanae

GC	Tuc	$00^h 24.1^m$	$-72° 05'$	50'
chart 24			m_v 4	

NGC 104, C106. High concentration of stars; very bright, large, and beautiful. Dimly visible to the naked eye. Per *NGC*, a (!!) very remarkable object. Extends 220 ly; distance 15,000 ly; S-S class 3.

Andromeda Galaxy

Gx	And	$00^h 42.7^m$	$+41° 16'$	$3.2° \times 1.0°$
chart 4			m_v 3.4	p.a. 35°

M31, NGC 224. Extremely bright, large, and elongated; readily visible to the naked eye. A poem by Rufus Festus Avienus in the mid-4th century alludes to M31 as one of the "clouds" binding Andromeda's arms to the sky. Resembles two cones or pyramids of light, opposed at the base (Messier). A pair of dark streaks near the brightest region are well seen in a 10-inch reflector (Denning). High magnifications will help bring out the tiny, starlike nucleus (Houston). Per *NGC*, a (!!!) most remarkable object. Modern astronomers have usually placed its distance at about 2.3 million ly, but recent studies using Hipparcos data seem to indicate 2.5 million ly. Hubble class Sb. This is the largest galaxy in the Local Group, about 25 percent larger than our own Milky Way. Its satellite galaxies include M32 and M110.

Antennae

See NGC 4038 and NGC 4039. Two galaxies in collision.

B8, 9, 11, 13

DN	Cam	$04^h 19.0^m$	$+55° 03'$	$2.5° \times 0.5°$
charts 1, 5				

High opacity, irregular shape. Located about 3° N of open cluster NGC 1528 and 2° NW of dark nebula B12.

B12

DN	Cam	$04^h 30.0^m$	$+54° 17'$	23'
charts 1, 5				

High opacity, irregular shape. Lies 2° SE of dark-nebula complex B8, 9, 11, 13, and about 3.5° NE of open cluster NGC 1528.

B29

DN	Aur	$05^h 06.2^m$	$+31° 44'$	10'
chart 5				

High opacity, circular shape.

B30-32

DN	Ori	$05^h 29.8^m$	$+12° 32'$	$80' \times 50'$
chart 11				

High opacity, irregular shape. Located about 3° NW of star λ Ori (Meissa); involved with dark nebula B225 to SW.

B33

See Horsehead Nebula.

B34

DN	Aur	$05^h 43.5^m$	$+32° 39'$	19'
chart 5				

Medium opacity, circular shape; includes globule. Located approx. 2° W of open cluster M37.

B35

DN	Ori	$05^h 45.5^m$	$+09° 03'$	$19' \times 9'$
chart 11				

High opacity, elliptical shape. Located on the sky midway between stars Betelgeuse (α Ori) and Meissa (λ Ori); involved in bright nebula Ced 59.

B40

DN	Sco	$16^h 14.7^m$	$-18° 59'$	15'
charts 15, 22				

Medium opacity, irregular shape. Located in NE part of bright nebula IC 4592 and approx. 1° NE of star ν Sco.

B41

DN	Sco	$16^h 22.5^m$	$-19° 40'$	$39' \times 30'$
charts 15, 22				

High opacity, elliptical shape. Located approx. 2.5° E of star ν Sco and 2° W of dark nebula B43.

B42

DN   Oph   16^h 25.0^m −23° 30′   19′
chart 22

High opacity, irregular shape. A narrow dark lane W of star ρ Oph, extending N–S. Located in bright nebula IC 4604.

B43

DN   Oph   16^h 31.0^m −19° 20′   2.3° × 1.2°
charts 15, 22

High opacity, irregular shape. Located about 4° E of star ν Sco and 2° E of dark nebula B41.

B44

DN   Oph   16^h 40.0^m −24° 20′   5.5° × 0.8°
chart 22

High opacity, irregular shape. A long, thick lane extending E. Lies approx. 2° NE of star Antares (α Sco).

B45

DN   Oph   16^h 38.0^m −22° 30′   6.0° × 1.3°
chart 22

High opacity, irregular shape. A Y-shaped narrow lane extending NE–SW; lies E of star ρ Oph.

B46

DN   Oph   16^h 57.3^m −22° 40′   19′
chart 22

High opacity; includes a globule. Located approx. 0.5° N of star 24 Oph.

B47

DN   Oph   17^h 01.0^m −22° 40′   30′
chart 22

Narrow lane in patchy region extending E of star 24 Oph. Located approx. 1° W of B51 and connected to it by two distinct lanes.

B48

DN   Sco   17^h 01.0^m −40° 47′   39′ × 14′
chart 22

High opacity, irregular shape. Located approx. 1° SE of bright nebula IC 4628.

B50

DN   Sco   17^h 03.0^m −34° 26′   15′
chart 22

High opacity, irregular shape. Located approx. 0.5° SW of star κ Sco.

B51

DN   Oph   17^h 04.0^m −22° 15′   60′ × 30′
chart 22

Narrow lane in patchy region extending NE of star 24 Oph. Approx. 1° E of dark nebula B47 and connected to it by two distinct lanes.

B53

DN   Sco   17^h 06.1^m −33° 15′   30′ × 10′
chart 22

Medium opacity. Crescent shaped; extending N–S. Lies approx. 0.5° NE of star κ Sco.

B55

DN   Sco   17^h 07.5^m −32° 00′   30′ × 10′
chart 22

High opacity, irregular shape.

B57

DN   Oph   17^h 08.3^m −22° 50′   5′
chart 22

High opacity, elliptical shape; elongated NE–SW. Includes globule. In patchy region lying E of dark nebula B44.

B58

DN   Sco   17^h 11.2^m −40° 25′   15′
chart 22

High opacity, oblong shape; extends N–S. Located about 2.7° E of bright nebula IC 4628.

B60

DN   Oph   17^h 11.8^m −22° 27′   30′ × 19′
chart 22

Medium opacity; extends NE–SW. In patchy region E of dark nebula B44. Dark nebula B246 lies about 10′ to SE.

B61

DN   Oph $17^h\,15.2^m$ −20° 21′   10′ × 4′
chart 22

High opacity, irregular or elongated shape. Lies approx. 1° N of dark nebula B63.

B62

DN   Oph $17^h\,16.2^m$ −20° 53′   25′ × 15′
chart 22

High opacity, irregular shape. Located approx. 0.5° N of dark nebula B63.

B63

DN   Oph $17^h\,16.0^m$ −21° 23′   1.7° × 0.3°
chart 22

Medium opacity; extends E–SW; a curving arc. Globule at W end. Located about 3° NNW of the star θ Oph.

B64

DN   Oph $17^h\,17.2^m$ −18° 32′   19′
charts 15, 22

High opacity, cometary shape. Globular cluster M9 located at E end.

B65, 6, 7

See Pipe Nebula.

B67a

DN   Oph $17^h\,22.5^m$ −21° 53′   15′
chart 22

High opacity, irregular shape; includes globule. Located approx. 1° SE of star ξ Oph.

B68

DN   Oph $17^h\,22.6^m$ −23° 44′   4′
chart 22

High opacity, kidney shaped; includes globule. Lies approximately 20′ SW of the Snake Nebula (dark nebula B72).

B69

DN   Oph $17^h\,22.9^m$ −23° 53′   4′
chart 22

High opacity, irregular shape. Lies about 20′ SSW of the Snake Nebula (dark nebula B72).

B70

DN   Oph $17^h\,23.5^m$ −23° 58′   4′
chart 22

Medium opacity, cometary shape. Located about 20′ S of the Snake Nebula (dark nebula B72).

B72

See Snake Nebula.

B74

DN   Oph $17^h\,25.2^m$ −24° 12′   15′ × 10′
chart 22

High opacity; irregular, slightly curved shape. Located about 15′ W of star 44 Oph.

B77

See Pipe Nebula.

B78

See Pipe Nebula.

B79

DN   Oph $17^h\,39.5^m$ −19° 47′   50′ × 30′
charts 15, 22

High opacity; narrow, curved lanes; NW extension of dark nebula B276.

B83a

DN   Sgr $17^h\,45.3^m$ −20° 00′   4′
charts 15, 22

High opacity, elliptical shape, with N–S elongation; includes globule. Located on a star cloud approx. 1.8° NE of star 58 Oph. Dark nebula B84 is adjacent to SE.

B84

DN   Sgr $17^h 46.5^m$ $-20° 11'$ $30' \times 15'$
charts 15, 22

High opacity, irregular or looped shape. Located approx. 1.7° NE of star 58 Oph; dark nebula B83a is nearby to NW.

B84a

DN   Sgr $17^h 57.5^m$ $-17° 40'$ $15'$
charts 15, 22

High opacity, cometary shape, with faint S extension. Lies approx. 1.5° N of open cluster M23.

B86

DN   Sgr $18^h 02.7^m$ $-27° 50'$ $4'$
chart 22

High opacity, irregular shape; includes globule. Located just W of open cluster NGC 6520.

B87

DN   Sgr $18^h 04.3^m$ $-32° 30'$ $12'$
chart 22

Parrot's Head. Medium opacity, cometary shape; includes globule. Located approx. 2° S of star γ Sgr (Alnasl).

B90

DN   Sgr $18^h 10.2^m$ $-28° 19'$ $10'$
chart 22

High opacity, irregular shape; elongated N–S; includes globule.

B91

DN   Sgr $18^h 10.0^m$ $-23° 39'$ $5' \times 2'$
chart 22

High opacity, kidney shape. Adjacent to bright nebulae IC 1274 and 1275; lies 1.5° E of the Lagoon Nebula (M8).

B92

DN   Sgr $18^h 15.5^m$ $-18° 11'$ $12' \times 6'$
charts 15, 16, 22

High opacity, elliptical shape; elongated N–S; includes globule. Located approx. 0.7° NW of the star cloud M24 (Milky Way Patch).

B93

DN   Sgr $18^h 16.9^m$ $-18° 04'$ $12' \times 2'$
charts 15, 16, 22

Medium opacity, cometary shape; includes globule. Located about 0.7° N of the star cloud M24 (Milky Way Patch).

B95

DN   Sct $18^h 25.6^m$ $-11° 45'$ $30'$
charts 15, 16

High opacity, cometary shape; includes globule. Located about 2.6° NE of the Eagle Nebula (M16).

B97

DN   Sct $18^h 29.1^m$ $-09° 56'$ $50'$
charts 15, 16

Medium opacity, irregular shape. Located approx. 2° SW of star α Sct.

B100, 1

DN   Sct $18^h 32.7^m$ $-09° 08'$ $39' \times 14'$
charts 15, 16

High opacity, crescent shape; includes globule. Located approx. 1° SW of star α Sct.

B103

DN   Sct $18^h 39.2^m$ $-06° 37'$ $39'$
charts 15, 16

High opacity, irregular shape. Located on NW side of the Scutum Star Cloud, approx. 2° NE of star α Sct.

B104

DN   Sct $18^h 47.3^m$ $-04° 32'$ $15' \times 1'$
charts 15, 16

High opacity, L shaped. Located about 20' N of the star β Sct.

B108

DN   Sct $18^h 49.6^m$ $-06° 19'$ $3'$
charts 15, 16

Medium opacity. Located approx. 0.5° W of the Wild Duck Cluster (M11).

B110

DN Sct $18^h 50.2^m$ $-04° 46'$ 8'
charts 15, 16

High opacity, irregular shape; includes globule.

B111

DN Sct $18^h 51.0^m$ $-05° 00'$ 2.0°
charts 15, 16

Medium opacity; consists of twin crescent-shaped areas just N of the Wild Duck Cluster (M11).

B112

DN Sct $18^h 51.2^m$ $-06° 40'$ 19'
charts 15, 16

Medium opacity, irregular shape. Lies approx. 0.5° S of the Wild Duck Cluster (M11).

B113

DN Sct $18^h 51.4^m$ $-04° 19'$ 10'
charts 15, 16

High opacity, irregular shape; includes globule.

B114-7

DN Sct $18^h 53.2^m$ $-07° 06'$ 50' × 5'
charts 15, 16

High opacity, irregular shape; extended N–S. Dark nebula B118 located at SE end.

B118

DN Sct $18^h 53.9^m$ $-07° 27'$ 50' × 5'
charts 15, 16

High opacity, cometary shape; includes globule. Dark-nebula complex B114-7 extends to NW.

B127, 29, 30

DN Aql $19^h 02.0^m$ $-05° 26'$ 19' × 5'
chart 16

High opacity, irregular or curved shape. These lie just NE of star 12 Aql.

B132

DN Aql $19^h 04.1^m$ $-04° 28'$ 15' × 7'
chart 16

High opacity, irregular shape; extends NE–SW. Located about 40' NW of star λ Aql.

B133

DN Aql $19^h 06.1^m$ $-06° 50'$ 10' × 3'
chart 16

High opacity, cometary shape; includes globule. Located approx. 2° S of star λ Aql.

B134

DN Aql $19^h 06.9^m$ $-06° 14'$ 6'
chart 16

High opacity, cometary shape; includes globule. Lies approx. 1.4° S of star λ Aql.

B135

DN Aql $19^h 07.7^m$ $-03° 55'$ 25'
chart 16

High opacity. Located approx. 1° NE of star λ Aql; dark nebula B136 lies just to SE.

B136

DN Aql $19^h 08.8^m$ $-04° 02'$ 5'
chart 16

High opacity. Located approx. 1° NE of star λ Aql; dark nebula B135 lies just to NW.

B139

DN Aql $19^h 18.1^m$ $-01° 28'$ 10' × 2'
chart 16

High opacity, elliptical shape. Planetary nebula NGC 6778 lies just to SE.

B142

DN Aql $19^h 40.7^m$ $+10° 30'$ 60' × 30'
chart 16

High opacity, irregular shape; extends SE–W. Located about 3° NW of star α Aql (Altair) and 1° WSW of star γ Aql (Tarazed).

B143

DN Aql 19^h 41.5^m +11° 00′ 50′ × 39′
chart 16

High opacity, irregular shape. Located about 3° NW of star α Aql (Altair) and 1° WNW of star γ Aql (Tarazed).

B145

DN Cyg 20^h 02.8^m +37° 40′ 34′ × 5′
charts 8, 9

Medium opacity, triangular shape; includes globule.

B146

DN Cyg 20^h 03.5^m +36° 02′ 1′
charts 8, 9

High opacity; very small.

B148, 9

DN Cep 20^h 49.1^m +59° 32′ 3′
chart 3

High opacity, cometary shape. Located approx. 0.5° SW of dark nebula B150.

B150

DN Cep 20^h 50.6^m +60° 18′ 60′ × 3′
chart 3

High opacity; irregular, curved shape; E part extends to NE. Located approx. 1.6° S of star η Cep and 0.5° NE of dark nebulae B148, 9.

B152

DN Cep 21^h 14.5^m +61° 45′ 15′ × 3′
chart 3

High opacity, irregular shape; elongated SE–NW. Located approx. 1° SW of star α Cep (Alderamin).

B157

DN Cyg 21^h 33.7^m +54° 40′ 4′
charts 3, 9

High opacity, cometary shape; includes globule. Dark nebula B364 lies approx. 0.7° to SW.

B160

DN Cep 21^h 38.0^m +56° 14′ 30′ × 15′
charts 3, 9

Medium opacity, kidney shaped. Lies approx. 1° S of bright nebula IC 1396.

B161

DN Cep 21^h 40.3^m +57° 49′ 12′ × 3′
chart 3

High opacity, cometary shape; includes globule. Located in NE part of bright nebula IC 1396.

B162

DN Cep 21^h 41.1^m +56° 19′ 12′ × 2′
charts 3, 9

Medium opacity, irregular shape; a curving dark strip. Located approx. 1° S of bright nebula IC 1396.

B163

DN Cep 21^h 42.2^m +56° 42′ 4′
charts 3, 9

Medium opacity; small, irregular shape pointing S; includes globule. Located in SE portion of bright nebula IC 1396.

B164

DN Cyg 21^h 46.5^m +51° 04′ 12′ × 6′
charts 3, 9

High opacity, kidney shaped; includes globule. Located approx. 0.8° E of star π^1 Cyg.

B168

DN Cyg 21^h 53.2^m +47° 12′ 1.7° × 0.2°
chart 9

High opacity, irregular shape; extends NE–SW with lane extending SSE toward the Cocoon Nebula (IC 5146).

B169, 70, 71

DN Cep 21^h 58.9^m +58° 45′ 80′
chart 3

High opacity, irregular shape. These objects run NE–SW with small NE extension. Located about 3° NE of bright nebula IC 1396; the much smaller dark nebulae B173, 4, are located about 0.5° to NE.

B173, 4

DN Cep $22^h 07.4^m$ +59° 20' 39' × 10'
chart 3

High opacity, elongated S shape. Lie approx. 0.5° NE of dark-nebula complex B169, 70, 71.

B225

DN Ori $05^h 29.8^m$ +12° 32' 80' × 50'
chart 11

High opacity, irregular shape; SW extension of dark-nebula complex B30-32. Located about 3° NW of star λ Ori (Meissa).

B228

DN Lup $15^h 45.5^m$ −34° 24' 4.0° × 0.3°
charts 21, 22

High opacity; long, thin lane extending NW–SE. Additional dark areas located several degrees to W.

B231

DN Sco $16^h 37.5^m$ −35° 12' 50' × 39'
chart 22

High opacity, irregular shape; sharper on W side. A 4.5-mag. star is located on W boundary. Dark nebula B233 lies approx. 1° to E.

B233

DN Sco $16^h 44.1^m$ −35° 21' 50' × 19'
chart 22

High opacity, irregular shape. Dark nebula B231 lies approx. 1° to W.

B238

DN Oph $16^h 52.5^m$ −23° 05' 17'
chart 22

High opacity, irregular shape. An 8th-mag. star lies on NW boundary.

B244

DN Oph $17^h 10.1^m$ −28° 24' 30' × 19'
chart 22

High opacity; shaped like an upside-down V. Located approx. 0.5° S of tip of the stem of the Pipe Nebula (dark nebulae B65, 6, 7).

B246

DN Oph $17^h 11.8^m$ −22° 27' 30' × 19'
chart 22

Medium opacity. Located in patchy region E of dark nebula B44. Dark nebula B60 lies about 10' to the NW.

B252

DN Sco $17^h 15.2^m$ −32° 13' 19' × 5'
chart 22

Triangular shape, high opacity.

B256

DN Oph $17^h 12.2^m$ −28° 51' 50' × 10'
chart 22

High opacity, irregular or curved shape; extends E–W. Lies approx. 1.5° S of the stem of the Pipe Nebula (dark nebulae B65, 6, 7).

B257

DN Sco $17^h 22.0^m$ −35° 35' 10' × 7'
chart 22

High opacity. A faint reflection nebula lies at the edge of this object.

B259

DN Oph $17^h 22.0^m$ −19° 19' 30'
charts 15, 22

Medium opacity, curved shape. Located approx. 1° SE of globular cluster M9.

B261

DN Oph $17^h 25.3^m$ −23° 00' 19'
chart 22

Medium opacity, irregular shape; extends E–W, diffused to S. Located approx. 0.5° NE of the Snake Nebula (dark nebula B72) and approx. 10' S of dark nebula B262.

B262

DN Oph $17^h 26.0^m$ $-22° 28'$ $60' \times 39'$
chart 22

Medium opacity, irregular shape with two thin extensions to W. Lies approx. 1° NE of the Snake Nebula (dark nebula B72) and approx. 10′ N of dark nebula B261.

B263

DN Sco $17^h 26.3^m$ $-42° 38'$ $30'$
chart 22

High opacity, irregular shape; extends NE–SW.

B268

DN Oph $17^h 31.0^m$ $-21° 00'$ $60' \times 50'$
chart 22

Medium opacity, irregular shape; extends N–S. Dark nebula B270 located in N part.

B270

DN Oph $17^h 32.0^m$ $-19° 40'$ $60'$
charts 15, 22

High opacity, roundish shape, Located in NE part of dark nebula B268.

B276

DN Oph $17^h 39.5^m$ $-19° 47'$ $50' \times 30'$
charts 15, 22

High opacity, irregular shape. Dark nebula B79 is a narrow, curved NW extension.

B283

DN Sco $17^h 51.3^m$ $-33° 53'$ $90' \times 60'$
chart 22

High opacity, irregular shape. Located approx. 1° NNW of open cluster M7.

B287

DN Sco $17^h 54.4^m$ $-35° 12'$ $25' \times 15'$
chart 22

High opacity, irregular shape. Lies approx. 0.5° SE of open cluster M7.

B303

DN Sgr $18^h 09.2^m$ $-24° 07'$ $1'$
chart 22

High opacity, S shaped. Located in the bright nebula IC 4685.

B312

DN Sct $18^h 30.9^m$ $-15° 08'$ $1.7° \times 0.5°$
charts 15, 16

Medium opacity, almost elliptical in shape. Located about 2.5° E of the Omega Nebula (M17).

B314

DN Sct $18^h 37.7^m$ $-09° 37'$ $34' \times 24'$
charts 15, 16

High opacity, irregular or curved shape; extends NE–SW. Located about 2° W of open cluster M26.

B320

DN Sct $18^h 53.0^m$ $-05° 50'$ $15'$
charts 15, 16

Medium opacity, irregular shape. Located at SE end of dark nebula B111.

B336

DN Aql $19^h 36.8^m$ $+12° 20'$ $5'$
chart 16

High opacity. Dark nebulae B337, 34, lie to NW.

B337, 34

DN Aql $19^h 36.0^m$ $+12° 25'$ $39' \times 5'$
chart 16

Medium opacity, extending NE–SW. Dark nebula B336 lies to SE.

B343

DN Cyg $20^h 13.5^m$ $+40° 16'$ $10' \times 5'$
charts 8, 9

High opacity, irregular shape; includes globule. Located in N part of bright nebula IC 1318 and approx. 1.7° W of star γ Cyg (Sadr).

B346

DN Cyg $20^h 26.7^m$ $+43° 45'$ $10' \times 4'$
charts 8, 9

High opacity, kidney shaped. Located about 3° SW of star α Cyg (Deneb).

B350

DN Cyg $20^h 49.1^m$ $+45° 53'$ $3'$
charts 8, 9

High opacity, cometary shape. Lies about 14′ S of star 55 Cyg.

B352

DN Cyg $20^h 57.1^m$ $+45° 54'$ $19' \times 9'$
chart 9

High opacity, irregular shape. Located in N part of the North America Nebula (NGC 7000).

B361

DN Cyg $21^h 12.9^m$ $+47° 22'$ $16'$
chart 9

Medium opacity, a cometary shape with faint W extension; includes globule. Located just S of open cluster IC 1369.

B362

DN Cyg $21^h 24.0^m$ $+50° 10'$ $15' \times 8'$
charts 3, 9

High opacity, elliptical shape; includes globule; extends NE–SW. A 9th-mag. star is located on the NE edge.

B364

DN Cyg $21^h 33.6^m$ $+54° 33'$ $39'$
charts 3, 9

High opacity, irregular shape with narrow lanes. Lies approx. 0.7° SW of dark nebula B157.

B365

DN Cep $21^h 34.9^m$ $+56° 43'$ $21' \times 3'$
charts 3, 9

Medium opacity, elliptical shape. Located in SW section of bright nebula IC 1396.

B367

DN Cep $21^h 44.4^m$ $+57° 12'$ $3'$
chart 3

High opacity, irregular shape. Lies in ESE part of bright nebula IC 1396.

Barbell Nebula

See M76.

Barnard's Galaxy

See NGC 6822.

Barnard's Loop

BNe Ori $05^h 20.0^m$ $-04° 00'$ $6.7° \times 0.7°$
chart 11

Sh2-276. Extremely large and very faint; a huge crescent of nebulosity; extended NW–SSE. Very difficult to detect visually, it shows up nicely on film. N part is brightest, possibly illuminated by 2.1-mag. star ζ Ori.

Bas 6

OC Cyg $20^h 06.8^m$ $+38° 21'$ $14'$
charts 8, 9 m_v 7.7

40 stars; mag. of brightest star 10.2; involved in nebulosity. Distance 6,800 ly.

Bas 11A

OC CMa $07^h 17.1^m$ $-13° 58'$ $8'$
chart 12 m_v 8.2

30 stars; mag. of brightest star 10.9. Distance 4,800 ly.

Bas 18

OC Cen $13^h 28.3^m$ $-62° 22'$ $4'$
chart 25 m_v 8.2

20 stars; mag. of brightest star 8.2. Distance 5,000 ly.

Be 86

OC Cyg $20^h 20.4^m$ $+38° 42'$ $7'$
charts 8, 9 m_v 7.9

30 stars; detached, strong concentration of stars;

large range in brightness; involved in nebulosity; mag. of brightest star 9.5; Trumpler class I 3 p n. Located about 40′ WNW of open cluster M29.

Be 135

DN Pup $07^h\,19.0^m$ $-44°\,35′$ $12′ \times 5′$
charts 19, 20

High opacity, elliptical shape; contains a small reflection nebula.

Be 145

DN Cir $14^h\,48.6^m$ $-65°\,15′$ $12′ \times 5′$
charts 25, 26

High opacity. Located near reflection nebula vdBH 63.

Be 146

DN Cen $13^h\,57.6^m$ $-40°\,00′$ $19′ \times 7′$
chart 21

High opacity, irregular shape. Located adjacent to bright nebula NGC 5367.

Be 149

DN Sco $16^h\,09.4^m$ $-39°\,08′$ $60′ \times 12′$
charts 21, 22

High opacity, irregular shape; contains faint reflection nebula.

Be 157

DN CrA $19^h\,02.9^m$ $-37°\,08′$ $50′ \times 17′$
chart 22

High opacity, irregular shape. Star γ CrA is located on E edge.

Bear Paw Galaxy

See NGC 2537.

Beehive Cluster

See Praesepe.

Bi 2

OC Cyg $20^h\,09.2^m$ $+35°\,29′$ $12′$
charts 8, 9 $m_v\,6.3$

15 stars; detached, no concentration of stars; moderate brightness range; mag. of brightest star 7.9; Trumpler class III 2 p.

Black Eye Galaxy

See M64.

Blanco 1

OC Scl $00^h\,04.3^m$ $-29°\,56′$ $90′$
charts 18, 23 $m_v\,4.5$

Zeta Sculptoris Cluster. 30 stars; detached, no concentration; moderate brightness range. Distance 815 ly; age 50 million years; Trumpler class III 2 m.

Blinking Planetary

See NGC 6826.

Blue Flash Nebula

See NGC 6905.

Blue Planetary Nebula

See NGC 3918.

Blue Snowball Nebula

See NGC 7662.

Bode's Nebulae

See M81 and M82.

Bowl of Pipe Nebula

See Pipe Nebula.

Box Nebula

See NGC 6309.

Brocchi's Cluster

OC Vul $19^h\,25.4^m$ $+20°\,11′$ $90′$
charts 8, 16 $m_v\,3.6$

Also the Coathanger or Cr 399. 10 stars of mag. 5.2 to 7.2. Recorded as "a little cloud" by the Arabic astronomer al-Ṣūfī more than 600 years before the invention of the telescope. Best seen in binoculars. Hipparcos measurements show that this is not a true cluster but a chance alignment of stars at very different distances.

Bubble Nebula

See NGC 7635.

Bug Nebula

PN　　Sco　　17^h 13.7^m　　−37° 06′　　44″
chart 22　　　　　　　　m$_v$ 9.6

NGC 6302, C69. Pretty bright; an elongated figure eight; phot. mag. 12.8; central star fainter than phot. mag. 21. Distance 2,000 ly; expansion velocity 8 km/sec (5 miles/sec); V-V class 6.

Butterfly Cluster

See M6.

Butterfly Wing Nebula

See NGC 2346.

C1

See NGC 188.

C2

See NGC 40.

C3

See NGC 4236.

C4

See NGC 7023.

C5

See IC 349.

C6

See NGC 6543.

C7

See NGC 2403.

C8

See NGC 559.

C9

See Sh2-155.

C10

See NGC 663.

C11

See NGC 7635.

C12

See NGC 6946.

C13

See NGC 457.

C14

See Double Cluster.

C15

See NGC 6826.

C16

See NGC 7243.

C17

See NGC 147.

C18

See NGC 185.

C19

See IC 5146.

C20

See North America Nebula.

C21

See NGC 4449.

C22

See NGC 7662.

C23

See NGC 891.

C24

See NGC 1275.

C25

See NGC 2419.

C26

See NGC 4244.

C27
See NGC 6888.

C28
See NGC 752.

C29
See NGC 5005.

C30
See NGC 7331.

C31
See IC 405.

C32
See NGC 4631.

C33
See NGC 6992 and NGC 6995.

C34
See NGC 6960.

C35
See NGC 4889.

C36
See NGC 4559.

C37
See NGC 6885.

C38
See NGC 4565.

C39
See NGC 2392.

C40
See NGC 3626.

C41
See Hyades.

C42
See NGC 7006.

C43
See NGC 7814.

C44
See NGC 7479.

C45
See NGC 5248.

C46
See Hubble's Variable Nebula.

C47
See NGC 6934.

C48
See NGC 2775.

C49
See Rosette Nebula.

C50
See NGC 2244.

C51
See IC 1613.

C52
See NGC 4697.

C53
See NGC 3115.

C54
See NGC 2506.

C55
See Saturn Nebula.

C56
See NGC 246.

C57
See NGC 6822.

C58
See NGC 2360.

C59

See NGC 3242.

C60

See NGC 4038.

C61

See NGC 4039.

C62

See NGC 247.

C63

See Helix Nebula.

C64

See NGC 2362.

C65

See NGC 253.

C66

See NGC 5694.

C67

See NGC 1097.

C68

See NGC 6729.

C69

See Bug Nebula.

C70

See NGC 300.

C71

See NGC 2477.

C72

See NGC 55.

C73

See NGC 1851.

C74

See NGC 3132.

C75

See NGC 6124.

C76

See NGC 6231.

C77

See NGC 5128.

C78

See NGC 6541.

C79

See NGC 3201.

C80

See Omega Centauri.

C81

See NGC 6352.

C82

See NGC 6193.

C83

See NGC 4945.

C84

See NGC 5286.

C85

See IC 2391.

C86

See NGC 6397.

C87

See NGC 1261.

C88

See NGC 5823.

C89

See NGC 6087.

C90

See NGC 2867.

C91

See NGC 3532.

C92
See Eta Carinae Nebula.

C93
See NGC 6752.

C94
See Jewel Box Cluster.

C95
See NGC 6025.

C96
See NGC 2516.

C97
See NGC 3766.

C98
See NGC 4609.

C99
See Coal Sack.

C100
See IC 2944, 48.

C101
See NGC 6744.

C102
See Southern Pleiades.

C103
See Tarantula Nebula.

C104
See NGC 362.

C105
See NGC 4833.

C106
See 47 Tucanae.

C107
See NGC 6101.

C108
See NGC 4372.

C109
See NGC 3195.

California Nebula
BNe Per $04^h\,00.7^m$ $+36°\,37'$ $2.3° \times 0.7°$ charts 4, 5

NGC 1499. Huge; very faint and very elongated; shaped like the state of California. Visible in low-power telescopes and binoculars under very dark skies; an H-beta or O III filter helps greatly. Illuminated by 4.0-mag. star ξ Per.

Campbell's Star
See PK64+5.1.

Cassiopeia Nebula
See IC 59 and IC 63.

Cat's Eye Nebula
See NGC 6543.

Cave Nebula
See Sh2-155.

CBS Eye Nebula
See NGC 3242.

Ced 33
BNr Tau $04^h\,27.1^m$ $+26°\,06'$ $5' \times 2'$ chart 5

Very faint and diffuse; illuminated by a 10th-mag. star.

Ced 34

    BNr   Tau $04^h\,27.2^m$ $+22°\,57'$ $10'\times6'$
    chart 5

Bright; illuminated by a 5.5-mag. star.

Ced 59

    BNer Ori $05^h\,45.3^m$ $+09°\,04'$ $3'\times2'$
    chart 11

Bright; illuminated by a 12th-mag. star. Located in dark nebula B35.

Ced 62

    BNr   Ori $06^h\,07.8^m$ $+18°\,41'$ $3'\times2'$
    charts 5, 11, 12

Bright; consists of two symmetrical fans extending N–S; illuminated by a 13th-mag. star.

Ced 90

    BNer CMa $07^h\,05.2^m$ $-12°\,20'$ $10'$
    chart 12

Bright; irregular, amorphous, or filamentary structure; illuminated by an 8.5-mag. star. Located at S tip of bright nebula IC 2177.

Ced 122

    BNe   Cen $13^h\,25.4^m$ $-64°\,01'$ $2.5°$
    chart 25

Large; illuminated by a 5.5-mag. star; visible to naked eye. Located E of the Coal Sack dark nebula.

Ced 174

    BNe   Cyg $20^h\,02.8^m$ $+36°\,58'$ $15'\times5'$
    charts 8, 9

Medium brightness.

Ced 211

    BNe   Aqr $23^h\,43.8^m$ $-15°\,17'$ $2'\times1'$
    charts 17, 23

Irregular, filamentary shape; illuminated by a 6.4-mag. star.

Ced 214

    BNer Cep $00^h\,04.7^m$ $+67°\,10'$ $50'\times39'$
    charts 1, 3

Medium brightness; brightest toward NE. Located approx. 1° S of bright nebula NGC 7822.

Centaurus A

See NGC 5128.

Centaurus Cluster

See IC 2944, 48.

Christmas Tree Cluster

See NGC 2264.

Circinus Galaxy

See E 97-13.

Cirrus Nebula

See Veil Nebula.

Clown Face Nebula

See NGC 2392.

Coal Sack

    DN   Cru $12^h\,53.0^m$ $-63°\,00'$ $6.7°\times5.0°$
    chart 25

C99. A dark patch in the Milky Way adjacent to and E of the Southern Cross. The most famous naked-eye dark nebula, a nearly starless spot. A single faint star is visible within this lagoon of darkness (Serviss). Extends 60 ly; distance 500 ly.

Coathanger

See Brocchi's Cluster.

Cocoon Nebula

See IC 5146.

Coddington's Nebula

See IC 2574.

Cone Nebula

See NGC 2264.

Cork Nebula

See M76.

Cr 21

OC	Tri	01^h 50.2^m	+27° 05′	6′
chart 4			m$_p$ 8.2	

20 stars; detached, weak concentration; medium brightness range. Might be an asterism and not a true cluster. Trumpler class IV 2 p.

Cr 62

OC	Aur	05$_p$ 22.5^m	+41° 00′	28′
chart 5			m$_p$ 4.2	

Few stars; not well detached; large brightness range; Trumpler class IV 3 p.

Cr 69

OC	Ori	05^h 35.1^m	+09° 56′	60′
chart 11			m$_p$ 2.8	

20 stars; detached, weak concentration; large brightness range; involved in nebulosity. Distance 1,600 ly; Trumpler class II 3 p n. Star λ Ori (Meissa) is located at cluster's center.

Cr 89

OC	Gem	06^h 18.0^m	+23° 38′	34′
chart 5			m$_p$ 5.7	

15 stars; not well detached; moderate brightness range; involved in nebulosity. Distance 4,200 ly; Trumpler class IV 2 p n.

Cr 91

OC	Mon	06^h 21.7^m	+02° 22′	16′
charts 11, 12			m$_p$ 6.4	

20 stars; not well detached; moderate brightness range; Trumpler class IV 2 p.

Cr 92

OC	Mon	06^h 22.9^m	+05° 07′	11′
charts 11, 12			m$_p$ 8.6	

Probably not a true cluster. Located about 2° W of the Rosette Nebula (NGC 2237-9).

Cr 96

OC	Mon	06^h 30.3^m	+02° 52′	7′
charts 11, 12			m$_v$ 7.3	

15 stars; not well detached; moderate brightness range; mag. of brightest star 8.8. Distance 3,600 ly; Trumpler class IV 2 p. Located about 2° SSW of the Rosette Nebula (NGC 2237-9).

Cr 97

OC	Mon	06^h 31.3^m	+05° 55′	20′
charts 11, 12			m$_p$ 5.4	

15 stars; not well detached; large brightness range; Trumpler class IV 3 p.

Cr 106

OC	Mon	06^h 37.1^m	+05° 57′	44′
charts 11, 12			m$_p$ 4.6	

20 stars; detached, no concentration; large brightness range; Trumpler class III 3 p. Located approx. 1° NE of the Rosette Nebula (NGC 2237-9).

Cr 107

OC	Mon	06^h 37.7^m	+04° 44′	34′
charts 11, 12			m$_v$ 5.1	

15 stars; not well detached; large brightness range. Distance 5,500 ly; Trumpler class IV 3 p. Located approx. 1° ESE of the Rosette Nebula (NGC 2237-9).

Cr 111

OC	Mon	06^h 38.7^m	+06° 54′	3.2′
charts 11, 12			m$_p$ 7.0	

Probably an asterism and not a true cluster.

Cr 121

OC	CMa	06^h 54.1^m	−24° 11′	50′
chart 19			m$_v$ 2.6	

20 stars; detached, no concentration; large brightness range. The brightest star (mag. 3.9) is 16 o^1 CMa, located at the cluster's center. Distance 2,300 ly; Trumpler class III 3 p.

Cr 132

OC	CMa	07^h 13.6^m	−30° 50′	1.7°
chart 19			m$_v$ 3.6	

25 stars; detached, no concentration; large bright-

ness range; mag. of brightest star 5.3. A large cluster. Trumpler class III 3 p.

Cr 135

OC	Pup	07^h 17.0^m	−36° 50′	50′
chart 19			m$_v$ 2.1	

Not well detached; moderate brightness range; poor cluster; mag. of brightest star 2.7; Trumpler class IV 2 p.

Cr 140

OC	CMa	07^h 23.9^m	−32° 00′	41′
chart 19			m$_v$ 3.5	

Also called the Tuft in the Tail of the Dog. 30 stars; detached, no concentration; large brightness range; mag. of brightest star 5.4. Distance 1,000 ly; Trumpler class III 3 p.

Cr 185

OC	Pup	08^h 22.5^m	−36° 10′	8′
charts 19, 20			m$_v$ 7.8	

35 stars; detached, no concentration; moderate brightness range; mag. of brightest star 10.1. Distance 8,100 ly; Trumpler class III 2 p.

Cr 197

OC	Vel	08^h 44.7^m	−41° 22′	16′
chart 20			m$_v$ 6.7	

40 stars; not well detached; moderate brightness range; involved in nebulosity (bright nebula Gum 15 is located on N part of cluster); mag. of brightest star 7.4. Distance 3,200 ly; Trumpler class IV 2 p n.

Cr 228

OC	Car	10^h 43.0^h	−60° 01′	15′
chart 25			m$_v$ 4.4	

Mag. of brightest star 6.3. Distance 8,500 ly. Located within bright nebula Eta Carinae.

Cr 236

OC	Car	10^h 57.0^m	−61° 02′	7′
chart 25			m$_p$ 7.7	

20 stars; detached, no concentration; moderate brightness range; Trumpler class III 2 p.

Cr 240

OC	Car	11^h 11.2^m	−60° 17′	25′
chart 25			m$_v$ 3.9	

30 stars; detached, no concentration; small brightness range; involved in nebulosity; mag. of brightest star 4.6; Trumpler class III 1 p n.

Cr 272

OC	Cen	13^h 30.6^m	−61° 16′	8′
chart 25			m$_v$ 7.7	

40 stars; detached, no concentration; small brightness range; mag. of brightest star 10.5. A very young cluster, only 2 million years old; distance 9,500 ly; Trumpler class III 1 m. Adjacent to open cluster Ho 16.

Cr 292

OC	Nor	15^h 50.7^m	−57° 40′	15′
charts 25, 26			m$_p$ 7.9	

50 stars; detached, no concentration; moderate brightness range; Trumpler class III 2 m.

Cr 299

OC	Nor	16^h 18.4^m	−55° 07′	19′
charts 25, 26			m$_p$ 6.9	

Detached, no concentration; moderate brightness range; poor cluster; Trumpler class III 2 p. In a rich field.

Cr 338

OC	Sco	17^h 38.2^m	−37° 34′	25′
chart 22			m$_p$ 8.0	

40 stars; detached, no concentration; moderate brightness range; Trumpler class III 2 p.

Cr 350

OC	Oph	17^h 48.1^m	+01° 18′	44′
charts 15, A1			m$_p$ 6.1	

20 stars; not well detached; moderate brightness range; Trumpler class IV 2 p.

Cr 367

OC	Sgr	$18^h 09.6^m$	$-23°59'$	$37'$
chart 22			$m_p 6.4$	

30 stars; not well detached; large brightness range; involved in nebulosity (bright nebulae IC 1274, IC 1275, IC 4685, and dark nebulae B91 and B303); Trumpler class IV 3 p n. Located approx. 1° ENE of Lagoon Nebula (M8).

Cr 394

OC	Sgr	$18^h 52.5^m$	$-20°23'$	$21'$
chart 22			$m_p 6.3$	

Not well detached; moderate brightness range; moderately rich; Trumpler class IV 2 m.

Cr 399

See Brocchi's Cluster.

Cr 401

OC	Aql	$19^h 38.4^m$	$+00°20'$	$1'$
chart 16			$m_p 7.0$	

Not well detached; moderate brightness range; moderately rich; Trumpler class IV 2 m.

Cr 419

OC	Cyg	$20^h 18.1^m$	$+40°43'$	$5'$
charts 8, 9			$m_p 5.4$	

Not well detached; moderate brightness range; poor cluster; Trumpler class IV 2 p. Located in SE portion of bright nebula IC 1311.

Cr 463

OC	Cas	$01^h 47.7^m$	$+71°46'$	$36'$
charts 1, 3			$m_v 5.7$	

40 stars; detached, no concentration; moderate brightness range; mag. of brightest star 8.5. Age 150 million years; distance 2,000 ly; Trumpler class III 2 p.

Crab Nebula

SNR	Tau	$05^h 34.5^m$	$+22°01'$	$6' \times 4'$
chart 5			$m_v 8.4{:}$	

M1, NGC 1952. Created by the supernova of A.D. 1054, which became as bright as Venus and was visible in the daytime sky. The nebula was originally discovered by John Bevis in 1731, then independently by Messier in 1758. Does not contain a single star; it is a whitish glow, elongated like a candle flame (Messier). Expansion velocity 1,600 km/sec (990 miles/sec); distance 6,000 ly.

Crescent Nebula

See NGC 6888.

Cygnus Nebula

See IC 1318.

Delta Lyrae Cluster

See Steph 1.

Do 25

OC	Mon	$06^h 45.1^m$	$+00°18'$	$23'$
charts 11, 12			$m_v 7.6$	

50 stars; not well detached; moderate brightness range; involved in nebulosity; mag. of brightest star 8.9. Distance 17,000 ly; Trumpler class IV 2 p n.

Donut Nebula

See Ring Nebula.

Double Cluster

OC	Per	$02^h 20.5^m$	$+57°08'$	$60' \times 30'$
chart 1			$m_v 5$	

C14, consisting of NGC 869 and NGC 884. A pair of very close, very bright open clusters, each with well over 100 stars. Faintly visible to the naked eye as an unresolved glow, they were described as a misty star almost 2,000 years ago in Ptolemy's *Almagest*. A glittering, dual starburst in binoculars (Mullaney). Like two swarms of bees encountering in midair (Serviss). When viewed together with a good low-power telescope they are a truly wonderful sight!

Draco Dwarf

Gx	Dra	$17^h 20.2^m$	$+57°55'$	$37' \times 23'$
chart 3			$m_v 9.9$	p.a. 84°

UGC 10822. Distance 3 million ly; Hubble class Ep.

Dumbbell Nebula

PN	Vul	$19^h 59.6^m$	$+22°\ 43'$	$8' \times 4'$
charts 8, 9			m_v 7.3:	

M27, NGC 6853, PK60–3.1, also the Hourglass Nebula. Shaped like a dumbbell or hourglass with a faint outer halo; very bright and very large; central star mag. 13.9. Quite possibly the finest planetary nebula. Two or three minute stars in it can be picked out with an 8-inch reflector (Webb). Seven stars seen with a 10-inch reflector (Denning). Looks like a comfortable pillow (Copeland). Age 20,000 years.

E 15-8

Gx	Men	$04^h 07.2^m$	$-82°\ 17'$	$1.0' \times 0.4'$
charts 24, 25			m_p 12.8	p.a. 177°

Star superimposed.

E 18-2

Gx	Cha	$08^h 19.2^m$	$-78°\ 42'$	$2.2' \times 1.1'$
charts 24, 25			m_p 12.9	p.a. 150°

Distance 230 million ly; Hubble class Sb.

E 19-3

Gx	Cha	$10^h 38.0^m$	$-81°\ 06'$	$1.6' \times 0.9'$
chart 25			m_p 13.0	p.a. 137°

Distance 79 million ly; Hubble class SBc.

E 27-1

Gx	Oct	$21^h 52.5^m$	$-81°\ 32'$	$3.0' \times 2.7'$
charts 24, 26			m_p 12.1	p.a. 175°

Distance 100 million ly; Hubble class SBb.

E 27-8

Gx	Oct	$22^h 23.0^m$	$-80°\ 00'$	$2.8' \times 1.0'$
charts 24, 26			m_p 12.8	p.a. 144°

Distance 100 million ly; Hubble class SBc.

E 37-10

Gx	Car	$10^h 04.3^m$	$-75°\ 29'$	$2.5' \times 2.3'$
chart 25			m_p 13.0	p.a. 140°

Distance 75 million ly; Hubble class SBc.

E 54-21

Gx	Hyi	$03^h 49.8^m$	$-71°\ 38'$	$4.7' \times 2.5'$
chart 24			m_p 12.7	p.a. 93°

Distance 55 million ly; Hubble class SBd.

E 60-19

Gx	Vol	$08^h 57.5^m$	$-69°\ 04'$	$3.2' \times 1.2'$
chart 25			m_p 12.7	p.a. 157°

Distance 53 million ly; Hubble class SBc.

E 69-14

Gx	TrA	$16^h 52.3^m$	$-69°\ 08'$	$1.5' \times 1.1'$
chart 26			m_p 12.4	p.a. 80°

Distance 200 million ly; Hubble class E/S0.

E 91-3

Gx	Car	$09^h 13.5^m$	$-63°\ 38'$	$2.1' \times 1.4'$
chart 25			m_p 12.9	p.a. 74°

Distance 73 million ly; Hubble class Sa.

E 97-13

Gx	Cir	$14^h 13.2^m$	$-65°\ 20'$	$6.0' \times 3.0'$
chart 25			m_v 10.1	p.a. 40°

Also Circinus Galaxy. Distance 11 million ly; Hubble class Sb.

E 101-14

Gx	Ara	$16^h 48.7^m$	$-62°\ 36'$	$1.7'$
chart 26			m_p 13.0	

Distance 100 million ly; Hubble class E/S0.

E 107-4

Gx	Pav	$21^h 03.5^m$	$-67°\ 11'$	$1.7' \times 1.4'$
chart 26			m_p 12.9	p.a. 136°

Distance 130 million ly; Hubble class E/S0.

E 115-21

Gx	Hor	$02^h 37.8^m$	$-61°\ 20'$	$7.0' \times 1.0'$
chart 24			m_p 13.0	p.a. 44°

Distance 16 million ly; Hubble class SBc.

E 116-12
Gx	Hor	03^h 13.1^m	−57° 21′	3.6′ × 1.2′
chart 24			m$_p$ 12.9	p.a. 25°

Distance 43 million ly; Hubble class SBc.

E 121-6
Gx	Pic	06^h 07.5^m	−61° 48′	3.9′ × 0.7′
chart 24			m$_p$ 10.7	p.a. 41°

Distance 44 million ly; Hubble class Sc.

E 121-26
Gx	Pic	06^h 21.6^m	−59° 44′	3.4′ × 2.2′
chart 24			m$_p$ 12.5	p.a. 115°

Distance 89 million ly; Hubble class SBa.

E 137-10
Gx	TrA	16^h 15.8^m	−60° 48′	2.3′ × 1.4′
charts 25, 26			m$_p$ 12.4	p.a. 167°

Distance 140 million ly; Hubble class S0.

E 137-12
Gx	TrA	16^h 16.3^m	−61° 18′	1.2′ × 0.5′
charts 25, 26			m$_p$ 13.0	p.a. 15°

Distance 270 million ly; Hubble class S0.

E 137-18
Gx	TrA	16^h 21.0^m	−60° 29′	3.3′ × 1.2′
chart 26			m$_p$ 12.1	p.a. 30°

Distance 21 million ly; Hubble class Sc.

E 137-34
Gx	Nor	16^h 35.2^m	−58° 05′	2.4′ × 1.8′
chart 26			m$_p$ 12.1	p.a. 10°

Distance 110 million ly; Hubble class SBa.

E 137-38
Gx	Ara	16^h 40.9^m	−60° 24′	2.6′ × 0.9′
chart 26			m$_p$ 12.6	p.a. 103°

Distance 210 million ly; Hubble class SBc.

E 138-5
Gx	Ara	16^h 53.9^m	−58° 47′	1.7′ × 1.3′
chart 26			m$_p$ 12.8	p.a. 140°

Distance 120 million ly; Hubble class E/S0.

E 138-10
Gx	Ara	16^h 59.1^m	−60° 13′	5.0′ × 4.2′
chart 26			m$_p$ 11.6	p.a. 55°

Distance 45 million ly; Hubble class Sc.

E 138-29
Gx	Ara	17^h 29.2^m	−62° 27′	2.3′ × 1.4′
chart 26			m$_p$ 12.7	p.a. 50°

Distance 200 million ly; Hubble class S0.

E 151-36A
Gx	Phe	01^h 14.3^m	−55° 24′	1.5′ × 1.0′
charts 18, 24			m$_p$ 13.0	p.a. 140°

Hubble class S0.

E 154-23
Gx	Hor	02^h 56.9^m	−54° 34′	8.0′ × 1.5′
charts 18, 24			m$_v$ 12.2	p.a. 39°

Distance 19 million ly; Hubble class SBc.

E 183-30
Gx	Tel	18^h 56.9^m	−54° 33′	2.0′ × 1.6′
charts 22, 26			m$_v$ 11.8	p.a. 16°

Distance 120 million ly; Hubble class S0p. Galaxy IC 4797 located 20′ to NW.

E 185-54
Gx	Tel	20^h 03.5^m	−55° 57′	3.4′ × 2.3′
chart 26			m$_p$ 12.0	p.a. 122°

Distance 190 million ly; Hubble class E.

E 186-62
Gx	Ind	20^h 34.0^m	−52° 59′	2.1′
charts 23, 26			m$_p$ 12.9	

Distance 110 million ly; Hubble class SBc.

E 208-21
Gx   Pup $07^h\,33.9^m$ $-50°\,27'$ $3.0' \times 2.2'$
charts 19, 20, 24 m_p 12.2   p.a. 110°

Distance 35 million ly; Hubble class E/S0.

E 209-9
Gx   Pup $07^h\,58.2^m$ $-49°\,51'$ $6.0' \times 0.9'$
charts 19, 20, 24 m_p 12.4   p.a. 152°

Distance 39 million ly; Hubble class SBc.

E 213-11
Gx   Vel $10^h\,16.9^m$ $-48°\,53'$ $3.6' \times 2.4'$
chart 20 m_p 12.0   p.a. 7°

Distance 110 million ly; Hubble class Sc.

E 219-21
Gx   Cen $13^h\,02.3^m$ $-50°\,20'$ $5.0' \times 1.2'$
charts 21, 25 m_p 12.7   p.a. 33°

Distance 52 million ly; Hubble class SBc.

E 219-41
Gx   Cen $13^h\,14.0^m$ $-49°\,29'$ $3.0' \times 0.9'$
charts 21, 25 m_p 12.9   p.a. 80°

Distance 190 million ly; Hubble class Sb.

E 221-6
Gx   Cen $13^h\,50.4^m$ $-48°\,23'$ $2.3'$
charts 21, 25 m_p 13.0

Distance 180 million ly; Hubble class Sc.

E 221-10
Gx   Cen $13^h\,51.0^m$ $-49°\,03'$ $1.3'$
charts 21, 25 m_p 13.0

Distance 130 million ly; Hubble class Sc.

E 221-26
Gx   Cen $14^h\,08.4^m$ $-47°\,58'$ $2.8' \times 1.8'$
charts 21, 25 m_p 12.0   p.a. 1°

Distance 120 million ly; Hubble class E. Open cluster NGC 5460 is located approx. 0.5° to S.

E 221-32
Gx   Cen $14^h\,12.2^m$ $-49°\,23'$ $2.1'$
charts 21, 25 m_p 12.7   p.a. 4°

Distance 120 million ly; Hubble class Sc.

E 221-34A
Gx   Cen $14^h\,16.1^m$ $-48°\,08'$ $5.0' \times 3.1'$
charts 21, 25 m_p 12.6   p.a. 175°

Distance 180 million ly; Hubble class S0.

E 235-55
Gx   Ind $21^h\,05.9^m$ $-48°\,12'$ $3.5' \times 2.7'$
charts 23, 26 m_p 12.7   p.a. 60°

Distance 220 million ly; Hubble class SBb.

E 240-10
Gx   Phe $23^h\,37.7^m$ $-47°\,30'$ $3.5' \times 1.8'$
charts 18, 23 m_p 12.6   p.a. 133°

Distance 140 million ly; Hubble class S0.

E 245-5
Gx   Phe $01^h\,45.1^m$ $-43°\,36'$ $3.8' \times 3.4'$
chart 18 m_p 12.7   p.a. 122°

Distance 14 million ly; Hubble class SBm.

E 249-31B
Gx   Eri $03^h\,55.8^m$ $-42°\,22'$ $3.7' \times 1.7'$
charts 18, 19 m_p 12.3   p.a. 40°

Distance 28 million ly.

E 263-48
Gx   Vel $10^h\,31.2^m$ $-46°\,15'$ $2.7' \times 1.5'$
chart 20 m_p 12.6   p.a. 168°

Distance 120 million ly; Hubble class S0.

E 265-7
Gx   Cen $11^h\,07.8^m$ $-46°\,31'$ $3.8' \times 1.3'$
charts 20, 21 m_p 12.4   p.a. 141°

Distance 37 million ly; Hubble class SBc.

E 266-15

Gx   Cen $11^h 40.9^m$ −44° 29′   1.7′ × 1.0′
charts 20, 21 m_p 13.0   p.a. 140°

Distance 130 million ly; Hubble class Sb.

E 269-57

Gx   Cen $13^h 10.1^m$ −46° 26′   3.3′ × 2.5′
charts 21, 25 m_p 12.5   p.a. 54°

Distance 130 million ly; Hubble class SBa.

E 269-85

Gx   Cen $13^h 20.0^m$ −47° 17′   2.5′ × 1.6′
charts 21, 25 m_p 12.7   p.a. 53°

Distance 120 million ly; Hubble class Sc.

E 270-17

Gx   Cen $13^h 34.7^m$ −45° 32′   12′ × 2′
charts 21, 25 m_p 11.8   p.a. 118°

Distance 29 million ly; Hubble class SBc.

E 271-10

Gx   Cen $14^h 00.8^m$ −45° 25′   2.1′ × 1.9′
charts 21, 25 m_v 12.1   p.a. 42°

Distance 59 million ly; Hubble class SBd.

E 273-14

Gx   Lup $14^h 58.4^m$ −47° 42′   3.8′ × 2.8′
charts 21, 25 m_p 12.9   p.a. 130°

Distance 40 million ly; Hubble class Ir.

E 274-1

Gx   Lup $15^h 14.2^m$ −46° 49′   11′ × 2′
charts 21, 22 m_p 11.7   p.a. 38°

Distance 18 million ly; Hubble class Sc.

E 300-14

Gx   Eri $03^h 09.6^m$ −41° 02′   4.9′ × 2.3′
chart 18 m_p 13.0   p.a. 166°

Distance 36 million ly; Hubble class SBc.

E 311-12

Gx   Pup $07^h 47.6^m$ −41° 27′   3.5′ × 0.5′
charts 19, 20 m_p 12.9   p.a. 14°

Distance 39 million ly; Hubble class Sa.

E 320-26

Gx   Cen $11^h 49.8^m$ −38° 47′   2.4′ × 0.9′
charts 20, 21 m_p 12.8   p.a. 163°

Distance 120 million ly; Hubble class Sb.

E 321-25

Gx   Cen $12^h 21.7^m$ −39° 46′   2.2′ × 1.1′
charts 20, 21 m_p 12.8   p.a. 15°

Distance 85 million ly; Hubble class SBc.

E 323-34

Gx   Cen $12^h 53.4^m$ −41° 12′   2.0′ × 1.0′
charts 21, 25 m_p 12.9   p.a. 165°

Distance 180 million ly; Hubble class E.

E 324-24

Gx   Cen $13^h 27.6^m$ −41° 29′   3.3′ × 2.5′
charts 21, 25 m_p 12.9   p.a. 50°

Distance 15 million ly; Hubble class Ir.

E 342-27

Gx   Mic $21^h 16.9^m$ −42° 16′   2.0′ × 1.5′
charts 23, 26 m_p 12.9   p.a. 52°

Distance 230 million ly; Hubble class E/S0.

E 342-50

Gx   Gru $21^h 28.3^m$ −37° 52′   2.3′ × 1.3′
charts 23, 26 m_p 12.9   p.a. 19°

Distance 120 million ly; Hubble class Sc.

E 351-30

See Sculptor Dwarf Galaxy.

E 356-4

See Fornax Dwarf Galaxy.

E 358-63

Gx For 03^h 46.3^m −34° 57′ 5.0′ × 1.3′
charts 18, 19 m$_p$ 12.6 p.a. 133°

Distance 78 million ly; Hubble class Sc.

E 362-11

Gx Col 05^h 16.7^m −37° 06′ 4.7′ × 0.8′
charts 19, 24 m$_p$ 13.0 p.a. 76°

Distance 51 million ly; Hubble class Sb.

E 371-16

Gx Pyx 08^h 47.1^m −33° 46′ 3.5′ × 3.1′
chart 20 m$_p$ 13.0 p.a. 90°

Distance 88 million ly; Hubble class Sba.

E 373-5

Gx Ant 09^h 30.9^m −35° 41′ 3.0′
chart 20 m$_p$ 12.9

Distance 94 million ly; Hubble class Sc.

E 373-8

Gx Ant 09^h 33.3^m −33° 02′ 5.0′ × 0.9′
chart 20 m$_p$ 12.7 p.a. 89°

Distance 31 million ly; Hubble class Sc.

E 380-1

Gx Hya 12^h 14.7^m −35° 31′ 3.2′ × 2.0′
charts 20, 21 m$_p$ 12.9 p.a. 8°

Distance 110 million ly; Hubble class SBb.

E 380-6

Gx Hya 12^h 15.6^m −35° 38′ 3.8′ × 1.6′
charts 20, 21 m$_p$ 12.6 p.a. 79°

Distance 120 million ly; Hubble class Sb.

E 383-76

Gx Cen 13^h 47.5^m −32° 52′ 2.5′ × 1.1′
chart 21 m$_p$ 13.0 p.a. 3°

Distance 490 million ly; Hubble class S0.

E 383-87

Gx Cen 13^h 49.3^m −36° 04′ 4.6′ × 3.8′
chart 21 m$_p$ 11.4 p.a. 95°

Distance 8 million ly; Hubble class SBc.

E 384-2

Gx Cen 13^h 51.3^m −33° 49′ 5.0′ × 3.0′
chart 21 m$_p$ 12.6 p.a. 125°

Distance 55 million ly; Hubble class SBd.

E 385-30

Gx Cen 14^h 29.3^m −33° 27′ 2.3′ × 1.3′
chart 21 m$_p$ 13.0 p.a. 15°

Distance 120 million ly; Hubble class S0.

E 404-12

Gx PsA 21^h 57.1^m −34° 35′ 2.3′ × 1.9′
chart 23 m$_p$ 12.8 p.a. 140°

Distance 120 million ly; Hubble class SBb.

E 428-11

Gx CMa 07^h 15.5^m −29° 21′ 1.6′ × 1.3′
chart 19 m$_p$ 12.9 p.a. 20°

Hubble class S0.

E 436-27

Gx Ant 10^h 28.9^m −31° 37′ 4.1′ × 2.1′
chart 20 m$_p$ 12.6 p.a. 0°

Distance 180 million ly; Hubble class S0.

E 442-26

Gx Hya 12^h 52.2^m −29° 51′ 3.0′ × 1.0′
chart 21 m$_p$ 12.6 p.a. 8°

Distance 120 million ly; Hubble class S0.

E 443-24

Gx Cen 13^h 01.0^m −32° 26′ 1.9′ × 1.5′
chart 21 m$_p$ 12.9 p.a. 167°

Distance 220 million ly; Hubble class E/S0.

E 445-2

Gx Cen 13^h 39.4^m −30° 47′ 2.3′ × 1.9′
chart 21 m$_p$ 12.8 p.a. 100°

Distance 190 million ly; Hubble class E/S0.

E 462-15

Gx Sgr 20^h 23.2^m −27° 43′ 1.8′ × 1.4′
chart 23 m$_p$ 12.9 p.a. 166°

Distance 260 million ly; Hubble class E.

E 479-4

Gx For 02^h 26.4^m −24° 18′ 2.7′ × 1.4′
chart 18 m$_p$ 12.9 p.a. 55°

Distance 63 million ly; Hubble class SBd.

E 490-37

Gx CMa 06^h 44.4^m −26° 07′ 2.1′ × 1.3′
chart 19 m$_p$ 13.0 p.a. 168°

Distance 100 million ly; Hubble class Sa.

E 492-2

Gx CMa 07^h 11.7^m −26° 42′ 2.1′ × 1.4′
chart 19 m$_p$ 13.0 p.a. 143°

Distance 110 million ly; Hubble class SBb.

E 494-26

Gx Pup 08^h 06.2^m −27° 31′ 4.9′ × 3.3′
charts 19, 20 m$_p$ 12.5 p.a. 155°

Distance 33 million ly; Hubble class SBb.

E 495-21

Gx Pyx 08^h 36.3^m −26° 25′ 1.8′
chart 20 m$_p$ 12.5

Distance 29 million ly; Hubble class S0.

E 499-23

Gx Hya 09^h 56.4^m −26° 06′ 2.0′ × 1.2′
chart 20 m$_p$ 12.8 p.a. 109°

Hubble class E/S0.

E 501-51

Gx Hya 10^h 37.5^m −26° 19′ 2.8′ × 1.6′
chart 20 m$_p$ 12.9 p.a. 117°

Distance 140 million ly; Hubble class S0.

E 507-25

Gx Hya 12^h 51.5^m −26° 27′ 2.4′ × 1.9′
chart 21 m$_p$ 12.6 p.a. 102°

Distance 130 million ly; Hubble class E/S0.

E 507-45

Gx Hya 12^h 55.6^m −26° 49′ 1.9′ × 1.4′
chart 21 m$_p$ 13.0 p.a. 165°

Distance 200 million ly; Hubble class S0.

E 548-81

Gx Eri 03^h 42.1^m −21° 15′ 1.7′
chart 18 m$_p$ 12.9 p.a. 146°

Distance 180 million ly; Hubble class SBa.

E 550-24

Gx Eri 04^h 21.2^m −21° 51′ 5.0′ × 2.1′
chart 19 m$_p$ 12.7 p.a. 132°

Distance 34 million ly; Hubble class SBc.

E 556-15

Gx CMa 06^h 21.1^m −20° 03′ 2.8′ × 2.0′
charts 11, 12, 19 m$_p$ 12.7 p.a. 141°

Distance 79 million ly; Hubble class SBb.

E 562-23

Gx Pyx 08^h 36.6^m −20° 28′ 2.2′ × 0.5′
chart 20 m$_p$ 13.0 p.a. 170°

Hubble class S0.

E 563-17

Gx Pyx 08^h 44.5^m −20° 21′ 1.9′ × 1.1′
chart 20 m$_p$ 13.0 p.a. 29°

Distance 140 million ly; Hubble class Sa.

E 563-31

Gx	Hya	$08^h 52.3^m$	$-17° 45'$	$1.7' \times 1.4'$
charts 12, 20			m_p 13.0	p.a. 133°

Hubble class S0.

Eagle Nebula

BNe+C	Ser	$18^h 18.8^m$	$-13° 47'$	$34' \times 27'$
charts 15, 16			m_v 6	

M16, NGC 6611, also the Star Queen Nebula. Large, bright nebula containing a 7′ open cluster; brightest star mag. 8.1. Cluster of small stars, mingled with a feeble glow (Messier). A grand cluster (Webb). Age 5.5 million years; distance 5,900 ly. Located in Serpens Cauda.

Egg Nebula

See PK80–6.1.

Eight Burst Nebula

See NGC 3132.

Embryo Nebula

See IC 1848.

Eskimo Nebula

See NGC 2392.

ET Cluster

See NGC 457.

Eta Carinae Nebula

BNe	Car	$10^h 43.8^m$	$-59° 52'$	2.0°
chart 25				

NGC 3372, C92, also the Keyhole Nebula. A very large, very bright, irregularly shaped nebula; naked-eye object. Extends 200 ly; distance 6,000 ly; even more impressive than the Orion Nebula (M42) but not visible from midnorthern latitudes. Illuminated by variable star η Car, which has hovered around mag. 6 in recent years but became as bright as mag. −0.8 in 1843.

Eyes

See NGC 4435 and NGC 4438.

F 591

Gx	Ind	$21^h 48.3^m$	$-54° 59'$	$2.0' \times 0.3'$
charts 23, 26			m_p 12.8	p.a. 102°

Distance 550 million ly; Hubble class Sb.

Filamentary Nebula

See NGC 6960.

Flaming Star Nebula

See IC 405.

Footprint Nebula

See M1-92.

Fornax A

See NGC 1316.

Fornax Dwarf Galaxy

Gx	For	$02^h 40.0^m$	$-34° 27'$	$60' \times 48'$
chart 18			m_p 9.0	p.a. 60°

E 356-4. Distance 2 million ly; Hubble class E.

Ghost of Jupiter Nebula

See NGC 3242.

Great Orion Nebula

BNer	Ori	$05^h 35.4^m$	$-05° 27'$	$70' \times 60'$
charts 11, B2			m_v 4	

M42, NGC 1976, also the Orion Nebula or Great Nebula in Orion. Extremely bright and large; a stellar nursery. The most spectacular nebula visible from north of latitude +40°. An irregular branching mass of greenish haze (Webb). One of the grandest objects in the heavens (Denning). Illuminated by the four stars of the Trapezium. Extends 30 ly; distance 1,500 ly; per *NGC*, a (!!!) most remarkable object.

Grus Quartet

See NGC 7552, NGC 7582, NGC 7590, and NGC 7599. These four galaxies form an impressive arc spanning less than 1°.

Gum 1

BNer   Mon $07^h 04.3^m$ $-10° 28'$ 19'
chart 12

Small, bright nebula. Located approx. 0.5° NW of bright nebula IC 2177.

Gum 15

BNe   Vel $08^h 44.6^m$ $-41° 17'$ 19'
chart 20

Irregular or roundish shape; mottled appearance; medium brightness. Part of the Vela Supernova Remnant. Open cluster Cr 197 lies at S end.

Gum 17

BNe   Vel $08^h 50.5^m$ $-42° 07'$ 1.7° × 1.0'
chart 20

Bright knot with dark lane; medium brightness; dark nebula SL 4 in E section. Part of the Vela Supernova Remnant.

Gum 23

BNe   Vel $08^h 59.7^m$ $-47° 27'$ 19' × 9'
chart 20

Bright; appears to consist of one small and three larger nebulosities. Part of the Vela Supernova Remnant.

Gum 25

BNe   Vel $09^h 02.4^m$ $-48° 42'$ 7' × 6'
chart 20

Bright; roundish shape. Part of the Vela Supernova Remnant.

Gum 32

BNe   Car $10^h 46.3^m$ $-58° 39'$ 7'
chart 25

Brighter part is broad, irregular, and crescent shaped; illuminated by a 10th-mag. star. Located approx. 0.5° NNE of bright nebula Eta Carinae.

Gum 39

BNe   Cen $11^h 28.9^m$ $-62° 41'$ 19' × 9'
chart 25

Medium brightness; irregular shape with slight E–W extension; illuminated by an 8.6-mag. star. Located approx. 0.5° NW of bright nebula IC 2948 and open cluster IC 2944.

Gum 41

BNer   Cen $11^h 30.4^m$ $-63° 50'$ 15'
chart 25

Located approx. 1° WSW of bright nebula IC 2948 and open cluster IC 2944.

H5

OC   Cru $12^h 27.3^m$ $-60° 46'$ 6'
chart 25 m_v 7.1

Few stars; detached, slight concentration of stars; large range in brightness; mag. of brightest star 8.4; Trumpler class II 3 p.

H20

OC   Sge $19^h 53.1^m$ $+18° 20'$ 7'
charts 8, 16 m_v 7.7

15 stars; detached, no concentration of stars; moderate range in brightness; mag. of brightest star 9.8; Trumpler class III 2 p. Located approx. 0.5° SSW of globular cluster M71.

Heart Nebula

See IC 1805.

Helix Galaxy

See NGC 2685.

Helix Nebula

PN   Aqr $22^h 29.6^m$ $-20° 48'$ 16' × 12'
chart 23 m_v 7.3

NGC 7293, PK36–57.1, C63; also known as the Helical Nebula. Central star mag. 13.5. The nebula is dimly visible in binoculars and rich-field telescopes at low magnification. Discovered by the German astronomer K. L. Harding in 1824, but first classified as a planetary nebula by H. D. Curtis in 1918. One of the nearest of all planetaries, situated about 300 ly from our solar system; expansion velocity 13 km/sec (8 miles/sec); V-V class 4 + 3.

Hercules Cluster

See M13.

Hind's Variable Nebula

BNr	Tau	$04^h 21.8^m$	+19° 32′	0.5′
charts 5, 11				

NGC 1554, 5. A very faint, small, roundish nebula associated with the star T Tau; per *NGC*, a (!!!) most remarkable object. After being discovered by J. R. Hind in 1852, the nebula faded from view in 1868 and did not reappear until 1890.

Ho 16

OC	Cen	$13^h 29.3^m$	−61° 12′	4′
chart 25			m_v 8.4	

10 stars; detached, weak concentration; large brightness range; mag. of brightest star 11.0. Age 160 million years; distance 6,800 ly; Trumpler class II 3 p. Adjacent to open cluster Cr 272.

Ho 17

OC	Cen	$14^h 33.7^m$	−61° 23′	7′
charts 25, A3			m_v 8.3	

10 stars; detached, weak concentration; large brightness range; mag. of brightest star 9.6. Age 180 million years; distance 4,200 ly; Trumpler class II 3 p. Located approx. 1° SW of star α Cen (Rigil Kent).

Ho 18

OC	Lup	$14^h 50.7^m$	−52° 15′	3′
charts 21, 25			m_v 8	

15 stars; detached, strong concentration; large brightness range; mag. of brightest star 8.8. Age 50 million years; distance 3,600 ly; Trumpler class I 3 p.

Ho 22

OC	Ara	$16^h 46.7^m$	−47° 06′	1.5′
charts 21, 22			m_v 6.7	

8 stars; not well detached; large brightness range; mag. of brightest star 7.3. Distance 9,100 ly; Trumpler class IV 3 p. Located adjacent to open cluster NGC 6204.

Hole in a Cluster

See NGC 6811.

Horsehead Nebula

DN	Ori	$05^h 40.9^m$	−02° 28′	6′ × 4′
charts 11, B2				

B33. High opacity. A very difficult object to detect visually, usually requiring more than a 12-inch aperture and very dark skies. An H-beta filter is very helpful. This dark nebula is seen in silhouette against the bright nebula IC 434, and if it lies at the same distance (800 ly) it is about 1 ly across.

Hourglass Nebula

See Dumbbell Nebula.

Hubble's Variable Nebula

BNer	Mon	$06^h 39.2^m$	+08° 44′	2′ × 1′
charts 11, 12				

NGC 2261, C46. Bright and fan shaped; apex at S end. Distance 3,000 ly. Envelopes the variable star R Mon.

Hyades

OC	Tau	$04^h 27.0^m$	+15° 50′	5.5°
charts 5, 11			m_v 0.5	

Also known as the Taurus Moving Cluster, C41, Cr 50, and Mel 25. Has at least 400 member stars; detached, weak central condensation of stars; large range in brightness. Brightest star is 3.4-mag. 78 θ^2 Tau. Mentioned in the works of Homer, Virgil, and other early writers. Trumpler class II 3 m. One of the nearest open clusters, lying at 151 ly according to Hipparcos measurements. The 0.9-mag. star Aldebaran (α Tau) is well in the foreground at 65 ly and is not a cluster member.

IC 10

Gx	Cas	$00^h 20.4^m$	+59° 18′	5.1′ × 4.3′
charts 1, 3			m_v 11.3	p.a. 120°

Distance 4 million ly; Hubble class S0. A member of the Local Group.

IC 59

BNer   Cas   00^h 56.7^m +61° 04'   10' × 5'
chart 1

Part of the Cassiopeia Nebula (along with IC 63). Pretty faint, large, nebulous patch. Illuminated by the bright star γ Cas, which varies unpredictably between about mag. 1.6 and 3.0.

IC 63

BNer   Cas   00^h 59.5^m +60° 49'   10' × 3'
chart 1

Part of the Cassiopeia Nebula (with IC 59). Pretty faint, large, fan-shaped patch of nebulosity. Illuminated by the variable star γ Cas.

IC 239

Gx   And   02^h 36.5^m +38° 58'   5'
charts 1, 4 m_v 11.1

Low surface brightness; very small, bright nucleus; two main arms. Distance 44 million ly; Hubble class S(B)c. Similar to M101 (but much smaller angular size).

IC 284

Gx   Per   03^h 06.2^m +42° 22'   4.1' × 2.1'
charts 1, 4 m_v 11.5   p.a. 13°

Faint, stellar nucleus; large number of very thin arms. Distance 120 million ly; Hubble class Sm.

IC 289

PN   Cas   03^h 10.3^m +61° 19'   35"
chart 1 m_v 13.2

PK138+02.1, Hubble 1. Ring structure with smooth disk; pretty bright, pretty large; located between two very faint stars; phot. mag. 12.3; central star mag. fainter than 15.9. Distance 3,900 ly; expansion velocity 22 km/sec (14 miles/sec); V-V class 4 + 2.

IC 334

Gx   Cam   03^h 45.3^m +76° 38'   3.7' × 1.9'
charts 1, 3 m_v 11.3   p.a. 80°

Bright, complex nuclear region; diffuse envelope. Distance 120 million ly; Hubble class P.

IC 335

Gx   For   03^h 35.5^m −34° 27'   2.6' × 0.7'
charts 18, 19 m_p 12.9   p.a. 84°

Distance 66 million ly; Hubble class Sa. Located approx. 1° N of the Fornax Galaxy Cluster.

IC 342

Gx   Cam   03^h 46.8^m +68° 06'   21'
chart 1 m_v 8.4

Faint, very large, and round; very small, bright nucleus. Distance 10 million ly; Hubble class S(B)c.

IC 348

BNr+C   Per   03^h 44.6^m +32° 09'   10'
charts 4, 5

Pretty bright and large reflection nebula. The 7' cluster within it has a total magnitude of 7.3 and contains 20 stars; not well detached from surrounding star field; mag. of brightest star 8.5. Age 130 million years; distance 1,300 ly.

IC 349

BN   Tau   03^h 46.3^m +23° 56'
charts 4, A2

C5, also called the Merope Nebula or Pleiades Reflection Nebula. A large and extremely faint patch of luminosity surrounding the star Merope (23 Tau), located in the Pleiades (M45).

IC 351

PN   Per   03^h 47.5^m +35° 03'   7"
charts 4, 5 m_v 11.9

PK159−15.1. Just barely nonstellar; phot. mag. 12.4; central star mag. 15.8. Expansion velocity 15 km/sec (9 miles/sec); distance 13,000 ly; V-V class 2a. Located near a triangle of stars.

IC 353

BNr(?)   Tau   03^h 55.0^m +25° 29'   3.0° × 0.5°
charts 4, 5, A2

Extremely large, very diffuse nebula; very faint filamentary structure; illuminated by a 6.4-mag. star. This is the E portion of bright nebula IC 1995, just NE of the Pleiades.

IC 356

| Gx | Cam | $04^h\,07.8^m$ | $+69°\,49'$ | $4.6' \times 3.5'$ |
| chart 1 | | | $m_v\,10.5$ | p.a. 90° |

Pretty faint and pretty large; brighter toward middle. Distance 44 million ly; Hubble class Sbp.

IC 382

| Gx | Eri | $04^h\,37.9^m$ | $-09°\,31'$ | $2.3' \times 1.4'$ |
| chart 11 | | | $m_v\,12.2$ | p.a. 170° |

Pretty bright, pretty large and roundish; has a small nucleus. Distance 210 million ly; Hubble class Sb.

IC 391

| Gx | Cam | $04^h\,57.4^m$ | $+78°\,11'$ | $1.3'$ |
| chart 1 | | | $m_v\,12.7$ | |

Very bright main body, then extremely diffuse. Distance 73 million ly; Hubble class Sc.

IC 405

| BNer | Aur | $05^h\,16.2^m$ | $+34°\,16'$ | $30'$ |
| chart 5 | | | | |

C31, Flaming Star Nebula. Very large, very faint; use an O III or H-beta filter if possible. Illuminated by a 6.0-mag. star.

IC 410

| BNe | Aur | $05^h\,22.6^m$ | $+33°\,31'$ | $40' \times 30'$ |
| chart 5 | | | | |

Very faint, large emission nebula surrounding the large (10′) open cluster NGC 1893 (the Letter Y Cluster).

IC 417

| BNe | Aur | $05^h\,28.1^m$ | $+34°\,26'$ | $12' \times 9'$ |
| chart 5 | | | | |

Very large, diffuse, faint nebula; includes wedge-shaped filaments.

IC 418

| PN | Lep | $05^h\,27.5^m$ | $-12°\,42'$ | $12''$ |
| chart 11 | | | $m_v\,9.3$ | |

PK215–24.1. Ring structure; very small, bright; phot. mag. 10.7; central star mag. 10.2. Expansion velocity 22 km/sec (14 miles/sec); distance 1,300 ly; V-V class 4.

IC 423

| BNr | Ori | $05^h\,33.4^m$ | $-00°\,37'$ | $6' \times 4'$ |
| charts 11, B2 | | | | |

Very faint, comet-shaped nebula; possibly a planetary nebula.

IC 426

| BNr | Ori | $05^h\,36.8^m$ | $-00°\,15'$ | $5'$ |
| charts 11, B2 | | | | |

Very faint, fan-shaped nebula; illuminated by a 9th-mag. star.

IC 430

| BNr | Ori | $05^h\,38.5^m$ | $-07°\,05'$ | $10'$ |
| charts 11, B2 | | | | |

Very large, very faint, fan-shaped nebula; illuminated by a 4.8-mag. star (49 Ori).

IC 431

| BNr | Ori | $05^h\,40.3^m$ | $-01°\,27'$ | $5' \times 3'$ |
| charts 11, B2 | | | | |

Appears as a haze around several bright stars.

IC 432

| BNr | Ori | $05^h\,40.9^m$ | $-01°\,29'$ | $7' \times 4'$ |
| charts 11, B2 | | | | |

Appears as a slightly elongated haze around some bright stars.

IC 434

| BNe | Ori | $05^h\,41.0^m$ | $-02°\,24'$ | $60' \times 9'$ |
| charts 11, B2 | | | | |

Large, faint nebula; very difficult visual object. Extends 14 by 2 ly; distance about 800 ly. Probably excited by ζ Ori, this nebula's glow is what silhouettes the Horsehead Nebula (B33).

IC 435

BNr Ori 05^h 43.0^m −02° 19′ 5′ × 3′
charts 11, B2

Faint nebulosity surrounding an 8.3-mag. star.

IC 438

Gx Lep 05^h 53.0^m −17° 53′ 2.8′ × 2.2′
charts 11, 19 m$_p$ 12.8 p.a. 55°

Extremely faint, pretty small, and elongated. Distance approximately 130 million ly; Hubble class SBc.

IC 443

BNe Gem 06^h 16.9^m +22° 47′ 50′ × 39′
chart 5

Faint, crescent shaped; appears filamentary and wispy; possibly a supernova remnant. Illuminated by an 8.8-mag. star.

IC 444

BNr Gem 06^h 19.4^m +23° 16′ 7′ × 4′
chart 5

Large, faint, glowing nebula; illuminated by a 7.0-mag. star (12 Gem).

IC 446

BNr Mon 06^h 31.0^m +10° 27′ 5′ × 4′
charts 11, 12

A faint, easily seen patch of nebulosity; involved with a 9.5-mag. star.

IC 448

BNr Mon 06^h 32.7^m +07° 19′ 15′ × 10′
charts 11, 12

Large and faint; illuminated by light from a 4.5-mag. star (13 Mon).

IC 456

Gx CMa 07^h 00.3^m −30° 10′ 2.2′ × 1.4′
chart 19 m$_p$ 13.0 p.a. 110°

Very faint, pretty small, and roundish; Hubble class S0. Bright stars are located at NE and NW.

IC 466

BNer Mon 07^h 08.6^m −04° 19′ 1′
chart 12

Very small, very faint; contains an 11.5-mag. star.

IC 520

Gx Cam 08^h 53.7^m +73° 29′ 2.1′ × 1.7′
charts 1, 2 m$_v$ 11.7 p.a. 0°

Small and bright; diffuse nucleus. Distance 160 million ly; Hubble class S(B)b.

IC 529

Gx Cam 09^h 18.5^m +73° 46′ 3.6′ × 1.7′
charts 1, 2 m$_v$ 11.9 p.a. 145°

Distance 100 million ly; Hubble class Sc.

IC 600

Gx Sex 10^h 17.2^m −03° 30′ 2.3′ × 1.2′
chart 13 m$_p$ 13.0 p.a. 25°

Faint, pretty small, and roundish; gradually brighter toward the middle. Distance 51 million ly; Hubble class SBc.

IC 630

Gx Sex 10^h 38.5^m −07° 10′ 2.1′ × 1.7′
chart 13 m$_p$ 12.9 p.a. 130°

Faint, extremely small; appears almost stellar; Hubble class S0. A 9.5-mag. star is located approx. 1.5° SW.

IC 651

Gx Sex 10^h 51.0^m −02° 09′ 0.8′
chart 13 m$_p$ 12.7

Pretty bright, pretty small, and roundish; gradually brighter toward middle. Distance 190 million ly; Hubble class Sb.

IC 694

Gx UMa 11^h 28.5^m +58° 33′ 1.3′
chart 2 m$_p$ 12.0

Very small. Distance 140 million ly; Hubble class S. Paired with galaxy NGC 3690.

IC 749

| Gx | UMa | $11^h 58.6^m$ | +42° 44′ | 2.3′ × 1.9′ |
| charts 6, 7 | | | m_v 12.4 | p.a. 150° |

Large and pretty bright. Distance 36 million ly; Hubble class Sc. Paired with galaxy IC 750.

IC 750

| Gx | UMa | $11^h 58.9^m$ | +42° 43′ | 2.8′ × 1.1′ |
| charts 6, 7 | | | m_v 11.9 | p.a. 43° |

Large, pretty bright, and elongated. Distance 33 million ly; Hubble class Sb. Paired with galaxy IC 749.

IC 764

| Gx | Hya | $12^h 10.2^m$ | −29° 44′ | 4.9′ × 1.4′ |
| charts 20, 21 | | | m_v 12.3 | p.a. 177° |

Pretty large, elongated, and extremely faint; faint nucleus. Distance 83 million ly; Hubble class S.

IC 983

| Gx | Boo | $14^h 10.1^m$ | +17° 44′ | 5.0′ × 4.8′ |
| charts 7, 14 | | | m_p 12.5 | p.a. 120° |

Extremely small and roundish. Distance 240 million ly; Hubble class SBa.

IC 1029

| Gx | Boo | $14^h 32.5^m$ | +49° 54′ | 2.6′ × 0.5′ |
| charts 2, 7 | | | m_p 12.2 | p.a. 152° |

Small, very faint, and elongated; much brighter toward middle. Distance 110 million ly; Hubble class Sb.

IC 1055

| Gx | Lib | $14^h 47.4^m$ | −13° 43′ | 2.1′ × 0.7′ |
| chart 14 | | | m_p 12.7 | p.a. 5° |

Pretty large and faint; elongated N–S. Distance 120 million ly; Hubble class Sb.

IC 1274

| BNe | Sgr | $18^h 09.5^m$ | −23° 44′ | 8′ × 7′ |
| chart 22 | | | | |

Nebulosity includes IC 1275, IC 4685, and NGC 6559. Adjacent to dark nebulae B91 and B303; located approx. 1° ENE of the Lagoon Nebula (M8).

IC 1275

| BNe | Sgr | $18^h 10.0^m$ | −23° 50′ | 10′ × 6′ |
| chart 22 | | | | |

Nebulosity includes IC 1274, IC 4685, and NGC 6559. Adjacent to dark nebulae B91 and B303; located approx. 1° ENE of the Lagoon Nebula (M8).

IC 1276

| GC | Ser | $18^h 10.7^m$ | −07° 12′ | 7′ |
| charts 15, 16 | | | m_p 10.3: | |

Palomar 7. Very low concentration of stars. Distance 42,000 ly.

IC 1283, 84

| BNer | Sgr | $18^h 17.8^m$ | −19° 40′ | 16′ × 15′ |
| charts 15, 16, 22 | | | | |

Irregularly shaped. IC 1283 is nebulosity surrounding a 7.6-mag. star; IC 1284 is the similar nebulosity surrounding a 9.3-mag. star.

IC 1287

| BNr | Sct | $18^h 31.3^m$ | −10° 50′ | 43′ × 33′ |
| charts 15, 16 | | | | |

Very large, elongated; illuminated by a 5.5-mag. star.

IC 1297

| PN | CrA | $19^h 17.4^m$ | −39° 37′ | 7″ |
| chart 22 | | | m_v 10.7 | |

PK358−21.1. Distance 17,000 ly. Central star is RU CrA, about mag. 13.

IC 1311

| BNe+C | Cyg | $20^h 10.8^m$ | +41° 11′ | 60′ × 20′ |
| charts 8, 9 | | | | |

Large, extremely faint emission nebula. Contains a 9′ open cluster with 60 stars; detached, weak concentration of stars; large range in brightness.

IC 1318

| BNe | Cyg | $20^h 21.0^m$ | +39° 54′ | 3.3° × 2.3° |
| charts 8, 9 | | | | |

Cygnus Nebula. Large patch of faint nebulosity surrounding the star γ Cyg (Sadr).

IC 1340

BNe　Cyg　　$20^h 56.2^m$　+31° 04'　25' × 19'
chart 9

Located in S part of the E segment of the Veil Nebula.

IC 1369

OC　Cyg　　$21^h 12.1^m$　+47° 44'　　4'
chart 9　　　　　　　　　m_v 6.8

40 stars; detached, strong concentration of stars; small brightness range; small, faint cluster; mag. of brightest star 12.1. Distance 4,900 ly; age 1.3 billion years; Trumpler class I 1 m. Located just N of dark nebula B361.

IC 1392

Gx　Cyg　　$21^h 35.5^m$　+35° 24'　1.6' × 1.3'
chart 9　　　　　　　　　m_p 12.9　p.a. 75°

Pretty bright; much brighter toward middle. Distance 200 million ly; Hubble class E/S0.

IC 1396

BNe+C　Cep　$21^h 39.1^m$　+57° 30'　2.8° × 2.3°
charts 3, 9

Faint and very large emission nebula. Contains an open cluster with 50 stars; detached, weak concentration of stars; large range in brightness; total mag. of cluster 3.5; mag. of brightest star 3.8. Distance 2,600 ly; Trumpler class II 3 m. Dark nebula B160 lies approx. 1° to N.

IC 1438

Gx　Aqr　　$22^h 16.5^m$　−21° 26'　2.4' × 2.0'
chart 23　　　　　　　　m_p 12.7　p.a. 120°

Faint; two nuclei. Distance 120 million ly; Hubble class SBa.

IC 1447

Gx　Aqr　　$22^h 30.0^m$　−05° 07'　1.4' × 0.7'
chart 17　　　　　　　　m_p 12.9　p.a. 80°

Extremely faint, pretty small, and roundish. Distance 130 million ly; Hubble class Sb. A 9th-mag. star lies about 3° to N.

IC 1459

Gx　Gru　　$22^h 57.2^m$　−36° 28'　4.9' × 3.5'
chart 23　　　　　　　　m_v 10　p.a. 40°

Faint and pretty small; cometary appearance. Distance 71 million ly; Hubble class E3.

IC 1470

BNe　Cep　　$23^h 05.2^m$　+60° 15'　15' × 1'
chart 3

Very faint and comet shaped; possibly a planetary nebula.

IC 1558

Gx　Scl　　$00^h 35.8^m$　−25° 22'　3.4' × 2.5'
chart 18　　　　　　　　m_p 12.8　p.a. 150°

Distance 68 million ly; Hubble class SBd.

IC 1613

Gx　Cet　　$01^h 04.9^m$　+02° 08'　16' × 14'
chart 10　　　　　　　　m_v 9.2　p.a. 50°

C51. Dwarf; very low surface brightness. Distance 5.4 million ly; Hubble class Ir+. Member of the Local Group.

IC 1625

Gx　Phe　　$01^h 07.7^m$　−46° 55'　1.7' × 1.3'
chart 18　　　　　　　　m_p 13.0　p.a. 7°

Extremely faint, very small, and roundish. Distance 290 million ly; Hubble class E/S0.

IC 1633

Gx　Phe　　$01^h 09.9^m$　−45° 56'　2.8' × 2.4'
chart 18　　　　　　　　m_p 12.5　p.a. 120°

Very faint, small, and roundish. Distance 310 million ly; Hubble class E. Very faint star to E.

IC 1660

BN　Tuc　　$01^h 12.6^m$　−71° 45'
chart 24

Extremely faint, very small, and roundish; appears almost stellar.

IC 1727

| Gx | Tri | 01^h 47.5^m | +27° 20′ | 6.0′ × 2.6′ |
| chart 4 | | | m$_v$ 11.5 | p.a. 150° |

Low surface brightness; emission patch at SE end. Distance 23 million ly; Hubble class Sbm.

IC 1747

| PN | Cas | 01^h 57.6^m | +63° 19′ | 13″ |
| chart 1 | | | m$_v$ 12.1 | |

PK130+01.1. Irregular disk; traces of ring structure; phot. mag. 13.6; central star mag. 15.4. Distance 6,800 ly; expansion velocity 30 km/sec (19 miles/sec); V-V class 3b.

IC 1788

| Gx | For | 02^h 15.8^m | −31° 12′ | 2.6′ × 1.1′ |
| chart 18 | | | m$_v$ 12.4 | p.a. 27° |

Thin, knotty arms. Distance 140 million ly; Hubble class Sa. Located approx. 1° SE of star μ For.

IC 1805

| BNe+C | Cas | 02^h 33.4^m | +61° 26′ | 60′ |
| chart 1 | | | | |

Heart Nebula. Faint, very large patch of nebulosity surrounding large (21′) open cluster. 30 stars; detached, no concentration; large brightness range; total mag. of cluster 6.5; mag. of brightest star 7.9. Distance 6,900 ly; Trumpler class III 3 p.

IC 1848

| BNe+C | Cas | 02^h 51.3^m | +60° 25′ | 60′ × 30′ |
| chart 1 | | | | |

Embryo Nebula. Faint and very large; irregular shape; filamentary structure. Nebulosity is brightest in NE portion. Contains a 12′ open cluster of 10 stars; not well detached from surrounding star field; large range in brightness; mag. of brightest star 7.1. Distance 7,200 ly; a young cluster, only 1 million years old; Trumpler class IV 3 p n.

IC 1870

| Gx | Eri | 02^h 57.9^m | −02° 21′ | 2.8′ × 1.5′ |
| chart 10 | | | m$_p$ 12.6 | p.a. 132° |

Very faint and roundish; very gradually brighter toward middle. Distance 67 million ly; Hubble class SBd.

IC 1871

| BNe | Cas | 03^h 03.2^m | +60° 29′ | 4′ |
| chart 1 | | | | |

Faint patch of nebulosity surrounding a 10th-mag. star.

IC 1933

| Gx | Hor | 03^h 25.7^m | −52° 47′ | 2.2′ × 1.1′ |
| charts 18, 24 | | | m$_v$ 12.8 | p.a. 55° |

Faint, small, and slightly elongated; very patchy arms; brighter toward center. Distance 38 million ly; Hubble class Sc.

IC 1953

| Gx | Eri | 03^h 33.7^m | −21° 29′ | 2.9′ × 2.1′ |
| chart 18 | | | m$_v$ 11.7 | p.a. 121° |

Faint and small; very faint nucleus. Distance 83 million ly; Hubble class Sc.

IC 1954

| Gx | Hor | 03^h 31.5^m | −51° 54′ | 3.1′ × 1.6′ |
| charts 18, 19, 24 | | | m$_v$ 11.4 | p.a. 66° |

Faint, pretty large and round; two bright arms; small, with a bright nucleus. Distance 40 million ly; Hubble class SBb. Located in a group of six galaxies.

IC 1970

| Gx | Hor | 03^h 36.5^m | −43° 57′ | 3.2′ × 0.9′ |
| charts 18, 19 | | | m$_p$ 12.9 | p.a. 75° |

Very small, extremely faint, and extremely elongated. Distance 48 million ly; Hubble class Sb.

IC 1993

| Gx | For | 03^h 47.1^m | −33° 42′ | 2.5′ |
| charts 18, 19 | | | m$_p$ 12.5 | |

Large and extremely faint. Distance 37 million ly; Hubble class SBb.

IC 1995

| BN | Tau | 03^h 50.3^m | +25° 35′ | |
| charts 4, 5, A2 | | | | |

A very faint, extremely large, filamentary nebula with a 6th-mag. star involved. Located in the W portion of bright nebula IC 353.

IC 2003

PN	Per	03^h 56.4^m	+33° 52′	9″
charts 4, 5			m$_v$ 11.5	

PK161–14.1. Pretty bright, extremely small; smooth disk; brighter toward center; expansion velocity 18 km/sec (11 miles/sec); nebula has phot. mag. 12.6; central star is about mag. 15.0 (var.). Distance 14,000 ly; V-V class 2.

IC 2006

Gx	Eri	03^h 54.5^m	–35° 58′	2.0′ × 1.7′
charts 18, 19			m$_v$ 11.4	p.a. 35°

Pretty bright, small, and roundish. Distance 53 million ly; Hubble class E1. This is a member of Fornax Galaxy Cluster.

IC 2035

Gx	Hor	04^h 09.0^m	–45° 31′	1.2′ × 0.9′
charts 18, 19, 24			m$_v$ 11.5	p.a. 86°

Faint, very small, and roundish. Distance 55 million ly; Hubble class SB0p.

IC 2051

Gx	Men	03^h 52.0^m	–83° 50′	2.6′ × 1.6′
charts 24, 25, A5			m$_p$ 12.1	p.a. 67°

Very faint and very small; stellar nucleus with elliptical ring. This is a remarkable (!) object per *NGC*. Distance 67 million ly; Hubble class SBb.

IC 2056

Gx	Ret	04^h 16.4^m	–60° 12′	1.9′ × 1.6′
chart 24			m$_v$ 11.6	p.a. 8°

Small, with a bright nucleus. Distance 38 million ly; Hubble class SB0p.

IC 2087

BNr	Tau	04^h 40.0^m	+25° 44′	4′
chart 5				

Small, extremely faint nebula.

IC 2118

BNr	Eri	05^h 06.9^m	–07° 13′	3.0° × 1.0°
chart 11				

Witch's Head Nebula. Very large, extremely faint nebula; possibly illuminated by star Rigel (β Ori).

IC 2149

PN	Aur	05^h 56.3^m	+46° 07′	9″
chart 5			m$_v$ 10.7	

PK166+10.1. Irregular disk; traces of ring structure; very small, very bright; phot. mag. 11.6; central star mag. 11.6. Distance 3,300 ly; expansion velocity 20 km/sec (12 miles/sec); V-V class 3b + 2. Located approx. 1° NNW of star β Aur.

IC 2157

OC	Gem	06^h 05.0^m	+24° 00′	7′
chart 5			m$_v$ 8.4	

20 stars; detached, no concentration of stars; moderate range in brightness; small; mag. of brightest star 11.1. Distance 6,500 ly; Trumpler class III 2 p. Located approx. 1° WSW of open cluster M35.

IC 2158

Gx	Col	06^h 05.3^m	–27° 51′	1.6′ × 1.3′
chart 19			m$_p$ 12.9	p.a. 100°

Distance 60 million ly; Hubble class SBa.

IC 2162

BNe	Ori	06^h 12.9^m	+17° 59′	15′ × 4′
charts 5, 11, 12				

Pretty large, very faint nebula; extended E–W.

IC 2163

Gx	CMa	06^h 16.5^m	–21° 23′	2.8′ × 1.1′
chart 19			m$_p$ 12.4	p.a. 98°

Extremely faint and pretty small. Distance 110 million ly; Hubble class SBc. Possibly interacting with galaxy NGC 2207.

IC 2165

PN	CMa	06^h 21.7^m	–12° 59′	9″
charts 11, 12			m$_v$ 10.5	

PK221–12.1. Very small ring; nearly stellar; phot. mag. 12.9; central star mag. 17.9. Distance 10,000 ly; expansion velocity 20 km/sec (12 miles/sec).

IC 2169

BNr   Mon $06^h\,31.2^m$ $+09°\,54'$ $25'\times19'$
charts 11, 12

Contains a small open cluster; large absorption patch in NE part.

IC 2177

BNe   Mon $07^h\,05.1^m$ $-10°\,42'$ $2.0°\times0.7°$
chart 12

Pretty bright, extremely large, very diffuse; extends N–S; illuminated by a 6.2-mag. star. Bright nebula Ced 90 located at S tip.

IC 2233

Gx   Lyn $08^h\,14.0^m$ $+45°\,45'$ $4.6'\times0.5'$
charts 5, 6 m_v 12.6   p.a. 172°

Distance 26 million ly; Hubble class SBd. Located about 20′ S of the Bear Paw Galaxy (NGC 2537).

IC 2311

Gx   Pup $08^h\,18.8^m$ $-25°\,22'$ $2.0'$
chart 20 m_p 12.5

Pretty bright, very small, and roundish. Distance 98 million ly; Hubble class E. Located about 6′ N of galaxy NGC 2566.

IC 2367

Gx   Pup $08^h\,24.2^m$ $-18°\,47'$ $2.4'\times1.7'$
charts 12, 20 m_p 12.5   p.a. 55°

Pretty bright and small. Distance 98 million ly; Hubble class SBb.

IC 2391

OC   Vel $08^h\,40.2^m$ $-53°\,04'$ $50'$
charts 20, 25 m_v 2.5

C85, also known as the Omicron Velorum Cluster. 30 stars; detached, weak concentration of stars; large range in brightness; very large; bright; mag. of brightest star 3.6. Distance 600 ly; age 36 million years; Trumpler class II 3 p.

IC 2395

OC   Vel $08^h\,41.1^m$ $-48°\,12'$ $7'$
charts 19, 20, 25 m_v 4.6

40 stars; detached, weak concentration of stars; large range in brightness; large; mag. of brightest star 5.5. Distance 2,800 ly; age 16 million years; Trumpler class II 3 p.

IC 2448

PN   Car $09^h\,07.1^m$ $-69°\,57'$ $9''$
chart 25 m_v 10.4

PK285–14.1. Very small, ring shaped; nearly stellar; phot. mag. 11.5; central star mag. 14.2. Distance 9,000 ly; V-V class 2b. Located approx. 0.5° SW of star β Car.

IC 2469

Gx   Pyx $09^h\,23.0^m$ $-32°\,27'$ $5.0'\times1.2'$
chart 20 m_p 11.3   p.a. 36°

Pretty faint, quite small, and very elongated. Distance 63 million ly; Hubble class SBa.

IC 2482

Gx   Hya $09^h\,27.0^m$ $-12°\,07'$ $2.6'\times1.8'$
charts 12, 13 m_p 12.5   p.a. 145°

Faint and very small; Hubble class E.

IC 2488

OC   Vel $09^h\,27.6^m$ $-56°\,59'$ $15'$
chart 25 m_p 7.4

70 stars; detached, weak concentration of stars; moderate range in brightness; large; brightest star is phot. mag. 10; Trumpler class II 2 m.

IC 2501

PN   Car $09^h\,38.8^m$ $-60°\,06'$ $2''$
chart 25 m_v 10.4

PK281–5.1. Appears stellar; phot. mag. 11.3; central star mag. 14.5. Distance 5,500 ly; V-V class 1.

IC 2511

Gx	Ant	09^h 49.4^m	−32° 51′	2.9′ × 0.6′
chart 20			m$_p$ 13.0	p.a. 38°

Pretty bright, pretty small, and extremely elongated. Distance 120 million ly; Hubble class SBb.

IC 2522

Gx	Ant	09^h 55.2^m	−33° 08′	3.2′ × 2.0′
chart 20			m$_v$ 11.9	p.a. 0°

Very faint, elongated, and quite large; distorted, narrow outer arms. Distance 120 million ly; Hubble class Sc. Paired with galaxy IC 2523.

IC 2531

Gx	Ant	09^h 59.9^m	−29° 37′	6.0′ × 0.6′
chart 20			m$_p$ 12.9	p.a. 75°

Extremely faint, extremely elongated, and pretty small. Distance 98 million ly; Hubble class Sc.

IC 2533

Gx	Ant	10^h 00.5^m	−31° 15′	1.9′ × 1.4′
chart 20			m$_p$ 12.9	p.a. 1°

Quite bright, small, and roundish; Hubble class either E or S0.

IC 2537

Gx	Ant	10^h 03.9^m	−27° 34′	2.3′ × 1.7′
chart 20			m$_v$ 12.2	p.a. 26°

Very faint, large, and elongated; filamentary arms; small, bright nucleus. Distance 110 million ly; Hubble class S–.

IC 2552

Gx	Ant	10^h 10.8^m	−34° 51′	1.6′ × 1.5′
chart 20			m$_p$ 13.0	p.a. 85°

Quite bright, small, and round; bright middle. Distance 120 million ly; Hubble class E/S0.

IC 2553

PN	Car	10^h 09.4^m	−62° 37′	9″
chart 25			m$_v$ 10.3	

PK285−5.1. Appears stellar; phot. mag. 13.0; central star mag. 15.5. Distance 9,100 ly; V-V class 1.

IC 2554

Gx	Car	10^h 08.9^m	−67° 02′	3.1′ × 1.3′
chart 25			m$_p$ 12.5	p.a. 175°

Small, quite faint, and quite elongated. Distance 50 million ly; Hubble class SBb.

IC 2560

Gx	Ant	10^h 16.3^m	−33° 34′	3.3′ × 1.9′
chart 20			m$_p$ 12.6	p.a. 45°

Extremely faint and pretty small. Distance 120 million ly; Hubble class SBa. Located among four stars.

IC 2574

Gx	UMa	10^h 28.4^m	+68° 25′	13′ × 6′
chart 2			m$_v$ 10.4	p.a. 50°

Coddington's Nebula (but really a galaxy), in the M81 galaxy group. Very large, extremely faint, and elongated. Distance 8 million ly; Hubble class S+. A stellar group is at one edge.

IC 2581

OC	Car	10^h 27.4^m	−57° 38′	7′
chart 25			m$_v$ 4.3	

25 stars; detached, strong concentration of stars; large range in brightness; mag. of brightest star 4.6. Distance 6,500 ly; age 10 million years; Trumpler class I 3 m. Located about 3° NW of bright nebula Eta Carinae.

IC 2597

Gx	Hya	10^h 37.8^m	−27° 05′	1.8′ × 1.3′
chart 20			m$_p$ 12.9	p.a. 4°

Pretty bright and pretty small. Distance 90 million ly; Hubble class E. Double star located to W.

IC 2602

See Southern Pleiades.

IC 2627

Gx	Crt	11^h 09.9^m	−23° 44′	2.6′
chart 20			m$_v$ 12	

Faint, large, and round; two main arms; small, bright nucleus. Distance 80 million ly; Hubble class Sb.

IC 2714

OC	Car	$11^h 17.5^m$	$-62° 46'$	$12'$
chart 25			m_p 8.2	

100 stars; detached, weak concentration of stars; large range in brightness; large cluster; brightest star is phot. mag. 10. Distance 3,900 ly; Trumpler class II 3 m.

IC 2872

BNe	Cen	$11^h 29.0^m$	$-62° 57'$	$12' \times 4'$
chart 25				

Faint and pretty large; contains few stars; extended NNE–SSW. Illuminated by a 9.3-mag. star.

IC 2944, 48

BNe+C	Cen	$11^h 36.6^m$	$-63° 02'$	$80' \times 50'$
chart 25				

C100, Running Chicken Nebula. NW portion is mottled; S portion is brighter. Contains an open cluster (the Centaurus Cluster) with 30 stars; detached, weak concentration of stars; small brightness range; mag. of brightest star 6.4. Distance 6,900 ly; age 10 million years; Trumpler class II 1 p n.

IC 2966

BNr	Mus	$11^h 50.4^m$	$-64° 54'$	$3'$
chart 25				

Small, fairly well-defined nebula; extends E–W; illuminated by an 11.5-mag. star.

IC 2995

Gx	Hya	$12^h 05.8^m$	$-27° 56'$	$3.0' \times 0.8'$
charts 20, 21			m_v 12.4	p.a. 117°

Very faint, large, and quite elongated; patchy arms; very faint nucleus. Distance 70 million ly; Hubble class S.

IC 3253

Gx	Cen	$12^h 23.8^m$	$-34° 37'$	$2.8' \times 1.1'$
charts 20, 21			m_v 12.1	p.a. 23°

Extremely faint, elongated, and very large; bright, patchy arms. Distance 110 million ly; Hubble class S.

IC 3370

Gx	Cen	$12^h 27.6^m$	$-39° 20'$	$3.1' \times 2.5'$
charts 20, 21			m_v 11	p.a. 45°

Pretty bright, pretty large, and round; bright, patchy arms; bright, rectangular nucleus. Distance 120 million ly; Hubble class E2.

IC 3568

PN	Cam	$12^h 32.9^m$	$+82° 33'$	$10''$
chart 2			m_v 10.6	

PK123+34.1. Smooth disk; phot. mag. 11.6; central star mag. 13.5. Distance 6,900 ly; expansion velocity 8 km/sec (5 miles/sec); V-V class 2 + 2a.

IC 3896

Gx	Cen	$12^h 56.7^m$	$-50° 21'$	$2.6' \times 1.9'$
charts 20, 21, 25			m_v 11.5	p.a. 10°

Bright middle. Distance 86 million ly; Hubble class E1. Galaxy IC 3896A is located approximately $20'$ to NW.

IC 3896A

Gx	Cen	$12^h 55.5^m$	$-50° 04'$	$2.6' \times 2.0'$
charts 20, 21, 25			m_p 13.0	p.a. 12°

Distance 90 million ly; Hubble class SB. Galaxy IC 3896 located about $20'$ to SE.

IC 4182

Gx	CVn	$13^h 05.8^m$	$+37° 36'$	$6.0'$
chart 7			m_v 12.5	

Quite faint and very large. Distance 17 million ly; Hubble class Sm.

IC 4191

PN	Mus	$13^h 08.8^m$	$-67° 39'$	$5''$
chart 25			m_v 10.6	

PK304–4.1. Smooth disk; phot. mag. 12.0; central star mag. 16.4. Expansion velocity 14 km/sec (9 miles/sec); distance 7,500 ly; V-V class 2.

IC 4214

Gx	Cen	$13^h 17.7^m$	$-32° 06'$	$2.3' \times 1.3'$
chart 21			m_p 12.3	p.a. 176°

Pretty bright, pretty small, and roundish. Distance 94 million ly; Hubble class SBa. A 9th-mag. star lies to SE.

IC 4296

Gx	Cen	$13^h 36.6^m$	$-33° 58'$	$3.2'$
chart 21			m_v 10.5	

Pretty faint, pretty small, and roundish; a Seyfert galaxy. Distance 150 million ly; Hubble class E0. Brightest member of the Centaurus Galaxy Cluster.

IC 4329

Gx	Cen	$13^h 49.1^m$	$-30° 18'$	$3.5' \times 2.1'$
chart 21			m_v 11	p.a. 63°

Pretty faint, small, and elongated; bright nucleus. Distance 180 million ly; Hubble class E3. The brightest in a group of galaxies that includes NGC 5357.

IC 4351

Gx	Hya	$13^h 57.9^m$	$-29° 19'$	$5.0' \times 1.1'$
chart 21			m_v 11.8	p.a. 17°

Large, faint, very elongated, edge-on spiral; bright nucleus with dark lane. Distance 110 million ly; Hubble class Sb.

IC 4367

Gx	Cen	$14^h 05.6^m$	$-39° 12'$	$1.7' \times 1.5'$
chart 21			m_p 13.0	p.a. 60°

Extremely faint, pretty small, and roundish. Distance 170 million ly; Hubble class SBb. Located between two stars.

IC 4386

Gx	Cen	$14^h 15.1^m$	$-43° 58'$	$3.2' \times 1.8'$
chart 21			m_p 13.0	p.a. 147°

Faint, very small, and roundish. Distance 76 million ly; Hubble class SBc.

IC 4402

Gx	Lup	$14^h 21.2^m$	$-46° 18'$	$4.9' \times 0.9'$
charts 21, 25			m_p 12.1	p.a. 127°

Large and extremely elongated with pointed ends. Distance 66 million ly; Hubble class Sc.

IC 4406

PN	Lup	$14^h 22.4^m$	$-44° 09'$	35″
chart 21			m_v 10.2	

PK319+15.1. Ring structure and irregular disk; phot. mag. 10.6; central star mag. 14.7. Distance 4,900 ly; expansion velocity 6 km/sec (4 miles/sec); V-V class 4 + 3.

IC 4444

Gx	Lup	$14^h 31.7^m$	$-43° 25'$	$1.7' \times 1.3'$
chart 21			m_v 11.4	p.a. 75°

Very faint and very small; very thick arms; nucleus appears starlike. Distance 77 million ly; Hubble class S(B)b.

IC 4499

GC	Aps	$15^h 00.3^m$	$-82° 13'$	$7'$
charts 25, 26			m_v 10.1	

Low concentration of stars; very faint. Distance 61,000 ly; S-S class 11.

IC 4538

Gx	Lib	$15^h 21.2^m$	$-23° 39'$	$2.6' \times 2.0'$
chart 21			m_p 12.9	p.a. 50°

Extremely faint and very large. Distance 120 million ly; Hubble class SBc.

IC 4585

Gx	TrA	$16^h 00.3^m$	$-66° 19'$	$2.2' \times 0.8'$
charts 25, 26			m_p 13.0	p.a. 45°

Small and extremely faint; irregular shape. Distance 160 million ly; Hubble class SBb.

IC 4592

BNr	Sco	$16^h 12.0^m$	$-19° 28'$	$2.5° \times 1.0°$
charts 15, 22				

Very large, elongated, and extremely faint; outer region is red due to light from star Antares (α Sco); brightest in region surrounding star 14 ν Sco. Dark nebula B40 located in NE portion.

IC 4593

PN	Her	$16^h 12.2^m$	$+12°\,04'$	$13''$
chart 15			$m_v\,10.7$	

PK25+40.1, the White-Eyed Pea. Small, faint, and stellar; surrounded by much larger, fainter disk; phot. mag. 10.9; central star mag. 11.2. Distance 6,500 ly; expansion velocity 13 km/sec (8 miles/sec); V-V class 2 + 2.

IC 4595

Gx	TrA	$16^h 20.7^m$	$-70°\,09'$	$3.0' \times 0.6'$
chart 26			$m_p\,12.7$	p.a. 60°

Faint, small, and extremely elongated. Distance 140 million ly; Hubble class Sc.

IC 4601

BNr	Sco	$16^h 20.0^m$	$-20°\,02'$	$19' \times 9'$
chart 22				

Stars of mag. 6 and 7 involved in a large, diffuse nebula consisting of several components. Located between dark nebula B41 and the SE part of bright nebula IC 4592.

IC 4603

BNr	Oph	$16^h 25.6^m$	$-24°\,28'$	$19' \times 9'$
charts 15, 22				

Very large, extremely faint, and diffuse.

IC 4604

BNr	Oph	$16^h 25.6^m$	$-23°\,26'$	$60' \times 25'$
chart 22				

Also called the Rho Ophiuchi Nebula. Extremely large, very faint; contains dark nebula B42; illuminated by the 4.6-mag. star ρ Oph.

IC 4605

BNr	Sco	$16^h 30.2^m$	$-25°\,06'$	$30'$
chart 22				

Very large, extremely faint; illuminated by a 4.8-mag. star (22 Sco). Located about 1° N of star Antares (α Sco) and immediately SW of dark nebula B44.

IC 4606

BN	Sco	$16^h 31.6^m$	$-26°\,03'$	$60' \times 40'$
chart 22				

Nebulosity surrounding star Antares (α Sco).

IC 4618

Gx	Aps	$16^h 57.9^m$	$-77°\,00'$	$1.8' \times 1.3'$
chart 26			$m_p\,12.7$	p.a. 118°

Extremely faint and extremely small; two-branched spiral; per *NGC*, a (!!) very remarkable object; Hubble class SBb.

IC 4628

BN	Sco	$16^h 57.0^m$	$-40°\,20'$	$90' \times 60'$
chart 22				

Elongated; very irregular shape. Located in N part of open cluster Tr 24.

IC 4633

Gx	Aps	$17^h 13.8^m$	$-77°\,32'$	$4.1' \times 3.1'$
chart 26			$m_p\,12.4$	p.a. 145°

Very faint and quite large; brighter toward middle. Distance 120 million ly; Hubble class Sc.

IC 4634

PN	Oph	$17^h 01.6^m$	$-21°\,50'$	$8''$
chart 22			$m_v\,10.9$	

PK0+12.1. Bright, extremely small, blue disk with irregular outer disk; phot. mag. 10.7; central star mag. 13.9. Distance 9,500 ly; expansion velocity 15 km/sec (9 miles/sec); V-V class 2a + 3. Lies about 0.5° N of dark nebulae B47 and B51.

IC 4637

PN	Sco	$17^h 05.2^m$	$-40°\,53'$	$18''$
chart 22			$m_v\,12.5$:	

PK345+00.1. Faint and small, nearly stellar; phot. mag. 13.6; central star mag. 12.5. Distance 5,000 ly; V-V class 3.

IC 4642

PN	Ara	$17^h 11.8^m$	$-55°\,24'$	$16''$
charts 22, 26			$m_v\,12.4$	

PK334–9.1. Ring structure; central star mag. 15.7. Distance 11,000 ly; V-V class 4.

IC 4646

Gx	Ara	$17^h\ 23.9^m$	–60° 00′	3.2′ × 2.2′
chart 26			m_p 12.6	p.a. 0°

Faint and pretty large; spiral structure. Distance 130 million ly; Hubble class Sc.

IC 4651

OC	Ara	$17^h\ 25.0^m$	–49° 57′	12′
charts 22, 26			m_v 6.9	

80 stars; detached, weak concentration of stars; large range in brightness; large cluster; mag. of brightest star 9.0. Distance 2,500 ly; age 2.4 billion years; Trumpler class II 3 m.

IC 4654

Gx	Aps	$17^h\ 37.1^m$	–74° 23′	1.7′ × 1.1′
chart 26			m_p 13.0	p.a. 102°

Extremely faint, extremely small, and roundish; much brighter toward middle. Distance 220 million ly; Hubble class SBb.

IC 4662

Gx	Pav	$17^h\ 47.1^m$	–64° 38′	2.8′ × 1.5′
chart 26			m_v 11.3	p.a. 105°

Faint, pretty small, and slightly elongated; very faint nucleus. Distance 7 million ly; Hubble class Ir+. Lies just NE of star η Pav.

IC 4663

PN	Sco	$17^h\ 45.5^m$	–44° 54′	13″
chart 22			m_v 12.3	

PK346–8.1. Very faint and small; ring structure; phot. mag. 13.1; central star mag. 15.2. Expansion velocity 15 km/sec (9 miles/sec); distance 9,800 ly; V-V class 4.

IC 4665

OC	Oph	$17^h\ 46.3^m$	+05° 43′	40′
charts 15, A1			m_v 4.2	

30 stars; detached, no concentration of stars; moderate range in brightness; very large; mag. of brightest star 6.9. Distance 1,400 ly; age 36 million years; Trumpler class III 2 p.

IC 4679

Gx	Tel	$18^h\ 11.4^m$	–56° 15′	2.3′ × 0.9′
charts 22, 26			m_p 12.8	p.a. 99°

Quite faint; small and roundish; much brighter toward middle. Distance 160 million ly; Hubble class SBc.

IC 4684

BNr	Sgr	$18^h\ 09.1^m$	–23° 25′	3′ × 2′
chart 22				

A 9th-mag. star involved in a small patch of nebulosity.

IC 4685

BNr	Sgr	$18^h\ 09.3^m$	–23° 59′	10′ × 7′
chart 22				

Large, faint, diffuse nebula in a group with dark nebulae B91 and B303 and bright nebulae IC 1274, IC 1275, and NGC 6559.

IC 4699

PN	Tel	$18^h\ 18.5^m$	–45° 59′	5″
chart 22			m_v 13.3:	

PK348–13.1. Small, faint, appears stellar; phot. mag. 11.9; mag. of central star 15.1. Expansion velocity 10 km/sec (6 miles/sec); distance 17,000 ly; V-V class 2.

IC 4703

BNe	Sgr	$18^h\ 18.6^m$	–13° 58′	35′ × 28′
charts 15, 16				

A small patch of nebulosity associated with the Eagle Nebula or Star Queen Nebula (M16). Brightest area is to the SE and contains a conspicuous absorption patch that on photographs resembles a horse and rider.

IC 4704

Gx	Pav	$18^h\ 27.9^m$	–71° 37′	2.8′ × 2.2′
chart 26			m_p 13.0	p.a. 165°

Quite bright; bright middle. Distance 150 million ly; Hubble class E/S0.

IC 4706

BN　　Sgr　　$18^h\,19.6^m$　　$-16°\,01'$
charts 15, 16

A 9.2-mag. star involved in a small patch of nebulosity. Connected with the Swan Nebula or Omega Nebula (M17).

IC 4710

Gx　　Pav　　$18^h\,28.6^m$　　$-66°\,59'$　　$3.3'\times2.8'$
chart 26　　　　　　　　$m_v\,12$　　　p.a. 5°

Small and very faint; several superimposed stars; Hubble class SBm.

IC 4712

Gx　　Pav　　$18^h\,31.1^m$　　$-71°\,42'$　　$2.4'\times1.3'$
chart 26　　　　　　　　$m_p\,13.0$　　p.a. 60°

Very faint, very small, and roundish; bright middle. Distance 150 million ly; Hubble class Sb.

IC 4721

Gx　　Pav　　$18^h\,34.4^m$　　$-58°\,30'$　　$5.0'\times1.6'$
chart 26　　　　　　　　$m_v\,11.9$　　p.a. 146°

Very elongated; very small and bright nucleus. Distance 250 million ly; Hubble class SBc. Paired with galaxy IC 4720, which lies about 8′ WNW.

IC 4725

See M25.

IC 4732

PN　　Sgr　　$18^h\,33.9^m$　　$-22°\,39'$　　　　4″
chart 22　　　　　　　　$m_v\,12.1$

PK10–6.1. Appears stellar; phot. mag. 13.3; central star mag. >16.2. Distance 9,000 ly; V-V class 1. Located approx. 1.5° WNW of birght globular cluster M22.

IC 4742

Gx　　Pav　　$18^h\,41.9^m$　　$-63°\,52'$　　$1.6'\times1.3'$
chart 26　　　　　　　　$m_p\,13.0$　　p.a. 20°

Quite faint, very small, and roundish; bright toward middle. Distance 190 million ly; Hubble class E.

IC 4756

OC　　Ser　　$18^h\,39.0^m$　　$+05°\,27'$　　50′
charts 15, 16　　　　　　$m_v\,4.6$

80 stars; detached, no concentration of stars; moderate range in brightness; a very large cluster; mag. of brightest star 8.7. Can be seen with naked eye under very dark skies. Distance 1,300 ly; age 580 million years. Trumpler class III 2 m. Located in Serpens Cauda.

IC 4765

Gx　　Pav　　$18^h\,47.3^m$　　$-63°\,20'$　　$3.2'\times1.9'$
chart 26　　　　　　　　$m_p\,12.4$　　p.a. 115°

Very faint, small, and roundish; bright middle. Distance 190 million ly; Hubble class E.

IC 4776

PN　　Sgr　　$18^h\,45.9^m$　　$-33°\,21'$　　　　8″
chart 22　　　　　　　　$m_v\,10.4$

PK2–13.1. Smooth disk with brighter central region; phot. mag. 11.7; central star mag. 14.1. Distance 11,000 ly; V-V class 2a.

IC 4785

Gx　　Pav　　$18^h\,52.9^m$　　$-59°\,15'$　　$3.4'\times1.5'$
chart 26　　　　　　　　$m_p\,13.0$　　p.a. 140°

Extremely faint and very small, with irregular shape; stellar nucleus. Distance 160 million ly; Hubble class SBb.

IC 4797

Gx　　Tel　　$18^h\,56.5^m$　　$-54°\,18'$　　$2.7'\times1.3'$
charts 22, 26　　　　　　$m_v\,11.3$　　p.a. 146°

Bright nucleus. Distance 110 million ly; Hubble class E5. Paired with galaxy IC 4796, which is one mag. fainter and 6′ north. Galaxy E 183-30 is located about 20′ SSE.

IC 4808

Gx　　CrA　　$19^h\,01.1^m$　　$-45°\,19'$　　$2.0'\times0.8'$
chart 22　　　　　　　　$m_p\,12.9$　　p.a. 45°

Very faint, quite large, and quite elongated; a little brighter toward middle. Distance 220 million ly; Hubble class Sb.

IC 4812

BNr	CrA	$19^h 01.1^m$	$-37° 04'$	$10' \times 7'$
chart 22				

Patch of nebulosity surrounding two 6.5-mag. stars.

IC 4831

Gx	Pav	$19^h 14.7^m$	$-62° 16'$	$3.5' \times 1.0'$
chart 26			m_p 12.7	p.a. 111°

Quite faint, very small, and very elongated. Distance 190 million ly; Hubble class Sa.

IC 4837

Gx	Tel	$19^h 15.2^m$	$-54° 40'$	$2.3' \times 1.2'$
charts 22, 26			m_v 12.4	p.a. 8°

Faint, quite small; faint nucleus. Distance 110 million ly; Hubble class SBcp. Galaxy IC 4837A located approx. 0.5° to N.

IC 4837A

Gx	Tel	$19^h 15.3^m$	$-54° 08'$	$4.1' \times 0.7'$
charts 22, 26			m_p 12.5	p.a. 165°

Distance 120 million ly; Hubble class Sa. Galaxy IC 4837 located approx. 0.5° to S.

IC 4845

Gx	Pav	$19^h 20.4^m$	$-60° 23'$	$1.9' \times 1.6'$
chart 26			m_p 12.4	p.a. 87°

Distance 170 million ly; Hubble class SBa. Located about 15′ E of the galaxy pair NGC 6769 and NGC 6770.

IC 4846

PN	Aql	$19^h 16.5^m$	$-09° 03'$	2″
chart 16			m_v 11.9	

PK27–9.1. Smooth disk; phot. mag. 12.7; central star mag. 15.2. Distance 9,000 ly; expansion velocity 19 km/sec (12 miles/sec); V-V class 2.

IC 4889

Gx	Tel	$19^h 45.3^m$	$-54° 21'$	$3.0' \times 1.9'$
charts 22, 26			m_v 11.3	p.a. 0°

Hexagonal shape. Distance 100 million ly; Hubble class E5. Faint star located approx. 6″ from nucleus.

IC 4901

Gx	Pav	$19^h 54.4^m$	$-58° 43'$	$4.9' \times 3.3'$
chart 26			m_p 12.1	p.a. 135°

Quite faint, small, and slightly elongated. Distance 90 million ly; Hubble class Sc.

IC 4931

Gx	Sgr	$20^h 00.8^m$	$-38° 34'$	$2.3' \times 1.9'$
charts 22, 23			m_p 12.9	p.a. 120°

Extremely faint, pretty small, and roundish. Distance 260 million ly; Hubble class E/S0. An 8th-mag. star is located about 20″ to E.

IC 4933

Gx	Tel	$20^h 03.5^m$	$-54° 59'$	$2.3' \times 1.9'$
chart 26			m_p 13.0	p.a. 0°

Extremely faint, extremely small, two-branched spiral. Distance is roughly 210 million ly; Hubble class SBb.

IC 4946

Gx	Sgr	$20^h 24.0^m$	$-44° 00'$	$2.5' \times 1.0'$
charts 22, 23, 26			m_p 12.6	p.a. 68°

Extremely faint, small, and roundish. Distance 130 million ly; Hubble class SBa.

IC 4954, 5

BNr	Vul	$20^h 04.8^m$	$+29° 15'$	25′
charts 8, 9				

A pair of detached nebulosities separated by about 3.5′ and aligned NW–SE.

IC 4991

Gx	Sgr	$20^h 18.4^m$	$-41° 03'$	$2.5' \times 1.6'$
charts 22, 23			m_p 12.8	p.a. 145°

Very faint, quite small, and roundish. Distance 250 million ly; Hubble class S0.

IC 4996

OC	Cyg	$20^h 16.5^m$	$+37° 38'$	6′
charts 8, 9			m_v 7.3	

15 stars; detached, strong concentration of stars; large range in brightness; small; mag. of brightest

star 8.5; involved in nebulosity. Distance 5,300 ly; age 10 million years; Trumpler class I 3 p n.

IC 4997

PN	Sge	$20^h\,20.2^m$	$+16°\,44'$	1.6″
charts 9, 16			m_v 10.5	

PK58–10.1. Appears stellar; phot. mag. 11.6; central star mag. 14.4. Distance 5,000 ly; expansion velocity 15 km/sec (9 miles/sec); V-V class 1.

IC 5013

Gx	Mic	$20^h\,28.6^m$	$-36°\,02'$	$2.2' \times 1.1'$
charts 22, 23			m_p 12.6	p.a. 19°

Extremely small and elongated. Distance 100 million ly; Hubble class S0.

IC 5020

Gx	Mic	$20^h\,30.6^m$	$-33°\,29'$	$2.9' \times 2.0'$
chart 23			m_v 12.1	p.a. 153°

Pretty faint, pretty small, and elongated. Distance 140 million ly; Hubble class Sb.

IC 5052

Gx	Pav	$20^h\,52.1^m$	$-69°\,12'$	$5.0' \times 0.9'$
chart 26			m_v 11	p.a. 143°

Faint, large, and extremely elongated; very faint nucleus. Distance 34 million ly; Hubble class SBd.

IC 5063

Gx	Ind	$20^h\,52.0^m$	$-57°\,04'$	$2.5' \times 1.7'$
chart 26			m_v 11.9	p.a. 116°

Strong dust lane with bright bulge; small, bright nucleus. Distance about 150 million ly; Hubble class S0.

IC 5067

BN	Cyg	$20^h\,47.8^m$	$+44°\,22'$	$25' \times 10'$
charts 8, 9				

W portion of the Pelican Nebula. Distance 4,000 ly.

IC 5068

BNe	Cyg	$20^h\,50.8^m$	$+42°\,31'$	$80' \times 29'$
chart 9				

Large, very faint nebula SW of North America.

IC 5070

BNe	Cyg	$20^h\,50.8^m$	$+44°\,21'$	$80' \times 70'$
chart 9				

E portion of the Pelican Nebula. Very large and faint. Distance 4,000 ly.

IC 5076

BNr	Cyg	$20^h\,55.9^m$	$+47°\,25'$	$8' \times 5'$
charts 8, 9				

Very faint nebula. Bright star located in NE portion.

IC 5092

Gx	Pav	$21^h\,16.2^m$	$-64°\,28'$	$2.7' \times 2.2'$
chart 26			m_p 13.0	p.a. 8°

Pretty large and elongated; brighter toward middle. Distance 140 million ly; Hubble class SBc.

IC 5105

Gx	Mic	$21^h\,24.4^m$	$-40°\,32'$	$2.8' \times 1.7'$
charts 23, 26			m_v 11.4	p.a. 40°

Very faint, very small, and round. Distance 230 million ly; Hubble class E4. Stars are located at opposite ends.

IC 5117

PN	Cyg	$21^h\,32.5^m$	$+44°\,35'$	1.2″
chart 9			m_v 11.5	

PK89–5.1. Smooth disk; phot. mag. 13.3; central star mag. 16.7. Distance 7,200 ly; expansion velocity 18 km/sec (11 miles/sec); V-V class 2.

IC 5146

BNe+C	Cyg	$21^h\,53.4^m$	$+47°\,16'$	12′
chart 9				

C19, Cocoon Nebula. Large, pretty bright; several absorption patches in interior; small, detached patch located 10′ WSW of center. Contains an open cluster of 20 stars; not well detached from surrounding star field; moderate range in brightness; total mag. 7.2.

IC 5148

PN	Gru	21^h 59.5^m	−39° 23′		1.9′
chart 23			m$_v$ 11:		

PK2–52.1; central star mag. 16.5; V-V class 4. IC 5150 is probably the same object.

IC 5152

Gx	Ind	22^h 02.7^m	−51° 18′		5.0′ × 3.7′
charts 23, 26			m$_v$ 10.6		p.a. 100°

Fairly bright center; a dwarf. Distance 5 million ly; extends 7,000 ly; Hubble class Ir+. A member of the Local Group.

IC 5156

Gx	PsA	22^h 03.2^m	−33° 50′		2.3′ × 0.8′
chart 23			m$_v$ 12.1		p.a. 175°

Pretty faint, pretty small, and elongated; with patchy spiral arms. Distance 110 million ly; Hubble class Sb.

IC 5179

Gx	Gru	22^h 16.1^m	−36° 51′		2.5′ × 1.2′
chart 23			m$_p$ 12.3		p.a. 57°

Large, very faint and roundish. Distance 150 million ly; Hubble class Sb. A star is located at S, and a bright star to SW.

IC 5181

Gx	Gru	22^h 13.4^m	−46° 01′		2.5′ × 0.9′
charts 23, 26			m$_v$ 11.4		p.a. 74°

Bright, smooth, edge on; extremely bright nucleus. Distance 89 million ly; Hubble class S0. Paired with galaxy NGC 7232A.

IC 5186

Gx	Gru	22^h 18.8^m	−36° 48′		1.7′ × 1.2′
chart 23			m$_p$ 12.6		p.a. 112°

Extremely faint, small, and roundish. Distance 210 million ly; Hubble class SBb. Faint stars to W.

IC 5201

Gx	Gru	22^h 21.0^m	−46° 02′		8.0′ × 4.1′
charts 23, 26			m$_v$ 11.1		p.a. 33°

Large and very faint; short, bright, knotty bar; filamentary arms. Distance 91 million ly; Hubble class SBc. Rich cluster of galaxies nearby.

IC 5217

PN	Lac	22^h 23.9^m	+50° 58′		7″
charts 3, 9			m$_v$ 11.3		

PK100–5.1. Smooth disk; phot. mag. 12.6; central star mag. 15.5. Expansion velocity 18 km/sec (11 miles/sec); distance 10,000 ly; V-V class 2.

IC 5240

Gx	Gru	22^h 41.9^m	−44° 46′		3.0′ × 1.9′
chart 23			m$_v$ 11.4		p.a. 100°

Small, bright nucleus; has two superimposed stars on bright inner ring. Distance 64 million ly; Hubble class SBa.

IC 5250

Gx	Tuc	22^h 47.3^m	−65° 03′		3.2′ × 2.9′
charts 24, 26			m$_p$ 12.1		p.a. 175°

Quite bright, small, and roundish. Distance 140 million ly; Hubble class S0. Faint stars located approx. 0.5° E. Paired with galaxy IC 5250A.

IC 5250A

Gx	Tuc	22^h 47.4^m	−65° 03′		3.3′ × 3.0′
charts 24, 26			m$_p$ 12.2		p.a. 135°

Distance 140 million ly; Hubble class S0. Paired with galaxy IC 5250.

IC 5267

Gx	Gru	22^h 57.2^m	−43° 24′		5.0′ × 3.7′
chart 23			m$_v$ 10.3		p.a. 140°

Pretty bright, small, and roundish; bright nucleus. Distance 74 million ly; Hubble class S0.

IC 5270

Gx	PsA	22^h 57.9^m	−35° 51′		3.2′ × 0.6′
chart 23			m$_p$ 12.9		p.a. 103°

Very faint, pretty small, and very elongated. Distance 84 million ly; Hubble class SBc.

IC 5271

Gx	PsA	22ʰ 58.0ᵐ	−33° 45′	2.6′ × 0.9′
chart 23			m_v 11.7	p.a. 138°

Pretty faint, pretty small, slightly elongated; patchy outer arms. Distance 76 million ly; Hubble class Sb.

IC 5273

Gx	Gru	22ʰ 59.4ᵐ	−37° 42′	2.6′ × 1.7′
chart 23			m_v 11.1	p.a. 56°

Very faint, quite large, and slightly elongated; small, bright nucleus. Distance about 57 million ly; Hubble class SBc.

IC 5325

Gx	Phe	23ʰ 28.7ᵐ	−41° 20′	2.8′
charts 18, 23			m_v 11.1	p.a. 8°

Faint, small, and round; small, bright nucleus. Distance 65 million ly; Hubble class S(B)b. Bright star located approx. 1.1′ to SW.

IC 5328

Gx	Phe	23ʰ 33.3ᵐ	−45° 01′	2.6′ × 1.6′
charts 18, 23			m_v 11.2	p.a. 40°

Very faint and small; bright, diffuse nucleus. Distance 130 million ly; Hubble class E4. Paired with galaxy IC 5328A.

IC 5332

Gx	Scl	23ʰ 34.5ᵐ	−36° 06′	9′
charts 18, 23			m_v 10.3	

Extremely faint and very large; very small, bright nucleus. Distance 31 million ly; Hubble class Sd. Located between two stars.

II Zw 5

Gx	Cet	02ʰ 41.3ᵐ	+04° 13′	0.2′
chart 10			m_p 11.8	

Faint, small, and round.

III Zw 66

Gx	Com	12ʰ 27.2ᵐ	+14° 07′	0.4′
charts 13, 14, B1			m_p 12.3	

Very small and round.

Intergalactic Wanderer

See NGC 2419.

J320

PN	Ori	05ʰ 05.6ᵐ	+10° 42′	6″
chart 11			m_v 12	

PK190−17.1. Smooth disk involved in a larger, fainter halo of nebulosity; obviously nonstellar; phot. mag. 12.9; central star mag. 14.4. Expansion velocity 18 km/sec (11 miles/sec); distance 7,500 ly; V-V class 2 + 4.

J900

PN	Gem	06ʰ 26.0ᵐ	+17° 48′	9″
charts 5, 11, 12			m_v 11.7	

PK194+02.1. Bright, very small; appears almost stellar; irregular shape with traces of ring structure; involved in a larger, fainter disk of smooth nebulosity; phot. mag. 12.4; central star mag. 17.8. Distance 7,500 ly; expansion velocity 18 km/sec (11 miles/sec); V-V class 3b + 2.

Jewel Box Cluster

OC	Cru	12ʰ 53.6ᵐ	−60° 20′	10′
chart 25			m_v 4.2	

NGC 4755, C94. Detached, strong concentration of stars; large range in brightness; very large and rich; one of the youngest open clusters; mag. of brightest star 5.8. Ranks among the finest objects in the sky's southern hemisphere. A gorgeous piece of fancy jewelry (John Herschel). Distance 7,600 ly; age 7.1 million years; Trumpler class I 3 r.

K14

OC	Cas	00ʰ 31.9ᵐ	+63° 10′	7′
charts 1, 3			m_v 8.5	

20 stars; detached, no concentration; small brightness range; mag. of brightest star 11.3. Distance 900 ly; age 16 million years; Trumpler class III 2p.

Kemble's Cascade

—	Cam	03ʰ 57.0ᵐ	+63° 00′	3.0°
chart 1				

Running NW of open cluster NGC 1502, this chain

Keyhole Nebula

See Eta Carinae Nebula.

Lagoon Nebula

BNe Sgr 18^h 03.8^m −24° 23′ 90′ × 39′
chart 22 m$_v$ 5.8:

M8, NGC 6523. Very large, very bright nebula; contains open cluster NGC 6530; per *NGC*, a (!!!) most remarkable object. A large, floating nebulous patch crossed by a great, curving dark lane (Mullaney). Located approx. 1.5° SSE of the Trifid Nebula (M20).

Large Magellanic Cloud

Gx Dor 05^h 23.6^m −69° 45′ 11° × 9°
charts 24, 25 m$_v$ 0.1

Irregular galaxy in the Local Group; possible satellite of the Milky Way. Prominently visible to the naked eye in the south circumpolar sky. Bright, central bar and vague spiral structure suggest Hubble class SBm. Its Hipparcos-calibrated distance is 180,000 ly.

LBN 1036

BNe Mon 07^h 16.0^m −10° 40′ 60′ × 10′
chart 12

Very faint and diffuse; the S-shaped half is brightest. Open cluster NGC 2353 is located in NW part.

LDN 134

DN Lib 15^h 53.6^m −04° 39′ 21′ × 12′
charts 14, 15

Elliptical shape, high opacity; includes globule.

LDN 557

DN Ser 18^h 38.6^m −01° 47′ 60′ × 9′
charts 15, 16

Irregular or crescent shape, high opacity.

LDN 582

DN Aql 18^h 52.6^m −01° 56′ 50′ × 10′
charts 15, 16

Irregular shape, high opacity.

LDN 617

DN Aql 18^h 57.5^m +01° 04′ 3.0° × 0.3°
charts 15, 16

Irregular shape, high opacity.

LDN 663

DN Aql 19^h 36.9^m +07° 34′ 4.0′
chart 16

Elliptical shape, high opacity; includes globule.

LDN 673

DN Aql 19^h 20.9^m +11° 16′ 50′ × 15′
chart 16

High opacity, irregular shape; filamentary appearance. Lies approx. 1° S of dark nebula LDN 684.

LDN 684

DN Aql 19^h 21.8^m +12° 26′ 50′ × 10′
chart 16

High opacity, irregular shape; lies approx. 1° N of dark nebula LDN 673.

LDN 889

DN Cyg 20^h 24.8^m +40° 10′ 90′ × 19′
charts 8, 9

Medium opacity; a wide lane E of star γ Cyg (Sadr).

LDN 935

DN Cyg 20^h 56.8^m +43° 52′ 2.5° × 0.7°
chart 9

Medium opacity; a wide lane separating the North America Nebula (NGC 7000) and the Pelican Nebula (IC 5067 and IC 5070).

LDN 1616

DN Ori 05^h 06.5^m −03° 30′ 10′ × 8′
chart 11

[continued from previous page] of stars stretches for almost 3°. While not labeled in *Sky Atlas 2000.0*, its brightest stars are plotted. Best viewed with binoculars. A celestial waterfall of dozens of 9th- and 10th-mag. stars (Houston).

High opacity, circular shape; apparently involved with bright nebula NGC 1788, which is located in NE part.

LDN 1710

DN   Sco   17^h 20.7^m –31° 57′   60′ × 9′
chart 22

High opacity, irregular shape.

LDN 1773

See Pipe Nebula.

Leo I

Gx   Leo   10^h 08.5^m +12° 18′   10′ × 7′
chart 13   m$_v$ 9.8   p.a. 80°

UGC 5470. Low surface brightness. Were it not for glare from Regulus, this dwarf galaxy would be an easy target for a 10-inch (Houston). Extends 2,200 ly; distance 750,000 ly; Hubble class dE3. A member of the Local Group.

Leo II

See UGC 6253.

Leo's Triplet

Gx   Com   11^h 20.0^m +13° 20′   0.5°
chart 13   m$_v$ 9

Three galaxies (M65, M66 and NGC 3628) located within 0.5° of each other. Use low power to get all three in the field; a very nice sight.

Letter S Cluster

See NGC 663.

Letter Y Cluster

See NGC 1893.

LH 120

BNe   Dor   05^h 44.0^m –67° 50′   7′
chart 24

Faint ring with several stars involved in nebulosity; possibly a supernova remnant. Located approx. 1° NE of the Tarantula Nebula (NGC 2070).

Little Dumbbell Nebula

See M76.

Little Gem Nebula

See NGC 6818.

Little Ghost Nebula

See NGC 6369.

Lost Galaxy

See NGC 4526.

Lower's Nebula

See Sh2-261.

Ly 2

OC   Cen   14^h 24.5^m –61° 20′   12′
charts 25, A3   m$_v$ 6.4

30 stars; not well detached; moderate brightness range; mag. of brightest star 7.7. Distance 3,600 ly; age 32 million years; Trumpler class IV 2 p.

M1

See Crab Nebula.

M2

GC   Aqr   21^h 33.5^m –00° 49′   15′
charts 16, 17   m$_v$ 6.5

NGC 7089. High concentration of stars; very bright, very large, very well resolved. Beautiful, large, and round, showing with 3.7-inch a granulated aspect (Webb). Per *NGC*, a (!!) very remarkable object. Extends 240 ly; distance 55,000 ly; S-S class 2.

M3

GC   CVn   13^h 42.2^m +28° 23′   16′
chart 7   m$_v$ 6.4

NGC 5272. Medium concentration of stars; extremely bright and very large. In a 4-inch at 72× the finely resolved cluster has a three-dimensional quality (O'Meara). Per *NGC*, a (!!) very remarkable object. Contains 45,000 stars; extends 220 ly; distance 48,000 ly; age 6.5 billion years. S-S class 6.

M4

GC	Sco	16^h 23.6^m	−26° 32′	34′
chart 22			m$_v$ 5.9	

NGC 6121. Low concentration of stars; very well resolved; interior appears to contain a line of stars. This object is elongated vertically (Smyth). In a 4-inch this cluster is a well-defined glow with a brighter center, the outer parts broken up into faint stars (Mallas). Extends 100 ly; distance 10,000 ly; S-S class 9. Located approx. 1° W of the star Antares (α Sco).

M5

GC	Ser	15^h 18.6^m	+02° 05′	21′
charts 14, 15			m$_v$ 5.8	

NGC 5904. Medium concentration of stars; very bright, very large, and extremely rich; slightly oval shaped in NE–SW direction. Per *NGC*, a (!!) very remarkable object. Extends 150 ly; distance 24,000 ly; S-S class 5. Located in Serpens Caput.

M6

OC	Sco	17^h 40.1^m	−32° 13′	15′
chart 22			m$_v$ 4.2	

NGC 6405, the Butterfly Cluster. 80 stars; detached, no concentration of stars; moderate range in brightness; mag. of brightest star 6.2. Extends 6 ly; distance 1,300 ly; age 51 million years; Trumpler class III 2 p.

M7

OC	Sco	17^h 53.9^m	−34° 49′	80′
chart 22			m$_v$ 3.3	

NGC 6475. 80 stars; detached, weak concentration of stars; moderate range in brightness; very large, very bright; mag. of brightest star 5.6. A sparkling appearance with an opera glass (Serviss). Viewed in binoculars, a gem-studded cross inside a double halo of similarly bright stars (O'Meara). Extends 20 ly; distance 800 ly; age 220 million years; Trumpler class II 2 r.

M8

See Lagoon Nebula.

M9

GC	Oph	17^h 19.2^m	−18° 31′	10′
charts 15, 22			m$_v$ 7.9:	

NGC 6333. Medium concentration of stars; bright, large, round, and very well resolved. A myriad of minute stars, wonderfully aggregated, with numerous outliers seen by glimpses (Smyth). Extends 75 ly; distance 26,000 ly; S-S class 8. Located at E end of dark nebula B64.

M10

GC	Oph	16^h 57.1^m	−04° 06′	19′
chart 15			m$_v$ 6.6	

NGC 6254. Medium concentration of stars; very large, round, bright; very well resolved in apertures greater than 10-inch. Look for a yellow "spark" at the very center (O'Meara). Along with near-twin M12, this is the best of the great swarm of globulars in Ophiuchus (Mullaney). Extends 90 ly; distance 16,000 ly; S-S class 7.

M11

OC	Sct	18^h 51.1^m	−06° 16′	14′
charts 15, 16			m$_v$ 5.8	

NGC 6705, the Wild Duck Cluster. About 200 stars; detached, strong concentration; moderate range in brightness; very bright and large. Resembles a flight of wild ducks in shape (Smyth). Noble fan-shaped cluster (Webb). Per *NGC*, a (!!) very remarkable object. Extends 22 ly; distance 5,500 ly; age 220 million years; Trumpler class I 2 r. Discovered by Gottfried Kirch in 1681. The brightest star (mag. 8.0) near the cluster's center may be in the foreground. A pair of 9th-mag. stars lie just to SE.

M12

GC	Oph	16^h 47.2^m	−01° 57′	14′
chart 15			m$_v$ 6.6	

NGC 6218. Low concentration of stars; very bright and very large. At low power there is a rocket-shaped asterism just N of M12, and the globular looks like a puff of smoke from the rocket's exhaust (O'Meara). Per *NGC*, a (!!) very remarkable object. Extends 80 ly; distance 19,000 ly; S-S class 9.

M13

| GC | Her | 16^h 41.7^m | +36° 28' | 20' |
| chart 8 | | | m$_v$ 5.9 | |

NGC 6205, the Hercules Cluster. Extremely bright, very large, round, and very rich. The best globular cluster north of the celestial equator; a naked-eye object under very dark skies. Discovered by Edmond Halley in 1714. While Messier never saw its individual stars, even a 3-inch telescope begins to show them at the cluster's edge. An extensive and magnificent mass of stars (Smyth). Three dark rifts radiate outward from near the center, like a dark "propeller" (Houston). Per *NGC*, a (!!) very remarkable object. Contains 500,000 stars; extends 150 ly; distance 26,000 ly; extremely old, about 14 billion years; S-S class 5. Galaxy NGC 6207 lies approx. 0.5° to NE.

M14

| GC | Oph | 17^h 37.6^m | –03° 15' | 10' |
| chart 15 | | | m$_v$ 7.6 | |

NGC 6402. Bright, very large, round, and extremely rich. Per *NGC*, a (!) remarkable object. Extends 70 ly; distance 24,000 ly.

M15

| GC | Peg | 21^h 30.0^m | +12° 10' | 18' |
| charts 16, 17 | | | m$_v$ 6.4 | |

NGC 7078. High concentration of stars; very bright, very large, and very well resolved. Noble cluster, not exactly round (Smyth). Per *NGC*, a (!) remarkable object. Extends 250 ly; distance 50,000 ly; S-S class 4. Contains a very small, very faint (phot. mag. 15.5) planetary nebula, PK65–27.1.

M16

See Eagle Nebula.

M17

See Omega Nebula.

M18

| OC | Sgr | 18^h 19.9^m | –17° 08' | 8' |
| charts 15, 16, 22 | | | m$_v$ 6.9 | |

NGC 6613. 20 stars; detached, weak concentration of stars; large range in brightness; seems to have a nebulous glow; mag. of brightest star 8.7. Glorious field in very rich vicinity (Webb). Extends 9 ly; distance 3,900 ly; age 32 million years; Trumpler class II 3 p n. Located approx. 1° SSW of M17.

M19

| GC | Oph | 17^h 02.6^m | –26° 16' | 14' |
| chart 22 | | | m$_v$ 7.2 | |

NGC 6273. Medium concentration of stars; very bright, large, round, and very well resolved. Extends 90 ly; distance 22,000 ly; S-S class 8.

M20

See Trifid Nebula.

M21

| OC | Sgr | 18^h 04.6^m | –22° 30' | 12' |
| chart 22 | | | m$_v$ 5.9 | |

NGC 6531. 70 stars; detached, strong concentration of stars; large range in brightness; mag. of brightest star 7.3. Extends 15 ly; distance 4,300 ly; age 4.6 million years; Trumpler class I 3 m. Located approx. 0.5° NE of the Trifid Nebula (M20).

M22

| GC | Sgr | 18^h 36.4^m | –23° 54' | 33' |
| chart 22 | | | m$_v$ 5.1 | |

NGC 6656. Medium concentration of stars; very bright, very large, rich, round, and well resolved. A sparkling fuzzy ball in a 3-inch glass at 45×, and 75× breaks it up into pinpoints of light (Mullaney). A dark gash runs SW to NE across the core (O'Meara). Per *NGC*, a (!!) very remarkable object. Contains about 70,000 stars; extends 75 ly; distance 7,800 ly; S-S class 7.

M23

| OC | Sgr | 17^h 56.8^m | –19° 01' | 26' |
| charts 15, 22 | | | m$_v$ 5.5 | |

NGC 6494. 150 stars; detached, no concentration of

stars; small range in brightness; bright, very large; mag. of brightest star 9.2. The stars of this cluster are very near each other (Messier). Elegant sprinkling of telescopic stars over the whole field (Smyth). Grand low-power field (Webb). Extends 16 ly; age 220 million years; distance 2,200 ly; Trumpler class III 1 m.

M24

—	Sgr	$18^h 16.9^m$	$-18° 29'$	$95' \times 35'$
charts 15, 16, 22			m_v 4.5:	

Also the Milky Way Star Cloud or Milky Way Patch. Not a true cluster, but a detached portion of the Milky Way; visible to naked eye on a clear night. Two or three times as broad as the full Moon (Serviss). Sometimes M24 is incorrectly equated with the open cluster NGC 6603, which it contains.

M25

OC	Sgr	$18^h 31.6^m$	$-19° 15'$	$32'$
charts 15, 16, 22			m_v 4.6	

IC 4725. 30 stars; detached, strong concentration of stars; moderate range in brightness; large; mag. of brightest star 6.7. Extends 18 ly; distance 1,900 ly; age 89 million years; Trumpler class I 2 p.

M26

OC	Sct	$18^h 45.2^m$	$-09° 24'$	$15'$
charts 15, 16			m_v 8	

NGC 6694. 30 stars; detached, strong concentration of stars; small range in brightness; large; mag. of brightest star 10.3. Coarse cluster (Webb). Extends 21 ly; distance 4,900 ly; age 89 million years; Trumpler class I 1 m.

M27

See Dumbbell Nebula.

M28

GC	Sgr	$18^h 24.5^m$	$-24° 52'$	$11'$
chart 22			m_v 6.8	

NGC 6626. High concentration of stars; very bright, large, and very well resolved. Per *NGC*, a (!) remarkable object. Extends 60 ly; distance 19,000 ly; S-S class 4.

M29

OC	Cyg	$20^h 23.9^m$	$+38° 32'$	$7'$
charts 8, 9			m_v 6.6	

NGC 6913. 50 stars; detached, no concentration of stars; large range in brightness; mag. of brightest star 8.6. Best viewed at very low magnification, as each increase in power reduces the cluster's beauty (Mallas). Traces of nebulosity are seen on photographs. Extends 8 ly; distance 4,000 ly; age 10 million years; Trumpler class III 3 p n.

M30

GC	Cap	$21^h 40.4^m$	$-23° 11'$	$12'$
chart 23			m_v 7.5	

NGC 7099. Medium concentration of stars; bright, large, slightly elongated. At high power the entire cluster seems to be served on a plate of fainter stars (O'Meara). Per *NGC*, a (!) remarkable object. Extends 140 ly; distance 41,000 ly; S-S class 5.

M31

See Andromeda Galaxy.

M32

Gx	And	$00^h 42.7^m$	$+40° 52'$	$7.6' \times 5.8'$
chart 4			m_v 8.2	p.a. 170°

NGC 221. Very bright, large, and round. Easy to mistake for a bright star in binoculars (O'Meara). Per *NGC*, a (!) remarkable object. Extends 6,000 ly; distance about 2.5 million ly; Hubble class E2. A Local Group member and a satellite of the Andromeda Galaxy (M31).

M33

Gx	Tri	$01^h 33.9^m$	$+30° 39'$	$70' \times 41'$
chart 4			m_v 5.7	p.a. 23°

NGC 598, the Triangulum Galaxy or Pinwheel Galaxy. Very large and very faint; visible to the naked eye in exceptionally dark skies. A whitish, nearly uniform glow, but a little brighter across two-thirds of its diameter (Messier). Spiral arms may be seen in large telescopes. Per *NGC*, a (!) remarkable object. Extends 50,000 ly; distance 2.4 million ly; Hubble class Sc. A member of the Local Group.

M34

OC	Per	02^h 42.0^m	+42° 47′	34′
charts 1, 4			m$_v$ 5.2	

NGC 1039. 60 stars; detached, weak concentration of stars; large range in brightness; bright, very large cluster; mag. of brightest star 7.3. Appears with an opera glass like a faint comet (Serviss). A fine group, chiefly of 9th-mag. stars (Denning). Distance 1,400 ly; extends 14 ly; age 190 million years; Trumpler class II 3 m.

M35

OC	Gem	06^h 08.9^m	+24° 20′	28′
chart 5			m$_v$ 5.1	

NGC 2168. About 200 stars; detached, no concentration of stars; moderate range in brightness; very large; mag. of brightest star 8.2. With an opera glass, like a piece of frosted silver over which a twinkling light is playing (Serviss). A big, splashy stellar jewel box (Mullaney). My personal favorite open cluster (Houston). Extends 23 ly; distance 2,800 ly; age 110 million years; Trumpler class III 2 m.

M36

OC	Aur	05^h 36.5^m	+34° 08′	12′
chart 5			m$_v$ 6	

NGC 1960. 60 stars; detached, weak concentration of stars; large range in brightness; bright, very large; mag. of brightest star 8.9. Extends 13 ly; distance 3,700 ly; age 25 million years; Trumpler class II 3 m.

M37

OC	Aur	05^h 52.4^m	+32° 33′	23′
chart 5			m$_v$ 5.6	

NGC 2099; 150 stars; detached, weak concentration of stars; small range in brightness. A magnificent object, the whole field being strewed as it were with sparkling gold dust (Smyth). Has a dark area near center. Brightest star is mag. 9.2, a topaz jewel surrounded by a pear-shaped cluster of scintillating diamonds (O'Meara). Distance 3,600 ly; age 300 million years; extends 24 ly. Trumpler class II 1 r.

M38

OC	Aur	05^h 28.7^m	+35° 50′	20′
chart 5			m$_v$ 6.4	

NGC 1912. 100 stars; detached, no concentration of stars; moderate range in brightness; very large and bright; mag. of brightest star 9.5. Of square shape, and contains no nebulosity (Messier). Noble cluster arranged as an oblique cross, with a pair of larger stars in each arm (Webb). Extends 16 ly; distance 2,800 ly; age 220 million years; Trumpler class III 2 m.

M39

OC	Cyg	21^h 32.2^m	+48° 26′	32′
chart 9			m$_p$ 4.6	

NGC 7092. 30 stars; detached, no concentration of stars; moderate range in brightness; very large; mag. of brightest star 6.8. Extends 7 ly; distance 800 ly; age 270 million years; Trumpler class III 2 p.

M40

—	UMa	12^h 22.2$_v$	+58° 05′	
chart 2			m$_v$ 9.1	

Winnecke 4. Two stars very near each other and very small (Messier). They have always appeared as a very wide pair to subsequent observers with better telescopes. The components are mag. 9.6 and 10.1, separated by 52″.

M41

OC	CMa	06^m 46.0^m	−20° 44′	38′
chart 19			m$_p$ 4.5	

NGC 2287. 80 stars; detached, weak concentration of stars; large range in brightness; very large, bright. Has a bright orange star (mag. 6.9) near the center. Extends 18 ly; distance 1,600 ly; age 190 million years; Trumpler class II 3 m.

M42

See Great Orion Nebula.

M43

BNer	Ori	05^h 35.6^m	−05° 16′	19′ × 15′
charts 11, B2			m$_v$ 9	

NGC 1982. Very bright, large nebula. A small star surrounded by nebulosity N of the Orion Nebula (Messier). Per *NGC*, a (!) remarkable object. Extends 8 ly; distance 1,500 ly. Illuminated by a 6.9-mag. star.

M44

See Praesepe.

M45

See Pleiades.

M46

| OC | Pup | 07^h 41.8^m | –14° 49′ | 26′ |
| charts 12, 19 | | | m$_v$ 6.1 | |

NGC 2437. 100 stars; detached, no concentration of stars; moderate range in brightness; very bright, very large; mag. of brightest star 8.7. Beautiful circular cloud of small stars, best with low power (Webb). Per *NGC*, a (!) remarkable object. Extends 24 ly; distance 3,200 ly; age 300 million years; Trumpler class III 2 m. A planetary nebula (NGC 2438) lies in front of the cluster.

M47

| OC | Pup | 07^h 36.6^m | –14° 30′ | 30′ |
| charts 12, 19 | | | m$_v$ 4.4 | |

NGC 2422. 30 stars; detached, no concentration of stars; moderate range in brightness; bright, very large; mag. of brightest star 5.7. Grand broad group, visible to naked eye (Webb). Extends 16 ly; distance 1,800 ly; age 78 million years; Trumpler class III 2 m.

M48

| OC | Hya | 08^h 13.8^m | –05° 48′ | 50′ |
| chart 12 | | | m$_v$ 5.8 | |

NGC 2548. 80 stars; detached, strong concentration of stars; moderate range in brightness; very large; mag. of brightest star 8.2. A tremendously pleasing cluster, a perfect arrowhead of bright stars with a tight, off-axis core (O'Meara). Extends 22 ly, distance 1,500 ly; age 300 million years; Trumpler class I 2 m.

M49

| Gx | Vir | 12^h 29.8^m | +08° 00′ | 8.9′ × 7.4′ |
| charts 13, 14, B1 | | | m$_v$ 8.4 | p.a. 155° |

NGC 4472. Very bright, large, and round; very bright nucleus. Obvious even in a 2-inch glass (Mullaney). Extends 110,000 ly; distance 41 million ly; Hubble class E4. A member of the Virgo Galaxy Cluster; in a group with galaxies NGC 4467, NGC 4470, and a faint irregular dwarf galaxy. A 13th-mag. star lies just to E.

M50

| OC | Mon | 07^h 02.8^m | –08° 23′ | 15′ |
| chart 12 | | | m$_v$ 5.9 | |

NGC 2323. 80 stars; detached, weak concentration of stars; large range in brightness; very large; mag. of brightest star 7.9. A red star is toward the southern verge and a pretty little equilateral triangle just north of it (Smyth). Per *NGC*, a (!) remarkable object. Extends 13 ly; distance 3,000 ly; age 78 million years; Trumpler class II 3 m.

M51

See Whirlpool Galaxy.

M52

| OC | Cas | 23^h 24.2^m | +61° 35′ | 12′ |
| chart 3 | | | m$_v$ 6.9 | |

NGC 7654. Detached, strong concentration of stars; moderate range in brightness; large. Rich sparkling group of at least 100 suns (Mullaney). An 8.2-mag. topaz field star on the SW edge all but leaps out at you, as if trying to steal the show (O'Meara). Cluster extends 10 ly; distance 3,000 ly; age 35 million years; Trumpler class I 2 r.

M53

| GC | Com | 13^h 12.9^m | +18° 10′ | 12′ |
| charts 7, 14 | | | m$_v$ 7.7 | |

NGC 5024. Medium concentration of stars; a large, bright cluster. Brilliant mass of minute stars, blazing in center (Webb). Per *NGC*, a (!) remarkable object. Extends 240 ly; distance 70,000 ly; S-S class 5.

M54

| GC | Sgr | 18^h 55.1^m | –30° 29′ | 12′ |
| chart 22 | | | m$_v$ 7.7 | |

NGC 6715. High concentration of stars; very bright, large, and round. At the cluster's N edge two arms of very faint stars appear to swirl counterclockwise around the nucleus, and an orange star is superposed on one of these arms (O'Meara). Extends 170 ly; distance 49,000 ly; S-S class 3.

M55

GC	Sgr	19^h 40.0^m	−30° 58′	19′
charts 22, 23			m$_v$ 7	

NGC 6809. Low concentration of stars; pretty bright, large, round, and very rich. Several dark lanes infiltrate the cluster, and a bright star lies in a dark hole near the center (O'Meara). Extends 110 ly; distance 20,000 ly; S-S class 11.

M56

GC	Lyr	19^h 16.6^m	+30° 11′	7′
chart 8			m$_v$ 8.2	

NGC 6779. Low concentration of stars; large, bright, and very well resolved at high power. Unlike most globulars, this one has no bright core (Mallas). With its delicate spherical halo, my favorite noncomet (O'Meara). Extends 90 ly; distance 46,000 ly; S-S class 8.

M57

See Ring Nebula.

M58

Gx	Vir	12^h 37.7^m	+11° 49′	5.4′ × 4.4′
charts 13, 14, B1			m$_v$ 9.8	p.a. 95°

NGC 4579. Bright, large, irregularly round; small, bright, diffuse nucleus. Extends 60,000 ly; distance 41 million ly; Hubble class Sb. Member of the Virgo Galaxy Cluster.

M59

Gx	Vir	12^h 42.0^m	+11° 39′	5.1′ × 3.4′
charts 13, 14, B1			m$_v$ 9.8	p.a. 165°

NGC 4621. Bright, pretty large, and slightly elongated; bright, diffuse nucleus. Extends 60,000 ly; distance 41 million ly; Hubble class E3. Member of the Virgo Galaxy Cluster.

M60

Gx	Vir	12^h 43.7^m	+11° 33′	7.2′ × 6.2′
charts 13, 14, B1			m$_v$ 8.8	p.a. 105°

NGC 4649. Very bright, pretty large, and round. Extends 80,000 ly; distance 41 million ly; Hubble class E1. Member of the Virgo Galaxy Cluster. About 2.5′ to the NW is galaxy NGC 4647.

M61

Gx	Vir	12^h 21.9^m	+04° 28′	6.0′ × 5.5′
charts 13, 14, B1			m$_v$ 9.7	p.a. 162°

NGC 4303. A very bright, very large, face-on spiral; extremely bright center. The starlike nucleus is sourrounded by a mottled, diamond-shaped inner core (O'Meara). Extends 70,000 ly; distance 41 million ly; Hubble class Sc. Paired with galaxy NGC 4303A.

M62

GC	Oph	17^h 01.2^m	−30° 07′	10′
chart 22			m$_v$ 6.6	

NGC 6266. High concentration of stars; large, very bright, and very well resolved. Waves of starlight seem to ripple out from the ruddy nucleus (O'Meara). Per *NGC*, a (!) remarkable object. Extends 60 ly; distance 22,000 ly; S-S class 4.

M63

Gx	CVn	13^h 15.8^m	+42° 02′	12.3′ × 7.6′
chart 7			m$_v$ 8.6	p.a. 105°

NGC 5055, the Sunflower Galaxy. Very bright and large; very small, bright nucleus. Extends 80,000 ly; distance 24 million ly; Hubble class Sb. A star is located at one end.

M64

Gx	Com	12^h 56.7^m	+21° 41′	9.3′ × 5.4′
chart 7			m$_v$ 8.5	p.a. 115°

NGC 4826, the Black Eye Galaxy. Very bright, very large and very elongated; extremely bright nucleus. Dark dust lane (the black eye) is obvious on photographs but visually difficult in small telescopes. Per *NGC*, a (!) remarkable object. The galaxy extends 35,000 ly; distance 12 million ly; Hubble class Sb.

M65

Gx	Leo	11^h 18.9^m	+13° 05′	10.0′ × 3.3′
chart 13			m$_v$ 9.3	p.a. 174°

NGC 3623. bright, very large, and very elongated; small, diffuse, very bright nucleus. Extends 80,000 ly; distance 29 million ly; Hubble class Sb. A member of Leo's Triplet (along with galaxies M66 and NGC 3628).

M66

Gx	Leo	11^h 20.2^m	+12° 59'	8.7' × 4.4'
chart 13			m$_v$ 9	p.a. 173°

NGC 3627. Bright, very large, and very elongated; small, very bright nucleus. Extends 70,000 ly; distance 25 million ly; Hubble class Sb. A member of Leo's Triplet (along with galaxies M65 and NGC 3628).

M67

OC	Cnc	08^h 51.4^m	+11° 49'	30'
chart 12			m$_v$ 6.9	

NGC 2682. 200 stars; detached, weak concentration of stars; moderate range in brightness; very bright, very large; mag. of brightest star 9.7. As a whole, its stars form a reverse S-shaped pattern, like a king cobra that has just swallowed a meal (O'Meara). Per *NGC*, a (!) remarkable object. Extends 24 ly; distance 2,700 ly; age 3.2 billion yearsTrumpler class II 2 m.

M68

GC	Hya	12^h 39.5^m	−26° 45'	10'
chart 21			m$_v$ 8.2	

NGC 4590. Low concentration of stars; large, very rich, and very well resolved telescopically. A cinch in binoculars, unresolved yet obviously not a star (O'Meara). Extends 100 ly; distance 36,000 ly; S-S class 10.

M69

GC	Sgr	18^h 31.4^m	−32° 21'	10'
chart 22			m$_v$ 7.7	

NGC 6637. Medium concentration of stars; bright, large, round, and very well resolved. Extends 70 ly; distance 24,000 ly; S-S class 5.

M70

GC	Sgr	18^h 43.2^m	−32° 18'	7'
chart 22			m$_v$ 8.1	

NGC 6681, Medium concentration of stars; bright, pretty large, and round. Extends 130 ly; distance 65,000 ly; S-S class 5.

M71

GC	Sge	19^h 53.8^m	+18° 47'	7'
charts 8, 16			m$_v$ 8.3	

NGC 6838. Large and very rich. Visually an oval, with the brighter side forming a curving V (Mallas). Extends 30 ly; distance 13,000 ly.

M72

GC	Aqr	20^h 53.5^m	−12° 32'	6'
chart 16			m$_v$ 9.4	

NGC 6981. Low concentration of stars; pretty bright, pretty large, round, and very well resolved in large telescopes. In a 4-inch, faint speckles pop in and out of view in an otherwise foggy moor of starlight (O'Meara). Extends 110 ly; distance 62,000 ly; S-S class 9.

M73

—	Aqr	20^h 59.0^m	−12° 38'	1.1'
chart 16			m$_v$ 9.3	

NGC 6994. A cluster of three or four little stars, which looks like a nebula at first glance (Messier). The four individual stars have magnitudes of 10.4 and fainter, according to the Tycho detectors on the Hipparcos spacecraft, but the mission failed to obtain conclusive distances to any of the four. Whether these stars are true neighbors in space or a chance alignment is still unknown.

M74

Gx	Psc	01^h 36.7^m	+15° 47'	10.2' × 9.5'
chart 10			m$_v$ 9.2	p.a. 25°

NGC 628. A fine, face-on spiral; round and very large; very small nucleus; two major arms that can be partially resolved with large instruments. Although bright in total light, it is so spread out that its contrast with the sky background is very low. No other Messier object has proven more troublesome and elusive to amateurs with small telescopes (O'Meara). Extends 80,000 ly; distance 26 million ly; Hubble class Sc.

M75

GC	Sgr	20^h 06.1^m	−21° 55'	7'
charts 22, 23			m$_v$ 8.6	

NGC 6864. High concentration of stars; bright, round, pretty large, partially resolved. Extends 150 ly; distance 80,000 ly; S-S class 1.

M76

PN	Per	01^h 42.4^m	+51° 34′	67″
charts 1, 4			m$_v$ 10.1	

NGC 650-1, PK130–10.1; the Little Dumbbell, Cork, or Barbell Nebula. Consists of two lobes, one brighter and larger; phot. mag. of nebula is 12.2; central star's visual mag. is 15.9. Faintest of the Messier objects, but easier to see than some others because it is so compact. Dimly perceptible within a diamond of faint stars (Copeland). Has a heart of darkness (O'Meara). Nebula extends 3 ly; distance 8,200 ly, expansion velocity 42 km/sec (26 miles/sec); V-V class 3 + 6.

M77

Gx	Cet	02^h 42.7^m	−00° 01′	6.9′ × 5.9′
chart 10			m$_v$ 8.8	p.a. 10°

NGC 1068. Very bright, pretty large, irregularly round; very bright nucleus. A Seyfert galaxy. Distance 52 million ly; extends 110,000 ly; Hubble class Sbp. Brightest in a group of galaxies.

M78

BNr	Ori	05^h 46.7^m	+00° 03′	7′ × 6′
charts 11, B2			m$_v$ 8	

NGC 2068. Bright, large, wispy nebula; contains two 10th-mag. stars. Resembles a comet with a split nucleus, parabolic hood, and a tail that fans from S to E (O'Meara). Extends 3 ly; distance 1,600 ly. In a group with nebulae NGC 2064, NGC 2067 and NGC 2071.

M79

GC	Lep	05^h 24.5^m	−24° 33′	6′
chart 19			m$_v$ 8	

NGC 1904. Medium concentration of stars; pretty large, extremely rich, very well resolved; very faint stars. Extends 75 ly; distance 43,000 ly; S-S class 5.

M80

GC	Sco	16^h 17.0^m	−22° 59′	8′
chart 22			m$_v$ 7.2	

NGC 6093. High concentration of stars; large, very bright, well resolved, extremely rich. A field glass shows it as a mere wisp of light (Serviss). The outer hood seems to scintillate like flecks of mica (O'Meara). Per NGC, a (!!) very remarkable object. Extends 80 ly; distance 36,000 ly; S-S class 2.

M81

Gx	UMa	09^h 55.6^m	+69° 04′	26′ × 14′
charts 1, 2			m$_v$ 6.8	p.a. 157°

NGC 3031, one of Bode's Nebulae (along with galaxy M82). Extremely large, elongated, and extremely bright; bright nucleus. The nuclear hub and spiral-arm halo of M81 make a lovely contrast with cigar-shaped M82 and its dust lanes (Mullaney). Per NGC, a (!) remarkable object. Distance 8.5 million ly; extends 60,000 ly; Hubble class Sb. About 37′ to N is M82.

M82

Gx	UMa	09^h 55.8^m	+69° 41′	11.2′ × 4.6′
charts 1, 2			m$_v$ 8.4	p.a. 65°

NGC 3034, one of Bode's Nebulae (along with galaxy M81). Very bright, large, and very elongated; spindle-shaped. Like a ghostly starship, cracked and floating through space (O'Meara). Distance 8.5 million ly; extends 25,000 ly; Hubble class P. M81 lies about 37′ to S.

M83

Gx	Hya	13^h 37.0^m	−29° 52′	11.2′ × 10.2′
chart 21			m$_v$ 7.6	p.a. 44°

NGC 5236. Very bright, large, and face on; two main arms; extremely bright nucleus. This nucleus is tiny, looking like a star trapped in a maelstrom of shimmering light (O'Meara). A delightful spiral for small telescopes (Houston). Per NGC, a (!!) very remarkable object. Extends 40,000 ly; distance 8.5 million ly; Hubble class Sc.

M84

Gx	Vir	12^h 25.1^m	+12° 53′	5.0′ × 4.4′
charts 13, 14, B1			m$_v$ 9.3	p.a. 135°

NGC 4374. Very bright, round, and pretty large. Extends 70,000 ly; distance 41 million ly; Hubble class E1. Paired with galaxy NGC 4387. A member of the Virgo Galaxy Cluster.

M85

| Gx | Com | 12ʰ 25.4ᵐ | +18° 11′ | 7.1′ × 5.2′ |
| charts 7, 13, 14, B1 | | | m_v 9.2 | p.a. 178° |

NGC 4382. Very bright, round, and pretty large; extremely bright, diffuse nucleus. Extends 80,000 ly; distance 41 million ly; Hubble class Ep. Paired with galaxy NGC 4394. A member of the Virgo Galaxy Cluster.

M86

Gx Vir 12ʰ 26.2ᵐ +12° 57′ 7.4′ × 5.5′
charts 13, 14, B1 m_v 9.2   p.a. 130°

NGC 4406. Large, very bright, and round. Hubble class E3. Paired with galaxy NGC 4402. A member of the Virgo Galaxy Cluster.

M87

Gx Vir 12ʰ 30.8ᵐ +12° 24′ 7.2′ × 6.8′
charts 13, 14, B1 m_v 8.6   p.a. 159°

NGC 4486, the Virgo A radio source. Very bright, very large, and round. Visible in 20-inch and larger telescopes is a faint ray or optical jet extending from the galaxy's NW side. This feature seems to be related to M87's powerful radio and X-ray emissions and to the supposed black hole in its nucleus. The galaxy extends 90,000 ly, and the jet is about 4,000 ly long. Distance 41 million ly; Hubble class E1. Member of the Virgo Galaxy Cluster.

M88

Gx Com 12ʰ 32.0ᵐ +14° 25′ 6.9′ × 3.9′
charts 13, 14, B1 m_v 9.5   p.a. 140°

NGC 4501. Bright, very large, and quite elongated; bright, very small nucleus. Two 12th-mag. stars touch the SE extremity of the nebulous glow (O'Meara). Extends 70,000 ly; distance 41 million ly; Hubble class Sb. A member of the Virgo Galaxy Cluster.

M89

Gx Vir 12ʰ 35.7ᵐ +12° 33′ 5.0′ × 4.6′
charts 13, 14, B1 m_v 9.8   p.a. 100°

NGC 4552. Pretty bright, pretty small, and round. Extends 60,000 ly; distance 41 million ly; Hubble class E0. Member of the Virgo Galaxy Cluster.

M90

Gx Vir 12ʰ 36.8ᵐ +13° 10′ 9.5′ × 4.7′
charts 13, 14, B1 m_v 9.5   p.a. 23°

NGC 4569. Pretty large and elongated; extremely bright, very small nucleus. Extends 120,000 ly; distance 41 million ly; Hubble class Sb. A member of the Virgo Galaxy Cluster. Possibly interacting with dwarf galaxy IC 3583.

M91

Gx Com 12ʰ 35.4ᵐ +14° 30′ 5.4′ × 4.4′
charts 13, 14, B1 m_v 10.2   p.a. 150°

NGC 4548. Bright, large, and slightly elongated; bright, diffuse nucleus, with a central bar dimly visible in large amateur telescopes. Galaxy extends 50,000 ly; distance 37 million ly; Hubble class SBb. Paired with galaxy NGC 4571, and a member of the Virgo Galaxy Cluster.

M92

GC Her 17ʰ 17.1ᵐ +43° 08′ 14′
chart 8 m_v 6.5

NGC 6341. High concentration of stars; large and very bright. A mass of stars and stardust (Denning). Extends 110 ly; distance 28,000 ly; S-S class 3.

M93

OC Pup 07ʰ 44.6ᵐ −23° 52′ 21′
chart 19 m_v 6.2:

NGC 2447. 80 stars; not well detached from the surrounding star field; large; small range in brightness; mag. of brightest star 8.2. Bright cluster in a rich neighborhood (Webb). The core has a distinct arrowhead shape (O'Meara). Extends 22 ly; distance 3,600 ly; age 98 million years; Trumpler class IV 1 p.

M94

Gx CVn 12ʰ 50.9ᵐ +41° 07′ 11.0′ × 9.1′
chart 7 m_v 8.1   p.a. 105°

NGC 4736. Large, very bright, irregularly round; extremely bright nucleus; no spiral arms; very faint outer ring (about 15′). Seems to peer back at you, like a hypnotic eye (O'Meara). Easy to find and see

M95

Gx	Leo	10^h 44.0^m	+11° 42′	7.4′ × 5.1′
chart 13			m$_v$ 9.7	p.a. 13°

NGC 3351. Large, bright, and round; extremely bright nucleus; internal ring with bar. Extends 60,000 ly; distance 29 million ly; Hubble class S(B)b. Member of the Leo galaxy group.

M96

Gx	Leo	10^h 46.8^m	+11° 49′	7.1′ × 5.1′
chart 13			m$_v$ 9.2	p.a. 5°

NGC 3368. Very bright, very large, and slightly elongated; small, bright nucleus; has visible dark lanes. Extends 60,000 ly; distance 29 million ly; Hubble class Sbp. Brightest in the Leo galaxy group.

M97

See Owl Nebula.

M98

Gx	Com	12^h 13.8^m	+14° 54′	9.5′ × 3.2′
charts 7, 13, 14, B1			m$_v$ 10.1	p.a. 155°

NGC 4192. Bright, very large, and very elongated; very small, extremely bright nucleus. Extends 90,000 ly; distance 36 million ly; Hubble class Sb. Member of the Virgo Galaxy Cluster; in a group with galaxies NGC 4198 and Holmberg 348c.

M99

Gx	Com	12^h 18.8^m	+14° 25′	5.4′ × 4.8′
charts 7, 13, 14, B1			m$_v$ 9.8	p.a. 60°

NGC 4254. Bright, large, and round; face on; small, very bright nucleus. Similar in appearance to M33 but much smaller in angular size. Fine example of a three-branched spiral (Denning). Per *NGC*, a (!) remarkable object. Extends 60,000 ly; distance 41 million ly; Hubble class Sc. Member of the Virgo Galaxy Cluster.

M100

Gx	Com	12^h 22.9^m	+15° 49′	6.9′ × 6.2′
charts 7, 13, 14, B1			m$_v$ 9.4	p.a. 30°

NGC 4321. Pretty faint, very large, round, face on; very bright nucleus. Per *NGC*, a (!!) very remarkable object. Extends 80,000 ly; distance 41 million ly; Hubble class Sc. Brightest spiral in the Virgo Galaxy Cluster. Paired with galaxy NGC 4312.

M101

Gx	UMa	14^h 03.2^m	+54° 21′	27′ × 26′
charts 2, 7			m$_v$ 7.7	p.a. 65°

NGC 5457. Pretty bright and very large; face on; faint spiral arms; small, bright nucleus. Distance 27 million ly according to Hipparcos studies; extends 200,000 ly; Hubble class Sc. Galaxies NGC 5422, NGC 5473, NGC 5474, and NGC 5485 all lie within about 1° of M101.

M102

Same object as M101. According to discoverer Pierre Méchain, the confusion was due to an error in his star chart. Many sources incorrectly equate M102 and the galaxy NGC 5866.

M103

OC	Cas	01^h 33.2^m	+60° 42′	6′
chart 1			m$_v$ 7.4:	

NGC 581. 25 stars; detached, no concentration of stars; moderate range in brightness; mag. of brightest star 10.6. Arrowhead or fan-shaped small group of stars (Mullaney). Distance 8,500 ly; age 22 million years; extends 15 ly; Trumpler class III 2 p.

M104

See Sombrero Galaxy.

M105

Gx	Leo	10^h 47.8^m	+12° 35′	4.5′ × 4.0′
chart 13			m$_v$ 9.3	p.a. 71°

NGC 3379. Very bright, large, and round; very bright nucleus. Extends 30,000 ly; distance 22 million ly; Hubble class E1. Member of the Leo galaxy group; paired with galaxy NGC 3384.

M106

Gx	CVn	12^h 19.0^m	+47° 18′	18′ × 8′
charts 2, 6, 7			m$_v$ 8.3	p.a. 150°

NGC 4258. Very bright, very large, and very elongated; small, bright nucleus in bright bulge. Distance 20 million ly; extends 100,000 ly; Hubble class Sbp. Added to Messier list in 1947 by Helen Sawyer Hogg. Galaxy NGC 4248 is located about 20′ to NW.

M107

GC	Oph	16^h 32.5^m	−13° 03′	12′
chart 15			m$_v$ 8.1	

NGC 6171. Low concentration of stars; large, very rich, very well resolved. There are five telescopic stars around it, so placed as to form a crucifix (Smyth). Extends about 70 ly; distance 20,000 ly; S-S class 10.

M108

Gx	UMa	11^h 11.5^m	+55° 40′	8.3′ × 2.5′
charts 2, 6			m$_v$ 10	p.a. 80°

NGC 3556. Bright, very large, and very elongated; nearly edge on; no visible nucleus. Distance 24 million ly; extends 60,000 ly; Hubble class Sc. A foreground star, nearly centered on this galaxy, is easy to mistake for a supernova. The Owl Nebula (M97) is located about 50′ to SE.

M109

Gx	UMa	11^h 57.6^m	+53° 23′	7.6′ × 4.9′
charts 2, 6, 7			m$_v$ 9.8	p.a. 68°

NGC 3992. Quite bright, very large, and elongated; spiral arms; diffuse, very bright nucleus. Extends 50,000 ly; distance 27 million ly; Hubble class S(B)b. Added to the Messier list by Owen Gingerich in 1953. A 12th-mag. foreground star lies less than 1′ N of nucleus on one of the spiral arms.

M110

Gx	And	00^h 40.4^m	+41° 41′	17′ × 10′
charts 4, 9			m$_v$ 8	p.a. 170°

NGC 205. Very bright, very large, and somewhat elongated. Discovered by Caroline Herschel in 1783; added to the Messier catalog by K. Glyn Jones in 1967. Extends 14,000 ly; distance about 2.5 million ly; Hubble class E6. A member of the Local Group and satellite of the Andromeda Galaxy (M31).

M1-92

BNr	Cyg	19^h 36.3^m	+29° 33′	0.2′ × 0.1′
chart 8				

The Footprint Nebula.

M-0-27-5

Gx	Sex	10^h 23.9^m	−03° 11′	1.8′ × 1.1′
chart 13			m$_p$ 13.0	p.a. 150°

Distance 240 million ly; Hubble class SBc.

M-1-3-85

Gx	Cet	01^h 05.1^m	−06° 13′	4.4′ × 3.2′
chart 10			m$_p$ 12.4	p.a. 70°

Distance 50 million ly; Hubble class SBc.

M-1-23-19

Gx	Hya	09^h 06.6^m	−07° 14′	1.5′ × 1.0′
charts 12, 13			m$_p$ 12.9	p.a. 80°

Hubble class E/S0.

M-1-24-1

Gx	Hya	09^h 10.9^m	−08° 53′	4.3′ × 1.0′
charts 12, 13			m$_p$ 11.9	p.a. 33°

Distance 72 million ly; Hubble class SBb.

M-1-26-30

Gx	Sex	10^h 11.0^m	−04° 43′	5.0′ × 4.8′
chart 13			m$_p$ 11.7	p.a. 135°

Distance 7 million ly; Hubble class Ir.

M-1-29-15

Gx	Crt	11^h 22.7^m	−07° 41′	2.3′ × 1.6′
chart 13			m$_p$ 13.0	p.a. 125°

Distance 310 million ly; Hubble class S0.

M-1-32-19

Gx	Vir	12^h 30.3^m	−08° 24′	2.3′
charts 13, 14			m$_p$ 12.5	p.a. 132°

Distance 230 million ly; Hubble class SBb.

M-1-32-28

| Gx | Vir | 12ʰ 35.6ᵐ | −07° 53′ | 4.2′ × 2.2′ |
| charts 13, 14 | | | m_p 12.6 | p.a. 55° |

Distance 38 million ly; Hubble class SBc.

M-1-33-1

Gx Vir 12ʰ 44.1ᵐ −05° 41′ 3.2′ × 2.6′
charts 13, 14   m_p 12.9   p.a. 40°

Distance 58 million ly; Hubble class Sd.

M-1-33-3

Gx Vir 12ʰ 45.7ᵐ −06° 04′ 3.2′ × 2.3′
charts 13, 14   m_p 13.0   p.a. 45°

Distance 60 million ly; Hubble class SBd.

M-1-33-27

Gx Vir 12ʰ 51.2ᵐ −06° 34′ 1.3′ × 1.1′
charts 13, 14   m_p 12.4   p.a. 30°

Distance 61 million ly; Hubble class Ir.

M-1-57-18

Gx Aqr 22ʰ 38.9ᵐ −05° 51′ 2.6′ × 0.9′
chart 17   m_p 12.9   p.a. 83°

Distance 130 million ly; Hubble class Sc.

M-2-9-36

Gx Eri 03ʰ 22.9ᵐ −11° 12′ 2.5′ × 1.8′
charts 10, 11   m_p 13.0   p.a. 165°

Distance 120 million ly; Hubble class SBc.

M-2-15-11

Gx Lep 05ʰ 50.9ᵐ −14° 47′ 3.0′ × 2.3′
chart 11   m_p 11.7   p.a. 160°

Distance 33 million ly; Hubble class SBc.

M-2-25-6

Gx Hya 09ʰ 36.5ᵐ −11° 20′ 1.9′ × 1.5′
charts 12, 13   m_p 12.6   p.a. 0°

Distance 230 million ly.

M-2-30-14

Gx Crt 11ʰ 40.6ᵐ −10° 05′ 2.7′ × 2.2′
chart 13   m_p 12.6   p.a. 5°

Distance 69 million ly; Hubble class SBc.

M-2-33-15

Gx Vir 12ʰ 49.4ᵐ −10° 07′ 4.2′ × 3.2′
charts 13, 14   m_v 11.6   p.a. 70°

Distance 52 million ly; Hubble class SBd.

M-2-33-17

Gx Crv 12ʰ 50.1ᵐ −14° 44′ 2.5′ × 2.3′
charts 13, 14   m_p 12.9   p.a. 35°

Distance 160 million ly; Hubble class Sa.

M-2-34-6

Gx Vir 13ʰ 09.8ᵐ −10° 20′ 3.0′ × 2.6′
chart 14   m_p 11.0   p.a. 160°

Distance 48 million ly; Hubble class Sc.

M-2-34-54

Gx Vir 13ʰ 27.9ᵐ −13° 25′ 1.7′ × 1.2′
chart 14   m_p 13.0   p.a. 150°

Distance 170 million ly; Hubble class SBb.

M-2-35-10

Gx Vir 13ʰ 38.2ᵐ −09° 48′ 2.3′ × 1.9′
chart 14   m_p 12.8   p.a. 85°

Distance 53 million ly; Hubble class SBd.

M-2-41-1

Gx Sco 16ʰ 17.3ᵐ −11° 44′ 2.4′ × 1.7′
chart 15   m_p 13.0   p.a. 85°

Distance 43 million ly; Hubble class Sb.

M-3-1-15

Gx Cet 00ʰ 01.9ᵐ −15° 27′ 11′ × 4′
charts 10, 17   m_p 11.1   p.a. 4°

Distance 2.6 million ly; Hubble class S.

M-3-10-42

Gx Eri $03^h 44.0^m$ $-14° 22'$ $2.0' \times 0.7'$
charts 10, 11, 18 m_p 13.0 p.a. 138°

Distance 66 million ly; Hubble class Sb.

M-3-10-45

Gx Eri $03^h 46.6^m$ $-16° 33'$ $1.4' \times 0.7'$
charts 10, 11, 18 m_p 13.0 p.a. 40°

Distance 50 million ly; Hubble class S.

M-3-25-4

Gx Hya $09^h 33.3^m$ $-16° 46'$ $2.1' \times 1.6'$
charts 12, 13 m_p 12.6 p.a. 35°

Distance 84 million ly; Hubble class SBb.

M-3-34-14

Gx Vir $13^h 12.6^m$ $-17° 32'$ $2.5' \times 0.8'$
charts 14, 21 m_p 12.8 p.a. 130°

Distance 120 million ly; Hubble class SBc.

M 4-28-110

Gx Com $11^h 58.7^m$ $+25° 03'$ $1.4' \times 0.9'$
charts 6, 7 m_p 12.8 p.a. 125°

Distance 190 million ly; Hubble class Sa.

M-6-20-4

Gx Vel $08^h 57.5^m$ $-39° 16'$ $0.9'$
chart 20 m_p 12.9

Hubble class E.

M-6-51-14

Gx Scl $23^h 37.1^m$ $-37° 43'$ $1.3' \times 1.1'$
charts 18, 23 m_p 12.9 p.a. 75°

Hubble class Sb.

Magnificent Cluster

See NGC 7789.

Manger

See Praesepe.

Medusa Nebula

See PK205+14.1.

Mel 66

OC Pup $07^h 26.3^m$ $-47° 44'$ $10'$
charts 19, 20 m_v 7.8

200 stars; detached, slight concentration; small brightness range; mag. of brightest star 11.4. Distance 8,200 ly; age 6 billion years; Trumpler class II 2 m.

Mel 71

OC Pup $07^h 37.5^m$ $-12° 04'$ $8'$
chart 12 m_v 7.1

80 stars; detached, weak concentration of stars; small range in brightness; mag. of brightest star 10.2. Distance 9,000 ly; age 420 million years; Trumpler class II 1 m.

Mel 101

OC Car $10^h 42.1^m$ $-65° 06'$ $14'$
chart 25 m_v 8.0:

50 stars; detached, weak concentration of stars; large range in brightness; mag. of brightest star 9.7. Distance 6,900 ly; Trumpler class II 3 m. Located approx. 0.5° SSW of the Southern Pleiades (IC 2602).

Mel 105

OC Car $11^h 19.5^m$ $-63° 30'$ $4'$
chart 25 m_v 8.5

70 stars; detached, strong concentration of stars; moderate range in brightness; mag. of brightest star 11.1. Distance 6,900 ly; age 59 million years; Trumpler class I 2 m.

Mel 227

OC Oct $20^h 18.0^m$ $-79° 09'$ $50'$
charts 24, 26 m_p 5.3

40 stars; detached, no concentration; large brightness range; Trumpler class II 2 p.

Merope Nebula

See IC 349.

Milky Way Star Cloud (or Patch)

See M24.

Mrk 6

OC	Cas	02^h 29.6^m	+60° 39′	4.5′
chart 1			m$_v$ 7.1	

6 stars; detached, no concentration; small brightness range; mag. of brightest star 8.4. Distance 2,000 ly; Trumpler class IV 2 p.

Mrk 18

OC	Vel	09^h 00.6^m	−48° 59′	2′
charts 19, 20			m$_v$ 7.8	

30 stars; detached, strong concentration; moderately rich in faint stars; mag. of brightest star 9.3. Distance 5,200 ly; Trumpler class I 3 p.

Mrk 50

OC	Cep	23^h 15.3^m	+60° 28′	5′
chart 3			m$_v$ 8.5	

5 stars; detached, no concentration; small brightness range; involved in nebulosity; mag. of brightest star 9.8. Distance 7,300 ly; age 10 million years; Trumpler class I 2 p n.

Mrk 1236

Gx	Sex	09^h 49.9^m	+00° 37′	2.1′ × 1.1′
charts 12, 13			m$_p$ 13.0	p.a. 85°

A double system with galaxy NGC 3023.

Network Nebula

See NGC 6992.

NGC 14

Gx	Peg	00^h 08.8^m	+15° 49′	2.8′ × 2.2′
charts 4, 10, 17			m$_v$ 12.1	p.a. 25°

Bright central region. Distance 46 million ly; Hubble class Irp+.

NGC 16

Gx	Peg	00^h 09.1^m	+27° 44′	1.8′ × 1.0′
charts 4, 9			m$_v$ 12	p.a. 16°

Pretty bright, small, and round; faint, very small nucleus. Distance 140 million ly; Hubble class E3.

NGC 23

Gx	Peg	00^h 09.9^m	+25° 55′	2.1′ × 1.4′
charts 4, 9			m$_v$ 12	p.a. 8°

Very small and bright nucleus. Distance 210 million ly; Hubble class Sbp.

NGC 24

Gx	Scl	00^h 09.9^m	−24° 58′	6.0′ × 1.3′
charts 18, 23			m$_v$ 11.3	p.a. 46°

Very faint, quite large, and very elongated; patchy arms. Distance 26 million ly; Hubble class Sb. Paired with galaxy NGC 45.

NGC 40

PN	Cep	00^h 13.0^m	+72° 31′	48″
charts 1, 3			m$_v$ 12.4	

PK120-9.1, C2. Faint, very small; traces of ring structure; phot. mag. 10.7; central star mag. 11.6. In a relatively sparse region and not easy to find, but well worth the search (Mullaney). Distance 2,900 ly; expansion velocity 28 km/sec (18 miles/sec); V-V class 3b +3.

NGC 45

Gx	Cet	00^h 14.1^m	−23° 11′	8.1′ × 5.8′
chart 18			m$_v$ 10.8	p.a. 142°

Large and extremely faint; very faint nucleus; many knots on faint arms; extremely low surface brightness. Distance 22 million ly; Hubble class S−. Paired with galaxy NGC 25.

NGC 50

Gx	Cet	00^h 14.7^m	−07° 21′	2.6′ × 1.9′
charts 10, 17			m$_v$ 11.6	p.a. 155°

Very faint. Distance 250 million ly; Hubble class S0p.

NGC 55

Gx	Scl	00^h 15.1^m	−39° 13′	31′ × 6′
charts 18, 23			m$_v$ 8.1	p.a. 108°

C72. Very bright, very large, very elongated. Most amateur telescopes can trace this slender, spindle-shaped form across 20′ of sky (Houston). Distance 4 million ly; Hubble class SBm. Brightest galaxy in Sculptor galaxy group.

NGC 57

Gx	Psc	00^h 15.5^m	+17° 20′	2.4′ × 2.0′
charts 4, 10, 17			m$_v$ 11.6	p.a. 40°

Faint, small, and roundish; suddenly brighter toward middle. Distance 240 million ly; Hubble class E.

NGC 63

Gx	Psc	00^h 17.8^m	+11° 27′	1.7′ × 1.1′
charts 10, 17			m$_v$ 11.6	p.a. 108°

Pretty faint, small, and roundish; suddenly brighter toward middle. Distance 59 million ly; Hubble class E5.

NGC 104

See 47 Tucanae.

NGC 121

GC	Tuc	00^h 26.8^m	−71° 32′	1.5′
chart 24			m$_v$ 10.6	

Pretty bright and small; slightly elongated. Located NNE of globular cluster 47 Tucanae (NGC 104) and W of the Small Magellanic Cloud.

NGC 128

Gx	Psc	00^h 29.3^m	+02° 52′	2.8′ × 0.9′
charts 10, 17			m$_v$ 11.8	p.a. 1°

Pretty bright, pretty small, elongated; very bright, box-shaped nucleus. Distance 190 million ly; Hubble class S0p. Brightest in a group of galaxies that includes NGC 125, NGC 126, NGC 127, and NGC 130.

NGC 129

OC	Cas	00^h 29.9^m	+60° 14′	20′
charts 1, 3			m$_v$ 6.5	

35 stars; not well detached from surrounding star field; moderate range in brightness; very large cluster; mag. of brightest star 8.6. Distance 5,200 ly; age 150 million years; Trumpler class IV 2 p.

NGC 134

Gx	Scl	00^h 30.4^m	−33° 15′	8.0′ × 1.7′
chart 18			m$_v$ 10.4	p.a. 50°

Very bright, large, and elongated; nearly edge on; very small, bright nucleus with dark lanes. Distance 67 million ly; Hubble class S(B)b. Paired with galaxy NGC 131.

NGC 147

Gx	Cas	00^h 33.2^m	+48° 30′	12.9′ × 8.1′
charts 4, 9			m$_v$ 9.5	p.a. 25°

C17. Very faint, very large, irregularly round; bright, extremely small nucleus. Distance 2.4 million ly; Hubble class dE4. A member of the Local Group and a satellite of galaxy M31; physically paired with galaxy NGC 185, which is approx. 1° to ESE.

NGC 150

Gx	Scl	00^h 34.3^m	−27° 48′	3.9′ × 1.8′
chart 18			m$_v$ 11.3	p.a. 118°

Pretty faint, pretty small, round; extremely small, very bright nucleus. Distance 70 million ly; Hubble class Sc.

NGC 151

Gx	Cet	00^h 34.0^m	−09° 42′	3.7′ × 1.5′
charts 10, 17			m$_v$ 11.6	p.a. 75°

Pretty faint, pretty large, slightly elongated; small, bright nucleus with dark area. Distance 160 million ly; Hubble class Sb.

NGC 157

Gx	Cet	00^h 34.8^m	−08° 24′	4.2′ × 2.7′
charts 10, 17			m$_v$ 10.4	p.a. 40°

Pretty bright, large, and elongated; very small, very bright nucleus. Distance 76 million ly; Hubble class Sc. Located between two quite bright stars.

NGC 175

Gx	Cet	00^h 37.4^m	−19° 56′	2.1′ × 1.9′
charts 10, 17, 18			m$_v$ 12.1	p.a. 120°

Pretty bright and pretty large; mottled appearance; very bright nucleus. Distance 75 million ly; Hubble class SBb.

NGC 185

| Gx | Cas | $00^h\,39.0^m$ | $+48°\,20'$ | $12' \times 10'$ |
| charts 4, 9 | | | $m_v\,9.2$ | p.a. 35° |

C18. Pretty bright, very large, and round; a dwarf galaxy of low surface brightness; very faint nucleus. A small dust patch is silhouetted against the nucleus. Distance 2.2 million ly; Hubble class dE0. A member of the Local Group and satellite of the Andromeda Galaxy (M31). Physically paired with galaxy NGC 147, which is approx. 1° to WNW.

NGC 188

| OC | Cep | $00^h\,44.4^m$ | $+85°\,20'$ | $14'$ |
| charts 1, 3, A4 | | | $m_v\,8.1$ | |

C1. 120 stars; detached, weak concentration of stars; moderate range in brightness; large cluster; mag. of brightest star 12.1. Distance 5,000 ly; age 5 billion years; Trumpler class II 2 r.

NGC 198

| Gx | Psc | $00^h\,39.4^m$ | $+02°\,48'$ | $1.1'$ |
| charts 10, 17 | | | $m_v\,13.2$ | |

Very small and very bright nucleus. Distance 220 million ly; Hubble class Sc.

NGC 205

See M110.

NGC 206

| BN | And | $00^h\,40.6^m$ | $+40°\,44'$ | $4.2' \times 1.5'$ |
| chart 4 | | | | |

Extends N–S; located in the SW part of the Andromeda Galaxy (M31).

NGC 210

| Gx | Cet | $00^h\,40.6^m$ | $-13°\,52'$ | $4.7' \times 3.1'$ |
| charts 10, 17 | | | $m_v\,10.9$ | p.a. 160° |

Bright, pretty small, and round; extremely bright nucleus. Distance 74 million ly; Hubble class Sb. Brightest in a group of galaxies.

NGC 214

| Gx | And | $00^h\,41.5^m$ | $+25°\,30'$ | $1.9' \times 1.5'$ |
| chart 4 | | | $m_v\,12.2$ | p.a. 35° |

Pretty faint, pretty small and round; filamentary arms; very small, bright nucleus. Distance 200 million ly; Hubble class Sb.

NGC 221

See M32.

NGC 224

See Andromeda Galaxy.

NGC 225

| OC | Cas | $00^h\,43.4^m$ | $+61°\,47'$ | $12'$ |
| charts 1, 3 | | | $m_v\,7$ | |

15 stars; detached, no concentration of stars; small brightness range; large cluster; involved in nebulosity; mag. of brightest star 9.3. Distance 2,000 ly; age 140 million years; Trumpler class III 1 p n.

NGC 245

| Gx | Cet | $00^h\,46.1^m$ | $-01°\,43'$ | $1.3' \times 1.1'$ |
| charts 10, 17 | | | $m_v\,12.2$ | p.a. 145° |

Faint, pretty small, and very irregularly shaped; very small, very bright nucleus. Distance 190 million ly; Hubble class Ir.

NGC 246

| PN | Cet | $00^h\,47.1^m$ | $-11°\,52'$ | $4.1'$ |
| charts 10, 17 | | | $m_v\,10.9$ | |

PK118–74.1, C56. Large, pretty bright, with traces of ring structure; phot. mag. 8.0. The central star's mag. is 12.0 and several other stars are superimposed. Distance 1,300 ly; age 15,000 years; expansion velocity 38 km/sec (24 miles/sec); V-V class 3b.

NGC 247

| Gx | Cet | $00^h\,47.1^m$ | $-20°\,46'$ | $21' \times 5'$ |
| chart 18 | | | $m_v\,9.2$ | p.a. 174° |

C62. Faint, extremely large, and very elongated; patchy arms; very bright, extremely small nucleus. Distance 8 million ly; Hubble class S–. Located in Sculptor galaxy group.

NGC 248

BNe Tuc $00^h 45.4^m$ $-73° 23'$
chart 24

Located in the S part of the Small Magellanic Cloud.

NGC 249

BNe Tuc $00^h 45.4^m$ $-73° 05'$
chart 24

Small and round; located in a rich star field in the S part of the Small Magellanic Cloud.

NGC 253

Gx Scl $00^h 47.6^m$ $-25° 17'$ $26' \times 6'$
chart 18 m_v 7.6 p.a. 52°

C65, Sculptor Galaxy. Extremely bright, extremely large, and very elongated; has complex dust lanes; extremely small, bright nucleus. Very, very large and bright (Denning). Per *NGC*, a (!!) very remarkable object. A beautiful galaxy! Extends about 80,000 ly; distance 11 million ly; Hubble class Scp.

NGC 254

Gx Scl $00^h 47.5^m$ $-31° 25'$ $2.6' \times 1.7'$
chart 18 m_v 11.6 p.a. 137°

Very bright and pretty small; faint outer arms. Distance 62 million ly; Hubble class Sa.

NGC 255

Gx Cet $00^h 47.8^m$ $-11° 28'$ $2.9' \times 2.6'$
charts 10, 17 m_v 11.9 p.a. 15°

Faint, pretty small and round; very bright, small, and elongated nucleus. Distance 81 million ly; Hubble class Sb.

NGC 261

BN Tuc $00^h 46.5^m$ $-73° 07'$
chart 24

Lies in a rich star field in the S portion of the Small Magellanic Cloud; includes a central star. This object is larger and more diffuse than nebula NGC 249.

NGC 266

Gx Psc $00^h 49.8^m$ $+32° 17'$ $3.0' \times 2.9'$
chart 4 m_v 11.6 p.a. 95°

Pretty bright and pretty small; pretty suddenly brighter toward middle. Distance 210 million ly; Hubble class SBb.

NGC 271

Gx Cet $00^h 50.7^m$ $-01° 55'$ $2.2' \times 1.7'$
charts 10, 17 m_v 12 p.a. 130°

Pretty faint, small and slightly elongated; pretty suddenly brighter toward middle. Distance 190 million ly; Hubble class SBb.

NGC 274

Gx Cet $00^h 51.0^m$ $-07° 03'$ $2.0' \times 1.4'$
charts 10, 17 m_v 11.8 p.a. 155°

Pretty bright and pretty small; extremely bright, very small nucleus. Distance 79 million ly; Hubble class E1. Interacting with galaxy NGC 275.

NGC 275

Gx Cet $00^h 51.1^m$ $-07° 04'$ $1.4' \times 1.1'$
charts 10, 17 m_v 12.5 p.a. 40°

Very faint, small, and round. Distance 80 million ly; Hubble class SBcp. Interacting with galaxy NGC 274.

NGC 278

Gx Cas $00^h 52.1^m$ $+47° 33'$ $2.4'$
charts 4, 9 m_v 10.8

Quite bright, pretty large, and round; several strong arms; large, bright nucleus. Distance 38 million ly; Hubble class E0p.

NGC 281

BNe+C Cas $00^h 52.8^m$ $+56° 37'$ $34' \times 29'$
chart 1

Pretty faint emission nebula containing a 4' clump of four stars; brightest star is phot. mag. 9.

NGC 288

GC Scl $00^h 52.8^m$ $-26° 35'$ $12'$
chart 18 m_v 8.1:

Low concentration of stars; large, bright, slightly elongated. Distance 27,000 ly; S-S class 10.

NGC 289

Gx   Scl $00^h 52.7^m$ $-31° 12'$ $5.0' \times 4.2'$
chart 18 m_v 10.6   p.a. 130°

Very bright, large, and slightly elongated; four arms; small, bright nucleus. Distance 78 million ly; Hubble class Sb.

NGC 294

BN   Tuc $00^h 52.1^m$ $-73° 21'$
chart 24

Very faint, roundish, gradually brighter toward middle. Located in the S part of the Small Magellanic Cloud.

NGC 300

Gx   Scl $00^h 54.9^m$ $-37° 41'$ $19' \times 13'$
chart 18 m_v 8.1   p.a. 111°

C70. Very bright, very large, elongated, S shaped; contains dark lanes; bright, extremely small nucleus. Distance 4 million ly; Hubble class SBmp. Member of Sculptor galaxy group.

NGC 309

Gx   Cet $00^h 56.7^m$ $-09° 55'$ $3.0' \times 2.3'$
charts 10, 17 m_v 11.9   p.a. 175°

Pretty bright and pretty large; many arms; extremely small, bright nucleus. Good example of a spiral galaxy. Distance 250 million ly; Hubble class Sd.

NGC 315

Gx   Psc $00^h 57.8^m$ $+30° 21'$ $3.2' \times 2.5'$
chart 4 m_v 11.2   p.a. 40°

Pretty bright, pretty large, roundish, gradually brighter toward middle. Distance 230 million ly; Hubble class S0.

NGC 337

Gx   Cet $00^h 59.8^m$ $-07° 35'$ $2.9' \times 1.8'$
charts 10, 17 m_v 11.6   p.a. 120°

Pretty faint, large, and elongated. Distance 77 million ly; Hubble class Sc.

NGC 337A

Gx   Cet $01^h 01.6^m$ $-07° 35'$ $5.8' \times 4.6'$
chart 10 m_v 12.2   p.a. 10°

Distance 49 million ly; Hubble class S(B)m.

NGC 346

BNe+C   Tuc $00^h 59.1^m$ $-72° 11'$ $14' \times 11'$
chart 24

Nebula containing a 10.3-mag. star cluster. Cluster size is 5.2'.

NGC 362

GC   Tuc $01^h 03.2^m$ $-70° 51'$ 12'
chart 24 m_v 6.8

C104. High concentration of stars; very bright, very large. Distance 29,000 ly; S-S class 3.

NGC 371

BN+C   Tuc $01^h 03.3^m$ $-72° 05'$ 7.5'
chart 24 m_v 9.3

Henize N76. Nebulosity involved with cluster in the Small Magellanic Cloud.

NGC 404

Gx   And $01^h 09.5^m$ $+35° 43'$ 4.4'
chart 4 m_v 10.3

Pretty bright, quite large, round; very small, extremely bright nucleus; semicircular dust lane. This galaxy lies just 7' NW of β And (mag. 2.1) and would be much easier to see without the star's glare. Because of this proximity, NGC 404 is omitted from many atlases and is often reported as a new comet. Distance 8 million ly; Hubble class E0. Possible member of the Local Group.

NGC 406

Gx   Tuc $01^h 07.4^m$ $-69° 53'$ $3.0' \times 1.3'$
chart 24 m_v 12.3   p.a. 160°

Faint, very large, and round; knotty arms. Distance 56 million ly; Hubble class Sc. It is seen through the outer N part of the Small Magellanic Cloud.

NGC 410

Gx	Psc	01^h 11.0^m	+33° 09′	2.3′ × 1.8′
chart 4			m$_v$ 11.5	p.a. 30°

Pretty bright and pretty large. Distance 240 million ly; Hubble class SB0.

NGC 416

GC	Tuc	01^h 08.0^m	−72° 21′	1.1′
chart 24			m$_v$ 11	

Faint, pretty small, and roundish; gradually brighter toward middle. Located in E portion of the Small Magellanic Cloud.

NGC 419

GC	Tuc	01^h 08.3^m	−72° 53′	2.6′
chart 24			m$_v$ 10	

Pretty bright, pretty large, roundish; gradually brighter toward middle. Located in E portion of the Small Magellanic Cloud.

NGC 428

Gx	Cet	01^h 12.9^m	+00° 59′	3.8′ × 3.0′
chart 10			m$_v$ 11.5	p.a. 120°

Faint, large, and round; low surface brightness. Distance 55 million ly; Hubble class Scp.

NGC 434

Gx	Tuc	01^h 12.2^m	−58° 15′	2.2′ × 1.2′
chart 24			m$_v$ 12.1	p.a. 6°

Bright, small, and round; very bright center; several diffuse arms. Distance 200 million ly; Hubble class S(B)b. Possibly interacting with nearby galaxy NGC 434A.

NGC 439

Gx	Scl	01^h 13.8^m	−31° 45′	2.5′ × 1.6′
chart 18			m$_v$ 11.4	p.a. 156°

Pretty bright, small, and round; hexagon-shaped nucleus. Distance 240 million ly; Hubble class E3.

NGC 448

Gx	Cet	01^h 15.3^m	−01° 38′	1.8′ × 0.9′
chart 10			m$_v$ 12.1	p.a. 116°

Pretty bright, very small, and slightly elongated. Distance 86 million ly; Hubble class E/S0.

NGC 450

Gx	Cet	01^h 15.5^m	−00° 52′	3.0′ × 2.3′
chart 10			m$_v$ 11.5	p.a. 72°

Very faint and large; two main knotty arms; bright, very small nucleus. Distance 81 million ly; Hubble class S.

NGC 457

OC	Cas	01^h 19.1^m	+58° 20′	12′
chart 1			m$_v$ 6.4	

C13, also the Owl Cluster or ET Cluster. 80 stars; detached, strong concentration of stars; large brightness range; large, bright cluster; mag. of brightest star 8.6. Resembles an owl's face or a stick figure with two very bright eyes. Distance 9,000 ly; age 25 million years; Trumpler class I 3 r.

NGC 460

BN	Tuc	01^h 14.8^m	−72° 18′
chart 24			

Nebulosity surrounding a small star cluster. Located just E of the Small Magellanic Cloud.

NGC 467

Gx	Psc	01^h 19.2^m	+03° 18′	2.2′
chart 10			m$_v$ 11.8	

Has a bright, diffuse nucleus. Distance 250 million ly; Hubble class S0p. Westernmost of a group of three galaxies that includes NGC 470 and NGC 474.

NGC 470

Gx	Psc	01^h 19.8^m	+03° 25′	2.8′ × 1.5′
chart 10			m$_v$ 11.8	p.a. 155°

Pretty bright, large, irregularly round; small, extremely bright nucleus. Distance 120 million ly; Hubble class Sc. In a group with galaxies NGC 467 and NGC 474.

NGC 474

Gx	Psc	01^h 20.1^m	+03° 25′	7.9′ × 7.2′
chart 10			m$_v$ 11.5	p.a. 75°

Pretty bright and small; very bright, small, diffuse nucleus; very faint rings. Distance 100 million ly; Hubble class S0. In a group with galaxies NGC 467 and NGC 470.

NGC 488

Gx	Psc	$01^h 21.8^m$	$+05° 15'$	$5.0' \times 3.9'$
chart 10			m_v 10.3	p.a. 15°

Pretty bright, large, round; many arms; very bright, diffuse nucleus. Distance 100 million ly; Hubble class Sb.

NGC 493

Gx	Cet	$01^h 22.1^m$	$+00° 57'$	$3.5' \times 1.0'$
chart 10			m_v 12.3	p.a. 58°

Very faint and large; slightly brighter toward middle. Distance 110 million ly; Hubble class SBc.

NGC 507

Gx	Psc	$01^h 23.7^m$	$+33° 15'$	$4.0' \times 2.6'$
chart 4			m_v 11.2	p.a. 90°

Very faint, pretty large, and roundish; brighter toward middle. Distance 220 million ly; Hubble class S0.

NGC 514

Gx	Psc	$01^h 24.1^m$	$+12° 55'$	$3.6' \times 3.0'$
chart 10			m_v 11.7	p.a. 110°

Faint, large, and slightly elongated; several arms; very small, bright nucleus. Distance 110 million ly; Hubble class Sc.

NGC 520

Gx	Psc	$01^h 24.6^m$	$+03° 48'$	$3.7' \times 1.4'$
chart 10			m_v 11.4	p.a. 130°

Faint and quite large; viewed nearly edge on. Distance 99 million ly; Hubble class P. Interacting with a pair of spiral galaxies.

NGC 521

Gx	Cet	$01^h 24.6^m$	$+01° 44'$	$3.1' \times 2.8'$
chart 10			m_v 11.7	p.a. 20°

Faint, pretty large, and round; very small, bright nucleus. Distance 220 million ly; Hubble class S(B)b.

NGC 524

Gx	Psc	$01^h 24.8^m$	$+09° 32'$	3.3'
chart 10			m_v 10.2	

Very bright and pretty large; extremely bright nucleus. Four stars near (Webb). Distance 120 million ly; Hubble class E1.

NGC 530

Gx	Cet	$01^h 24.7^m$	$-01° 35'$	$1.5' \times 0.3'$
chart 10			m_v 13	p.a. 135°

Extremely faint, small, and moderately elongated. Distance 220 million ly; Hubble class SB0.

NGC 533

Gx	Cet	$01^h 25.5^m$	$+01° 46'$	$3.9' \times 2.8'$
chart 10			m_v 11.4	p.a. 50°

Pretty bright, pretty large, and round. Distance 220 million ly; Hubble class E2. Many small galaxies nearby.

NGC 559

OC	Cas	$01^h 29.5^m$	$+63° 18'$	4.4'
chart 1			m_v 9.5	

C8. 60 stars; detached, weak concentration of stars; moderate range in brightness; mag. of brightest star 10.6. Distance 3,000 ly; over 1 billion years old; Trumpler class II 2 m.

NGC 578

Gx	Cet	$01^h 30.5^m$	$-22° 40'$	$4.8' \times 3.2'$
chart 18			m_v 11	p.a. 110°

Bright, large, and slightly elongated; three arms; very small, bright nucleus. Distance 73 million ly; Hubble class Sc.

NGC 581

See M103.

NGC 584

Gx	Cet	$01^h 31.3^m$	$-06° 52'$	$3.5' \times 2.1'$
chart 10			m_v 10.5	p.a. 68°

Very bright, pretty large, and round; small, very bright, diffuse nucleus. Distance 81 million ly; Hubble class E4. Located in a group of galaxies.

NGC 596

Gx   Cet $01^h 32.9^m$ $-07°02'$ $3.0' \times 2.3'$
chart 10 m_v 10.9   p.a. 50°

Pretty bright and roundish; distorted with a faint tail. Distance 92 million ly; Hubble class E2.

NGC 598

See M33.

NGC 600

Gx   Cet $01^h 33.1^m$ $-07°19'$ $3.7' \times 3.6'$
chart 10 m_v 12.4   p.a. 85°

Extremely faint. Distance 82 million ly; Hubble class SBd.

NGC 613

Gx   Scl $01^h 34.3^m$ $-29°25'$ $5.0' \times 4.4'$
chart 18 m_v 10   p.a. 120°

Very bright, very large, and elongated; small, extremely bright nucleus with dark lane. Distance 64 million ly; Hubble class S(B)b.

NGC 615

Gx   Cet $01^h 35.1^m$ $-07°20'$ $3.4' \times 1.3'$
chart 10 m_v 11.6   p.a. 155°

Pretty bright, pretty large, and slightly elongated; double arms; very bright nucleus. Distance 83 million ly; Hubble class Sb.

NGC 625

Gx   Phe $01^h 35.1^m$ $-41°26'$ $6.0' \times 1.8'$
chart 18 m_v 11   p.a. 92°

Patchy, large, very faint, and very elongated. Distance 14 million ly; Hubble class SBm.

NGC 628

See M74.

NGC 636

Gx   Cet $01^h 39.1^m$ $-07°31'$ $2.7' \times 2.2'$
chart 10 m_v 11.5   p.a. 140°

Pretty bright, very small, and round. Distance 87 million ly; Hubble class E1.

NGC 637

OC   Cas $01^h 42.9^m$ $+64°00'$ 3.5'
chart 1 m_v 8.2

20 stars; detached, strong concentration of stars; large range in brightness; pretty small, bright cluster; mag. of brightest star 10.0. Distance 6,900 ly; Trumpler class I 3 p.

NGC 650-1

See M76.

NGC 654

OC   Cas $01^h 44.1^m$ $+61°53'$ 5'
chart 1 m_v 6.5

60 stars; detached, weak concentration of stars; large range in brightness; mag. of brightest star 7.4. Distance 8,200 ly; age 15 million years; Trumpler class II 3 m.

NGC 659

OC   Cas $01^h 44.2^m$ $+60°42'$ 5'
chart 1 m_v 7.9

40 stars; detached, no concentration of stars; small range in brightness; mag. of brightest star 10.4. Distance 6,900 ly; age 20 million years; Trumpler class III 1 p.

NGC 660

Gx   Psc $01^h 43.0^m$ $+13°39'$ $9.1' \times 4.1'$
chart 10 m_v 11.2   p.a. 170°

Pretty bright, pretty large, and quite elongated, brighter toward middle. Distance 42 million ly; Hubble class SBap.

NGC 663

OC   Cas $01^h 46.0^m$ $+61°15'$ 15'
chart 1 m_v 7.1

C10, Letter S Cluster. 80 stars; appears to contain a group of stars arranged in an S shape; detached, no concentration of stars; large, bright; moderate range in brightness; mag. of brightest star 8.4. Distance 7,200 ly; age 22 million years; Trumpler class III 2 m.

NGC 672

| Gx | Tri | $01^h 47.9^m$ | +27° 26′ | 6.6′ × 2.7′ |
| chart 4 | | | m_v 10.9 | p.a. 65° |

Faint, small, and moderately elongated; very faint nucleus. Distance 25 million ly; Hubble class SBc. Brightest in a group of galaxies; possibly interacting with galaxy IC 1727.

NGC 676

| Gx | Psc | $01^h 49.0^m$ | +05° 54′ | 4.3′ × 1.6′ |
| chart 10 | | | m_v 9.6 | p.a. 172° |

Very small and bright nucleus; has a bright star superimposed. Distance 70 million ly; Hubble class Sa.

NGC 680

| Gx | Ari | $01^h 49.8^m$ | +21° 58′ | 2.0′ × 1.8′ |
| chart 4 | | | m_v 11.9 | p.a. 140° |

Pretty bright, small, and irregularly roundish; much brighter toward middle. Distance 130 million ly; Hubble class S(B)c.

NGC 681

| Gx | Cet | $01^h 49.2^m$ | −10° 26′ | 2.7′ × 1.8′ |
| chart 10 | | | m_v 12 | p.a. 68° |

Pretty faint, quite large, round; with dust lane; very bright, very small nucleus. Resembles a miniature M104. Distance 76 million ly; Hubble class Sb.

NGC 685

| Gx | Eri | $01^h 47.7^m$ | −52° 46′ | 3.8′ × 3.2′ |
| charts 18, 24 | | | m_v 11.3 | p.a. 90° |

Faint, very large, and round; thick arms; pretty large, pretty bright nucleus. Distance 55 million ly; Hubble class S(B)c.

NGC 691

| Gx | Ari | $01^h 50.7^m$ | +21° 46′ | 3.4′ × 2.5′ |
| chart 4 | | | m_v 11.4 | p.a. 95° |

Faint and quite elongated; very gradually brighter toward middle. Distance 120 million ly; Hubble class Sb.

NGC 692

| Gx | Phe | $01^h 48.7^m$ | −48° 39′ | 2.1′ × 1.8′ |
| chart 18 | | | m_p 13.0 | p.a. 80° |

Bright, small, and roundish; gradually brighter toward middle. Distance 270 million ly; Hubble class SBc.

NGC 697

| Gx | Ari | $01^h 51.3^m$ | +22° 22′ | 4.4′ × 1.4′ |
| chart 4 | | | m_v 12 | p.a. 105° |

Faint, quite large, and elongated. Distance 140 million ly; Hubble class S(B)b. Brightest in a group of galaxies.

NGC 701

| Gx | Cet | $01^h 51.1^m$ | −09° 42′ | 2.6′ × 1.3′ |
| chart 10 | | | m_v 12.2 | p.a. 40° |

Faint, pretty large, and elongated; two main arms; very bright bar. Distance 81 million ly; Hubble class SBc. Paired with galaxy IC 1738.

NGC 718

| Gx | Psc | $01^h 53.2^m$ | +04° 12′ | 2.3′ × 2.1′ |
| chart 10 | | | m_v 11.7 | p.a. 45° |

Pretty bright and small; three pairs of spiral arms, each pair having a different curvature; has an extremely bright nucleus. Distance 76 million ly; Hubble class Sb.

NGC 720

| Gx | Cet | $01^h 53.0^m$ | −13° 44′ | 4.7′ × 2.6′ |
| chart 10 | | | m_v 10.2 | p.a. 135° |

Quite bright, pretty large, and slightly elongated. Distance 79 million ly; Hubble class E3.

NGC 731

| Gx | Cet | $01^h 54.9^m$ | −09° 01′ | 1.6′ |
| chart 10 | | | m_v 12.1 | |

Extremely faint; appears stellar. Distance 170 million ly; Hubble class E0.

NGC 741

Gx	Psc	01ʰ 56.3ᵐ	+05° 38′	3.0′ × 2.7′
chart 10			m_v 11.2	p.a. 90°

Pretty faint, small, and round; stellar nucleus. Distance 240 million ly; Hubble class E1. Brightest in a group of galaxies; shares a common halo with galaxy NGC 742.

NGC 744

OC	Per	01ʰ 58.4ᵐ	+55° 29′	10′
charts 1, 4			m_v 7.9	

20 stars; not well detached from surrounding star field; moderate range in brightness; large cluster; mag. of brightest star 10.4. Distance 4,900 ly; age 39 million years; Trumpler class IV 2 p.

NGC 750

Gx	Tri	01ʰ 57.5ᵐ	+33° 12′	1.8′ × 1.7′
chart 4			m_v 11.9	p.a. 170°

Quite bright, pretty large, and round; very bright nucleus. Distance 230 million ly; Hubble class E. Interacting with galaxy NGC 751.

NGC 752

OC	And	01ʰ 57.8ᵐ	+37° 41′	50′
chart 4			m_v 5.7	

C28. 60 stars; detached, no concentration of stars; small range in brightness; extremely large cluster; brightest star is mag. 9.0. Distance about 1,300 ly; over 1 billion years old; Trumpler class III 1 m.

NGC 753

Gx	And	01ʰ 57.7ᵐ	+35° 55′	2.5′ × 2.0′
chart 4			m_v 12.3	p.a. 125°

Pretty bright, pretty large, and round; very small, very bright nucleus. Distance 220 million ly; Hubble class Sc.

NGC 772

Gx	Ari	01ʰ 59.3ᵐ	+19° 00′	7.1′ × 4.5′
charts 4, 10			m_v 10.3	p.a. 130°

Bright, quite large and round; small, very bright, diffuse nucleus. Distance 110 million ly; Hubble class Sb. Spiral arms seem to bend around nearby galaxy NGC 770.

NGC 777

Gx	Tri	02ʰ 00.3ᵐ	+31° 26′	2.8′ × 2.2′
chart 4			m_v 11.4	p.a. 155°

Pretty bright, pretty large, and round. Distance 230 million ly; Hubble class E2. Paired with fainter galaxy NGC 778, 9′ to S.

NGC 779

Gx	Cet	01ʰ 59.7ᵐ	−05° 58′	4.1′ × 1.2′
chart 10			m_v 11.2	p.a. 162°

Quite bright, large, and very elongated; an edge-on spiral; very bright nucleus. Distance 64 million ly; Hubble class Sb.

NGC 782

Gx	Eri	01ʰ 57.6ᵐ	−57° 47′	2.4′ × 2.0′
chart 24			m_v 11.8	p.a. 15°

Pretty bright, pretty large, and slightly elongated; thin, faint arms; small, bright nucleus. Distance 250 million ly; Hubble class SBb.

NGC 784

Gx	Tri	02ʰ 01.3ᵐ	+28° 51′	6.0′ × 1.5′
chart 4			m_v 11.7	p.a. 0°

Very faint, large, and very elongated in N–S direction; Hubble class S(B).

NGC 788

Gx	Cet	02ʰ 01.1ᵐ	−06° 49′	2.1′ × 1.5′
chart 10			m_v 12.1	p.a. 75°

Pretty faint, pretty small and round; bright nucleus. Distance 180 million ly; Hubble class Sa. Paired with galaxy IC 184.

NGC 803

Gx	Ari	02ʰ 03.8ᵐ	+16° 02′	3.0′ × 1.3′
chart 10			m_v 12.6	p.a. 8°

Very faint and small; gradually brighter toward middle. Distance 96 million ly; Hubble class Sb.

NGC 812

| Gx | And | 02^{h}06.9^m | +44° 34′ | 3.1′ × 1.3′ |
| charts 1, 4 | | | m$_v$ 11.2 | p.a. 160° |

Extremely faint and pretty large. Distance 240 million ly; Hubble class Sp.

NGC 821

| Gx | Ari | 02^{h}08.4^m | +11° 00′ | 2.8′ × 1.9′ |
| chart 10 | | | m$_v$ 10.7 | p.a. 25° |

Pretty bright, very small, and slightly elongated. Distance 82 million ly; Hubble class E2.

NGC 835

| Gx | Cet | 02^{h}09.4^m | −10° 08′ | 0.9′ × 0.8′ |
| chart 10 | | | m$_v$ 12.1 | p.a. 80° |

Faint, small, and roundish. Distance 180 million ly; Hubble class S(B)bp.

NGC 864

| Gx | Cet | 02^{h}15.5^m | +06° 00′ | 4.7′ × 3.2′ |
| chart 10 | | | m$_v$ 10.8 | p.a. 20° |

Extremely faint, quite large, and round; very small, very bright nucleus. Distance 71 million ly; Hubble class Sa.

NGC 869

| OC | Per | 02^{h}19.0^m | +57° 09′ | 30′ |
| chart 1 | | | m$_v$ 5.3 | |

The W member of the Double Cluster, which includes NGC 884. About 200 stars; detached, strong concentration of stars; large range in brightness; very large; mag. of brightest star 6.6. Per *NGC,* a (!) remarkable object. Distance 7,200 ly; age 5.6 million years; extends roughly 60 ly; Trumpler class I 3 r.

NGC 873

| Gx | Cet | 02^{h}16.5^m | −11° 21′ | 1.5′ × 1.2′ |
| chart 10 | | | m$_v$ 12.4 | p.a. 145° |

Faint, pretty large, and roundish; very gradually brighter toward middle, distance 180 million ly; Hubble class Sb.

NGC 877

| Gx | Ari | 02^{h}18.0^m | +14° 33′ | 2.4′ × 1.9′ |
| chart 10 | | | m$_v$ 11.9 | p.a. 140° |

Pretty faint, pretty large, slightly elongated; two arms; very small, very bright nucleus. Distance 180 million ly; Hubble class SBc. Brightest in a group with galaxies NGC 870, NGC 871, and NGC 876.

NGC 884

| OC | Per | 02^{h}22.4^m | +57° 07′ | 30′ |
| chart 1 | | | m$_v$ 6.1 | |

The E member of the Double Cluster, which includes NGC 869. 115 stars; detached, strong concentration of stars; large range in brightness; very large; mag. of brightest star 8.1. Per *NGC,* a (!) remarkable object. Distance 7,500 ly; age 3.2 million years; extends 70 ly.

NGC 887

| Gx | Cet | 02^{h}19.5^m | −16° 04′ | 1.9′ × 1.5′ |
| chart 10 | | | m$_v$ 12 | p.a. 5° |

Faint, small, and irregularly roundish; pretty gradually brighter toward middle, distance 190 million ly; Hubble class SBc.

NGC 890

| Gx | Tri | 02^{h}22.0^m | +33° 16′ | 2.9′ × 2.3′ |
| chart 4 | | | m$_v$ 11.2 | p.a. 60° |

Bright, small, and round; very small, bright nucleus. Distance 180 million ly; Hubble class E4.

NGC 891

| Gx | And | 02^{h}22.6^m | +42° 21′ | 13.5′ × 2.8′ |
| charts 1, 4 | | | m$_v$ 9.9 | p.a. 22° |

C23. Bright, very large, and very elongated; broad dust lane. Discovered by Caroline Herschel in 1783 with a small 30× reflector. Brighter at the edges than along the central part (Smyth). Most picturesque edge-on spiral in the entire sky (Mullaney). Per *NGC,* a (!) remarkable object. Distance 30 million ly; Hubble class Sb. Part of the NGC 1023 galaxy group.

NGC 895

Gx	Cet	02^h 21.6^m	−05° 31′	3.5′ × 2.6′
chart 10			m$_v$ 11.7	p.a. 110°

Faint, very large, and irregularly round; two main arms; very small, bright nucleus. Distance 100 million ly; Hubble class Sb. Joined to fainter galaxy NGC 894.

NGC 896

BNe	Cas	02^h 24.8^m	+61° 54′	26′ × 12′
chart 1				

Extremely faint and pretty large; irregular shape. Has a fairly prominent absorption patch in SW portion.

NGC 897

Gx	For	02^h 21.1^m	−33° 43′	2.1′ × 1.3′
chart 18			m$_v$ 12.2	p.a. 17°

Pretty bright, small, and roundish; pretty suddenly brighter toward middle. Distance 210 million ly; Hubble class Sa. Three faint galaxies nearby.

NGC 908

Gx	Cet	02^h 23.1^m	−21° 14′	5.5′ × 2.8′
chart 18			m$_v$ 10.4	p.a. 75°

Quite bright, very large, and elongated; several filamentary arms; very small, bright nucleus. Distance 64 million ly; Hubble class Sc.

NGC 922

Gx	For	02^h 25.1^m	−24° 47′	1.9′
chart 18			m$_v$ 12	p.a. 172°

Quite faint, pretty large, and round; several filamentary arms; bright nucleus. Distance 130 million ly; Hubble class Scp. Possibly interacting with an anonymous barred-spiral galaxy.

NGC 925

Gx	Tri	02^h 27.3^m	+33° 35′	9.8′ × 6.0′
chart 4			m$_v$ 10.1	p.a. 102°

Quite faint, quite large, and elongated; very faint nucleus. Distance 31 million ly; Hubble class S(B)c. Belongs to the NGC 1023 galaxy group.

NGC 936

Gx	Cet	02^h 27.6^m	−01° 09′	4.5′ × 3.7′
chart 10			m$_v$ 10.2	p.a. 135°

Very bright, very large, and round; diffuse, very bright nucleus. Bluish white and pale, but very distinct (Smyth). Distance 59 million ly; Hubble class SBa. Paired with galaxy NGC 941.

NGC 941

Gx	Cet	02^h 28.5^m	−01° 09′	2.7′ × 1.9′
chart 10			m$_v$ 12.4	p.a. 170°

Very faint, quite large, and roundish; very small, bright nucleus. Distance 68 million ly; Hubble class Sc. Paired with galaxy NGC 936.

NGC 945

Gx	Cet	02^h 28.6^m	−10° 32′	2.3′ × 2.2′
chart 10			m$_v$ 12.1	p.a. 65°

Very faint, large, irregularly roundish; gradually brighter toward middle. Distance 190 million ly; Hubble class SBc.

NGC 949

Gx	Tri	02^h 30.8^m	+37° 08′	2.8′ × 1.7′
chart 4			m$_v$ 11.8	p.a. 145°

Quite bright, small and elongated. Distance 34 million ly; Hubble class S.

NGC 955

Gx	Cet	02^h 30.6^m	−01° 06′	2.8′ × 0.8′
chart 10			m$_v$ 12	p.a. 19°

Pretty bright, small, elongated. Distance 68 million ly; Hubble class Sb.

NGC 957

OC	Per	02^h 33.6^m	+57° 32′	10′
chart 1			m$_v$ 7.6	

30 stars; detached, no concentration of stars; moderate range in brightness; pretty large; mag. of brightest star 9.5. Distance 7,200 ly; age 15 million years; Trumpler class III 2 p.

NGC 958

Gx	Cet	02ʰ 30.7ᵐ	−02° 56′	3.0′ × 0.9′
chart 10			m_v 12.2	p.a. 12°

Pretty faint; a nearly edge-on spiral. Distance 250 million ly; Hubble class S(B)b.

NGC 959

Gx	Tri	02ʰ 32.4ᵐ	+35° 30′	2.3′ × 1.4′
chart 4			m_v 12.4	p.a. 65°

Extremely faint, pretty large, and slightly elongated; slightly brighter toward middle. Distance 33 million ly; Hubble class Sm.

NGC 972

Gx	Ari	02ʰ 34.2ᵐ	+29° 19′	3.2′ × 1.6′
chart 4			m_v 11.4	p.a. 152°

Pretty bright, quite large, and slightly elongated; small, faint nucleus. Distance 73 million ly; Hubble class Sc.

NGC 986

Gx	For	02ʰ 33.6ᵐ	−39° 03′	3.9′ × 3.1′
chart 18			m_v 10.9	p.a. 150°

Pretty bright, pretty large, and quite elongated; very bright nucleus with dark lanes. Distance 85 million ly; Hubble class SBb.

NGC 988

Gx	Cet	02ʰ 35.5ᵐ	−09° 22′	4.5′ × 2.2′
chart 10			m_v 11	p.a. 112°

Distance 63 million ly; Hubble class SBc. A 7.5-mag. star is almost superimposed.

NGC 991

Gx	Cet	02ʰ 35.5ᵐ	−07° 09′	3.0′ × 2.6′
chart 10			m_v 11.7	p.a. 60°

Very faint and quite large; very small, bright nucleus. Distance 69 million ly; Hubble class S. In NGC 1052 galaxy group.

NGC 1003

Gx	Per	02ʰ 39.3ᵐ	+40° 52′	5.4′ × 2.1′
charts 1, 4			m_v 11.4	p.a. 97°

Pretty faint, large, and elongated. Distance 34 million ly; Hubble class Sa. Located in NGC 1023 galaxy group; cluster of spiral galaxies in background.

NGC 1012

Gx	Ari	02ʰ 39.3ᵐ	+30° 09′	2.6′ × 1.1′
chart 4			m_v 12	p.a. 24°

Large, nuclear bulge. Distance 49 million ly; Hubble class P.

NGC 1015

Gx	Cet	02ʰ 38.2ᵐ	−01° 19′	2.6′ × 2.6′
chart 10			m_v 12.1	p.a. 10°

Very faint and small. Distance 120 million ly; Hubble class SBa.

NGC 1016

Gx	Cet	02ʰ 38.3ᵐ	+02° 07′	2.3′
chart 10			m_v 11.6	

Faint, small, and roundish; bright middle. Distance 290 million ly; Hubble class E0.

NGC 1022

Gx	Cet	02ʰ 38.5ᵐ	−06° 41′	2.6′
chart 10			m_v 11.3	

Quite bright, pretty large, and round; small, very bright nucleus. Distance 65 million ly; Hubble class Sb. In NGC 1052 galaxy group.

NGC 1023

Gx	Per	02ʰ 40.4ᵐ	+39° 04′	8.7′ × 3.3′
charts 1, 4			m_v 9.3	p.a. 87°

Very bright, very large, and very elongated; extremely bright nucleus. Distance 34 million ly; Hubble class E7p. Brightest in a group of galaxies including NGC 925 and NGC 1003.

NGC 1027

OC	Cas	02ʰ 42.7ᵐ	+61° 33′	19′
chart 1			m_v 6.7	

40 stars; detached, no concentration of stars; moderate range in brightness; large; mag. of brightest

star 9.3. Distance 3,300 ly; age 350 million years; Trumpler class III 2 p.

NGC 1032

Gx	Cet	02^h 39.4^m	+01° 06'	3.4' × 1.0'
chart 10			m$_v$ 11.6	p.a. 68°

Pretty bright and slightly elongated. Distance 120 million ly; Hubble class Sa. Forms a trapezoid with three stars.

NGC 1035

Gx	Cet	02^h 39.5^m	−08° 08'	2.2' × 0.6'
chart 10			m$_v$ 12.2	p.a. 150°

Pretty faint, large, and quite elongated; a nearly edge-on spiral. Distance 57 million ly; Hubble class P. In NGC 1052 galaxy group.

NGC 1039

See M34.

NGC 1042

Gx	Cet	02^h 40.4^m	−08° 26'	4.7' × 3.9'
chart 10			m$_v$ 11	p.a. 15°

Faint, large, and round; two well-defined spiral arms; very small, very bright nucleus. Distance 59 million ly; Hubble class Sc. In NGC 1052 galaxy group.

NGC 1045

Gx	Cet	02^h 40.5^m	−11° 17'	1.9' × 1.3'
chart 10			m$_v$ 12.1	p.a. 55°

Faint, small, and roundish; brighter toward middle. Distance 20 million ly; Hubble class E/S0.

NGC 1049

GC	For	02^h 39.7^m	−34° 17'	0.4'
chart 18			m$_v$ 12.9	

The brightest globular cluster in the Fornax Dwarf Galaxy.

NGC 1052

Gx	Cet	02^h 41.1^m	−08° 15'	2.9' × 2.1'
chart 10			m$_v$ 10.5	p.a. 120°

Bright, pretty large, and round; very bright, diffuse center. This galaxy is an active radio source. Distance 62 million ly; Hubble class E2. Brightest in a group of galaxies.

NGC 1055

Gx	Cet	02^h 41.7^m	+00° 27'	7.6' × 3.0'
chart 10			m$_v$ 10.6	p.a. 105°

Pretty faint and quite large; a nearly edge-on spiral with dust lane. Distance 47 million ly; Hubble class Sb. In NGC 1068 galaxy group.

NGC 1058

Gx	Per	02^h 43.5^m	+37° 21'	3.2'
charts 1, 4			m$_v$ 11.2	

Pretty faint and round; many arms; small, very bright nucleus. Distance 29 million ly; Hubble class Sc. In NGC 1023 galaxy group.

NGC 1060

Gx	Tri	02^h 43.3^m	+32° 25'	2.3' × 1.7'
chart 4			m$_v$ 11.8	p.a. 75°

Faint, pretty large, and roundish; slightly brighter toward middle. Distance 230 million ly; Hubble class E/S0.

NGC 1068

See M77.

NGC 1070

Gx	Cet	02^h 43.4^m	+04° 58'	2.4' × 2.0'
chart 10			m$_v$ 11.9	p.a. 175°

Pretty faint, small, irregularly roundish; gradually brighter toward middle. Distance 180 million ly; Hubble class Sb.

NGC 1073

Gx	Cet	02^h 43.7^m	+01° 23'	4.8' × 4.4'
chart 10			m$_v$ 11	p.a. 15°

Very faint and large; very small, bright nucleus. Distance 54 million ly; Hubble class S(B)c. In NGC 1068 galaxy group.

NGC 1079

| Gx | For | 02ʰ 43.7ᵐ | −29° 00′ | 3.9′ × 2.4′ |
| chart 18 | | | m_v 11.3 | p.a. 87° |

Bright, pretty large, and quite elongated; small, very bright nucleus. Distance 94 million ly; Hubble class S(B)ap.

NGC 1084

| Gx | Eri | 02ʰ 46.0ᵐ | −07° 35′ | 3.3′ × 2.1′ |
| chart 10 | | | m_v 10.7 | p.a. 37° |

Very bright, pretty large, and elongated; very small, bright nucleus. Pale though distinct (Smyth). Distance 61 million ly; Hubble class Sc.

NGC 1087

| Gx | Cet | 02ʰ 46.4ᵐ | −00° 30′ | 3.9′ × 2.3′ |
| chart 10 | | | m_v 10.9 | p.a. 5° |

Pretty bright, quite large, and slightly elongated. Distance 80 million ly; Hubble class Sc. In NGC 1068 galaxy group; paired with fainter galaxy NGC 1090.

NGC 1090

| Gx | Cet | 02ʰ 46.6ᵐ | −00° 15′ | 4.1′ × 1.8′ |
| chart 10 | | | m_v 11.8 | p.a. 102° |

Very faint, pretty large, irregularly round; bright nucleus in narrow bar. Distance 120 million ly; Hubble class S−. In NGC 1068 galaxy group; paired with galaxy NGC 1087.

NGC 1097

| Gx | For | 02ʰ 46.3ᵐ | −30° 16′ | 9.3′ × 6.6′ |
| chart 18 | | | m_v 9.2 | p.a. 130° |

C67. Very bright, large, and very elongated; two main arms. Has a small, extremely bright, peculiar nucleus with inner spiral structure. Distance 53 million ly; Hubble class S(B)b.

NGC 1122

| Gx | Per | 02ʰ 52.9ᵐ | +42° 12′ | 1.8′ × 1.3′ |
| charts 1, 4 | | | m_v 12.1 | p.a. 40° |

Very faint, pretty small, and roundish. Distance 160 million ly; Hubble class SBb.

NGC 1134

| Gx | Ari | 02ʰ 53.7ᵐ | +13° 01′ | 2.5′ × 0.9′ |
| chart 10 | | | m_v 12.1 | p.a. 148° |

Faint, small, irregularly roundish. Distance 160 million ly; Hubble class S.

NGC 1140

| Gx | Eri | 02ʰ 54.6ᵐ | −10° 02′ | 1.7′ × 1.0′ |
| chart 10 | | | m_v 12.5 | p.a. 6° |

Pretty bright, small, and round; appears stellar; very bright nucleus. Distance 65 million ly; Hubble class P.

NGC 1156

| Gx | Ari | 02ʰ 59.7ᵐ | +25° 14′ | 3.3′ × 2.8′ |
| chart 4 | | | m_v 11.7 | p.a. 25° |

Pretty bright, quite large, and quite elongated. Distance 21 million ly; Hubble class Ir+. Located between two stars.

NGC 1161

| Gx | Per | 03ʰ 01.2ᵐ | +44° 54′ | 3.0′ × 1.9′ |
| charts 1, 4 | | | m_v 11 | p.a. 23° |

Faint, small, and slightly elongated; suddenly brighter toward middle, distance 90 million ly; Hubble class S0.

NGC 1169

| Gx | Per | 03ʰ 03.6ᵐ | +46° 23′ | 4.6′ × 2.7′ |
| charts 1, 4 | | | m_v 11.2 | p.a. 28° |

Faint, small; bright, very small nucleus. Distance 110 million ly; Hubble class S(B)b. A foreground star is superimposed near nucleus.

NGC 1172

| Gx | Eri | 03ʰ 01.6ᵐ | −14° 50′ | 2.3′ × 1.7′ |
| charts 10, 11 | | | m_v 11.9 | p.a. 25° |

Pretty faint, pretty large, and round. Distance 71 million ly; Hubble class E1.

NGC 1179

| Gx | Eri | 03ʰ 02.6ᵐ | −18° 54′ | 4.7′ × 3.6′ |
| charts 10, 11, 18 | | | m_v 11.9 | p.a. 35° |

Extremely faint and pretty small; several faint arms;

very small, bright nucleus. Distance 75 million ly; Hubble class Sp.

NGC 1186

| Gx | Per | 03^h 05.5^m | +42° 50′ | 3.2′ × 1.2′ |
| charts 1, 4 | | | m$_v$ 11.4 | p.a. 122° |

Faint, with nebulous appendages. Distance 120 million ly; Hubble class SBb.

NGC 1187

| Gx | Eri | 03^h 02.6^m | −22° 52′ | 5.0′ × 3.9′ |
| chart 18 | | | m$_v$ 10.7 | p.a. 130° |

Pretty faint, quite large, and quite elongated; multiple arms; very small, very bright nucleus. Distance 58 million ly; Hubble class S(B)c.

NGC 1199

| Gx | Eri | 03^h 03.6^m | −15° 37′ | 1.8′ × 1.4′ |
| charts 10, 11 | | | m$_v$ 11.3 | p.a. 48° |

Quite bright, pretty small, irregularly round. Distance 110 million ly; Hubble class E2. Brightest in a group of galaxies.

NGC 1200

| Gx | Eri | 03^h 03.9^m | −12° 00′ | 2.9′ × 1.8′ |
| charts 10, 11 | | | m$_v$ 12.7 | p.a. 85° |

Pretty faint, quite large, irregularly roundish; bright middle. Distance 170 million ly; Hubble class E/S0.

NGC 1201

| Gx | For | 03^h 04.1^m | −26° 04′ | 3.4′ × 1.9′ |
| chart 18 | | | m$_v$ 10.8 | p.a. 7° |

Quite bright and pretty small; very bright, diffuse nucleus. Distance 71 million ly; Hubble class Sa. Star nearby.

NGC 1209

| Gx | Eri | 03^h 06.1^m | −15° 37′ | 2.3′ × 1.1′ |
| charts 10, 11 | | | m$_v$ 11.4 | p.a. 80° |

Bright, small, and quite elongated. Distance 110 million ly; Hubble class E5. In NGC 1199 galaxy group.

NGC 1232

| Gx | Eri | 03^h 09.8^m | −20° 35′ | 7.8′ × 6.9′ |
| chart 18 | | | m$_v$ 10 | p.a. 108° |

Pretty bright, quite large, and round; thin arms; very small, bright, diffuse nucleus. Distance 72 million ly; Hubble class Sc. Galaxy NGC 1232A is a satellite.

NGC 1241

| Gx | Eri | 03^h 11.2^m | −08° 55′ | 3.2′ × 2.0′ |
| charts 10, 11 | | | m$_v$ 12 | p.a. 145° |

Faint, pretty large, and round; two main arms; very small, very bright nucleus. Distance 93 million ly; Hubble class Sb. Paired with galaxy NGC 1242. A 9th-mag. star is nearby.

NGC 1245

| OC | Per | 03^h 14.7^m | +47° 15′ | 10′ |
| charts 1, 4, 5 | | | m$_v$ 8.4 | |

200 stars; detached, no concentration of stars; small range in brightness; mag. of brightest star 11.2; pretty large. An elegant sprinkle (Smyth). Low power shows a very faint cloud of minute stars, beautifully bordered by a foreshortened pentagon (Webb). Distance 7,500 ly; over 1 billion years old; Trumpler class III 1 r.

NGC 1249

| Gx | Hor | 03^h 10.0^m | −53° 20′ | 4.8′ × 2.3′ |
| charts 18, 24 | | | m$_v$ 11.5 | p.a. 86° |

Bright, large, and very elongated; patchy arms; very faint nucleus. Distance 36 million ly; Hubble class SBc.

NGC 1253

| Gx | Eri | 03^h 14.2^m | −02° 49′ | 5.0′ × 2.3′ |
| charts 10, 11 | | | m$_v$ 11.7 | p.a. 82° |

Distance 74 million ly; Hubble class S(B)c.

NGC 1255

| Gx | For | 03^h 13.5^m | −25° 44′ | 4.1′ × 2.6′ |
| chart 18 | | | m$_v$ 11 | p.a. 117° |

Faint and pretty large; two main arms; very bright, very small nucleus. Distance 76 million ly; Hubble class Sc.

NGC 1261

GC	Hor	03^h 12.3^m	−55° 13′	7′
charts 18, 24			m$_v$ 8.3	

C87. Bright, large, and round; partially resolved; high concentration of stars. Stars and stardust (Denning). Distance 44,000 ly; S-S class 2.

NGC 1272

Gx	Per	03^h 19.4^m	+41° 30′	2.2′ × 2.0′
charts 1, 4, 5			m$_v$ 11.8	p.a. 0°

Faint, small, and roundish. Distance 190 million ly; Hubble class S0.

NGC 1275

Gx	Per	03^h 19.8^m	+41° 31′	2.6′ × 1.9′
charts 1, 4, 5			m$_v$ 11.9	p.a. 110°

C24, the Perseus A radio source. A faint and small Seyfert galaxy; distance 230 million ly; Hubble class P. Brightest member of the Perseus Galaxy Cluster (Abell 426).

NGC 1288

Gx	For	03^h 17.2^m	−32° 35′	2.3′ × 1.9′
chart 18			m$_v$ 12	p.a. 0°

Very faint and large; thin arms. Distance 190 million ly; Hubble class Sb.

NGC 1291

Gx	Eri	03^h 17.3^m	−41° 06′	10.5′ × 9.1′
chart 18			m$_v$ 8.5	p.a. 80°

Very bright, pretty large, and round; very bright nucleus. Distance 29 million ly; Hubble class SBa.

NGC 1292

Gx	For	03^h 18.3^m	−27° 37′	3.0′ × 1.3′
chart 18			m$_v$ 11.9	p.a. 7°

Faint, pretty small, and slightly elongated; two bright arms; small, bright center. Distance 58 million ly; Hubble class Sc.

NGC 1300

Gx	Eri	03^h 19.7^m	−19° 25′	6.5′ × 4.3′
charts 10, 11, 18			m$_v$ 10.4	p.a. 106°

Quite bright, very large, and very elongated; very small, extremely bright nucleus. While photographs reveal a central bar with two thin but tightly wound spiral arms, small amateur instruments show only a blurred spindle (Houston). Distance 62 million ly; Hubble class SBb.

NGC 1302

Gx	For	03^h 19.8^m	−26° 04′	4.0′
chart 18			m$_v$ 10.9	p.a. 171°

Small and round; extremely bright nucleus. Distance 71 million ly; Hubble class S(B)a.

NGC 1305

Gx	Eri	03^h 21.4^m	−02° 19′	1.6′ × 1.1′
charts 10, 11			m$_v$ 13.3	p.a. 130°

Pretty bright, pretty small, and roundish. Distance 270 million ly; Hubble class E2.

NGC 1309

Gx	Eri	03^h 22.1^m	−15° 24′	2.3′ × 2.2′
charts 10, 11, 18			m$_v$ 11.5	p.a. 45°

Bright, large, irregularly round; two main arms; small, bright nucleus. Distance 95 million ly; Hubble class Sc.

NGC 1310

Gx	For	03^h 21.1^m	−37° 06′	1.9′ × 1.5′
chart 18			m$_v$ 12.3	p.a. 95°

Very faint, pretty large, and roundish; only slightly brighter toward middle; knotty arms. Distance 68 million ly; Hubble class Sc.

NGC 1313

Gx	Ret	03^h 18.3^m	−66° 30′	8.5′ × 6.6′
chart 24			m$_v$ 8.9	p.a. 38°

Pretty bright, large, and elongated; two main arms; very small, bright nucleus. Distance 10 million ly; Hubble class SBd.

NGC 1316

Gx	For	03^h 22.7^m	−37° 12′	7.1′ × 5.5′
chart 18			m$_v$ 8.2	p.a. 50°

Fornax A radio source, also an X-ray source. Visually quite large, slightly elongated, with diffuse center; extended envelope with faint loops. Distance 71 million ly; Hubble class S(B)0p. Member of the Fornax Galaxy Cluster.

NGC 1317

Gx	For	$03^h\,22.7^m$	$-37°\,06'$	$2.8' \times 2.3'$
chart 18			m_v 10.8	p.a. 78°

Pretty bright and pretty small; extremely bright nucleus. Distance 83 million ly; Hubble class S(B)a. Member of the Fornax Galaxy Cluster.

NGC 1325

Gx	Eri	$03^h\,24.4^m$	$-21°\,33'$	$4.7' \times 1.7'$
chart 18			m_v 11.5	p.a. 56°

Small, bright nucleus. Distance 67 million ly; Hubble class Sb.

NGC 1326

Gx	For	$03^h\,23.9^m$	$-36°\,28'$	$4.3' \times 2.9'$
chart 18			m_v 10.9	p.a. 77°

A pretty small galaxy with a very bright nucleus. Distance 51 million ly; Hubble class SB0. Member of the Fornax Galaxy Cluster.

NGC 1332

Gx	Eri	$03^h\,26.3^m$	$-21°\,20'$	$4.0' \times 1.5'$
chart 18			m_v 10.5	p.a. 148°

Very bright and small; very bright, diffuse nucleus. Distance 64 million ly; Hubble class E7. Paired with galaxy NGC 1331.

NGC 1333

BNr	Per	$03^h\,29.3^m$	$+31°\,25'$	$8' \times 6'$
chart 4				

Faint and large; extended NNE–SSW; brightest at the ends.

NGC 1337

Gx	Eri	$03^h\,28.1^m$	$-08°\,23'$	$6.8' \times 2.0'$
charts 10, 11			m_v 11.9	p.a. 145°

Extremely faint, very large, and quite elongated; nearly edge on; two spiral arms; faint nucleus with dark lanes. Distance 52 million ly; Hubble class S–.

NGC 1338

Gx	Eri	$03^h\,28.9^m$	$-12°\,09'$	$1.3' \times 1.2'$
charts 10, 11			m_v 12.7	p.a. 55°

Very faint, small, irregularly roundish; slightly brighter toward middle, distance 110 million ly; Hubble class Sb.

NGC 1339

Gx	For	$03^h\,28.1^m$	$-32°\,17'$	$1.9' \times 1.4'$
chart 18			m_v 11.6	p.a. 172°

Quite bright, pretty small, and round. Distance 52 million ly; Hubble class E2.

NGC 1341

Gx	For	$03^h\,28.0^m$	$-37°\,09'$	$1.6'$
charts 18, 19			m_v 12.1	p.a. 118°

Faint and small; high surface brightness. Distance 73 million ly; Hubble class SBa. Member of the Fornax Galaxy Cluster.

NGC 1342

OC	Per	$03^h\,31.6^m$	$+37°\,20'$	$14'$
charts 4, 5			m_v 6.7	

40 stars; detached, no concentration of stars; large range in brightness; very large; mag. of brightest star 8.8. Distance 1,800 ly; age 300 million years; Trumpler class III 3 p.

NGC 1344

Gx	For	$03^h\,28.3^m$	$-31°\,04'$	$5.0' \times 3.7'$
chart 18			m_v 10.2	p.a. 165°

Quite bright, pretty large, irregularly round; bright, elongated nucleus. Distance 47 million ly; Hubble class E3.

NGC 1350

Gx	For	$03^h\,31.1^m$	$-33°\,38'$	$5.0' \times 2.9'$
chart 18			m_v 10.3	p.a. 0°

Bright, large, and very elongated; pretty bright arms. Distance 72 million ly; Hubble class SBb. Member of the Fornax Galaxy Cluster.

NGC 1351

Gx	For	03^h 30.6^m	−34° 51′	3.2′ × 1.9′
chart 18			m$_v$ 11.3	p.a. 140°

Pretty bright, pretty small, and round; very bright nucleus. Distance 59 million ly; Hubble class S0. Member of the Fornax Galaxy Cluster.

NGC 1353

Gx	Eri	03^h 32.1^m	−20° 49′	3.5′ × 1.4′
chart 18			m$_v$ 11.4	p.a. 138°

Pretty bright, quite large, and slightly elongated; patchy arms; very small, very bright nucleus. Distance 70 million ly; Hubble class Sb.

NGC 1357

Gx	Eri	03^h 33.3^m	−13° 40′	3.0′ × 2.2′
charts 10, 11			m$_v$ 11.5	p.a. 85°

Pretty bright, pretty large, and round; small, bright, diffuse nucleus. Distance about 85 million ly; Hubble class Sa.

NGC 1359

Gx	Eri	03^h 33.8^m	−19° 29′	2.5′ × 1.5′
charts 10, 11, 18			m$_v$ 12.1	p.a. 139°

Faint, large, and round; twisted arms. Distance 78 million ly; Hubble class SBp+. Paired with an anonymous barred-spiral galaxy.

NGC 1365

Gx	For	03^h 33.6^m	−36° 08′	9.8′ × 5.5′
charts 18, 19			m$_v$ 9.3	p.a. 32°

Very bright, very large, and quite elongated; extremely bright nucleus. An excellent example of a barred spiral. Per *NGC*, a (!!) very remarkable object. Extends 200,000 ly; distance 65 million ly; Hubble class SBb. Possibly in the foreground of the Fornax Galaxy Cluster.

NGC 1366

Gx	For	03^h 33.9^m	−31° 12′	2.0′ × 0.9′
chart 18			m$_v$ 11.9	p.a. 2°

Very faint and small; flattened central bulge. Distance 49 million ly; Hubble class E6.

NGC 1371

Gx	For	03^h 35.0^m	−24° 56′	5.4′ × 4.0′
chart 18			m$_v$ 10.6	p.a. 135°

Pretty bright and pretty large; small, very bright nucleus. Distance 61 million ly; Hubble class S(B)a.

NGC 1374

Gx	For	03^h 35.3^m	−35° 14′	2.6′ × 2.4′
charts 18, 19			m$_v$ 11	p.a. 100°

Very bright and pretty large; bright nucleus. Distance 48 million ly; Hubble class E0. Paired with galaxy NGC 1375; member of the Fornax Galaxy Cluster.

NGC 1376

Gx	Eri	03^h 37.1^m	−05° 03′	2.0′ × 1.8′
charts 10, 11			m$_v$ 12.1	p.a. 95°

Extremely faint, pretty large, irregularly roundish; filamentary arms; small, bright nucleus. Distance 180 million ly; Hubble class Sc. Paired with an anonymous spiral galaxy.

NGC 1379

Gx	For	03^h 36.1^m	−35° 26′	2.6′
charts 18, 19			m$_v$ 11	p.a. 17°

Bright, pretty large, and round; bright nucleus. Distance 54 million ly; Hubble class E0. Member of the Fornax Galaxy Cluster.

NGC 1380

Gx	For	03^h 36.4^m	−34° 59′	4.8′ × 3.0′
charts 18, 19			m$_v$ 10	p.a. 7°

Very bright, large, and elongated; very bright nucleus. Distance 72 million ly; Hubble class S0. Member of the Fornax Galaxy Cluster.

NGC 1381

Gx	For	03^h 36.5^m	−35° 18′	2.7′ × 0.9′
charts 18, 19			m$_v$ 11.5	p.a. 139°

Faint, with a small, very bright nucleus. Distance 71 million ly; Hubble class S0. Member of the Fornax Galaxy Cluster.

NGC 1385

Gx	For	03ʰ 37.5ᵐ	−24° 30′	3.7′ × 2.3′
chart 18			m_v 10.7	p.a. 165°

Pretty bright, pretty small, and round; patchy arms. Distance 60 million ly; Hubble class Sc.

NGC 1386

Gx	Eri	03ʰ 36.8ᵐ	−36° 00′	3.5′ × 1.3′
charts 18, 19			m_v 11.2	p.a. 25°

Faint, with a small, bright nucleus. Distance 28 million ly; Hubble class Sb. Member of the Fornax Galaxy Cluster.

NGC 1387

Gx	For	03ʰ 37.0ᵐ	−35° 30′	3.2′
charts 18, 19			m_v 10.8	p.a. 119°

Very bright, pretty large, and round; very bright, elongated nucleus. Distance 48 million ly; Hubble class S0. Member of the Fornax Galaxy Cluster.

NGC 1389

Gx	Eri	03ʰ 37.2ᵐ	−35° 45′	2.6′ × 1.5′
charts 18, 19			m_v 11.4	p.a. 30°

Faint, with a bright nucleus. Distance 40 million ly; Hubble class E4. Member of the Fornax Galaxy Cluster.

NGC 1393

Gx	Eri	03ʰ 38.6ᵐ	−18° 26′	1.9′ × 1.2′
charts 10, 11, 18			m_v 12	p.a. 170°

Elongated nucleus. Distance 92 million ly; Hubble class S0.

NGC 1395

Gx	Eri	03ʰ 38.5ᵐ	−23° 02′	5.0′ × 4.6′
chart 18			m_v 9.7	p.a. 90°

Bright, very small, and elongated; has a very bright center. Distance about 69 million ly; Hubble class E3.

NGC 1398

Gx	For	03ʰ 38.9ᵐ	−26° 20′	6.6′ × 5.2′
chart 18			m_v 9.5	p.a. 100°

Quite bright, quite large, and round; large, bright nucleus; internal ring with bar. Distance 57 million ly; Hubble class S(B)b.

NGC 1399

Gx	For	03ʰ 38.5ᵐ	−35° 27′	6.0′
charts 18, 19			m_v 8.8	p.a. 29°

Very bright and pretty large; bright nucleus. Distance 56 million ly; Hubble class E1p. Member of the Fornax Galaxy Cluster; several faint galaxies nearby.

NGC 1400

Gx	Eri	03ʰ 39.5ᵐ	−18° 41′	2.5′ × 2.1′
charts 10, 11, 18			m_v 11	p.a. 40°

Quite bright, pretty small, and round; small, bright, diffuse nucleus. Distance 17 million ly; Hubble class E1. Several faint companion galaxies nearby; in NGC 1407 galaxy group.

NGC 1404

Gx	Eri	03ʰ 38.9ᵐ	−35° 36′	4.0′ × 3.3′
charts 18, 19			m_v 9.7	p.a. 170°

Very bright, pretty large, and roundish; bright nucleus. Distance 77 million ly; Hubble class E1. Member of the Fornax Galaxy Cluster. A bright star is superimposed; faint galaxies nearby.

NGC 1406

Gx	For	03ʰ 39.4ᵐ	−31° 19′	3.8′ × 0.7′
charts 18, 19			m_v 11.6	p.a. 15°

Faint, quite large, and very elongated; bright nucleus with dust lane. Distance 32 million ly; Hubble class Sb.

NGC 1407

Gx	Eri	03ʰ 40.2ᵐ	−18° 35′	5.0′ × 4.6′
charts 10, 11, 18			m_v 9.7	p.a. 35°

Very bright, large, and round; small, bright, diffuse nucleus. Nearly midway between and W of two stars, the three forming an obtuse triangle (Smyth). Distance 75 million ly; Hubble class E0. Paired with galaxy NGC 1400; brightest member of a group of galaxies.

NGC 1411

Gx	Hor	03^h 38.7^m	−44° 06′	2.3′ × 1.7′
charts 18, 19			m$_v$ 11.1	p.a. 6°

Bright, pretty small, and round; superimposed bright star; very bright nucleus. Distance 39 million ly; Hubble class S0.

NGC 1415

Gx	Eri	03^h 40.9^m	−22° 34′	3.6′ × 1.7′
chart 18			m$_v$ 11.5	p.a. 148°

Pretty bright, small, and slightly elongated; very bright nucleus. Distance 61 million ly; Hubble class Sb. Paired with galaxy NGC 1416.

NGC 1417

Gx	Eri	03^h 42.0^m	−04° 42′	2.6′ × 1.5′
charts 10, 11			m$_v$ 12.1	p.a. 5°

Pretty faint, pretty large, and slightly elongated; several bright arms; small, very bright nucleus. Distance 180 million ly; Hubble class Sb. In a group with galaxies NGC 1418 and NGC 1424.

NGC 1421

Gx	Eri	03^h 42.5^m	−13° 29′	3.5′ × 0.9′
charts 10, 11			m$_v$ 11.4	p.a. 179°

Faint, quite large, and quite elongated; nearly edge on; very small, bright nucleus with dark lane. Distance 90 million ly; Hubble class Sb.

NGC 1425

Gx	For	03^h 42.2^m	−29° 54′	5.4′ × 2.7′
charts 18, 19			m$_v$ 10.8	p.a. 129°

Faint, pretty large, irregularly round; very small, very bright nucleus. Distance 65 million ly; Hubble class Sb.

NGC 1426

Gx	Eri	03^h 42.8^m	−22° 07′	2.7′ × 1.7′
chart 18			m$_v$ 11.2	p.a. 111°

Pretty faint, small, and slightly elongated; small, bright nucleus. Distance 54 million ly; Hubble class E2. Paired with galaxy NGC 1439.

NGC 1427

Gx	For	03^h 42.3^m	−35° 24′	3.5′ × 2.5′
charts 18, 19			m$_v$ 10.9	p.a. 76°

Pretty faint, small, and roundish; small, bright nucleus but otherwise featureless. Distance 62 million ly; Hubble class E3. Member of the Fornax Galaxy Cluster.

NGC 1433

Gx	Hor	03^h 42.0^m	−47° 13′	6′
charts 18, 19, 24			m$_v$ 10	p.a. 99°

Very bright, large, and quite elongated; small, extremely bright nucleus. Distance 35 million ly; Hubble class SBa.

NGC 1437

Gx	Eri	03^h 43.6^m	−35° 51′	3.1′ × 2.1′
charts 18, 19			m$_v$ 11.7	p.a. 150°

Faint, very large, and round; bright, very small nucleus; patchy arms. Distance 47 million ly; Hubble class SBa. Member of the Fornax Galaxy Cluster.

NGC 1439

Gx	Eri	03^h 44.8^m	−21° 55′	2.5′
chart 18			m$_v$ 11.2	p.a. 25°

Faint and pretty small; very small, bright nucleus. Distance 82 million ly; Hubble class E1. Paired with galaxy NGC 1426; several other faint galaxies lie nearby.

NGC 1440

Gx	Eri	03^h 45.1^m	−18° 16′	2.3′ × 1.7′
charts 10, 11, 18			m$_v$ 11.5	p.a. 28°

Pretty bright, pretty small, and roundish. Distance 63 million ly; Hubble class S(B)a. Paired with galaxy NGC 1452.

NGC 1444

OC	Per	03^h 49.4^m	+52° 40′	4′
charts 1, 4, 5			m$_v$ 6.6	

15 stars; not well detached from surrounding star field; small range in brightness; mag. of brightest star 6.8. Distance 3,300 ly; age 160 million years;

NGC 1448

Trumpler class IV 1 p. Nebula Sh2-205 is 2° E.

NGC 1448

| Gx | Hor | 03^h 44.5^m | −44° 39′ | 8.1′ × 1.8′ |
| charts 18, 19 | | m$_v$ 10.8 | p.a. 41° |

Pretty bright, large, and very elongated; nearly edge on; patchy arms; small, bright nucleus. Distance 44 million ly; Hubble class Sc.

NGC 1452

| Gx | Eri | 03^h 45.4^m | −18° 38′ | 2.3′ × 1.5′ |
| charts 10, 11, 18 | | m$_v$ 12.1 | p.a. 113° |

Faint and round; faint arms. Distance 78 million ly; Hubble class SBa. Paired with galaxy NGC 1440.

NGC 1453

| Gx | Eri | 03^h 46.5^m | −03° 58′ | 2.4′ |
| charts 10, 11 | | m$_v$ 11.5 | |

Pretty bright, small, and round; very bright nucleus. Distance 170 million ly; Hubble class S(B)c. Brightest in a group of galaxies.

NGC 1461

| Gx | Eri | 03^h 48.5^m | −16° 24′ | 3.0′ × 0.9′ |
| charts 10, 11, 18 | | m$_v$ 11.8 | p.a. 155° |

Pretty bright, small, and slightly elongated; very small, bright, diffuse nucleus. Distance 59 million ly; Hubble class Sa.

NGC 1487

| Gx | Eri | 03^h 55.8^m | −42° 22′ | 3.7′ × 2.3′ |
| charts 18, 19 | | m$_v$ 11.4 | p.a. 55° |

Pretty bright, pretty large, and round; bright nucleus. This is possibly a colliding pair of galaxies. Distance 23 million ly; Hubble class P. Forms a triangle with two stars.

NGC 1491

| BNe | Per | 04^h 03.4^m | +51° 19′ | 3′ |
| charts 1, 4, 5 | | | |

Small and very bright; has a fainter, extended envelope (about 25′). Two strong wisps form a narrow V shape with the NNE end open. Distance 2,500 ly.

NGC 1493

| Gx | Hor | 03^h 57.5^m | −46° 13′ | 3.8′ |
| charts 18, 19, 24 | | m$_v$ 11.2 | p.a. 155° |

Faint, quite large, and round; very small, bright nucleus. Distance about 36 million ly; Hubble class SBc.

NGC 1494

| Gx | Hor | 03^h 57.7^m | −48° 54′ | 3.3′ × 1.9′ |
| charts 18, 19, 24 | | m$_v$ 11.6 | p.a. 179° |

Faint, large, round; very faint nucleus. Distance 39 million ly; Hubble class Sd.

NGC 1499

See California Nebula.

NGC 1501

| PN | Cam | 04^h 07.0^m | +60° 55′ | 52″ |
| chart 1 | | m$_v$ 11.5 | |

PK144+06.1. Pretty bright, pretty small, irregular. A faint gray spot in amateur telescopes, but on photographs somewhat like a flower (Copeland). Phot. mag. 13.3; central star mag. 14.4. Expansion velocity 39 km/sec (24 miles/sec); distance 3,900 ly; V-V class 3.

NGC 1502

| OC | Cam | 04^h 07.7^m | +62° 20′ | 7′ |
| chart 1 | | m$_v$ 5.7 | |

Sometimes called the Golden Harp Cluster. 45 stars; detached, weak concentration of stars; large range in brightness; mag. of brightest star 6.9. Distance 3,000 ly; age 20 million years; Trumpler class II 3 p.

NGC 1507

| Gx | Eri | 04^h 04.5^m | −02° 11′ | 3.2′ × 0.9′ |
| chart 11 | | m$_v$ 12.3 | p.a. 11° |

Very faint, pretty large, and very elongated. Distance 37 million ly; Hubble class Sbm.

NGC 1511

| Gx | Hyi | 03ʰ 59.6ᵐ | –67° 38′ | 3.6′ × 1.3′ |
| chart 24 | | | m_v 11.1 | p.a. 125° |

Pretty bright, pretty small, and very elongated; has a very faint nucleus. Distance 70 million ly; Hubble class Scp.

NGC 1512

| Gx | Hor | 04ʰ 03.9ᵐ | –43° 21′ | 9′ × 5′ |
| charts 18, 19 | | | m_v 10.2 | p.a. 90° |

Bright and quite large; very bright nucleus. Distance 24 million ly; Hubble class SBa.

NGC 1513

| OC | Per | 04ʰ 10.0ᵐ | +49° 31′ | 8′ |
| charts 1, 4, 5 | | | m_v 8.4 | |

50 stars; detached, weak concentration of stars; small range in brightness; large; mag. of brightest star 11.2. A well-marked object, with a crown of larger stars around, somewhat in the form of the letter D (Smyth). Distance 2,700 ly; age 430 million years; Trumpler class III 1 m.

NGC 1514

| PN | Tau | 04ʰ 09.2ᵐ | +30° 47′ | 2.2′ |
| charts 4, 5 | | | m_v 10.9 | |

PK165–15.1. Irregular, smooth disk involved in large, faint nebulosity; phot. mag. 10.0. Appears more like a nebulous star than a planetary nebula because the 9.4-mag. central luminary dominates the view (Houston). Expansion velocity 25 km/sec (16 miles/sec); distance 2,000 ly.

NGC 1515

| Gx | Dor | 04ʰ 04.0ᵐ | –54° 06′ | 5.4′ × 1.3′ |
| charts 19, 24 | | | m_v 11.2 | p.a. 18° |

Bright, large, and very elongated; two main arms; very bright center. Distance 38 million ly; Hubble class S(B)b. Several small nearby galaxies.

NGC 1518

| Gx | Eri | 04ʰ 06.8ᵐ | –21° 11′ | 3.1′ × 1.4′ |
| chart 19 | | | m_v 11.7 | p.a. 35° |

Bright, large, and quite elongated; two main arms. Distance 35 million ly; Hubble class Scp. Paired with galaxy NGC 1521.

NGC 1521

| Gx | Eri | 04ʰ 08.3ᵐ | –21° 03′ | 3.2′ × 1.9′ |
| chart 19 | | | m_v 11.3 | p.a. 10° |

Pretty bright and round; elongated nucleus. Distance 180 million ly; Hubble class E3. Paired with galaxy NGC 1518.

NGC 1527

| Gx | Hor | 04ʰ 08.4ᵐ | –47° 54′ | 4.1′ × 1.7′ |
| charts 18, 19, 24 | | | m_v 10.7 | p.a. 78° |

Pretty bright, pretty small, and elongated; very bright nucleus. Distance 37 million ly; Hubble class S0. Located between two bright stars.

NGC 1528

| OC | Per | 04ʰ 15.4ᵐ | +51° 14′ | 23′ |
| charts 1, 4, 5 | | | m_v 6.4 | |

40 stars; detached, weak concentration of stars; moderate range in brightness; bright; mag. of brightest star 8.8. Good low-power object (Webb). Distance 26,000 ly; age 270 million years; Trumpler class II 2 m.

NGC 1530

| Gx | Cam | 04ʰ 23.5ᵐ | +75° 18′ | 4.6′ × 2.6′ |
| charts 1, 3 | | | m_v 11.4 | p.a. 50° |

Pretty bright and large; very bright, very small nucleus. Distance 120 million ly; Hubble class SBb.

NGC 1531

| Gx | Eri | 04ʰ 12.0ᵐ | –32° 51′ | 1.2′ × 0.9′ |
| charts 18, 19 | | | m_v 12.1 | p.a. 122° |

Pretty bright, pretty large, and round. Distance 47 million ly; Hubble class E6. Paired with much brighter galaxy NGC 1532.

NGC 1532

| Gx | Eri | 04ʰ 12.1ᵐ | –32° 52′ | 11.0′ × 3.4′ |
| charts 18, 19 | | | m_v 9.9 | p.a. 33° |

Bright, very large, and very elongated; patchy arms;

dust lane. Distance 44 million ly; Hubble class Sb. Paired with galaxy NGC 1531.

NGC 1533

Gx	Dor	04^h 09.9^m	−56° 07′	2.7′ × 2.3′
chart 24			m$_v$ 10.8	p.a. 151°

Very bright, very large and round; very bright nucleus. Distance 24 million ly; Hubble class SB0.

NGC 1535

PN	Eri	04^h 14.3^m	−12° 44′	21″
chart 11			m$_p$ 9.6	

PK206−40.1. Very bright, small, round; traces of ring structure; central star mag. 12.2. A splendid though not very conspicuous object (Smyth). Expansion velocity 19 km/sec (12 miles/sec); distance 5,000 ly.

NGC 1537

Gx	Eri	04^h 13.7^m	−31° 39′	3.9′ × 2.6′
charts 18, 19			m$_v$ 10.6	p.a. 98°

Very bright, pretty small, and slightly elongated. Distance 49 million ly; Hubble class E4.

NGC 1543

Gx	Ret	04^h 12.7^m	−57° 44′	5.0′ × 3.1′
chart 24			m$_v$ 9.7	p.a. 93°

Bright, pretty large, and elongated; very bright nucleus. Distance 51 million ly; Hubble class SB0.

NGC 1545

OC	Per	04^h 20.9^m	+50° 15′	18′
charts 1, 4, 5			m$_v$ 6.2	

20 stars; detached, weak concentration of stars; moderate range in brightness; mag. of brightest star 7.1. Distance 2,600 ly; age 190 million years; Trumpler class II 2 p.

NGC 1546

Gx	Dor	04^h 14.6^m	−56° 04′	3.4′ × 2.0′
chart 24			m$_v$ 11.3	p.a. 147°

Pretty bright and slightly elongated; very small, bright nucleus. Distance 41 million ly; Hubble class S0.

NGC 1549

Gx	Dor	04^h 15.8^m	−55° 36′	4.6′ × 3.7′
chart 24			m$_v$ 9.5	p.a. 135°

Bright, pretty small, and round; very bright nucleus. Distance 41 million ly; Hubble class E0. Paired with NGC 1553.

NGC 1553

Gx	Dor	04^h 16.2^m	−55° 47′	4.1′ × 2.8′
chart 24			m$_v$ 9.1	p.a. 150°

Very bright, pretty small, and round; extremely bright nucleus. Distance 46 million ly; Hubble class S0. Paired with galaxy NGC 1549; located among three stars.

NGC 1554, 5

See Hind's Variable Nebula.

NGC 1559

Gx	Ret	04^h 17.6^m	−62° 47′	3.6′ × 2.0′
chart 24			m$_v$ 10.4	p.a. 64°

Very bright, very large, and elongated; high surface brightness; patchy arms; very faint nucleus. Distance 45 million ly; Hubble class SBc.

NGC 1560

Gx	Cam	04^h 32.8^m	+71° 53′	9.8′ × 2.0′
chart 1			m$_v$ 11.4	p.a. 23°

Low surface brightness; brighter toward middle but no definite nucleus. Distance 7 million ly; Hubble class Sd.

NGC 1566

Gx	Dor	04^h 20.0^m	−54° 56′	7.6′ × 6.2′
charts 19, 24			m$_v$ 9.4	p.a. 60°

Bright and very large; small, very bright nucleus; patchy arms. A Seyfert galaxy. Distance 51 million ly; Hubble class S(B)b.

NGC 1569

Gx	Cam	04^h 30.8^m	+64° 51′	3.6′ × 1.9′
chart 1			m$_v$ 11	p.a. 120°

Pretty bright, small, and slightly elongated; very

bright nucleus. Distance about 4 million ly; Hubble class Irp+.

NGC 1573

Gx   Cam $04^h\,35.0^m$ $+73°\,16'$ $1.9'\times1.3'$
chart 1 $m_v\,11.7$ p.a. 35°

Very faint and small. Distance 190 million ly; Hubble class E3.

NGC 1574

Gx   Ret $04^h\,22.0^m$ $-56°\,58'$ $4.1'\times3.7'$
chart 24 $m_v\,10.2$ p.a. 35°

Pretty bright, small, and round; very bright nucleus. Distance 29 million ly; Hubble class S0.

NGC 1575

Gx   Eri $04^h\,26.3^m$ $-10°\,06'$ $1.5'\times1.0'$
chart 11 $m_v\,12.2$ p.a. 90°

Very faint, pretty small, and roundish; Hubble class Sb. A foreground star lies 2′ S.

NGC 1579

BNr   Per $04^h\,30.2^m$ $+35°\,16'$ $12'\times7'$
chart 5

Several stars involved in a pretty bright, very large nebula of irregular shape; extended N–S; brighter toward center; dark lanes. Stands nearly in the center of a trapezium of stars (William Herschel). Of a slight cream color (Smyth).

NGC 1582

OC   Per $04^h\,32.0^m$ $+43°\,51'$ 37′
charts 4, 5 $m_p\,7.0$

20 stars; not well detached; moderate brightness range; brightest star is phot. mag. 9.0; Trumpler class IV 2 p.

NGC 1587

Gx   Tau $04^h\,30.7^m$ $+00°\,40'$ $1.9'\times1.7'$
chart 11 $m_v\,11.7$ p.a. 144°

Faint, pretty small, and roundish. Distance 170 million ly; Hubble class E1p.

NGC 1589

Gx   Tau $04^h\,30.8^m$ $+00°\,52'$ $3.3'\times1.0'$
chart 11 $m_v\,11.8$ p.a. 160°

Faint, pretty large, and slightly elongated. Distance 160 million ly; Hubble class Sb.

NGC 1596

Gx   Dor $04^h\,27.6^m$ $-55°\,02'$ $3.9'\times1.0'$
charts 19, 24 $m_v\,11$ p.a. 20°

Bright, pretty large, and very elongated; slightly flattened bulge; bright nucleus. Distance 520 million ly; Hubble class S0. Paired with galaxy NGC 1602.

NGC 1600

Gx   Eri $04^h\,31.7^m$ $-05°\,05'$ $3.2'\times2.1'$
chart 11 $m_v\,10.9$ p.a. 15°

Pretty bright, pretty large, and round. Distance 210 million ly; Hubble class E2. Brightest in a group.

NGC 1615

Gx   Tau $04^h\,36.0^m$ $+19°\,57'$ $1.4'\times0.8'$
charts 5, 11 $m_v\,13.6$ p.a. 115°

Very faint and very small; slightly brighter toward middle. Distance 150 million ly; Hubble class S0.

NGC 1617

Gx   Dor $04^h\,31.7^m$ $-54°\,36'$ $4.8'\times2.3'$
charts 19, 24 $m_v\,10.5$ p.a. 107°

Bright, large, and very elongated; smooth arms; very bright nucleus. Distance 34 million ly; Hubble class SBa.

NGC 1620

Gx   Eri $04^h\,36.6^m$ $-00°\,09'$ $3.2'\times1.0'$
chart 11 $m_v\,12.3$ p.a. 25°

Very faint, pretty large, and moderately elongated. Distance 150 million ly; Hubble class Sb.

NGC 1624

BNe   Per $04^h\,40.5^m$ $+50°\,27'$ 5′
charts 1, 4, 5

Faint emission nebula containing a small (5′), faint cluster of 12 stars; mag. of brightest star 11.8.

NGC 1637

Gx	Eri	04^h 41.5^m	−02° 51′	3.7′ × 3.2′
chart 11			m$_v$ 10.8	p.a. 15°

Quite bright, large, and round; very small, very bright nucleus. Distance about 27 million ly; Hubble class Sc.

NGC 1638

Gx	Eri	04^h 41.6^m	−01° 49′	2.3′ × 1.7′
chart 11			m$_v$ 12	p.a. 70°

Faint, pretty large, and elongated; very bright nucleus. Distance 140 million ly; Hubble class E2.

NGC 1640

Gx	Eri	04^h 42.2^m	−20° 26′	2.6′ × 2.3′
chart 19			m$_v$ 11.7	p.a. 45°

Very faint, pretty small, and elongated; small, bright nucleus. Distance 61 million ly; Hubble class SBb.

NGC 1647

OC	Tau	04^h 46.0^m	+19° 04′	44′
charts 5, 11			m$_v$ 6.4	

200 stars; detached, weak concentration of stars; moderate range in brightness; mag. of brightest star 8.6. Distance 1,800 ly; age 210 million years; Trumpler class II 2 m.

NGC 1653

Gx	Eri	04^h 45.8^m	−02° 23′	1.8′
chart 11			m$_v$ 12	p.a. 100°

Faint, quite small, and round. Distance 180 million ly; Hubble class E0.

NGC 1662

OC	Ori	04^h 48.5^m	+10° 56′	19′
chart 11			m$_v$ 6.4	

35 stars; detached, strong concentration; a poor cluster with moderate brightness range; mag. of brightest star 8.3. Distance 1,300 ly; Trumpler class I 2 p.

NGC 1664

OC	Aur	04^h 51.1^m	+43° 42′	18′
chart 5			m$_v$ 7.6	

40 stars; detached, no concentration of stars; small range in brightness; pretty large; mag. of brightest star 10.6. Observers with 8-inch and larger telescopes mention numerous star chains (Houston). Distance 3,900 ly; age 300 million years; Trumpler class III 1 p.

NGC 1667

Gx	Eri	04^h 48.6^m	−06° 19′	1.8′ × 1.4′
chart 11			m$_v$ 12.1	p.a. 20°

Pretty faint, pretty small, and round; very small, bright nucleus. Distance 200 million ly; Hubble class SB0. Paired with galaxy NGC 1666.

NGC 1672

Gx	Dor	04^h 45.7^m	−59° 15′	4.8′ × 3.9′
chart 24			m$_v$ 9.8	p.a. 170°

Bright and large; very bright nucleus. Distance 47 million ly; Hubble class SBb.

NGC 1679

Gx	Cae	04^h 49.9^m	−31° 58′	2.8′ × 2.0′
chart 19			m$_v$ 11.5	p.a. 150°

Very patchy arms with many knots. Distance 39 million ly; Hubble class Sbm. Superimposed star.

NGC 1684

Gx	Ori	04^h 52.5^m	−03° 06′	2.9′ × 1.9′
chart 11			m$_v$ 11.7	p.a. 95°

Pretty faint, pretty small, and roundish; brighter toward middle. Distance 190 million ly; Hubble class E.

NGC 1688

Gx	Dor	04^h 48.4^m	−59° 48′	2.4′ × 1.9′
chart 24			m$_v$ 12	p.a. 177°

Pretty bright, pretty large, irregularly roundish; patchy arms. Distance 13 million ly; Hubble class SBc.

NGC 1691

Gx	Ori	04^h 54.6^m	+03° 16′	1.7′ × 1.5′
chart 11			m$_v$ 12	p.a. 85°

Faint and small. Distance 200 million ly; Hubble class SBa. With a superimposed 11th-mag. star.

NGC 1700

Gx	Eri	04^h 56.9^m	−04° 52′	3.0′ × 1.9′
chart 11			m$_v$ 11.2	p.a. 90°

Quite bright and small; very bright nucleus. Distance 170 million ly; Hubble class E1. Paired with galaxy NGC 1699.

NGC 1703

Gx	Dor	04^h 52.9^m	−59° 45′	3.1′
chart 24			m$_v$ 11.6	p.a. 139°

Bright arms with many knots; small, bright nucleus. Distance 58 million ly; Hubble class Sb.

NGC 1705

Gx	Pic	04^h 54.2^m	−53° 22′	1.9′ × 1.4′
charts 19, 24			m$_v$ 11.8	p.a. 50°

Pretty faint, small, and round; very bright nucleus. Distance 17 million ly; Hubble class S0p. Very faint star (mag. 14.5) is superimposed.

NGC 1723

Gx	Eri	04^h 59.4^m	−10° 59′	3.2′ × 2.0′
chart 11			m$_v$ 11.7	p.a. 95°

A faint galaxy. Distance 160 million ly; Hubble class SB. Located between two stars (at N and S).

NGC 1726

Gx	Eri	04^h 59.7^m	−07° 45′	2.2′ × 1.6′
chart 11			m$_v$ 11.7	p.a. 0°

Faint and round; very bright, small, diffuse nucleus; dark lane. Distance 170 million ly; Hubble class E2. In a group with galaxies NGC 1720 and IC 398.

NGC 1730

Gx	Lep	04^h 59.5^m	−15° 49′	2.2′ × 1.0′
chart 11			m$_v$ 12.3	p.a. 94°

Faint, pretty small, and slightly elongated. Distance 170 million ly; Hubble class S. Located between two faint stars.

NGC 1744

Gx	Lep	05^h 00.0^m	−26° 02′	6.8′ × 4.1′
chart 19			m$_v$ 11.3	p.a. 168°

Faint, very large, and very elongated; low surface brightness, distance 25 million ly; Hubble class S(B)c.

NGC 1746

OC	Tau	05^h 03.6^m	+23° 49′	41′
chart 5			m$_p$ 6.1	

20 stars; detached, no concentration of stars; small range in brightness; large; brightest star is phot. mag. 8. Distance 1,400 ly; Trumpler class III 1 p.

NGC 1763

BNe	Dor	04^h 56.8^m	−66° 24′	5′ × 3′
chart 24				

Bean shaped with ENE–WSW extensions; bright and well-defined; contains some stars.

NGC 1778

OC	Aur	05^h 08.1^m	+37° 03′	7′
chart 5			m$_v$ 7.7	

25 stars; detached, no concentration of stars; moderate range in brightness; mag. of brightest star 10.1. A loose cluster, having four brighter stars in a curve, of which the leader is double (Smyth). Distance 4,400 ly; age 160 million years; Trumpler class III 2 p.

NGC 1779

Gx	Eri	05^h 05.3^m	−09° 09′	2.7′ × 1.4′
chart 11			m$_v$ 12.1	p.a. 105°

Pretty bright, small, and roundish; gradually brighter toward middle. Distance 150 million ly; Hubble class Sba.

NGC 1784

Gx	Lep	05^h 05.5^m	−11° 52′	4.1′ × 2.5′
chart 11			m$_v$ 11.7	p.a. 105°

Pretty bright, pretty small, and very elongated; many arms; very small, bright nucleus. Distance 95 million ly; Hubble class S(B)c.

NGC 1786

GC	Dor	04^h 59.1^m	−67° 45′	1.2′
chart 24			m$_v$ 10.1	

Located on W edge of the Large Magellanic Cloud; high concentration of stars; S-S class 2. Relatively bright star field on NW edge, fainter star field to SW.

NGC 1788

BNr   Ori $05^h 06.9^m$ $-03° 21'$ $7' \times 5'$
chart 11

A 10th-mag. star involved in a bright, quite large patch of nebulosity. Flanked by the dark nebula LDN 1616.

NGC 1792

Gx   Col $05^h 05.3^m$ $-37° 59'$ $5.0' \times 2.9'$
charts 19, 24 m_v 9.9   p.a. 137°

Very bright, very large, and elongated; small, bright nucleus. Distance 43 million ly; Hubble class Sb.

NGC 1796

Gx   Dor $05^h 02.7^m$ $-61° 08'$ $1.9' \times 1.0'$
chart 24 m_v 12.3   p.a. 102°

Pretty faint, pretty small and quite elongated; very faint nucleus. Distance 34 million ly; Hubble class SBb.

NGC 1800

Gx   Col $05^h 06.4^m$ $-31° 57'$ $2.0' \times 1.0'$
chart 19 m_v 12.6   p.a. 113°

Pretty bright and elongated. Distance 23 million ly; Hubble class E6.

NGC 1807

OC   Tau $05^h 10.7^m$ $+16° 32'$ 16'
charts 5, 11 m_v 7

20 stars; detached, weak concentration of stars; moderate range in brightness; mag. of brightest star 8.6. Probably an asterism and not a real cluster. Trumpler class II 2 p. Located approx. 0.5° WSW of open cluster NGC 1817.

NGC 1808

Gx   Col $05^h 07.7^m$ $-37° 31'$ $7.2' \times 4.1'$
charts 19, 24 m_v 9.9   p.a. 133°

Bright, large, and elongated; several bright knots; many dark lanes; very bright nucleus. Distance 33 million ly; Hubble class S(B)a.

NGC 1817

OC   Tau $05^h 12.1^m$ $+16° 42'$ 15'
charts 5, 11 m_v 7.7

60 stars; detached, no concentration of stars; small range in brightness; large; mag. of brightest star 11.2. Distance 5,700 ly; age 790 million years; Trumpler class III 1 m. Located about 0.5° ENE of open cluster NGC 1807.

NGC 1824

Gx   Dor $05^h 06.9^m$ $-59° 43'$ $3.3' \times 0.9'$
chart 24 m_v 12.6   p.a. 160°

Bright, patchy bar; arms have many faint knots. Distance 46 million ly; Hubble class SBm.

NGC 1832

Gx   Lep $05^h 12.1^m$ $-15° 41'$ $2.3' \times 1.7'$
chart 11 m_v 11.3   p.a. 10°

Pretty bright, irregularly roundish; many arms; small, very bright nucleus. Distance 77 million ly; Hubble class Sc.

NGC 1835

GC   Dor $05^h 05.2^m$ $-69° 24'$ 1.2'
chart 24 m_v 9.8

Located on the W portion of the Large Magellanic Cloud. High concentration of stars; slightly oval shaped with E–W extension; S-S class 3.

NGC 1851

GC   Col $05^h 14.1^m$ $-40° 03'$ 10'
charts 19, 24 m_v 7.1

C73. High concentration of stars; very bright, very large, and well resolved. Fine object (Denning). Per *NGC*, this is a (!) remarkable object. Distance 35,000 ly; S-S class 2.

NGC 1857

OC   Aur $05^h 20.2^m$ $+39° 21'$ 6'
chart 5 m_v 7

40 stars; detached, weak concentration of stars; moderate range in brightness. A finder will show it as a hazy patch surrounding an orange 7.4-mag. star (Houston). Distance 6,200 ly.

NGC 1888

| Gx | Lep | 05^h 22.6^m | −11° 30′ | 3.2′ × 1.2′ |
| chart 11 | | | m$_v$ 11.9 | p.a. 145° |

Pretty bright and pretty large. Distance 100 million ly; Hubble class SBcp. Interacting with fainter galaxy NGC 1889 lying 1′ N.

NGC 1892

| Gx | Dor | 05^h 17.1^m | −64° 58′ | 2.9′ × 0.8′ |
| chart 24 | | | m$_v$ 12 | p.a. 74° |

Quite faint, pretty large, and elongated. Distance 51 million ly; Hubble class Sc. Many stars from the Large Magellanic Cloud are superimposed.

NGC 1893

| OC | Aur | 05^h 22.7^m | +33° 24′ | 10′ |
| chart 5 | | | m$_v$ 7.5 | |

The Letter Y Cluster. 60 stars; detached, weak concentration of stars; moderate brightness range; large; mag. of brightest star 9.3. Distance 13,000 ly; very young, only about 1 million years; Trumpler class II 2 m. Involved in a very large (40′ × 30′) emission nebula, IC 410.

NGC 1904

See M79.

NGC 1907

| OC | Aur | 05^h 28.0^m | +35° 19′ | 7′ |
| chart 5 | | | m$_v$ 8.2 | |

30 stars; detached, weak concentration of stars; small range in brightness; irregularly round; involved in nebulosity; mag. of brightest star 11.3. A compressed oval cluster in a splendid district of the heavens (Smyth). Distance 4,500 ly; age 440 million years; Trumpler class II 1 m n. Located about 0.5° SSW of open cluster M38.

NGC 1912

See M38.

NGC 1929, 34-6

| BNe | Dor | 05^h 22.0^m | −67° 58′ | 19′ × 15′ |
| chart 24 | | | | |

Group of four nebulae in a large complex in the N part of the Large Magellanic Cloud.

NGC 1931

| BNer+C | Aur | 05^h 31.4^m | +34° 15′ | 3′ |
| chart 5 | | | | |

Very bright; contains a very bright, large, and round cluster; mag. of brightest star in cluster is mag. 11.5. After intent gazing, under moderate power, a triangle of stars rises distinctly from the stardust (Smyth). Distance 4,000 ly.

NGC 1947

| Gx | Dor | 05^h 26.8^m | −63° 46′ | 2.4′ × 2.0′ |
| chart 24 | | | m$_v$ 10.8 | p.a. 119° |

Pretty bright, large, and round; bright, diffuse nucleus with dark lane. Distance 28 million ly; Hubble class S0p.

NGC 1952

See Crab Nebula.

NGC 1954

| Gx | Lep | 05^h 32.8^m | −14° 04′ | 4.2′ × 1.9′ |
| chart 11 | | | m$_v$ 11.8 | p.a. 135° |

Very faint; small and roundish; suddenly much brighter toward middle. Distance 130 million ly; Hubble class Sc.

NGC 1960

See M36.

NGC 1961

| Gx | Cam | 05^h 42.1^m | +69° 23′ | 4.6′ × 3.2′ |
| chart 1 | | | m$_v$ 11 | p.a. 85° |

Faint, large, and irregularly shaped; very small and bright nucleus; peculiar outer arm and streamers. Distance 180 million ly; Hubble class SBp.

NGC 1964

| Gx | Lep | 05^h 33.4^m | −21° 57′ | 6.2′ × 2.5′ |
| chart 19 | | | m$_v$ 10.7 | p.a. 32° |

Faint, large, and round; small, very bright nucleus.

Distance 65 million ly; Hubble class Sb. Superimposed bright star.

NGC 1966

BNe Dor $05^h 26.8^m$ $-68° 49'$ $12' \times 12'$
chart 24

Pretty bright and roundish; somewhat suddenly brighter toward middle, located near the center of the Large Magellanic Cloud.

NGC 1968

BN Dor $05^h 27.2^m$ $-67° 26'$ $13' \times 12'$
chart 24

Located in N part of the Large Magellanic Cloud. Contains an open cluster; mag. of brightest star in cluster is 12.6.

NGC 1973

BNer Ori $05^h 35.1^m$ $-04° 44'$ $5'$
charts 11, B2

Bright nebulosity associated with 8th- and 9th-mag. stars. Distance 1,500 ly. Connected to nebula NGC 1975.

NGC 1974

BN Dor $05^h 27.9^m$ $-67° 24'$
chart 24

Located in N part of the Large Magellanic Cloud; contains an open cluster.

NGC 1975

BNer Ori $05^h 35.4^m$ $-04° 41'$ $10' \times 5'$
charts 11, B2

Bright double star involved in a bright, large patch of nebulosity. Distance 1,500 ly. Connected to nebula NGC 1973.

NGC 1976

See Great Orion Nebula.

NGC 1977

BNer Ori $05^h 35.5^m$ $-04° 52'$ $19' \times 9'$
charts 11, B2

Sparse grouping of stars involved in a faint, large patch of nebulosity. Per NGC, a (!!) very remarkable object.

NGC 1978

GC Dor $05^h 28.6^m$ $-66° 14'$ $2.7'$
chart 24 m_v 9.9

Medium concentration of stars; oval shaped, extending NNW–SSE; S-S class 5. Located on N edge of the Large Magellanic Cloud.

NGC 1979

Gx Lep $05^h 34.0^m$ $-23° 19'$ $2.2' \times 1.8'$
chart 19 m_v 11.8 p.a. 10°

Pretty bright lens. Distance 69 million ly; Hubble class S0.

NGC 1981

OC Ori $05^h 35.2^m$ $-04° 26'$ $25'$
charts 11, B2 m_v 4.6

20 stars; detached, no concentration of stars; moderate range in brightness; very bright, large; mag. of brightest star 6.3; involved in nebulosity. Distance 1,300 ly; Trumpler class III 2 p n.

NGC 1982

See M43.

NGC 1999

BNer Ori $05^h 36.5^m$ $-06° 42'$ $15' \times 12'$
charts 11, B2

A 10th-mag. star involved in a bright, roundish patch of nebulosity.

NGC 2014

BN Dor $05^h 32.2^m$ $-67° 40'$
chart 24 m_p 8.0

Located within the Large Magellanic Cloud; a pretty large nebula containing a pretty large, fairly compressed open cluster.

NGC 2018

BNe Men $05^h 30.6^m$ $-71° 04'$ $25' \times 18'$
charts 24, 25

Pretty bright, large, and roundish; located in S portion of the Large Magellanic Cloud; contains an open cluster that is roundish and gradually brighter toward middle.

NGC 2022

PN　Ori　$05^h\,42.1^m$　$+09°\,05'$　$19''$
chart 11　　　　　$m_v\,11.9$

PK196–10.1. Pretty bright, very small, and round; phot. mag. 12.4; central star mag. 14.9. Expansion velocity 29 km/sec (18 miles/sec). Extends 0.6 ly; distance 6,900 ly; V-V class 4 + 2.

NGC 2023

BNer　Ori　$05^h\,41.6^m$　$-02°\,14'$　$10'$
charts 11, B2

A bright star involved in a large patch of nebulosity.

NGC 2024

BNe　Ori　$05^h\,41.9^m$　$-01°\,51'$　$30'$
charts 11, B2

Tank Trap Nebula. Bright, very large, and irregularly shaped; divided by dark lane extending N–S. The E section of nebula appears larger. Illuminated by 2.1-mag. star ζ Ori. Per *NGC*, a (!) remarkable object. Distance 1,500 ly.

NGC 2048

BNe　Dor　$05^h\,35.2^m$　$-69°\,46'$　$18'\times12'$
charts 24, 25

Several stars involved in nebulosity; located in E portion of the Large Magellanic Cloud and approx. 0.5° SSW of the Tarantula Nebula (NGC 2070).

NGC 2064

BNr　Ori　$05^h\,46.3^m$　$00°\,00'$　$12'\times2'$
charts 11, B2

Small, relatively bright patch of nebulosity extending N–S. Located in a group with nebulae NGC 2067, M78, and NGC 2071.

NGC 2067

BNr　Ori　$05^h\,46.5^m$　$+00°\,06'$　$7'\times3'$
charts 11, B2

Faint and pretty large; located in a group with nebulae NGC 2064, M78, and NGC 2071.

NGC 2068

See M78.

NGC 2070

See Tarantula Nebula.

NGC 2071

BNr　Ori　$05^h\,47.2^m$　$+00°\,18'$　$4'\times3'$
charts 11, B2

Small, very faint, and fuzzy. Located in a group with nebulae NGC 2064, NGC 2067, and M78.

NGC 2074

BNe　Dor　$05^h\,39.6^m$　$-69°\,27'$　$15'\times10'$
charts 24, 25

S shaped with many stars involved in nebulosity. Associated with nebula NGC 2081.

NGC 2077-80

BNe　Dor　$05^h\,40.4^m$　$-69°\,38'$　$15'\times11'$
charts 24, 25

Dumbbell shaped; extends SW–NE.

NGC 2081

BNe　Dor　$05^h\,39.6^m$　$-69°\,27'$　$15'\times10'$
charts 24, 25

Contains a small open cluster; associated with nebula NGC 2074.

NGC 2082

Gx　Dor　$05^h\,41.9^m$　$-64°\,18'$　$1.9'\times1.7'$
chart 24　　　　　$m_v\,12$　　p.a. 41°

Faint; strong arms; small, elongated nucleus. Distance 37 million ly; Hubble class Sb.

NGC 2089

Gx　Lep　$05^h\,47.9^m$　$-17°\,36'$　$1.9'\times1.1'$
charts 11, 19　　　$m_v\,11.9$　p.a. 39°

Elongated arms. Distance 120 million ly; Hubble

class E/S0. Several superimposed stars.

NGC 2090

Gx	Col	05^h 47.0^m	−34° 15′	5.0′ × 2.8′
chart 19			m$_v$ 11	p.a. 13°

Bright, pretty large, irregularly round; bright, small, diffuse nucleus; several knotty arms and dark lanes. Distance 69 million ly; Hubble class Sc.

NGC 2099

See M37.

NGC 2129

OC	Gem	06^h 01.0^m	+23° 18′	7′
chart 5			m$_v$ 6.7	

40 stars; detached, no concentration of stars; large range in brightness; pretty large; mag. of brightest star 7.4. Distance 6,500 ly; age 16 million years; Trumpler class III 3 p.

NGC 2139

Gx	Lep	06^h 01.1^m	−23° 41′	2.9′ × 2.0′
chart 19			m$_v$ 11.4	p.a. 140°

Faint and small; very bright arms; small, bright nucleus. Distance 69 million ly; Hubble class Scp.

NGC 2146

Gx	Cam	06^h 18.7^m	+78° 21′	5.0′ × 3.2′
chart 1			m$_v$ 10.6	p.a. 123°

Pretty bright, large, and slightly elongated. One arm has a dark lane. Distance 45 million ly; Hubble class SBbp. Possibly interacting with galaxy NGC 2146A.

NGC 2149

BNr	Mon	06^h 03.5^m	−09° 44′	3′ × 2′
charts 11, 12				

A 9.3-mag. star involved in a small, faint patch of nebulosity.

NGC 2158

OC	Gem	06^h 07.5^m	+24° 06′	5′
chart 5			m$_v$ 8.6	

Rich; detached, weak concentration; large range in brightness; mag. of brightest star 12.4. Generally too difficult for apertures less than 5 inches, this is a peculiar wedge-shaped cluster that resembles a diffuse nebula or comet (Houston). Distance about 16,000 ly; age 3.2 billion years; Trumpler class II 3 r. Lies only about 0.4° SW of M35.

NGC 2168

See M35.

NGC 2169

OC	Ori	06^h 08.4^m	+13° 57′	7′
charts 11, 12			m$_v$ 5.9	

30 stars; detached, strong concentration of stars; large range in brightness; small; mag. of brightest star 6.9. Consists of two distinct parts. Distance 3,600 ly; age 50 million years; Trumpler class I 3 p.

NGC 2170

BNr	Mon	06^h 07.5^m	−06° 24′	2′
charts 11, 12				

A 9.5-mag. star involved in a small, very faint patch of nebulosity.

NGC 2174

BNe	Ori	06^h 09.7^m	+20° 30′	39′ × 30′
chart 5				

Extremely faint, very large, and mottled patch of nebulosity surrounding a 7th-mag. star; several other stars involved.

NGC 2175

OC	Ori	06^h 09.8^m	+20° 19′	18′
chart 5			m$_v$ 6.8	

60 stars; not well detached from surrounding star field; large range in brightness; mag. of brightest star 7.6. Distance 6,400 ly; Trumpler class IV 3 p n. Involved in an extremely faint, large (39′ × 30′) emission nebula, NGC 2174.

NGC 2182

BNr	Mon	06^h 09.5^m	−06° 20′	3′
charts 11, 12				

A 9.0-mag. star involved in nebulosity.

NGC 2183, 85

BNr Mon $06^h\,11.0^m$ $-06°\,13'$ $3' \times 1'$
charts 11, 12

A small, bright patch.

NGC 2187A

Gx Dor $06^h\,03.7^m$ $-69°\,36'$ $2.3' \times 1.0'$
charts 24, 25 $m_p\,12.9$ p.a. 100°

Distance 160 million ly; Hubble class Sa.

NGC 2188

Gx Col $06^h\,10.2^m$ $-34°\,06'$ $4.7' \times 1.1'$
chart 19 $m_v\,11.6$ p.a. 175°

Pretty faint, pretty large, and very elongated; small, fairly bright nucleus. Distance 19 million ly; Hubble class SBm.

NGC 2194

OC Ori $06^h\,13.8^m$ $+12°\,48'$ $10'$
charts 11, 12 $m_v\,8.5$

80 stars; detached, no concentration of stars; small range in brightness; large; mag. of brightest star 12.1. Distance 5,200 ly; Trumpler class III 1 r.

NGC 2195

BN Ori $06^h\,14.4^m$ $+17°\,39'$ $<1'$
charts 5, 11, 12

A star involved in a small, nebulous patch.

NGC 2196

Gx Lep $06^h\,12.2^m$ $-21°\,48'$ $2.8' \times 2.2'$
chart 19 $m_v\,11.1$ p.a. 35°

Pretty faint, pretty small, and slightly elongated; bright, diffuse nucleus. Distance 90 million ly; Hubble class Sb.

NGC 2204

OC CMa $06^h\,15.7^m$ $-18°\,39'$ $13'$
charts 11, 19 $m_v\,8.6$

80 stars; detached, no concentration of stars; mag. of brightest star 12.2. Distance 14,500 ly; age 3 billion years; Trumpler class III 3 m.

NGC 2206

Gx CMa $06^h\,16.0^m$ $-26°\,46'$ $2.5' \times 1.3'$
chart 19 $m_v\,12.2$ p.a. 138°

Elongated nucleus; very bright, inner pseudo-ring; thin, knotty arms. Distance 260 million ly; Hubble class S(B)b.

NGC 2207

Gx CMa $06^h\,16.4^m$ $-21°\,22'$ $4.4' \times 2.6'$
chart 19 $m_v\,10.8$ p.a. 141°

Pretty bright, pretty large, and elongated; two main arms; very bright and elongated nucleus. Distance 110 million ly; Hubble class Sc. Possibly interacting with galaxy IC 2163.

NGC 2210

GC Dor $06^h\,11.5^m$ $-69°\,08'$ $1.7'$
charts 24, 25 $m_v\,10.2$

Very bright, pretty large, and roundish; much brighter toward middle; high concentration of stars.

NGC 2215

OC Mon $06^h\,21.0^m$ $-07°\,17'$ $10'$
charts 11, 12 $m_v\,8.4$

40 stars; detached, weak concentration of stars; moderate range in brightness; a large cluster; mag. of brightest star 10.5. Distance 3,300 ly; age 350 million years; Trumpler class II 2 p.

NGC 2217

Gx CMa $06^h\,21.7^m$ $-27°\,14'$ $4.8'$
chart 19 $m_v\,10.2$ p.a. 104°

Very bright, small, and round; very bright, diffuse nucleus. Distance 54 million ly; Hubble class SBa.

NGC 2223

Gx CMa $06^h\,24.6^m$ $-22°\,50'$ $3.0' \times 2.6'$
chart 19 $m_v\,11.8$ p.a. 175°

Faint, pretty large, round; patchy arms; small, bright, elongated nucleus. Distance 110 million ly; Hubble class S(B)b. Star nearby.

NGC 2232

OC Mon 06^h 28.0^m −04° 54′ 30′
charts 11, 12 m$_v$ 3.9

20 stars; not well detached from surrounding star field; large range in brightness; a large cluster; includes bright star 10 Mon (mag. 5.1). Distance 1,300 ly; age 22 million years; Trumpler class IV 3 p.

NGC 2236

OC Mon 06^h 29.7^m +06° 50′ 7′
charts 11, 12 m$_v$ 8.5

50 stars; detached, no concentration of stars; moderate range in brightness; mag. of brightest star 11.0. Distance 11,000 ly; Trumpler class III 2 p.

NGC 2237-39, 46

See Rosette Nebula.

NGC 2244

OC Mon 06^h 32.0^m +04° 55′ 23′
charts 11, 12 m$_v$ 4.8

C50, located in the Rosette Nebula (NGC 2237-39, 46). 100 stars; weak concentration of stars; large brightness range; large, bright; mag. of brightest star 5.8; involved in nebulosity. Distance 5,000 ly; Trumpler class II 3 p n. A little W and N nestles a beautiful, very small delta of stars, a delightful trigon (Copeland).

NGC 2245

BNr Mon 06^h 32.7^m +10° 10′ 5′ × 3′
charts 11, 12

Pretty large; cometlike appearance; 11th-mag. star involved in nebulosity.

NGC 2247

BNr Mon 06^h 33.2^m +10° 20′ 4′ × 3′
charts 11, 12

An 8.5-mag. star involved in the N part of an extremely faint patch of nebulosity.

NGC 2251

OC Mon 06^h 34.7^m +08° 22′ 10′
charts 11, 12 m$_v$ 7.3

30 stars; detached, no concentration of stars; moderate range in brightness; elongated and large; mag. of brightest star 9.1. Distance 5,000 ly; age 300 million years; Trumpler class III 2 p.

NGC 2252

OC Mon 06^h 35.0^m +05° 23′ 19′
charts 11, 12 m$_p$ 7.7

30 stars; not well detached; moderate brightness range; involved in nebulosity; brightest star is phot. mag. 9. Distance 5,000 ly; age 300 million years; Trumpler class IV 2 p n. Located on NE edge of the Rosette Nebula (NGC 2237-39, 46).

NGC 2258

Gx Cam 06^h 47.8^m +74° 29′ 2.3′ × 1.5′
chart 1 m$_v$ 11.9 p.a. 150°

Faint. Distance 180 million ly; Hubble class S0.

NGC 2261

See Hubble's Variable Nebula.

NGC 2263

Gx CMa 06^h 38.5^m −24° 51′ 2.8′ × 1.7′
chart 19 m$_v$ 11.9 p.a. 143°

A knotty inner ring, with very faint arms forming an outer ring. Distance 110 million ly; Hubble class SBb. Several superimposed stars.

NGC 2264

C+BNe Mon 06^h 41.1^m +09° 53′ 35′ × 15′
charts 11, 12 m$_v$ 3.9

Christmas Tree Cluster and Cone Nebula. 40 stars bathed in bright nebulosity; not well detached from surrounding star field; large range in brightness; large; mag. of brightest star 4.7. The Cone is the small dark nebula southeast of the cluster. Distance 2,400 ly; age 20 million years; Trumpler class IV 3 p n.

NGC 2266

OC Gem 06^h 43.2^m +26° 58′ 6′
chart 5 m$_p$ 9.5

50 stars; detached, weak concentration of stars; moderate range in brightness; pretty small; bright-

est star is phot. mag. 11. Distance 11,000 ly; Trumpler class II 2 m.

NGC 2268

Gx Cam 07^h 14.3^m +84° 23' 3.3' × 2.0'
charts 1, A4 m$_v$ 11.5 p.a. 63°

Pretty faint, pretty large, and slightly elongated; very bright, very small nucleus. Distance 110 million ly.

NGC 2272

Gx CMa 06^h 42.7^m −27° 28' 2.3' × 1.5'
chart 19 m$_v$ 11.7 p.a. 123°

Pretty faint, pretty small, and very slightly elongated; brighter toward middle; Hubble class S0.

NGC 2273

Gx Lyn 06^h 50.1^m +60° 51' 3.2' × 2.5'
chart 1 m$_v$ 11.7 p.a. 50°

Faint, small, and irregularly roundish; includes a ring of high surface brightness. Distance 90 million ly; Hubble class S(B)a.

NGC 2276

Gx Cep 07^h 27.2^m +85° 45' 2.8' × 2.6'
charts 1, A4 m$_v$ 11.4 p.a. 20°

Pretty large and faint; very bright, small nucleus. Many arms, one of which begins straight, then bends. Distance 110 million ly; Hubble class Sc. Interacting with galaxy NGC 2300.

NGC 2280

Gx CMa 06^h 44.8^m −27° 38' 5.6' × 3.2'
chart 19 m$_v$ 10.5 p.a. 163°

Pretty faint, pretty large, and slightly elongated; two main arms; small, bright, diffuse nucleus. Distance 72 million ly; Hubble class Sb.

NGC 2281

OC Aur 06^h 49.3^m +41° 04' 15'
chart 5 m$_v$ 5.4

30 stars; detached, strong concentration of stars; large range in brightness; mag. of brightest star 7.3. Distance 1,600 ly; age 300 million years; Trumpler class I 3 p.

NGC 2282

BNr Mon 06^h 46.9^m +01° 19' 3'
charts 11, 12

A 10th-mag. star involved in faint nebulosity in the SE part.

NGC 2283

Gx CMa 06^h 45.9^m −18° 13' 3.7' × 2.8'
charts 11, 12, 19 m$_v$ 12.2 p.a. 2°

Faint and small; patchy arms with three or four small stars superimposed. Distance 28 million ly; Hubble class SBc.

NGC 2286

OC Mon 06^h 47.6^m −03° 10' 15'
charts 11, 12 m$_v$ 7.5

50 stars; not well detached from surrounding star field; large range in brightness; mag. of brightest star 9.7. Curiously studded in pairs and triplets (Smyth). Distance 4,200 ly; Trumpler class IV 3 m.

NGC 2287

See M41.

NGC 2292

Gx CMa 06^h 47.7^m −26° 45' 3.9' × 3.4'
chart 19 m$_v$ 10.8 p.a. 1°

Extremely faint and roundish; gradually brighter toward middle. Distance 92 million ly; Hubble class E/S0. Interacting with galaxy NGC 2293.

NGC 2293

Gx CMa 06^h 47.7^m −26° 45' 3.9' × 3.2'
chart 19 m$_p$ 12.2 p.a. 125°

Pretty bright and roundish; gradually brighter toward middle. Distance 78 million ly; Hubble class S0. Interacting with galaxy NGC 2292.

NGC 2298

GC Pup 06^h 49.0^m −36° 00' 5'
chart 19 m$_v$ 9.3

Medium concentration of stars; bright and pretty large. Distance 40,000 ly; S-S class 6.

NGC 2300

Gx	Cep	07^h 32.3^m	+85° 43′	3.2′ × 2.5′
charts 1, A4			m$_v$ 11	p.a. 78°

Pretty bright, pretty large, and slightly elongated; bright, very small nucleus. Distance 94 million ly; Hubble class E1. Interacting with galaxy NGC 2276.

NGC 2301

OC	Mon	06^h 51.8^m	+00° 28′	12′
charts 11, 12			m$_v$ 6	

80 stars; detached, strong concentration of stars; large range in brightness; large; mag. of brightest star 8.0. A curving group topped with a flying wedge of suns (Copeland). Distance 2,500 ly; age 110 million years; Trumpler class I 3 m.

NGC 2305

Gx	Vol	06^h 48.6^m	−64° 16′	1.9′ × 1.5′
charts 24, 25			m$_p$ 12.5	p.a. 142°

Very faint, very small, and roundish. Distance 140 million ly; Hubble class E.

NGC 2307

Gx	Vol	06^h 48.9^m	−64° 20′	1.7′ × 1.6′
charts 24, 25			m$_p$ 13.0	p.a. 60°

Very faint and pretty small. Distance 160 million ly; Hubble class SBb.

NGC 2310

Gx	Pup	06^h 53.9^m	−40° 52′	4.2′ × 0.7′
chart 19			m$_v$ 11.7	p.a. 47°

Pretty bright, pretty large, and very elongated; small, bright nucleus; flattened bulge; dark lane. Distance 40 million ly; Hubble class S0.

NGC 2316

BNer	Mon	06^h 59.7^m	−07° 46′	4′ × 3′
charts 11, 12				

A double star involved in a pretty faint, small, and roundish patch of nebulosity.

NGC 2320

Gx	Lyn	07^h 05.7^m	+50° 35′	1.4′ × 0.8′
charts 1, 5, 6			m$_v$ 11.9	p.a. 140°

Pretty bright, small, and irregularly roundish, gradually brighter toward middle. Distance 250 million ly; Hubble class E.

NGC 2323

See M50.

NGC 2324

OC	Mon	07^h 04.2^m	+01° 03′	7′
chart 12			m$_v$ 8.4	

70 stars; detached, weak concentration of stars; moderate range in brightness; large; mag. of brightest star 10.4. Distance 9,500 ly; age 660 million years; Trumpler class II 2 r.

NGC 2325

Gx	CMa	07^h 02.7^m	−28° 42′	3.4′ × 2.1′
chart 19			m$_v$ 11.2	p.a. 6°

Pretty bright, pretty large, and slightly elongated. Distance 80 million ly; Hubble class E4. Several companion galaxies located nearby.

NGC 2331

OC	Gem	07^h 07.2^m	+27° 21′	18′
chart 5			m$_p$ 8.5	

30 stars; not well detached; small brightness range; brightest star is phot. mag. 9; Trumpler class IV 1 p.

NGC 2335

OC	Mon	07^h 06.6^m	−10° 05′	12′
chart 12			m$_v$ 7.2	

35 stars; detached, no concentration of stars; large range in brightness; large; involved in nebulosity; mag. of brightest star 9.5. Distance 3,300 ly; age 160 million years; Trumpler class III 3 m n.

NGC 2336

Gx	Cam	07^h 27.1^m	+80° 11′	6.9′ × 4.0′
chart 1			m$_v$ 10.4	p.a. 178°

Pretty bright, pretty large, and round; many weak, filamentary, knotty arms; faint, very small nucleus. Distance 100 million ly; Hubble class Sb. Paired with galaxy IC 467.

NGC 2339

Gx	Gem	$07^h 08.3^m$	$+18°\,47'$	$2.8' \times 2.0'$
charts 5, 12			$m_v\,11.8$	p.a. 175°

Pretty bright, pretty large, and round; very small, very bright nucleus. Distance 100 million ly; Hubble class Sc.

NGC 2340

Gx	Lyn	$07^h 11.2^m$	$+50°\,10'$	$1.8' \times 1.2'$
charts 1, 5, 6			$m_v\,11.7$	p.a. 8°

Pretty faint, small, and roundish; gradually brighter toward middle. Distance 260 million ly; Hubble class E.

NGC 2343

OC	Mon	$07^h 08.3_v$	$-10°\,39'$	7'
chart 12			$m_v\,6.7$	

20 stars; detached, no concentration of stars; large range in brightness; quite large; mag. of brightest star 8.4. Distance 3,300 ly; age 100 million years; Trumpler class III 3 p.

NGC 2344

Gx	Lyn	$07^h 12.5^m$	$+47°\,10'$	$1.8' \times 1.7'$
charts 5, 6			$m_p\,12$	p.a. 120°

Pretty bright, pretty small, and roundish; a little brighter toward middle, distance 43 million ly; Hubble class S(B)c.

NGC 2345

OC	CMa	$07^h 08.3^m$	$-13°\,10'$	12'
chart 12			$m_v\,7.7$	

70 stars; detached, strong concentration of stars; large range in brightness; mag. of brightest star 9.9. Distance 5,900 ly; age 79 million years; Trumpler class I 3 m.

NGC 2346

PN	Mon	$07^h 09.4^m$	$-00°\,48'$	52"
chart 12			$m_v\,11.6$	

PK215+03.1, the Butterfly Wing Nebula. A small, fairly faint, hazy, easily seen object; irregular disk with traces of ring structure; central star mag. 11.5 (var.). V-V class 3b + 6.

NGC 2353

OC	Mon	$07^h 14.6^m$	$-10°\,18'$	19'
chart 12			$m_v\,7.1$	

30 stars; detached, weak concentration of stars; moderate range in brightness; large; contains a very bright star. Distance 3,600 ly; age 13 million years; Trumpler class II 2 p. Located on N tip of bright nebula LBN 1036.

NGC 2354

OC	CMa	$07^h 14.3^m$	$-25°\,44'$	19'
chart 19			$m_v\,6.5$	

100 stars; detached, no concentration of stars; moderate range in brightness; mag. of brightest star 9.1. Distance 6,000 ly; age 180 million years; Trumpler class III 2 m.

NGC 2359

BNe	CMa	$07^h 18.6^m$	$-13°\,12'$	$7' \times 6'$
chart 12				

Thor's Helmet. A very faint, large, ring-shaped, filamentary nebula; S half is brightest. Per *NGC*, a (!!) very remarkable object. Distance 4,000 ly.

NGC 2360

OC	CMa	$07^h 17.8^m$	$-15°\,37'$	12'
charts 12, 19			$m_v\,7.2$	

C58, Opened Box Cluster. 80 stars; detached, weak concentration of stars; moderate range in brightness; very large; mag. of brightest star 10.4. A singular group of very lucid specks (Smyth). Distance 5,300 ly; age 1.3 billion years; Trumpler class II 2 m.

NGC 2362

OC	CMa	$07^h 18.8^m$	$-24°\,57'$	7'
chart 19			$m_v\,4.1$	

C64. 60 stars; detached, strong concentration of stars; large range in brightness; pretty large; mag. of brightest star 4.4. Distance 5,000 ly; age 25 million years; Trumpler class I 3. Involved in faint, very large (6° × 1.5°) emission nebula.

NGC 2366

Gx	Cam	07^h 28.9^m	+69° 13′	7.6′ × 3.5′
chart 1			m$_v$ 10.8	p.a. 25°

Very faint and pretty large; weak, curved arms. Distance 11 million ly; Hubble class Ir+. In M81 galaxy group.

NGC 2367

OC	CMa	07^h 20.1^m	−21° 56′	3.5′
chart 19			m$_v$ 7.9	

30 stars; detached, moderately rich in bright and faint stars; mag. of brightest star 9.4. Distance 6,500 ly.

NGC 2371, 2

PN	Gem	07^h 25.6^m	+29° 29′	44″
chart 5			m$_v$ 11.3	

PK189+19.1. Bright and small; an irregular roundish disk consisting of twin lobes; anomalous form. Phot. mag. 13.0; central star mag. 14.9. Two bright, round nebulae separated by an interval of 28″ (Denning). Distance 3,900 ly; age 10,000 years; V-V class 3a + 6.

NGC 2374

OC	CMa	07^h 24.0^m	−13° 16′	19′
chart 12			m$_v$ 8	

25 stars; detached, weak concentration of stars; large range in brightness; very large cluster; mag. of brightest star 10.8. Distance 4,200 ly; age 320 million years; Trumpler class II 3 p.

NGC 2380

Gx	CMa	07^h 23.9^m	−27° 32′	1.9′
chart 19			m$_v$ 11.5	p.a. 98°

Pretty faint, pretty small, and roundish; very suddenly much brighter toward middle, many superimposed stars; Hubble class S(B)0.

NGC 2383

OC	CMa	07^h 24.8^m	−20° 56′	6′
chart 19			m$_v$ 8.4:	

40 stars; detached, strong concentration of stars; large range in brightness; pretty small; mag. of brightest star 9.8. Distance 6,500 ly; Trumpler class I 3 m. Open cluster NGC 2384 directly to SE.

NGC 2384

OC	CMa	07^h 25.1^m	−21° 02′	2.5′
chart 19			m$_v$ 7.4:	

15 stars; not well detached; large brightness range; mag. of brightest star 8.6. Distance 6,500 ly; Trumpler class IV 3 p. Open cluster NGC 2383 directly to NW.

NGC 2392

PN	Gem	07^h 29.2^m	+20° 55′	19.5″
chart 5			m$_v$ 9.2	

PK197+17.1, C39, also the Eskimo Nebula or Clown Face Nebula. Irregular disk with traces of ring structure; phot. mag. 9.9; central star mag. 10.5. Has a faint halo and features resembling a human face. Bluish nebulosity quite like a telescopic comet (Webb). Expansion velocity 54 km/sec (34 miles/sec); distance 2,900 ly; V-V class 3b.

NGC 2395

OC	Gem	07^h 27.1^m	+13° 35′	12′
chart 12			m$_v$ 8	

30 stars; detached, no concentration of stars; small range in brightness; mag. of brightest star 10.0. Distance 3,900 ly; age 50 million years; Trumpler class III 1 p.

NGC 2396

OC	Pup	07^h 28.1^m	−11° 44′	10′
chart 12			m$_p$ 7.4	

30 stars; detached, no concentration; large brightness range; brightest star is phot. mag. 11. Trumpler class III 3 p.

NGC 2397

Gx	Vol	07^h 21.3^m	−69° 00′	2.5′ × 1.2′
charts 24, 25			m$_v$ 12	p.a. 123°

Pretty bright, quite large, and elongated. Distance

45 million ly; Hubble class S(B)b. Paired with galaxy NGC 2397A.

NGC 2403

Gx	Cam	$07^h 36.9^m$	$+65° 36'$	$23' \times 11'$
chart 1			m_v 8.5	p.a. 127°

C7. Bright, very elongated, very large; irregular arms; small, very faint nucleus. Mottled texture visible in 8-inch and larger telescopes (Mullaney). Easily missed if you look for too small an object (Houston). Per *NGC*, a (!!) very remarkable object. Distance 11 million ly; Hubble class Sc. In M81 galaxy group; similar in appearance to M33.

NGC 2414

OC	Pup	$07^h 33.3^m$	$-15° 27'$	4'
charts 12, 19			m_v 7.9	

35 stars; detached, strong concentration; moderately rich in bright and faint stars; mag. of brightest star 8.2. Distance 8,200 ly.

NGC 2415

Gx	Lyn	$07^h 36.9^m$	$+35° 15'$	0.9'
charts 5, 6			m_v 12.4	

Pretty bright and roundish; gradually brighter toward middle; high surface brightness. Distance 160 million ly; Hubble class Ir+.

NGC 2417

Gx	Car	$07^h 30.2^m$	$-62° 15'$	$3.2' \times 2.1'$
charts 24, 25			m_v 12.4	p.a. 81°

Very faint, large, and roundish; gradually brighter toward middle; small nucleus; narrow, knotty arms. Distance 130 million ly; Hubble class Sc.

NGC 2419

GC	Lyn	$07^h 38.2^m$	$+38° 53'$	4.7'
charts 5, 6			m_v 10.3	

C25, Intergalactic Wanderer. High concentration of stars; pretty bright, pretty large, and slightly elongated; S-S class 2. At 300,000 ly, this globular is more distant than the Magellanic Clouds and seemingly adrift in intergalactic space. In good seeing at the prime focus of the Palomar 200-inch reflector, René Racine and William Harris barely resolved the cluster's brightest stars visually.

NGC 2420

OC	Gem	$07^h 38.5^m$	$+21° 34'$	10'
chart 5			m_v 8.3	

100 stars; detached, strong concentration of stars; moderate range in brightness; quite large; mag. of brightest star 11.1. Distance 8,200 ly; age 4 billion years; Trumpler class I 2 r.

NGC 2421

OC	Pup	$07^h 36.3^m$	$-20° 37'$	10'
chart 19			m_v 8.3	

70 stars; detached, strong concentration of stars toward center; moderate range in brightness; mag. of brightest star 10.5. Distance 6,200; Trumpler class I 2 m.

NGC 2422

See M47.

NGC 2423

OC	Pup	$07^h 37.1^m$	$-13° 52'$	19'
chart 12			m_v 6.7	

40 stars; not well detached from surrounding star field; moderate range in brightness; very large; mag. of brightest star 9.0. Distance 2,800 ly; age 350 million years; Trumpler class IV 2 m.

NGC 2427

Gx	Pup	$07^h 36.5^m$	$-47° 38'$	$5.6' \times 2.6'$
charts 19, 20			m_v 11.5	p.a. 122°

Extremely faint, large, and elongated; very small, faint nucleus. Distance 30 million ly; Hubble class S(B)d.

NGC 2434

Gx	Vol	$07^h 34.9^m$	$-69° 17'$	2.2'
charts 24, 25			m_v 11.5	p.a. 133°

Pretty bright, small, and round. Distance 110 million ly; Hubble class E0. Bright star superimposed on edge.

NGC 2437

See M46.

NGC 2438

PN	Pup	07^h 41.8^m	−14° 44′	64″
charts 12, 19			m$_v$ 11	

PK231+04.2. Pretty bright, pretty small ring; phot. mag. 10.1; central star mag. 17.5. Distance 2,900 ly; V-V class 4 + 2. Lies in front of open cluster M46.

NGC 2439

OC	Pup	07^h 40.8^m	−31° 39′	10′
charts 19, 20			m$_v$ 6.9	

80 stars; detached, weak concentration of stars; large range in brightness; bright, pretty large; mag. of brightest star 8.9. Distance 5,300 ly; age 66 million years; Trumpler class II 3 m.

NGC 2440

PN	Pup	07^h 41.9^m	−18° 12′	16″
charts 12, 19			m$_v$ 9.4	

PK234+02.1. Quite bright; not very well defined, almost stellar); phot. mag. 10.8; central star mag. 17.7. Seen at 64× like a dull 8th-mag. star, but with more power surrounded by a little very faint haziness (Webb). Expansion velocity 24 km/sec (15 miles/sec); distance 3,600 ly; V-V class 5 + 3.

NGC 2441

Gx	Cam	07^h 51.9^m	+73° 01′	2.0′ × 1.8′
charts 1, 2			m$_v$ 12.2	p.a. 30°

Very faint and pretty small; several knotty arms; very small, bright nucleus. Distance 160 million ly; Hubble class Sc.

NGC 2442

Gx	Vol	07^h 36.4^m	−69° 32′	5′
charts 24, 25			m$_p$ 11.4	

Very faint, round; bright, very small nucleus; dark lane. Distance 54 million ly; Hubble class SBb. Connected to galaxy NGC 2443.

NGC 2447

See M93.

NGC 2451

OC	Pup	07^h 45.4^m	−37° 58′	44′
charts 19, 20			m$_v$ 2.8	

40 stars; detached, weak concentration of stars; moderate range in brightness; extremely large; mag. of brightest star 3.6. Distance 1,000 ly; age 36 million years; Trumpler class II 2 p.

NGC 2452

PN	Pup	07^h 47.4^m	−27° 20′	19″
charts 19, 20			m$_v$ 12	

PK243−1.1. A faint, small ring enclosed within a larger, fainter nebula; phot. mag. 12.6; central star mag. 16.1. Distance 8,800 ly; V-V class 4 + 2. Located immediately SW of the nice open cluster NGC 2453.

NGC 2453

OC	Pup	07^h 47.8^m	−27° 14′	5′
charts 19, 20			m$_v$ 8.3	

30 stars; detached, strong concentration of stars; moderate range in brightness; small; mag. of brightest star 9.5. Distance 4,900 ly; age 200 million years; Trumpler class I 2 p. Lies immediately NE of planetary nebula NGC 2452.

NGC 2460

Gx	Cam	07^h 56.9^m	+60° 21′	2.5′ × 1.9′
charts 1, 2			m$_v$ 11.8	p.a. 40°

Faint, small, and round; spiral structure; very bright nucleus. Distance 67 million ly; Hubble class Sb. Paired with galaxy IC 2209.

NGC 2467

C+BNe	Pup	07^h 52.6^m	−26° 23′	15′
charts 19, 20			m$_p$ 7.1	

50 stars; strong concentration of stars; pretty bright, very large, and round. Distance 11,000 ly; age 250 million years; involved in small (7′ × 6′), pretty bright emission nebula with faint streamers.

NGC 2477

OC	Pup	$07^h 52.3^m$	$-38° 33'$	26'
charts 19, 20			m_v 5.8	

160 stars; detached, strong concentration of stars; large range in brightness; very large, bright; mag. of brightest star 9.8. Perhaps the best of Puppis's amazing hoard of lovely star clusters (Mullaney). The richest open cluster (Harlow Shapley). Per *NGC*, a (!) remarkable object. Distance 4,200 ly; age 710 million years; Trumpler class I 3 r.

NGC 2482

OC	Pup	$07^h 54.9^m$	$-24° 18'$	12'
charts 19, 20			m_v 7.3	

Starfish Cluster. 40 stars; detached, no concentration of stars; small range in brightness; large; mag. of brightest star 10.0. Distance 2,600 ly; Trumpler class III 1 m.

NGC 2483

OC	Pup	$07^h 55.9^m$	$-27° 56'$	10'
charts 19, 20			m_v 7.6	

30 stars; detached, no concentration; moderate brightness range; mag. of brightest star 9.3. Probably not a true cluster.

NGC 2485

Gx	CMi	$07^h 56.8^m$	$+07° 29'$	1.6'
chart 12			m_v 12.2	

Distance 200 million ly; Hubble class Sa.

NGC 2489

OC	Pup	$07^h 56.2^m$	$-30° 04'$	7'
charts 19, 20			m_v 7.9	

45 stars; detached, weak concentration of stars; moderate range in brightness; pretty large; mag. of brightest star 11.1. Distance 3,900 ly; age 240 million years; Trumpler class II 2 m.

NGC 2493

Gx	Lyn	$08^h 00.4^m$	$+39° 50'$	2.0'
charts 5, 6			m_v 12	

Quite bright, small, and roundish; suddenly brighter toward middle. Distance 170 million ly; Hubble class SB0. Bright star 8' to NE.

NGC 2500

Gx	Lyn	$08^h 01.9^m$	$+50° 44'$	2.9' × 2.6'
charts 1, 2, 5, 6			m_v 11.6	p.a. 50°

Faint, large, and round; very small, bright nucleus. Distance 24 million ly; Hubble class S+. Located among stars.

NGC 2506

OC	Mon	$08^h 00.2^m$	$-10° 47'$	7'
chart 12			m_v 7.6	

C54. 150 stars; detached, strong concentration of stars; moderate range in brightness; mag. of brightest star 10.8; pretty large. Fine broad starry cloud, vicinity gorgeous (Webb). Distance 7,200 ly; age 4 billion years; Trumpler class I 2 r.

NGC 2513

Gx	Cnc	$08^h 02.4^m$	$+09° 25'$	2.5' × 2.0'
chart 12			m_v 11.6	p.a. 170°

Faint, small, and roundish; brighter toward middle. Distance 200 million ly; Hubble class E.

NGC 2516

OC	Car	$07^h 58.3^m$	$-60° 52'$	30'
charts 24, 25			m_v 3.8	

C96. 80 stars; detached, strong concentration of stars; large range in brightness; very bright, very large; brightest star is phot. mag. 7. Distance 1,300 ly; age 110 million years; Trumpler class I 3 r.

NGC 2517

Gx	Pup	$08^h 02.8^m$	$-12° 19'$	1.4' × 1.0'
chart 12			m_v 11.8	p.a. 70°

Faint, small, and roundish; Hubble class S(B)0. Located between two stars of mag. 13 or 14.

NGC 2523

Gx	Cam	$08^h 15.0^m$	$+73° 35'$	2.9' × 1.8'
charts 1, 2			m_v 11.9	p.a. 57°

Pretty bright, pretty large, and slightly elongated; has several filamentary, knotty arms; diffuse, bright, small nucleus. Distance 160 million ly; Hubble class SBb. Paired with galaxy NGC 2523B.

NGC 2525

| Gx | Pup | 08^h 05.6^m | −11° 26′ | 3.0′ × 2.0′ |
| chart 12 | | | m$_v$ 11.6 | p.a. 75° |

Pretty large, quite faint, and round; strong arms; extremely small, bright nucleus. Distance 57 million ly; Hubble class S(B)cp.

NGC 2527

| OC | Pup | 08^h 05.3^m | −28° 10′ | 21′ |
| charts 19, 20 | | | m$_v$ 6.5 | |

40 stars; detached, no concentration of stars; small range in brightness; mag. of brightest star 8.6; very large. Distance 2,000 ly; over 1 billion years old; Trumpler class III 1 p.

NGC 2532

| Gx | Lyn | 08^h 10.3^m | +33° 57′ | 2.0′ × 1.7′ |
| charts 5, 6 | | | m$_v$ 12.4 | p.a. 10° |

Pretty bright, pretty large, and roundish; slightly brighter toward middle. Distance 230 million ly; Hubble class S(B)c.

NGC 2533

| OC | Pup | 08^h 07.0^m | −29° 54′ | 3.5′ |
| charts 19, 20 | | | m$_v$ 7.6 | |

60 stars; detached, no concentration of stars; small range in brightness; mag. of brightest star 9.0; pretty large. Distance 5,500 ly; age 180 million years; Trumpler class III 1 p.

NGC 2537

| Gx | Lyn | 08^h 13.3^m | +45° 59′ | 3.1′ |
| charts 5, 6 | | | m$_v$ 11.7 | |

Bear Paw Galaxy. Pretty bright, pretty large, and round. Distance 19 million ly; Hubble class S. Paired with galaxy NGC 2537A.

NGC 2539

| OC | Pup | 08^h 10.7^m | −12° 50′ | 21′ |
| chart 12 | | | m$_v$ 6.5 | |

50 stars; detached, weak concentration of stars; small range in brightness; mag. of brightest star 9.2; very large. Loose cluster of stars (Webb). Distance 4,200 ly; age 660 million years; Trumpler class II 1 m.

NGC 2541

| Gx | Lyn | 08^h 14.7^m | +49° 04′ | 6.6′ × 3.5′ |
| charts 5, 6 | | | m$_v$ 11.8 | p.a. 165° |

Faint, large, and elongated; very small, faint nucleus. Distance 26 million ly; Hubble class S+.

NGC 2543

| Gx | Lyn | 08^h 13.0^m | +36° 15′ | 2.5′ × 1.3′ |
| charts 5, 6 | | | m$_v$ 11.9 | p.a. 45° |

Faint, pretty large, irregularly roundish; gradually brighter toward middle. Distance 100 million ly; Hubble class SBb.

NGC 2546

| OC | Pup | 08^h 12.4^m | −37° 38′ | 40′ |
| charts 19, 20 | | | m$_v$ 6.3 | |

40 stars; detached, no concentration of stars; moderate range in brightness; mag. of brightest star 8.2; bright, large. Distance 3,300 ly; age 42 million years; Trumpler class III 2 m.

NGC 2547

| OC | Vel | 08^h 10.7^m | −49° 16′ | 19′ |
| charts 19, 20, 25 | | | m$_v$ 4.7 | |

80 stars; detached, weak concentration of stars; moderate range in brightness; mag. of brightest star 6.5; bright, large; involved in nebulosity. Distance 1,300 ly; age 74 million years; Trumpler class II 2 p n.

NGC 2548

See M48.

NGC 2549

| Gx | Lyn | 08^h 19.0^m | +57° 48′ | 3.8′ × 1.2′ |
| charts 1, 2 | | | m$_v$ 11.2 | p.a. 177° |

Pretty bright, small, and very elongated; extremely bright, elongated nucleus. Distance 51 million ly; Hubble class E6.

NGC 2552

| Gx | Lyn | 08ʰ 19.3ᵐ | +50° 00′ | 3.5′ × 2.2′ |
| charts 2, 5, 6 | | | m_v 12.1 | p.a. 45° |

Extremely faint, quite large, and slightly elongated; very small, faint nucleus. Distance 24 million ly; Hubble class Ir+.

NGC 2554

| Gx | Cnc | 08ʰ 17.9ᵐ | +23° 28′ | 3.5′ × 2.6′ |
| chart 6 | | | m_v 12 | p.a. 150° |

Faint, small, and roundish; much brighter toward the middle. Distance about 180 million ly; Hubble class Sap.

NGC 2559

| Gx | Pup | 08ʰ 17.1ᵐ | −27° 27′ | 3.9′ × 1.8′ |
| chart 20 | | | m_v 10.9 | p.a. 6° |

Very bright and knotty inner arms; very faint outer arms. Distance 59 million ly; Hubble class SBb. Many superimposed stars.

NGC 2566

| Gx | Pup | 08ʰ 18.8ᵐ | −25° 30′ | 3.6′ × 2.5′ |
| chart 20 | | | m_v 11 | p.a. 110° |

Very knotty arms; inner pseudo-ring. Distance 62 million ly; Hubble class SBb.

NGC 2567

| OC | Pup | 08ʰ 18.6ᵐ | −30° 38′ | 10′ |
| charts 19, 20 | | | m_v 7.4 | |

40 stars; detached, no concentration of stars; moderate range in brightness; mag. of brightest star 10.1; pretty large. Distance 5,500 ly; age 68 million years; Trumpler class III 2 m.

NGC 2571

| OC | Pup | 08ʰ 18.9ᵐ | −29° 44′ | 12′ |
| chart 20 | | | m_v 7 | |

30 stars; not well detached from surrounding star field; small range in brightness; mag. of brightest star 8.8; very large. Distance 6,900 ly; age 22 million years; Trumpler class IV 1 p.

NGC 2579

| OC | Pup | 08ʰ 21.1ᵐ | −36° 11′ | 10′ |
| charts 19, 20 | | | m_v 7.5 | |

20 stars; not well detached; small brightness range; mag. of brightest star 9.5. Probably not a true cluster. Contains a small (2′), double, nebulous patch.

NGC 2591

| Gx | Cam | 08ʰ 37.4ᵐ | +78° 02′ | 3.0′ × 0.6′ |
| charts 1, 2 | | | m_v 12.2 | p.a. 32° |

Faint, small, and elongated; a little brighter toward the middle. Distance 66 million ly; Hubble class Sc.

NGC 2595

| Gx | Cnc | 08ʰ 27.7ᵐ | +21° 29′ | 3.1′ × 2.6′ |
| chart 6 | | | m_v 12.3 | p.a. 45° |

Very faint, pretty large, irregularly roundish; compact core. Distance 180 million ly; Hubble class S(B)c.

NGC 2608

| Gx | Cnc | 08ʰ 35.3ᵐ | +28° 28′ | 2.2′ × 1.4′ |
| chart 6 | | | m_v 12.3 | p.a. 60° |

Faint and slightly elongated; two bright arms; has a double nucleus or a superimposed star on the nucleus. Distance 89 million ly; Hubble class Sc.

NGC 2610

| PN | Hya | 08ʰ 33.4ᵐ | −16° 09′ | 38″ |
| charts 12, 20 | | | m_v 12.8 | |

PK239+13.1. A faint, small ring with a smooth disk; phot. mag. 13.6; central star mag. 15.9 (var.). Distance 5,500 ly; V-V class 4 + 2.

NGC 2612

| Gx | Hya | 08ʰ 33.8ᵐ | −13° 10′ | 3.2′ × 0.7′ |
| chart 12 | | | m_v 12.7 | p.a. 115° |

Faint, small, and elongated; brighter toward middle; Hubble class S0.

NGC 2613

| Gx | Pyx | 08ʰ 33.4ᵐ | −22° 58′ | 7.2′ × 2.1′ |
| chart 20 | | | m_v 10.5 | p.a. 113° |

Quite bright, large, and very elongated; multiple filamentary arms; small, very bright nucleus. Distance 63 million ly; Hubble class Sb.

NGC 2626

BNer	Vel	08^h 35.6^m	−40° 40′	5′
charts 19, 20				

A 10th-mag. star involved in the N part of a pretty bright, pretty large, roundish nebula.

NGC 2627

OC	Pyx	08^h 37.3^m	−29° 57′	10′
chart 20			m$_p$ 8.4	

60 stars; detached, no concentration of stars; moderate range in brightness; brightest star is phot. mag. 11; quite large. Trumpler class III 2 m.

NGC 2632

See Praesepe.

NGC 2633

Gx	Cam	08^h 48.1^m	+74° 06′	2.3′ × 1.4′
charts 1, 2			m$_v$ 12.2	p.a. 175°

Faint, small, and slightly elongated; very bright, very small nucleus; absorption lane crosses arm near nucleus. Distance 100 million ly; Hubble class SBbp. Brightest in a group of galaxies; paired with galaxy NGC 2634.

NGC 2634

Gx	Cam	08^h 48.4^m	+73° 58′	2.0′
charts 1, 2			m$_v$ 12	

Faint, small, and slightly elongated; Hubble class E1. Paired with galaxy NGC 2633.

NGC 2639

Gx	UMa	08^h 43.6^m	+50° 12′	1.8′ × 1.4′
charts 2, 5, 6			m$_v$ 11.7	p.a. 140°

Bright, small, and elongated; very small, very bright nucleus. Distance 150 million ly; Hubble class Sa.

NGC 2640

Gx	Car	08^h 37.4^m	−55° 07′	2.0′ × 1.7′
charts 20, 25			m$_p$ 12.7	p.a. 104°

Pretty bright, small, and roundish. Distance 36 million ly; Hubble class S0.

NGC 2648

Gx	Cnc	08^h 42.7^m	+14° 17′	3.2′ × 1.3′
chart 12			m$_v$ 11.8	p.a. 148°

Faint, small, and elongated. Distance 85 million ly; Hubble class Sa.

NGC 2649

Gx	Lyn	08^h 44.1^m	+34° 43′	1.5′ × 1.4′
chart 6			m$_v$ 12.3	p.a. 95°

Faint, large, and round. Distance 180 million ly; Hubble class S(B)b.

NGC 2654

Gx	UMa	08^h 49.2^m	+60° 13′	4.3′ × 0.8′
chart 2			m$_v$ 11.8	p.a. 63°

Small and pretty faint; faint stars in middle. Distance 63 million ly; Hubble class Sb.

NGC 2655

Gx	Cam	08^h 55.6^m	+78° 13′	5.1′ × 4.4′
charts 1, 2			m$_v$ 10.1	p.a. 90°

Very bright, quite large, and slightly elongated; very bright, very large nucleus with absorption lane on one side; very faint, diffuse outer arms. Distance 71 million ly; Hubble class S(B)a.

NGC 2659

OC	Vel	08^h 42.6^m	−44° 57′	2.7′
charts 19, 20			m$_v$ 8.6:	

80 stars; detached, no concentration of stars; large range of brightness; mag. of brightest star 9.7; Trumpler class III 3 m.

NGC 2663

Gx	Pyx	08^h 45.1^m	−33° 48′	3.5′ × 2.3′
chart 20			m$_v$ 10.9	p.a. 110°

Pretty faint, pretty small, and slightly elongated. Distance 82 million ly; Hubble class E/S0.

NGC 2665

| Gx | Hya | 08^h 46.0^m | −19° 18′ | 2.0′ × 1.5′ |
| charts 12, 20 | | | m$_v$ 12.7 | p.a. 144° |

Patchy, inner pseudo-ring; pretty faint outer ring; several superimposed stars. Distance 67 million ly; Hubble class SBa.

NGC 2669

| OC | Vel | 08^h 46.3^m | −52° 55′ | 12′ |
| charts 20, 25 | | | m$_v$ 6.1 | |

40 stars; detached, no concentration; large brightness range; mag. of brightest star 7.6. Distance about 3,300 ly.

NGC 2670

| OC | Vel | 08^h 45.5^m | −48° 47′ | 8′ |
| charts 19, 20, 25 | | | m$_v$ 7.8 | |

30 stars; detached, weak concentration of stars; moderate range in brightness; mag. of brightest star 9.3; pretty large. Distance 3,300 ly; age 95 million years; Trumpler class II 2 p.

NGC 2672

| Gx | Cnc | 08^h 49.4^m | +19° 05′ | 2.6′ × 2.4′ |
| charts 6, 12 | | | m$_v$ 11.7 | p.a. 120° |

Pretty bright and pretty large. Distance 180 million ly; Hubble class E1. Brightest in a group of galaxies; paired with galaxy NGC 2673.

NGC 2681

| Gx | UMa | 08^h 53.6^m | +51° 19′ | 3.6′ |
| charts 2, 6 | | | m$_v$ 10.3 | |

Very bright and very large; extremely bright nucleus. Distance 33 million ly; Hubble class Sa.

NGC 2682

See M67.

NGC 2683

| Gx | Lyn | 08^h 52.7^m | +33° 25′ | 9.3′ × 2.5′ |
| chart 6 | | | m$_v$ 9.8 | p.a. 44° |

Very bright, very large, and very elongated; very bright, small nucleus. A bright oval with a splendid center, closely followed by a star (Smyth). Scarcely worth the search with 3.7-inch (Webb). Distance 11 million ly; Hubble class Sb.

NGC 2685

| Gx | UMa | 08^h 55.6^m | +58° 44′ | 4.6′ × 2.3′ |
| chart 2 | | | m$_v$ 11.3 | p.a. 38° |

Helix Galaxy. Pretty faint, pretty large, and elongated; almost edge on; barlike core. Distance 42 million ly; Hubble class Sbp.

NGC 2693

| Gx | UMa | 08^h 57.0^m | +51° 21′ | 2.6′ × 1.9′ |
| charts 2, 6 | | | m$_v$ 11.9 | p.a. 160° |

Pretty bright and slightly elongated; bright nucleus. Distance 220 million ly; Hubble class Ep. Paired with galaxy NGC 2694.

NGC 2695

| Gx | Hya | 08^h 54.5^m | −03° 04′ | 1.8′ × 1.3′ |
| chart 12 | | | m$_v$ 11.9 | p.a. 175° |

Pretty faint, pretty small, and roundish. Distance 72 million ly; Hubble class S0. Lines up SE–NW with galaxies NGC 2698 and NGC 2708, all within 0.5° of each other.

NGC 2698

| Gx | Hya | 08^h 55.6^m | −03° 11′ | 1.5′ × 0.6′ |
| chart 12 | | | m$_v$ 12.6 | p.a. 96° |

Very faint, pretty small, and roundish; Hubble class Sa. Lines up SE–NW with galaxies NGC 2695 and NGC 2708, all within 0.5° of each other.

NGC 2701

| Gx | UMa | 08^h 59.1^m | +53° 46′ | 2.2′ × 1.7′ |
| charts 2, 6 | | | m$_v$ 12.3 | p.a. 23° |

Pretty bright, fan shaped; spiral arms; has a bright, very small nucleus. Distance 100 million ly; Hubble class Sc.

NGC 2708

| Gx | Hya | 08^h 56.1^m | −03° 22′ | 3.0′ × 1.3′ |
| chart 12 | | | m$_v$ 12 | p.a. 20° |

Pretty faint and pretty small. Distance 80 million ly;

Hubble class Sb. Lines up SE–NW with galaxies NGC 2695 and NGC 2698, all within 0.5° of each other.

NGC 2712

Gx	Lyn	08ʰ 59.5ᵐ	+44° 55′	2.9′ × 1.5′
chart 6			m_v 12.1	p.a. 178°

Large, pretty bright, and elongated; several knotty arms; very small, bright nucleus. Distance 81 million ly; Hubble class Sb.

NGC 2713

Gx	Hya	08ʰ 57.3ᵐ	+02° 55′	3.3′ × 1.3′
chart 12			m_v 11.8	p.a. 107°

Pretty bright; two main arms; small, very bright nucleus. Distance 160 million ly; Hubble class Sb. Paired with galaxy NGC 2716.

NGC 2715

Gx	Cam	09ʰ 08.1ᵐ	+78° 05′	4.8′ × 1.6′
charts 1, 2			m_v 11.2	p.a. 22°

Pretty bright, large, and elongated; several filamentary, knotty arms; very small, bright nucleus. Distance 57 million ly; Hubble class Sc.

NGC 2716

Gx	Hya	08ʰ 57.6ᵐ	+03° 05′	1.6′ × 1.2′
chart 12			m_v 11.8	p.a. 30°

Very small, very bright nucleus. Distance 150 million ly; Hubble class SB0. Paired with galaxy NGC 2713; double star nearby.

NGC 2721

Gx	Hya	08ʰ 58.9ᵐ	−04° 54′	2.3′ × 1.6′
chart 12			m_v 11.7	p.a. 30°

Quite faint, pretty large, and roundish; gradually brighter toward middle. Distance 150 million ly; Hubble class S.

NGC 2732

Gx	Cam	09ʰ 13.4ᵐ	+79° 11′	2.1′ × 0.9′
charts 1, 2			m_v 11.9	p.a. 67°

Pretty bright, small, and elongated; very bright, very small nucleus. Distance 96 million ly; Hubble class S. Paired with an anonymous spiral galaxy with a bright nucleus.

NGC 2742

Gx	UMa	09ʰ 07.6ᵐ	+60° 29′	3.0′ × 1.5′
chart 2			m_v 11.4	p.a. 87°

Bright, large, and elongated; several filamentary, knotty arms; small, bright nucleus. Distance 61 million ly; Hubble class Sc.

NGC 2748

Gx	Cam	09ʰ 13.7ᵐ	+76° 29′	3.2′ × 1.1′
charts 1, 2			m_v 11.7	p.a. 38°

Pretty bright, pretty large, and elongated; several knotty arms; small, faint nucleus. Distance 68 million ly; Hubble class Sc.

NGC 2749

Gx	Cnc	09ʰ 05.4ᵐ	+18° 19′	1.9′ × 1.7′
charts 6, 12, 13			m_v 11.8	p.a. 75°

Pretty faint, small, and round. Distance 180 million ly; Hubble class E2. Brightest in a group of galaxies.

NGC 2750

Gx	Cnc	09ʰ 05.8ᵐ	+25° 26′	2.1′ × 1.7′
chart 6			m_v 11.9	p.a. 85°

Very faint, quite large, and roundish; bright nucleus. Distance 110 million ly; Hubble class S(B)c.

NGC 2763

Gx	Hya	09ʰ 06.8ᵐ	−15° 30′	2.3′ × 1.9′
charts 12, 13			m_v 12	p.a. 120°

Very faint and pretty small; two main arms; faint, small nucleus. Distance about 70 million ly; Hubble class S–.

NGC 2768

Gx	UMa	09ʰ 11.6ᵐ	+60° 02′	6.3′ × 2.8′
chart 2			m_v 9.9	p.a. 95°

Quite bright, quite large, and elongated; with a bright, large nucleus. Distance 65 million ly; Hubble class E5.

NGC 2770

Gx	Lyn	09^h 09.6^m	+33° 07′	3.6′ × 1.1′
chart 6			m$_v$ 12.2	p.a. 148°

Very small, bright nucleus. Distance 83 million ly; Hubble class Sc.

NGC 2775

Gx	Cnc	09^h 10.3^m	+07° 02′	4.5′ × 3.5′
charts 12, 13			m$_v$ 10.1	p.a. 155°

C48; quite bright, quite large, and round; almost face on; many knotty arms; small, bright, diffuse nucleus. Distance 42 million ly; Hubble class Sa. Paired with galaxy NGC 2777.

NGC 2776

Gx	Lyn	09^h 12.2^m	+44° 57′	3.0′ × 2.8′
chart 6			m$_v$ 11.6	p.a. 80°

Pretty bright, large, and round; two main arms; small, very bright nucleus. Distance 110 million ly; Hubble class Sc.

NGC 2781

Gx	Hya	09^h 11.5^m	−14° 49′	3.1′ × 1.8′
charts 12, 13			m$_v$ 11.6	p.a. 75°

Bright, small, and slightly elongated; very small, bright nucleus. Distance about 80 million ly; Hubble class Sb.

NGC 2782

Gx	Lyn	09^h 14.1^m	+40° 07′	3.7′ × 2.6′
chart 6			m$_v$ 11.6	p.a. 75°

Quite bright and round; extremely bright nucleus with dark lane; diffuse outer arms. Distance 110 million ly; Hubble class Sb. Interacting with nearby spiral galaxy.

NGC 2784

Gx	Hya	09^h 12.3^m	−24° 10′	5.1′ × 2.3′
chart 20			m$_v$ 10	p.a. 73°

Bright, large, and very elongated; very small, very bright, diffuse nucleus. Distance 19 million ly; Hubble class S0. Several companion galaxies nearby at W and SE.

NGC 2787

Gx	UMa	09^h 19.3^m	+69° 12′	3.3′ × 2.1′
charts 1, 2			m$_v$ 10.8	p.a. 117°

Bright, pretty large, and slightly elongated; small, very bright nucleus. Distance 33 million ly; Hubble class Sap.

NGC 2792

PN	Vel	09^h 12.4^m	−42° 26′	13″
chart 20			m$_v$ 11.7	

PK265+04.1. Very small; has faint ring structure; phot. mag. 13.5; central star mag. 15.7. Per *NGC*, a (!) remarkable object. Distance 8,200 ly; V-V class 4.

NGC 2798

Gx	Lyn	09^h 17.4^m	+42° 00′	2.7′ × 0.9′
chart 6			m$_v$ 12.3	p.a. 160°

Pretty bright and small; extremely bright, small nucleus. Distance 74 million ly; Hubble class SBap. Interacting with galaxy NGC 2799.

NGC 2805

Gx	UMa	09^h 20.3^m	+64° 06′	6.3′ × 5.0′
chart 2			m$_v$ 11	p.a. 125°

Very small, very bright nucleus; many irregular, very faint and knotty arms, Distance 80 million ly; Hubble class S(B)d. Similar in appearance to M101.

NGC 2808

GC	Car	09^h 12.1^m	−64° 52′	12′
chart 25			m$_v$ 6.2	

Very large; high concentration of stars. Per *NGC*, a (!) remarkable object. Distance 30,000 ly; S-S class 1.

NGC 2811

Gx	Hya	09^h 16.2^m	−16° 19′	2.3′ × 0.8′
charts 12, 13			m$_v$ 11.3	p.a. 20°

Pretty bright, pretty small, and elongated; very bright nucleus. Distance 98 million ly; Hubble class Sb.

NGC 2815

Gx	Hya	09^h 16.3^m	−23° 38′	3.4′ × 1.1′
chart 20			m$_v$ 11.8	p.a. 10°

Faint, small, and slightly elongated; bright bulge in center; two main arms. Distance 99 million ly; Hubble class Sbp.

NGC 2817

| Gx | Hya | 09^h 17.2^m | –04° 45′ | 2.0′ × 1.7′ |
| charts 12, 13 | | | m$_v$ 12.6 | p.a. 140° |

Very faint, pretty small, and roundish. Distance 150 million ly; Hubble class SBc.

NGC 2818

| OC | Pyx | 09^h 16.0^m | –36° 37′ | 8′ |
| chart 20 | | | m$_v$ 8.2 | |

40 stars; detached, weak concentration of stars; moderate range in brightness; pretty bright, pretty large, round; mag. of brightest star 11.3. Involved in a large, hazy patch of nebulosity; has a very small planetary nebula (NGC 2818A) located on one side. Per *NGC*, a (!) remarkable object. Distance 10,000 ly; over 1 billion years old; Trumpler class II 2 m.

NGC 2818A

| PN | Pyx | 09^h 16.0^m | –36° 38′ | 50″ |
| chart 20 | | | m$_v$ 11.6 | |

PK261+08.1. A small, irregularly shaped, double-lobed planetary years; phot. mag. 13.0; central star phot. mag. 19.5. Located on one side of (and possibly belonging to) open cluster NGC 2818. Extends 2 ly; distance 10,000 ly; age 22,000; V-V class 3b.

NGC 2822

| Gx | Car | 09^h 13.8^m | –69° 39′ | 2.8′ × 1.9′ |
| chart 25 | | | m$_v$ 10.7 | p.a. 90° |

Pretty faint, very small, and roundish; gradually brighter toward middle; Hubble class E.

NGC 2832

| Gx | Lyn | 09^h 19.8^m | +33° 45′ | 3.2′ × 2.2′ |
| chart 6 | | | m$_v$ 11.9 | p.a. 160° |

Quite bright, quite large, and elongated. Distance 300 million ly; Hubble class E. Brightest member of Abell 779 galaxy cluster (which also includes NGC 2830 and NGC 2831).

NGC 2835

| Gx | Hya | 09^h 17.9^m | –22° 21′ | 6.3′ × 4.4′ |
| chart 20 | | | m$_v$ 10.4 | p.a. 8° |

Large and faint; a multiple-arm spiral; faint, very small nucleus. Distance 27 million ly; Hubble class Sp. Located between two 9th-mag. stars.

NGC 2836

| Gx | Car | 09^h 13.7^m | –69° 20′ | 2.8′ × 2.0′ |
| chart 25 | | | m$_v$ 11.8 | p.a. 118° |

Faint and formless; very faint arms; Hubble class Sb. Similar to M81 but much smaller.

NGC 2841

| Gx | UMa | 09^h 22.0^m | +50° 59′ | 8.1′ × 3.8′ |
| charts 2, 6 | | | m$_v$ 9.2 | p.a. 147° |

Very bright, large, and elongated; symmetrical, spiraled, knotty arms; bright, large nucleus. Distance 30 million ly; Hubble class Sb.

NGC 2845

| Gx | Vel | 09^h 18.6^m | –38° 01′ | 2.2′ × 1.1′ |
| chart 20 | | | m$_v$ 12.7 | p.a. 67° |

Elongated nucleus; several knots. Distance 100 million ly; Hubble class S0. A fairly bright star is located approx. 0.4′ to SE.

NGC 2848

| Gx | Hya | 09^h 20.2^m | –16° 32′ | 2.5′ × 1.5′ |
| charts 12, 13, 20 | | | m$_v$ 11.8 | p.a. 30° |

Very faint, quite large, and elongated; two main arms; bright center with star. Distance 78 million ly; Hubble class S–.

NGC 2855

| Gx | Hya | 09^h 21.5^m | –11° 55′ | 2.5′ × 1.9′ |
| charts 12, 13 | | | m$_v$ 11.7 | p.a. 130° |

Pretty bright, pretty large, and round; very bright, diffuse nucleus; prominent dust lanes. Distance 72 million ly; Hubble class E1.

NGC 2857

| Gx | UMa | 09^h 24.6^m | +49° 21′ | 2.2′ × 2.0′ |
| charts 2, 6 | | | m$_v$ 12.3 | p.a. 0° |

Very faint and pretty large; low surface brightness. Distance 210 million ly; Hubble class Sc.

NGC 2859

Gx	LMi	09^h 24.3^m	+34° 31′	4.1′ × 3.6′
chart 6			m$_v$ 10.9	p.a. 85°

Very bright, pretty large, and round; very bright nucleus; external ring. Distance 72 million ly; Hubble class SBa.

NGC 2865

Gx	Hya	09^h 23.5^m	−23° 10′	2.5′ × 2.1′
chart 20			m$_v$ 11.4	p.a. 155°

Bright, small, and round; small, bright, diffuse, elongated nucleus. Distance 110 million ly; Hubble class E4.

NGC 2867

PN	Car	09^h 21.4^m	−58° 19′	14″
chart 25			m$_v$ 9.7	

PK278−5.1, C90. Very small and round; ring structure; phot. mag. 9.7; central star mag. about 15.0. Per *NGC*, a (!!) very remarkable object. Distance 5,500 ly; V-V class 4.

NGC 2872

Gx	Leo	09^h 25.7^m	+11° 26′	1.8′ × 1.3′
charts 12, 13			m$_v$ 11.9	p.a. 22°

Pretty faint, pretty small, and roundish; bright middle. Distance 120 million ly; Hubble class E2.

NGC 2880

Gx	UMa	09^h 29.6^m	+62° 29′	2.4′ × 1.5′
chart 2			m$_v$ 11.5	p.a. 140°

Bright, small, and round; small, very bright nucleus. Distance 70 million ly; Hubble class E3. Located among stars.

NGC 2887

Gx	Car	09^h 23.4^m	−63° 49′	1.9′ × 1.4′
chart 25			m$_p$ 12.8	p.a. 78°

Faint, small, and roundish; brighter toward middle. Distance 120 million ly; Hubble class E. A star of mag. 8 is located about 8′ to NE.

NGC 2889

Gx	Hya	09^h 27.2^m	−11° 39′	2.2′ × 1.8′
charts 12, 13			m$_v$ 11.7	p.a. 65°

Pretty faint, pretty small, and slightly elongated; very small, bright nucleus; two main filamentary arms. Distance 140 million ly; Hubble class SB+. Paired with galaxy NGC 2884.

NGC 2903

Gx	Leo	09^h 32.2^m	+21° 30′	12.6′ × 6.6′
chart 6			m$_v$ 9	p.a. 17°

Quite bright, very large, and elongated; with a very bright nucleus; an excellent example of a multi-armed spiral galaxy. One of the best galaxies missed by Messier and a fine sight in small telescopes (Mullaney). Hangs like a misty jewel S of λ Leo (Houston). Distance 20 million ly; Hubble class Sb.

NGC 2906

Gx	Leo	09^h 32.1^m	+08° 27′	1.4′ × 0.9′
charts 12, 13			m$_v$ 12.7	p.a. 75°

Faint, pretty small, and slightly elongated; gradually brighter toward middle. Distance 88 million ly; Hubble class Sc.

NGC 2907

Gx	Hya	09^h 31.6^m	−16° 44′	2.2′ × 1.4′
charts 12, 13			m$_v$ 11.7	p.a. 115°

Pretty faint, small, and slightly elongated; nearly edge on; dust lane; very small, bright nucleus. Distance 79 million ly; Hubble class Sb. Paired with an anonymous spiral galaxy to SW.

NGC 2910

OC	Vel	09^h 30.4^m	−52° 54′	5′
charts 20, 25			m$_v$ 7.2	

30 stars; detached, strong concentration of stars; moderate range in brightness; mag. of brightest star 9.3; quite large. Distance 4,200 ly; age 40 million years; Trumpler class I 2 p.

NGC 2911
Gx	Leo	09^h 33.8^m	+10° 09'	4.0' × 3.0'
charts 12, 13			m$_v$ 11.5	p.a. 140°

Faint, pretty large, and round; small, very bright nucleus; dark lane on one side. Distance 130 million ly; Hubble class E2p. Brightest in a group of galaxies.

NGC 2916
Gx	Leo	09^h 35.0^m	+21° 42'	2.5' × 1.7'
chart 6			m$_v$ 12.1	p.a. 20°

Faint, small, and slightly elongated. Distance 190 million ly; Hubble class Sb.

NGC 2921
Gx	Hya	09^h 34.5^m	−20° 55'	3.0' × 1.0'
chart 20			m$_v$ 12	p.a. 83°

Very faint, pretty small, and slightly elongated; elongated nucleus, bright inner ring; gradually brighter toward middle; faint, smooth arms; Hubble class Sa.

NGC 2925
OC	Vel	09^h 33.7^m	−53° 26'	12'
charts 20, 25			m$_p$ 8.3	

40 stars; detached, no concentration of stars; small range in brightness. Distance 2,600 ly; age 79 million years; Trumpler class III 1 p.

NGC 2935
Gx	Hya	09^h 36.7^m	−21° 08'	3.9' × 2.9'
chart 20			m$_v$ 11.1	p.a. 0°

Pretty bright, pretty small, and slightly elongated; two main arms; extremely bright nucleus. Distance 84 million ly; Hubble class S(B)b.

NGC 2947
Gx	Hya	09^h 36.1^m	−12° 26'	1.5' × 1.3'
charts 12, 13			m$_v$ 12.4	p.a. 25°

Extremely faint, pretty large, irregularly roundish; gradually brighter toward middle. Distance 110 million ly; Hubble class Sb.

NGC 2950
Gx	UMa	09^h 42.6^m	+58° 51'	2.9' × 1.9'
chart 2			m$_v$ 10.9	p.a. 145°

Bright, pretty small, and round; extremely bright nucleus. Distance 63 million ly; Hubble class Sap.

NGC 2962
Gx	Hya	09^h 40.9^m	+05° 10'	2.6' × 1.9'
charts 12, 13			m$_v$ 11.9	p.a. 3°

Faint, very small, and slightly elongated; small, very bright, diffuse nucleus. Distance 78 million ly; Hubble class Sb.

NGC 2964
Gx	Leo	09^h 42.9^m	+31° 51'	3.0' × 1.7'
chart 6			m$_v$ 11.3	p.a. 97°

Bright, very large, and slightly elongated; small, bright nucleus; two massive, branching arms. Distance 55 million ly; Hubble class Sc. Brightest in NGC 2964 galaxy group, which includes NGC 2968 and NGC 2970.

NGC 2967
Gx	Sex	09^h 42.1^m	+00° 20'	2.7' × 2.6'
charts 12, 13			m$_v$ 11.6	p.a. 65°

Pretty faint, pretty large, and round; very small, bright nucleus; two large arms. Distance 85 million ly; Hubble class Sc.

NGC 2968
Gx	Leo	09^h 43.2^m	+31° 56'	2.2' × 1.5'
chart 6			m$_v$ 11.7	p.a. 45°

Pretty bright, pretty large, and slightly elongated; small, bright nucleus with dark lane. Distance 68 million ly; Hubble class P. In NGC 2964 galaxy group along with NGC 2970.

NGC 2974
Gx	Sex	09^h 42.6^m	−03° 42'	3.4' × 2.1'
charts 12, 13			m$_v$ 10.9	p.a. 42°

Bright, quite small, irregularly round; very bright nucleus. Distance 78 million ly; Hubble class Sa.

NGC 2976

Gx	UMa	$09^h\,47.3^m$	$+67°\,55'$	$4.9' \times 2.5'$
charts 1, 2			m_v 10.2	p.a. 143°

Bright, very large, and slightly elongated; very small, bright nucleus. Distance 8 million ly; Hubble class Scp. In M81 galaxy group.

NGC 2983

Gx	Hya	$09^h\,43.7^m$	$-20°\,29'$	$2.5' \times 1.4'$
chart 20			m_v 11.8	p.a. 95°

Faint, pretty small, and round; has a very small, very bright nucleus. Distance 76 million ly; Hubble class SBa.

NGC 2985

Gx	UMa	$09^h\,50.3^m$	$+72°\,17'$	$4.7' \times 3.5'$
charts 1, 2			m_v 10.4	p.a. 0°

Very bright, quite large, and round; small, very bright, diffuse nucleus. Distance 62 million ly; Hubble class Sb. Paired with galaxy NGC 3027.

NGC 2986

Gx	Hya	$09^h\,44.3^m$	$-21°\,17'$	$3.5' \times 3.0'$
chart 20			m_v 10.6	p.a. 105°

Pretty bright, pretty small, irregularly round; very bright center. Distance 93 million ly; Hubble class E1. Paired with an anonymous spiral galaxy.

NGC 2993

Gx	Hya	$09^h\,45.8^m$	$-14°\,22'$	$1.8' \times 1.2'$
charts 12, 13			m_v 12.6	p.a. 95°

Quite faint, small, and round; very bright nucleus. Distance 80 million ly; Hubble class P. Interacting with galaxy NGC 2992, which is distinctly cigar shaped.

NGC 2997

Gx	Ant	$09^h\,45.7^m$	$-31°\,11'$	$8.1' \times 6.5'$
chart 20			m_v 9.3	p.a. 110°

Very faint and very large; has a very small, very bright, complex nucleus. Per *NGC*, a (!) remarkable object. Distance 35 million ly; Hubble class Sc. A smaller version of M101.

NGC 3001

Gx	Ant	$09^h\,46.3^m$	$-30°\,26'$	$3.0' \times 2.1'$
chart 20			m_v 11.4	p.a. 6°

Faint, small, round; thin, patchy arms. Distance 95 million ly; Hubble class Sb. Has a faint, superimposed star.

NGC 3003

Gx	LMi	$09^h\,48.5^m$	$+33°\,25'$	$5.9' \times 1.7'$
chart 6			m_v 11.9	p.a. 79°

Quite bright, large, and very elongated; very faint nucleus. Per *NGC*, a (!) remarkable object. Distance 62 million ly; Hubble class S(B)c.

NGC 3020

Gx	Leo	$09^h\,50.1^m$	$+12°\,49'$	$3.0' \times 1.7'$
charts 12, 13			m_v 11.9	p.a. 105°

Extremely faint, pretty small, and slightly elongated. Distance 57 million ly; Hubble class SBc.

NGC 3021

Gx	LMi	$09^h\,51.0^m$	$+33°\,33'$	$1.4' \times 0.9'$
chart 6			m_v 12.1	p.a. 110°

Pretty bright, pretty small, and slightly elongated; very small, very bright nucleus. Distance 65 million ly; Hubble class Sb.

NGC 3023

Gx	Sex	$09^h\,49.9^m$	$+00°\,37'$	$2.8' \times 1.6'$
charts 12, 13			m_v 12.2	p.a. 70°

Distance 73 million ly; Hubble class S(B)cp. A double system with galaxy Mrk 1236.

NGC 3027

Gx	UMa	$09^h\,55.7^m$	$+72°\,12'$	$4.3' \times 1.9'$
charts 1, 2			m_v 11.8	p.a. 130°

Very faint, very large, and slightly elongated. Distance 53 million ly; Hubble class SBd. Paired with galaxy NGC 2985.

NGC 3031

See M81.

NGC 3034

See M82.

NGC 3038

Gx	Ant	09^h 51.3^m	−32° 45′	2.6′ × 1.4′
chart 20			m$_v$ 11.7	p.a. 130°

Pretty bright, pretty small, and roundish; thin arms. Distance 100 million ly; Hubble class Sa.

NGC 3041

Gx	Leo	09^h 53.1^m	+16° 41′	3.7′ × 2.4′
charts 12, 13			m$_v$ 11.5	p.a. 95°

Faint, large, and round; a small, bright nucleus. Distance 56 million ly; Hubble class Sc.

NGC 3044

Gx	Sex	09^h 53.7^m	+01° 35′	4.7′ × 0.7′
charts 12, 13			m$_v$ 11.9	p.a. 13°

Very faint, very large, and very elongated; nearly edge on; very faint nucleus. Distance 50 million ly; Hubble class Sc.

NGC 3051

Gx	Ant	09^h 54.0^m	−27° 17′	2.0′ × 1.9′
chart 20			m$_v$ 11.8	p.a. 55°

Pretty faint, small, and roundish; gradually brighter toward middle. Distance 100 million ly; Hubble class E/S0.

NGC 3052

Gx	Hya	09^h 54.5^m	−18° 38′	2.0′ × 1.3′
charts 12, 13, 20			m$_v$ 12.2	p.a. 102°

Faint, pretty large, and round; three arms; small, bright nucleus. Distance 140 million ly; Hubble class Sc. Paired with galaxy NGC 3045.

NGC 3054

Gx	Hya	09^h 54.5^m	−25° 42′	3.8′ × 2.4′
chart 20			m$_v$ 11.5	p.a. 118°

Pretty bright, large, irregularly round; several smooth arms; small, bright nucleus. Distance 84 million ly; Hubble class Sb.

NGC 3055

Gx	Sex	09^h 55.3^m	+04° 16′	2.1′ × 1.3′
charts 12, 13			m$_v$ 12.1	p.a. 63°

Faint, pretty large, and slightly elongated; two arms; very small, very bright nucleus. Distance 70 million ly; Hubble class Sc.

NGC 3056

Gx	Ant	09^h 54.6^m	−28° 18′	2.2′ × 1.3′
chart 20			m$_v$ 11.6	p.a. 16°

Pretty bright, small and round; bright, diffuse nucleus. Distance 33 million ly; Hubble class E3. Superimposed star.

NGC 3059

Gx	Car	09^h 50.1^m	−73° 55′	3.9′
chart 25			m$_v$ 10.8	

Faint, large, irregularly round. Distance 41 million ly; Hubble class SBb.

NGC 3067

Gx	Leo	09^h 58.4^m	+32° 22′	2.3′ × 0.9′
chart 6			m$_v$ 12.1	p.a. 105°

Pretty bright, pretty large, and elongated. Distance 61 million ly; Hubble class Sb.

NGC 3077

Gx	UMa	10^h 03.4^m	+68° 44′	4.6′ × 3.6′
charts 1, 2			m$_v$ 9.8	p.a. 45°

Quite bright, quite large, and round. Distance 10 million ly; Hubble class E2p. Located in the M81 galaxy group.

NGC 3078

Gx	Hya	09^h 58.4^m	−26° 56′	2.8′ × 2.3′
chart 20			m$_v$ 11	p.a. 177°

Pretty bright, small, and round; very bright nucleus. Distance 97 million ly; Hubble class E3.

NGC 3079

Gx	UMa	10^h 02.0^m	+55° 41′	7.6′ × 1.7′
charts 2, 6			m$_v$ 10.9	p.a. 165°

Very bright, large, and very elongated; almost edge

on; appears as a long streak. Has dark lanes on one side. Distance 53 million ly; Hubble class Sb.

NGC 3081

| Gx | Hya | 09^h 59.5^m | −22° 50′ | 2.2′ × 1.7′ |
| chart 20 | | | m$_v$ 12 | p.a. 158° |

Very faint and quite small; very bright nucleus. Distance 93 million ly; Hubble class Sb.

NGC 3087

| Gx | Ant | 09^h 59.1^m | −34° 13′ | 2.4′ × 2.1′ |
| chart 20 | | | m$_v$ 11.7 | p.a. 45° |

Pretty bright, small, and round. Distance 100 million ly; Hubble class E0. Located between two stars.

NGC 3091

| Gx | Hya | 10^h 00.2^m | −19° 38′ | 2.8′ × 1.8′ |
| charts 13, 20 | | | m$_v$ 11 | p.a. 149° |

Pretty bright, pretty small, irregularly round; bright nucleus. Distance 160 million ly; Hubble class E2. Paired with galaxy NGC 3096.

NGC 3095

| Gx | Ant | 10^h 00.1^m | −31° 33′ | 3.7′ × 2.1′ |
| chart 20 | | | m$_v$ 11.8 | p.a. 126° |

Faint, large, and elongated; very small, very bright nucleus; Hubble class S(B). Paired with galaxy NGC 3100 to SE.

NGC 3098

| Gx | Leo | 10^h 02.3^m | +24° 43′ | 2.2′ × 0.6′ |
| chart 6 | | | m$_v$ 12 | p.a. 90° |

Pretty bright, small, and elongated; small, bright nucleus. Distance 55 million ly; Hubble class E7.

NGC 3100

| Gx | Ant | 10^h 00.7^m | −31° 40′ | 3.4′ × 1.8′ |
| chart 20 | | | m$_v$ 11.2 | p.a. 154° |

Very bright nucleus; dark lane near nucleus. Distance 100 million ly; Hubble class S(B)0p. Two bright superimposed stars; paired with galaxy NGC 3095 to NW.

NGC 3108

| Gx | Ant | 10^h 02.5^m | −31° 41′ | 2.8′ × 1.9′ |
| chart 20 | | | m$_v$ 11.5 | p.a. 110° |

Very bright, very large bulge; strong dark lane. Distance 110 million ly; Hubble class S0. Bright star located approx. 1.5′ to SW.

NGC 3109

| Gx | Hya | 10^h 03.1^m | −26° 10′ | 14.5′ × 3.5′ |
| chart 20 | | | m$_v$ 9.8 | p.a. 93° |

Quite faint, very large, and extremely elongated; edge on. One of my favorites, a long, spindle-shaped galaxy whose ends appear squared off (Houston). Distance 6 million ly; Hubble class Ir+.

NGC 3114

| OC | Car | 10^h 02.7^m | −60° 07′ | 34′ |
| chart 25 | | | m$_v$ 4.2 | |

Detached, weak concentration of stars; large range in brightness; mag. of brightest star 7.3; extremely large. Distance 2,900 ly; age 110 million years; Trumpler class II 3 r.

NGC 3115

| Gx | Sex | 10^h 05.2^m | −07° 43′ | 8.3′ × 3.2′ |
| chart 13 | | | m$_v$ 8.9 | p.a. 43° |

C53, Spindle Galaxy. Very bright, large, and very elongated; very bright nucleus; almost spherical nuclear bulge with an extremely thin disk. Distance 21 million ly; Hubble class E6.

NGC 3124

| Gx | Hya | 10^h 06.7^m | −19° 13′ | 2.9′ × 2.5′ |
| charts 13, 20 | | | m$_v$ 12 | p.a. 165° |

Faint, pretty large, and round; face on; two main arms; very small, bright nucleus. Distance 130 million ly; Hubble class S(B)b.

NGC 3132

| PN | Vel | 10^h 07.0^m | −40° 26′ | 30″ |
| chart 20 | | | m$_v$ 9.7 | |

PK272+12.1, C74, the Eight Burst Nebula. Very bright, very large, slightly oblong ring and smooth disk; phot. mag. 8.2; central star mag. 10.1. Per

NGC, a (!!) very remarkable object. Distance 2,600 ly; expansion velocity 13 km/sec (8 miles/sec); V-V class 4 + 2.

NGC 3136

Gx	Car	$10^h 05.8^m$	−67° 23′	3.1′ × 2.2′
chart 25			m_v 10.6	p.a. 30°

Pretty bright, pretty small, and round; bright nucleus. Distance 61 million ly; Hubble class E4. Several superimposed stars.

NGC 3136B

Gx	Car	$10^h 10.2^m$	−67° 00′	1.3′ × 0.8′
chart 25			m_v 12.3	p.a. 30°

Pretty bright lens. Distance 68 million ly; Hubble class E. Several superimposed stars.

NGC 3137

Gx	Ant	$10^h 09.1^m$	−29° 04′	6.0′ × 2.3′
chart 20			m_v 11.5	p.a. 1°

Pretty bright nucleus; many knots; Hubble class Sc. Similar to M101 but with broader, patchier arms.

NGC 3145

Gx	Hya	$10^h 10.2^m$	−12° 26′	2.9′ × 1.4′
chart 13			m_v 11.7	p.a. 20°

Faint, pretty large, and very elongated; two main arms; small, very bright nucleus; internal ring. The trick is to pick it out from the glare of 3.6-mag. λ Hya; high magnification will certainly help (Houston). Distance 160 million ly; Hubble class Sb. Paired with galaxy NGC 3143.

NGC 3147

Gx	Dra	$10^h 16.9^m$	+73° 24′	4.3′ × 3.7′
charts 1, 2			m_v 10.6	p.a. 155°

Very bright, large, and round; filamentary arms; very bright nucleus. Distance 120 million ly; Hubble class Sb.

NGC 3149

Gx	Cha	$10^h 03.8^m$	−80° 25′	1.9′
chart 25			m_p 12.6	

Faint, small, and slightly elongated; slightly brighter toward middle. Distance 79 million ly; Hubble class Sb.

NGC 3158

Gx	LMi	$10^h 13.8^m$	+38° 46′	2.3′ × 2.2′
chart 6			m_v 11.9	p.a. 160°

Bright, small, and round. Distance 300 million ly; Hubble class E2. Brightest in a group of galaxies; paired with galaxy NGC 3160.

NGC 3162

Gx	Leo	$10^h 13.5^m$	+22° 44′	3.1′
chart 6			m_v 11.6	

Quite large, pretty faint, and round; two main arms; small, very bright nucleus. Distance 59 million ly; Hubble class Sc.

NGC 3166

Gx	Sex	$10^h 13.7^m$	+03° 26′	4.7′ × 2.1′
chart 13			m_v 10.4	p.a. 87°

Bright, pretty small, and round; very small, very bright nucleus. Distance 52 million ly; Hubble class S(B)a. Brightest in galaxy group; interacting with galaxy NGC 3169.

NGC 3169

Gx	Sex	$10^h 14.2^m$	+03° 28′	4.4′ × 2.5′
chart 13			m_v 10.2	p.a. 45°

Bright, pretty large, and elongated; small, very bright nucleus; dark lane on one side. Distance 46 million ly; Hubble class Sb. Interacting with galaxy NGC 3166.

NGC 3175

Gx	Ant	$10^h 14.7^m$	−28° 52′	4.9′ × 1.4′
chart 20			m_v 11.3	p.a. 56°

Quite bright, large, and very elongated; very small, bright nucleus with dark lanes. Distance 37 million ly; Hubble class Sb.

NGC 3177

Gx	Leo	$10^h 16.6^m$	+21° 07′	1.5′ × 1.2′
chart 6			m_v 12.4	p.a. 135°

Quite faint, small, and round; two main arms; very

small, very bright nucleus. Distance 49 million ly; Hubble class Sb.

NGC 3182

Gx	UMa	$10^h 19.6^m$	+58° 12′	2.1′ × 1.7′
chart 2			m_v 12.1	p.a. 155°

Quite bright, quite large, irregularly roundish; gradually brighter toward middle. Distance 95 million ly; Hubble class Sa.

NGC 3183

Gx	Dra	$10^h 21.8^m$	+74° 11′	2.3′ × 1.4′
charts 1, 2			m_p 12.7	p.a. 170°

Faint, pretty large, and elongated; slightly brighter toward middle. Distance 140 million ly; Hubble class SBb.

NGC 3184

Gx	UMa	$10^h 18.3^m$	+41° 25′	7′ × 7′
charts 2, 6			m_v 9.8	p.a. 135°

Pretty bright, very large, and round; two main knotty arms; small, very bright nucleus. Distance 39 million ly; Hubble class Sc.

NGC 3185

Gx	Leo	$10^h 17.6^m$	+21° 41′	1.7′ × 1.0′
chart 6			m_v 12.2	p.a. 130°

Pretty faint and pretty large; small, very bright nucleus. Distance 50 million ly; Hubble class S(B)b. A member of NGC 3190 galaxy group (along with NGC 3193).

NGC 3190

Gx	Leo	$10^h 18.1^m$	+21° 50′	3.5′ × 1.4′
chart 6			m_v 11.2	p.a. 125°

Bright, pretty small, and elongated. Distance 53 million ly; Hubble class Sb. Interacting with galaxy NGC 3187; brightest in NGC 3190 galaxy group (which includes NGC 3185 and NGC 3193).

NGC 3193

Gx	Leo	$10^h 18.4^m$	+21° 54′	2.2′
chart 6			m_v 10.9	

Bright, small, and slightly elongated; very bright center. Distance 56 million ly; Hubble class E0. A member of NGC 3190 galaxy group (along with NGC 3185).

NGC 3195

PN	Cha	$10^h 09.5^m$	−80° 52′	42″
chart 25			m_v 11.6	

PK296–20.1, C109. Central star is fainter than mag. 15.3. V-V class 3.

NGC 3198

Gx	UMa	$10^h 19.9^m$	+45° 33′	8.3′ × 3.7′
charts 2, 6			m_v 10.3	p.a. 35°

Pretty bright, very large, and very elongated; several arms; small, very bright nucleus. Distance 30 million ly; Hubble class Sc.

NGC 3199

BNe	Car	$10^h 17.1^m$	−57° 55′	20′ × 15′
chart 25				

Very bright, very large, and ring shaped. The brightest part forms a crescent; many stars involved in nebulosity. Per *NGC*, a (!) remarkable object.

NGC 3200

Gx	Hya	$10^h 18.6^m$	−17° 59′	4.3′ × 1.3′
charts 13, 20			m_v 12.2	p.a. 169°

Pretty bright and elongated; very small, bright nucleus. Distance 140 million ly; Hubble class Sb. Several faint galaxies nearby.

NGC 3201

GC	Vel	$10^h 17.6^m$	−46° 25′	20′
chart 20			m_v 6.9	

C79. Low concentration of stars; very large and irregularly round. Distance 16,300 ly; S-S class 10.

NGC 3206

Gx	UMa	$10^h 21.8^m$	+56° 56′	2.9′ × 1.9′
chart 2			m_v 11.9	p.a. 0°

Pretty bright, quite large, and elongated. Distance 54 million ly; Hubble class SBc.

NGC 3211

PN	Car	10^h 17.8^m	−62° 40′	16″
chart 25			m$_v$ 10.7	

PK286−4.1. Small and round; phot. mag. 11.8; central star mag. 15.5; V-V class 2b. Located among 150 stars.

NGC 3223

Gx	Ant	10^h 21.6^m	−34° 15′	4.0′ × 2.6′
chart 20			m$_v$ 11	p.a. 135°

Pretty bright, very large, and slightly elongated; patchy arms; bright, diffuse nucleus. Distance 110 million ly; Hubble class Sb.

NGC 3226

Gx	Leo	10^h 23.5^m	+19° 54′	2.6′ × 2.3′
charts 6, 13			m$_v$ 11.4	p.a. 15°

Pretty bright, quite large, and round; bright, diffuse nucleus. A large, very faint loop extends from nearby Seyfert galaxy NGC 3227 to this object. Distance 55 million ly; Hubble class E2.

NGC 3227

Gx	Leo	10^h 23.5^m	+19° 52′	5.6′ × 4.0′
charts 6, 13			m$_v$ 10.3	p.a. 155°

Pretty bright, quite large, and round; a Seyfert galaxy; two main arms; very small, extremely bright nucleus. A large, very faint loop extends to nearby galaxy NGC 3226. Distance 46 million ly; Hubble class Sb.

NGC 3228

OC	Vel	10^h 21.8^m	−51° 43′	18′
charts 20, 25			m$_v$ 6	

15 stars; detached, strong concentration of stars; small range in brightness; mag. of brightest star 7.9; large. Distance 1,600 ly; age 42 million years; Trumpler class I 1 p.

NGC 3239

Gx	Leo	10^h 25.1^m	+17° 10′	4.5′ × 2.4′
charts 6, 13			m$_v$ 11.3	p.a. 87°

Irregular shape; many knots; bright star superimposed. Distance 29 million ly; Hubble class Irp+.

NGC 3241

Gx	Ant	10^h 24.3^m	−32° 29′	2.3′ × 1.6′
chart 20			m$_v$ 12.1	p.a. 123°

Faint and quite elongated; thin arms; small, bright nucleus. Distance 110 million ly; Hubble class Sb.

NGC 3242

PN	Hya	10^h 24.8^m	−18° 38′	25″
charts 13, 20			m$_v$ 7.8	

PK261+32.1, C59, also the Ghost of Jupiter Nebula or the CBS Eye Nebula. Very bright, slightly elongated, pale-green ring; phot. mag. 8.6. Like a bluish egg (Copeland). Central star mag. 12.3, but detecting it is tough at all magnifications because of the nebula's brightness (Mullaney). Expansion velocity 23 km/sec (14 miles/sec); distance 2,600 ly; V-V class 4 + 3b.

NGC 3244

Gx	Ant	10^h 25.5^m	−39° 50′	2.0′ × 1.5′
chart 20			m$_v$ 12.4	p.a. 170°

Very faint, with faint outer arms, many knots. Distance 110 million ly; Hubble class Sc.

NGC 3245

Gx	LMi	10^h 27.3^m	+28° 30′	3.3′ × 2.2′
chart 6			m$_v$ 10.8	p.a. 177°

Very bright, pretty large, and elongated; extremely bright nucleus. Distance 52 million ly; Hubble class E5. Paired with galaxy NGC 3245A.

NGC 3247

C+BNe	Car	10^h 25.9^m	−57° 56′	7′
chart 25			m$_v$ 7.6	

20 stars; detached, weak concentration; moderate brightness range; mag. of brightest star 10.0; involved in mottled nebulosity. Distance 4,600 ly; age 50 million years; Trumpler class II 2 p n. Several detached, smaller nebulae in outlying areas.

NGC 3250

Gx	Ant	10^h 26.5^m	−39° 57′	2.6′ × 1.9′
chart 20			m$_v$ 11.3	p.a. 148°

Pretty bright, pretty large, and round; bright nucleus. Distance 110 million ly; Hubble class E4. In a group of five faint galaxies.

NGC 3254

| Gx | LMi | 10^h 29.3^m | +29° 29′ | 4.8′ × 1.5′ |
| chart 6 | | | m$_v$ 11.7 | p.a. 46° |

Quite bright, large, and very elongated; with a small, very bright nucleus. Distance 51 million ly; Hubble class Sb.

NGC 3256

| Gx | Vel | 10^h 27.9^m | −43° 54′ | 5.0′ × 2.6′ |
| chart 20 | | | m$_v$ 10.8 | p.a. 100° |

Quite bright, small, and roundish; a colliding system that includes two or three nuclei in contact, forming a very bright mass. Distance 110 million ly; Hubble class P.

NGC 3258

| Gx | Ant | 10^h 28.9^m | −35° 36′ | 3.0′ × 2.6′ |
| chart 20 | | | m$_v$ 11.5 | p.a. 75° |

Quite faint, small, and round; small, bright nucleus. Distance 110 million ly; Hubble class E1. In a group with galaxies NGC 3268, NGC 3271, and NGC 3281.

NGC 3259

| Gx | UMa | 10^h 32.6^m | +65° 02′ | 2.2′ × 1.1′ |
| chart 2 | | | m$_v$ 12.1 | p.a. 20° |

Faint, small, and round; bright, very small nucleus. Distance 81 million ly; Hubble class Irp+. Paired with galaxy NGC 3266.

NGC 3261

| Gx | Vel | 10^h 29.0^m | −44° 39′ | 3.7′ × 3.0′ |
| chart 20 | | | m$_v$ 11.2 | p.a. 85° |

Faint, small, and round; two main arms; very bright nucleus. Distance 99 million ly; Hubble class SBb. Located among stars.

NGC 3263

| Gx | Vel | 10^h 29.2^m | −44° 06′ | 6.0′ × 1.4′ |
| chart 20 | | | m$_v$ 10.8 | p.a. 97° |

Faint, small, and elongated; suddenly brighter toward middle, very patchy, distorted arms. Distance 120 million ly; Hubble class SBb.

NGC 3264

| Gx | UMa | 10^h 32.3^m | +56° 05′ | 3.0′ × 1.3′ |
| charts 2, 6 | | | m$_v$ 12 | p.a. 177° |

Extremely faint. Distance 45 million ly; Hubble class SBm. Located between two stars.

NGC 3268

| Gx | Ant | 10^h 30.0^m | −35° 20′ | 3.5′ × 2.6′ |
| chart 20 | | | m$_v$ 11.6 | p.a. 71° |

Faint, small, and round. Distance 110 million ly; Hubble class E2. In a group with galaxies NGC 3258, NGC 3271, and NGC 3281.

NGC 3271

| Gx | Ant | 10^h 30.4^m | −35° 22′ | 2.9′ × 1.4′ |
| chart 20 | | | m$_v$ 11.7 | p.a. 106° |

Pretty faint, small, and elongated; very bright, elongated nucleus. Distance 150 million ly; Hubble class SB0. In a group with galaxies NGC 3258, NGC 3268, and NGC 3281.

NGC 3275

| Gx | Ant | 10^h 30.9^m | −36° 44′ | 2.8′ × 2.2′ |
| chart 20 | | | m$_v$ 11.6 | p.a. 120° |

Faint, large, and slightly elongated; small, bright nucleus. Distance 130 million ly; Hubble class SBb. Paired with galaxy NGC 3275A.

NGC 3277

| Gx | LMi | 10^h 32.9^m | +28° 31′ | 2.1′ × 1.8′ |
| chart 6 | | | m$_v$ 11.7 | p.a. 170° |

Quite bright, considerably small and round; very small, very bright nucleus. Distance 61 million ly; Hubble class Sb.

NGC 3281

| Gx | Ant | 10^h 31.9^m | −34° 51′ | 3.2′ × 1.6′ |
| chart 20 | | | m$_v$ 11.7 | p.a. 140° |

Extremely faint, pretty large, and elongated. Distance 140 million ly; Hubble class Sb. In a group with the galaxies NGC 3258, NGC 3268, and NGC 3271.

NGC 3285

Gx	Hya	10^h 33.6^m	−27° 27′	2.8′ × 1.6′
chart 20			m$_v$ 12	p.a. 108°

Pretty bright, small, and slightly elongated; inner arms; small, extremely bright nucleus; Hubble class S. In a group with two other galaxies, NGC 3285A and NGC 3285B.

NGC 3287

Gx	Leo	10^h 34.8^m	+21° 39′	2.0′ × 0.9′
chart 6			m$_v$ 12.3	p.a. 20°

Faint and pretty large; two main arms; with a faint, very small nucleus. Distance 46 million ly; Hubble class S.

NGC 3293

C+BNer	Car	10^h 35.8^m	−58° 14′	40′
chart 25			m$_v$ 4.7	

Detached, strong concentration of stars in a 6′ region; large range in brightness; mag. of brightest star 6.5. Distance 8,500 ly; age 10 million years; Trumpler class I 3 r. Large nebular patch located at the S end.

NGC 3294

Gx	LMi	10^h 36.3^m	+37° 19′	3.5′ × 1.8′
charts 2, 6			m$_v$ 11.8	p.a. 122°

Quite bright, large, and elongated; very small, diffuse, bright nucleus. Distance 66 million ly; Hubble class S(B)b. Paired with galaxy NGC 3304.

NGC 3301

Gx	Leo	10^h 36.9^m	+21° 53′	3.6′ × 1.2′
chart 6			m$_v$ 11.4	p.a. 52°

Quite bright, small, and slightly elongated; extremely bright nucleus. Distance 54 million ly; Hubble class S0.

NGC 3309

Gx	Hya	10^h 36.6^m	−27° 31′	2.5′ × 2.1′
chart 20			m$_v$ 11	p.a. 40°

Bright, large, and round. Distance 160 million ly; Hubble class E0. Brightest in Hydra I Galaxy Cluster (Abell 1060); paired with galaxy NGC 3311.

NGC 3310

Gx	UMa	10^h 38.8^m	+53° 30′	2.8′
charts 2, 6			m$_v$ 10.8	

Quite bright, pretty large, and round; filamentary arms; very small, very bright nucleus. Distance 46 million ly; Hubble class S(B)bp.

NGC 3311

Gx	Hya	10^h 36.7^m	−27° 32′	3.6′
chart 20			m$_v$ 10.9	p.a. 19°

Bright, large, and round. Distance 140 million ly; Hubble class S0. Paired with galaxy NGC 3309.

NGC 3312

Gx	Hya	10^h 37.0^m	−27° 34′	3.2′ × 1.2′
chart 20			m$_v$ 11.8	p.a. 175°

Quite faint and elongated; twisting arms; extremely bright nucleus; dark lane. Distance 110 million ly; Hubble class Sbp.

NGC 3313

Gx	Hya	10^h 37.4^m	−25° 19′	4.3′ × 3.5′
chart 20			m$_v$ 11.6	p.a. 55°

Extremely faint, pretty small, irregularly roundish; gradually brighter toward middle. Distance 150 million ly; Hubble class SBb.

NGC 3318

Gx	Vel	10^h 37.3^m	−41° 38′	2.3′ × 1.2′
chart 20			m$_v$ 12	p.a. 78°

Quite faint, pretty large, and quite elongated; very bright nucleus. Distance 110 million ly; Hubble class S(B)b. Paired with galaxy NGC 3318B.

NGC 3319

Gx	UMa	10^h 39.2^m	+41° 41′	6.8′ × 3.9′
charts 2, 6			m$_v$ 11.1	p.a. 37°

Quite faint, large, and elongated. Distance 33 million ly; Hubble class S(B)c.

NGC 3320

Gx	UMa	$10^h 39.6^m$	$+47° 24'$	$2.2' \times 1.0'$
charts 2, 6			m_v 12.3	p.a. 20°

Faint, pretty small, and elongated; small, bright nucleus; filamentary, knotty arms. Distance 100 million ly; Hubble class Sc.

NGC 3324

BNer+C	Car	$10^h 37.3^m$	$-58° 38'$	$15' \times 13'$
chart 25				

Large, bright emission nebula containing a small (6′) open cluster; detached, strong concentration of stars; large range in brightness; mag. of brightest star 8.2. Pretty bright, large; total mag. of cluster 6.7. Distance 10,800 ly; age 2.2 million years; Trumpler class I 3 r n.

NGC 3329

Gx	Dra	$10^h 44.7^m$	$+76° 49'$	$1.9' \times 1.1'$
charts 1, 2			m_v 12.2	p.a. 140°

Small and slightly elongated; very bright nucleus. Distance 81 million ly; Hubble class S. Brightest in a group of 12 galaxies.

NGC 3330

OC	Vel	$10^h 38.6^m$	$-54° 09'$	7′
charts 20, 25			m_p 7.4	

30 stars; detached, no concentration; moderate brightness range; mag. of brightest star 8.8. Distance 4,500 ly; age 50 million years; Trumpler class II 2 p.

NGC 3336

Gx	Hya	$10^h 40.3^m$	$-27° 47'$	$1.9' \times 1.6'$
chart 20			m_v 12.3	p.a. 123°

Small, bright nucleus; very patchy arms with bright knots embedded in them. Distance 160 million ly; Hubble class SBc.

NGC 3338

Gx	Leo	$10^h 42.1^m$	$+13° 45'$	$5.5' \times 3.7'$
chart 13			m_v 11.1	p.a. 100°

Faint, quite large, and elongated; two main arms; small, very bright nucleus. Distance 52 million ly; Hubble class Sb. Member of the Leo galaxy group.

NGC 3344

Gx	LMi	$10^h 43.5^m$	$+24° 55'$	$6.9' \times 6.5'$
chart 6			m_v 9.9	

Quite bright and large; three arms; very small, bright nucleus. Distance 22 million ly; Hubble class Sc. A wide 9th-mag. double star just E of the galaxy's center interferes with the view (Houston). A fainter foreground star, even closer to the nucleus, is not a supernova.

NGC 3346

Gx	Leo	$10^h 43.6^m$	$+14° 52'$	$2.7' \times 2.5'$
chart 13			m_v 11.7	p.a. 90°

Quite faint, very large, and round. Distance 49 million ly; Hubble class Sc.

NGC 3347

Gx	Ant	$10^h 42.8^m$	$-36° 21'$	$3.3' \times 2.0'$
chart 20			m_v 11.4	p.a. 173°

Pretty faint, small, and elongated; thin arms, small, very bright nucleus. Distance 120 million ly; Hubble class SBb. In a group with galaxies NGC 3354 and NGC 3358.

NGC 3348

Gx	UMa	$10^v 47.2^m$	$+72° 50'$	2.0′
charts 1, 2			m_v 11.2	

Bright and small. Distance 130 million ly; Hubble class E1.

NGC 3351

See M95.

NGC 3358

Gx	Ant	$10^h 43.6^m$	$-36° 25'$	$3.2' \times 1.9'$
chart 20			m_v 11.5	p.a. 141°

Quite faint, very small, and slightly elongated; very bright nucleus. Distance 110 million ly; Hubble class S(B)a. In a group with galaxies NGC 3347 and NGC 3354.

NGC 3359

Gx UMa $10^h 46.6^m$ +63° 13' 6.8' × 4.3'
chart 2 m_v 10.6 p.a. 170°

Pretty bright, large, and elongated; very small, faint nucleus. Distance 49 million ly; Hubble class S(B)c.

NGC 3366

Gx Vel $10^h 35.1^m$ −43° 42' 2.2' × 1.0'
chart 20 m_v 12 p.a. 37°

Faint and elongated; gradually brighter toward middle. Distance 120 million ly; Hubble class S(B)0. Stars of mag. 6 or 7 very nearby.

NGC 3367

Gx Leo $10^h 46.6^m$ +13° 45' 2.5'
chart 13 m_v 11.5

Pretty bright, quite large, irregularly round; three arms; very small, extremely bright nucleus. Distance 130 million ly; Hubble class Sc.

NGC 3368

See M96.

NGC 3370

Gx Leo $10^h 47.1^m$ +17° 16' 3.0' × 1.7'
charts 6, 13 m_v 11.6 p.a. 148°

Quite bright, pretty large, and slightly elongated; small, bright nucleus. Distance 53 million ly; Hubble class Sc.

NGC 3372

See Eta Carinae Nebula.

NGC 3377

Gx Leo $10^h 47.7^m$ +13° 59' 4.4' × 2.6'
chart 13 m_v 10.4 p.a. 35°

Very bright, quite large, and slightly elongated; very bright center. Distance 26 million ly; Hubble class E5. Paired with galaxy NGC 3377A.

NGC 3379

See M105.

NGC 3381

Gx LMi $10^h 48.4^m$ +34° 43' 2.0'
charts 2, 6 m_v 11.7

Pretty faint, quite large, irregularly roundish; gradually brighter toward middle. Distance 70 million ly; Hubble class SBp.

NGC 3384

Gx Leo $10^h 48.3^m$ +12° 38' 5.9' × 2.6'
chart 13 m_v 9.9 p.a. 53°

Very bright, large, and round; extremely bright nucleus. Distance 28 million ly; Hubble class E7. Paired with galaxy M105; a member of the Leo galaxy group.

NGC 3389

Gx Leo $10^h 48.5^m$ +12° 32' 2.8' × 1.3'
chart 13 m_v 11.9 p.a. 112°

Faint, large, and elongated; very small, bright nucleus. Distance 50 million ly; Hubble class Sc.

NGC 3390

Gx Hya $10^h 48.1^m$ −31° 32' 3.5' × 0.6'
chart 20 m_v 12.4 p.a. 177°

Faint, small, and very elongated; nearly edge on; has a dark lane. Distance 110 million ly; Hubble class Sb.

NGC 3395

Gx LMi $10^h 49.8^m$ +32° 59' 1.7' × 0.9'
chart 6 m_v 12.1 p.a. 50°

Quite bright, pretty small, and slightly elongated; two main arms; very small, very bright nucleus. Distance 69 million ly; Hubble class Sc. Interacting with galaxy NGC 3396.

NGC 3396

Gx LMi $10^h 49.9^m$ +32° 59' 2.7' × 1.1'
chart 6 m_v 12.1 p.a. 100°

Pretty bright, pretty small, and slightly elongated. Distance 72 million ly; Hubble class P. Interacting with galaxy NGC 3395.

NGC 3411

Gx	Hya	$10^h 50.4^m$	$-12° 51'$	2.0'
chart 13			m_v 11.9	

Faint, small, and roundish; slightly brighter toward middle. Distance 190 million ly; Hubble class E.

NGC 3412

Gx	Leo	$10^h 50.9^m$	$+13° 25'$	$3.7' \times 2.2'$
chart 13			m_v 10.5	p.a. 155°

Bright, small, and slightly elongated; very bright nucleus. Distance 32 million ly; Hubble class E5.

NGC 3413

Gx	LMi	$10^h 51.4^m$	$+32° 46'$	$1.8' \times 0.8'$
chart 6			m_v 12.1	p.a. 178°

Faint and small. Distance 27 million ly; Hubble class S0. In a group with galaxies NGC 3430 and NGC 3424.

NGC 3414

Gx	LMi	$10^h 51.3^m$	$+27° 59'$	$3.4' \times 2.8'$
chart 6			m_v 11	p.a. 20°

Bright, pretty large, and round. Distance 61 million ly; Hubble class SBa. Brightest in a group of galaxies.

NGC 3423

Gx	Sex	$10^h 51.2^m$	$+05° 51'$	$3.9' \times 3.3'$
chart 13			m_v 11.1	p.a. 10°

Faint, very large, round; face on; several filamentary arms; very small, bright nucleus. Distance 37 million ly; Hubble class Sc.

NGC 3430

Gx	LMi	$10^h 52.2^m$	$+32° 57'$	$4.1' \times 2.2'$
chart 6			m_v 11.6	p.a. 30°

Pretty bright, large, and slightly elongated; filamentary arms; small, bright nucleus. Distance 67 million ly; Hubble class Sc. In a group with galaxies NGC 3413 and NGC 3424.

NGC 3432

Gx	LMi	$10^h 52.5^m$	$+36° 37'$	$6.2' \times 1.5'$
charts 2,6			m_v 11.2	p.a. 38°

Pretty bright, pretty large, and very elongated; nearly edge on; dwarf companion at one end. Distance 27 million ly; Hubble class SBm. Double star nearby.

NGC 3433

Gx	Leo	$10^h 52.1^m$	$+10° 09'$	$3.7' \times 3.3'$
chart 13			m_v 11.6	p.a. 50°

Very faint, very large, and round; two main arms; small, bright nucleus. Distance 110 million ly; Hubble class Sb. String of galaxies nearby.

NGC 3437

Gx	Leo	$10^h 52.6^m$	$+22° 56'$	$2.6' \times 0.8'$
chart 6			m_v 12.1	p.a. 122°

Pretty bright, pretty large, and slightly elongated; small, bright nucleus. Distance 45 million ly; Hubble class Sb.

NGC 3445

Gx	UMa	$10^h 54.6^m$	$+56° 59'$	1.5'
chart 2			m_v 12.6	

Quite bright, pretty large, and round; small, bright nucleus. Distance 90 million ly; Hubble class Sc. In a group with faint galaxies NGC 3440 and NGC 3458.

NGC 3448

Gx	UMa	$10^h 54.7^m$	$+54° 18'$	$5.4' \times 1.9'$
charts 2,6			m_v 12.1	p.a. 65°

Bright, pretty large, and very elongated; edge on. Distance 64 million ly; Hubble class P.

NGC 3449

Gx	Ant	$10^h 52.9^m$	$-32° 56'$	$3.3' \times 1.0'$
chart 20			m_v 12.1	p.a. 148°

Faint, small, and round. Distance 130 million ly; Hubble class Sa. Several faint galaxies nearby.

NGC 3450

Gx	Hya	$10^h 48.1^m$	$-20° 51'$	$2.6' \times 2.3'$
chart 20			m_v 11.9	p.a. 140°

Very faint, large, and roundish; very gradually brighter toward middle. Distance 170 million ly; Hubble class SBb.

NGC 3457
Gx	Leo	10^h 54.8^m	+17° 37'	0.9'
charts 6, 13			m$_v$ 12.6	

Distance 47 million ly; Hubble class S.

NGC 3464
Gx	Hya	10^h 54.7^m	−21° 04'	2.6' × 1.7'
chart 20			m$_v$ 12.5	p.a. 112°

Extremely faint, pretty large, and elongated; patchy arms; small, bright nucleus. Distance 160 million ly; Hubble class Sb.

NGC 3485
Gx	Leo	11^h 00.0^m	+14° 51'	2.3' × 2.2'
chart 13			m$_v$ 11.8	p.a. 100°

Faint, large, and round; several filamentary arms; small, very bright nucleus. Distance 59 million ly; Hubble class S(B)b.

NGC 3486
Gx	LMi	11^h 00.4^m	+28° 59'	6.9' × 5.4'
chart 6			m$_v$ 10.5	p.a. 80°

Quite bright, quite large, and round; many arms; bright, diffuse nucleus. Distance 29 million ly; Hubble class Sc.

NGC 3489
Gx	Leo	11^h 00.3^m	+13° 54'	3.5' × 2.2'
chart 13			m$_v$ 10.3	p.a. 70°

Very bright, pretty large, and slightly elongated; very small, very bright nucleus. Distance 25 million ly; Hubble class E6.

NGC 3495
Gx	Leo	11^h 01.3^m	+03° 38'	4.9' × 1.1'
chart 13			m$_v$ 11.8	p.a. 20°

Very faint, pretty large, and very elongated; nearly edge on; many arms and dark lanes visible; faint, very small nucleus. Distance 36 million ly; Hubble class Sb.

NGC 3496
OC	Car	10^h 59.8^m	−60° 20'	8'
chart 25			m$_v$ 8.2	

60 stars; detached, no concentration; small brightness range; mag. of brightest star 11.8. Distance 3,600 ly; age 230 million years; Trumpler class III 1 m.

NGC 3503
BNer	Car	11^h 01.3^m	−59° 51'	3'
chart 25				

Three 10th-mag. stars involved in a small, faint patch of nebulosity; surrounding field is rich in faint stars.

NGC 3504
Gx	LMi	11^h 03.2^m	+27° 58'	2.6' × 2.4'
chart 6			m$_v$ 10.9	p.a. 150°

Bright, large, and elongated; extremely bright nucleus with dark lanes. Distance 64 million ly; Hubble class Sb. In a group with galaxies NGC 3512 and NGC 3515.

NGC 3507
Gx	Leo	11^h 03.4^m	+18° 08'	3.3' × 2.9'
charts 6, 13			m$_v$ 10.9	p.a. 110°

Quite faint, pretty large, and roundish; bright nucleus. Distance 38 million ly; Hubble class SBb.

NGC 3510
Gx	LMi	11^h 03.7^m	+28° 53'	4.1' × 0.8'
chart 6			m$_v$ 12.2	p.a. 163°

Faint, large, and very elongated. Distance 29 million ly; Hubble class SBm.

NGC 3511
Gx	Crt	11^h 03.4^m	−23° 05'	5.4' × 2.2'
chart 20			m$_v$ 11	p.a. 76°

Very faint, very large, and very elongated; strong arms; faint, extremely small nucleus. Distance 42 million ly; Hubble class Sc. Paired with more compact galaxy NGC 3513.

NGC 3512
Gx	LMi	11^h 04.0^m	+28° 02'	1.7'
chart 6			m$_v$ 12.3	

Faint, pretty small, and round; two main arms; very bright, small nucleus. Distance 59 million ly;

Hubble class Sc. In a group with galaxies NGC 3504 and NGC 3515.

NGC 3513

Gx	Crt	$11^h 03.8^m$	$-23° 15'$	$3.0' \times 2.4'$
chart 20			m_v 11.5	p.a. 75°

Very faint and very large; an S-shaped spiral; two strong arms; small, extremely bright nucleus. Distance 41 million ly; Hubble class S(B)c. Paired with galaxy NGC 3511.

NGC 3516

Gx	UMa	$11^h 06.8^m$	$+72° 34'$	$1.9' \times 1.6'$
charts 1, 2			m_v 11.7	p.a. 60°

Pretty bright and very small; a Seyfert galaxy with extremely bright nucleus. Appears to contain a black hole into which matter is being drawn. Distance 120 million ly; Hubble class SB0.

NGC 3521

Gx	Leo	$11^h 05.8^m$	$-00° 02'$	$9.5' \times 5.0'$
chart 13			m_v 9	p.a. 163°

Quite bright, quite large, and elongated; many arms; very small, very bright nucleus. Distance 28 million ly; Hubble class Sb.

NGC 3528

Gx	Crt	$11^h 07.3^m$	$-19° 28'$	$2.7' \times 1.5'$
charts 13, 20			m_v 11.8	p.a. 59°

Faint, small, and round; bright nucleus; Hubble class S0.

NGC 3532

OC	Car	$11^h 06.4^m$	$-58° 40'$	50'
chart 25			m_v 3	

C91. 150 stars; detached, weak concentration of stars; small range in brightness; mag. of brightest star 7.1; extremely large, round. Distance 1,300 ly; age 270 million years; Trumpler class II 1 m.

NGC 3549

Gx	UMa	$11^h 10.9^m$	$+53° 23'$	$3.2' \times 1.0'$
charts 2, 6			m_v 12.1	p.a. 38°

Quite bright, quite large, and elongated; two main arms; very small, bright nucleus. Distance 130 million ly; Hubble class Sb.

NGC 3556

See M108.

NGC 3557

Gx	Cen	$11^h 10.0^m$	$-37° 32'$	$4.0' \times 3.1'$
chart 20			m_v 10.5	p.a. 30°

Bright, small, and round; bright nucleus. Distance 120 million ly; Hubble class E3. In a group with galaxies NGC 3564 and NGC 3568.

NGC 3568

Gx	Cen	$11^h 10.8^m$	$-37° 27'$	$2.5' \times 0.9'$
chart 20			m_v 12.3	p.a. 7°

No bright nucleus, but a bright knot off center; very patchy structure; Hubble class SBc. Four pretty bright stars to E; in a group with galaxies NGC 3564 and NGC 3557.

NGC 3572

OC	Car	$11^h 10.4^m$	$-60° 14'$	7'
chart 25			m_v 6.6	

35 stars; detached, strong concentration of stars; moderate range in brightness; mag. of brightest star 7.9. Distance 7,500 ly; age 13 million years; involved in a faint, large (20') emission nebula; Trumpler class I 2 m n.

NGC 3576, 79

BNe	Car	$11^h 12.0^m$	$-61° 12'$	$19' \times 15'$
chart 25				

Forms three knots along with nebula NGC 3603.

NGC 3583

Gx	UMa	$11^h 14.2^m$	$+48° 19'$	$2.5' \times 1.7'$
charts 2, 6, 7			m_v 11.1	p.a. 125°

Pretty bright, pretty large, and round; three arms; very small, very bright nucleus; dark lanes. Distance 96 million ly; Hubble class Sc. Paired with galaxy NGC 3577.

NGC 3585

| Gx | Hya | 11^h 13.3^m | −26° 45′ | 2.9′ × 1.6′ |
| chart 20 | | | m$_v$ 9.7 | p.a. 107° |

Bright, pretty large, and elongated. Distance 54 million ly; Hubble class E5. A string of five very faint galaxies is nearby.

NGC 3587

See Owl Nebula.

NGC 3590

| OC | Car | 11^h 12.9^m | −60° 47′ | 4′ |
| chart 25 | | | m$_v$ 8.2 | |

25 stars; detached, weak concentration of stars; small range in brightness; mag. of brightest star 10.3; slightly elongated. Distance 6,200 ly; age 50 million years; Trumpler class II 1 p.

NGC 3593

| Gx | Leo | 11^h 14.6^m | +12° 49′ | 5.8′ × 2.5′ |
| chart 13 | | | m$_v$ 10.4 | p.a. 92° |

Bright, quite large, and elongated; bright nucleus; large bulge and dark lane. Distance 24 million ly; Hubble class Sb.

NGC 3595

| Gx | UMa | 11^h 15.4^m | +47° 27′ | 1.6′ × 0.8′ |
| charts 2, 6, 7 | | | m$_v$ 12.1 | p.a. 176° |

Very faint, very small, and slightly elongated; appears stellar. Distance 100 million ly; Hubble class E.

NGC 3596

| Gx | Leo | 11^h 15.1^m | +14° 47′ | 4.1′ |
| chart 13 | | | m$_v$ 11.2 | |

Pretty faint, large, and round; two strong arms; small, very bright nucleus. Distance 45 million ly; Hubble class Sc.

NGC 3599

| Gx | Leo | 11^h 15.5^m | +18° 07′ | 2.5′ × 1.8′ |
| charts 6, 13 | | | m$_v$ 11.9 | p.a. 95° |

Bright, pretty small, and roundish; bright nucleus; Hubble class S0. In a group with galaxies NGC 3607 and NGC 3608.

NGC 3600

| Gx | UMa | 11^h 15.9^m | +41° 35′ | 4.1′ × 0.8′ |
| charts 2, 6 | | | m$_v$ 11.7 | p.a. 3° |

Pretty faint, small, and slightly elongated; bright nucleus. Distance 32 million ly; Hubble class Sa.

NGC 3603

| BN | Car | 11^h 15.0^m | −61° 12′ | 10′ |
| chart 25 | | | | |

Forms three knots along with nebulae NGC 3576 and NGC 3579.

NGC 3607

| Gx | Leo | 11^h 16.9^m | +18° 03′ | 4.6′ × 4.1′ |
| charts 6, 13 | | | m$_v$ 9.9 | p.a. 120° |

Very bright, large, and round; small, very bright nucleus. Distance 37 million ly; Hubble class E1. Brightest in a group of galaxies that includes NGC 3599 and NGC 3608.

NGC 3608

| Gx | Leo | 11^h 17.0^m | +18° 09′ | 3.2′ × 2.5′ |
| charts 6, 13 | | | m$_v$ 10.8 | p.a. 75° |

Bright, pretty large, and round; small, bright, diffuse nucleus. Distance 49 million ly; Hubble class E3. In a group of galaxies that includes NGC 3599 and NGC 3607.

NGC 3610

| Gx | UMa | 11^h 18.4^m | +58° 47′ | 2.7′ × 2.6′ |
| chart 2 | | | m$_v$ 10.8 | p.a. 130° |

Very bright, pretty small, and slightly elongated; very small, very bright nucleus. Distance 81 million ly; Hubble class E2.

NGC 3611

| Gx | Leo | 11^h 17.5^m | +04° 33′ | 2.2′ × 1.9′ |
| chart 13 | | | m$_v$ 12.2 | p.a. 20° |

Pretty faint, quite small, irregularly round; small, very bright nucleus. Distance 70 million ly; Hubble class Sa.

NGC 3613

Gx	UMa	11^h 18.6^m	+58° 00′	3.5′ × 1.9′
chart 2			m$_v$ 10.9	p.a. 102°

Very bright, quite large, and elongated; diffuse, small, bright nucleus. Distance 94 million ly; Hubble class E5. In a group with galaxies NGC 3619 and NGC 3625.

NGC 3614

Gx	UMa	11^h 18.3^m	+45° 45′	4.8′ × 2.9′
charts 2, 6, 7			m$_v$ 11.6	p.a. 80°

Faint, pretty large, and slightly elongated; two main arms; small, bright nucleus. Distance 100 million ly; Hubble class Sc. Paired with galaxy NGC 3614A.

NGC 3614A

Gx	UMa	11^h 18.3^m	+45° 43′	0.7′ × 0.3′
charts 2, 6, 7			m$_v$ 11.9	p.a. 98°

Diffuse and patchy; a dwarf galaxy; Hubble class SBm. Paired with galaxy NGC 3614.

NGC 3619

Gx	UMa	11^h 19.4^m	+57° 46′	3.5′
chart 2			m$_v$ 11.5	

Quite bright and quite large; very small, very bright nucleus. Distance 76 million ly; Hubble class Sa. In a group with galaxies NGC 3613 and NGC 3625.

NGC 3621

Gx	Hya	11^h 18.3^m	−32° 49′	10.0′ × 6.5′
chart 20			m$_v$ 8.9	p.a. 159°

Quite bright, very large, and elongated; knotty arms; very small, bright nucleus. Distance 20 million ly; Hubble class Sc.

NGC 3623

See M65.

NGC 3626

Gx	Leo	11^h 20.1^m	+18° 21′	3.2′ × 2.4′
charts 6, 13			m$_v$ 11	p.a. 157°

C40. Bright, small, and slightly elongated; small, extremely bright nucleus. Distance 59 million ly; Hubble class Sb.

NGC 3627

See M66.

NGC 3628

Gx	Leo	11^h 20.3^m	+13° 35′	14.8′ × 3.6′
chart 13			m$_v$ 9.5	p.a. 104°

A member of Leo's Triplet (along with galaxies M65 and M66); very large, pretty bright, and very elongated; edge on. Distance 32 million ly; Hubble class Sb.

NGC 3629

Gx	Leo	11^h 20.5^m	+26° 58′	2.0′ × 1.5′
chart 6			m$_v$ 12.1	p.a. 30°

Quite faint, large, and round; many arms; very small, bright nucleus. Distance 63 million ly; Hubble class Sc.

NGC 3630

Gx	Leo	11^h 20.3^m	+02° 58′	1.9′ × 0.8′
chart 13			m$_v$ 11.9	p.a. 37°

Pretty bright, small, and round; very bright nucleus. Distance 60 million ly; Hubble class E7. In a group with galaxies NGC 3640, NGC 3645, and others.

NGC 3631

Gx	UMa	11^h 21.0^m	+53° 10′	4.6′ × 4.1′
charts 2, 6			m$_v$ 10.4	p.a. 90°

Pretty bright, large, and round; two straight, large arms, with an absorption lane crossing one of them; small, bright nucleus. Distance 54 million ly; Hubble class Sc.

NGC 3640

Gx	Leo	11^h 21.1^m	+03° 14′	4.5′ × 4.0′
chart 13			m$_v$ 10.4	p.a. 100°

Bright, pretty large, and round; small, bright, diffuse nucleus. Distance 52 million ly; Hubble class E1. Paired with galaxy NGC 3641; also in a group with NGC 3630, NGC 3645, and others.

NGC 3642

Gx	UMa	11^h 22.3^m	+59° 05′	5.8′ × 4.9′
chart 2			m$_v$ 11.2	p.a. 105°

Pretty bright, pretty large, and round; many arms; very small, very bright nucleus. Distance 75 million ly; Hubble class Sc.

NGC 3646

Gx	Leo	$11^h 21.7^m$	+20° 10'	3.8' × 2.1'
chart 6			m_v 11.1	p.a. 50°

Quite faint, quite large, and slightly elongated; small, bright nucleus. Distance 180 million ly; Hubble class Sc. Paired with galaxy NGC 3649.

NGC 3652

Gx	UMa	$11^h 22.7^m$	+37° 46'	2.3' × 0.7'
chart 6			m_v 12.2	p.a. 150°

Pretty faint, quite large, and quite elongated; Z shaped. Distance 91 million ly; Hubble class SBcp. Companions close by to NNW.

NGC 3655

Gx	Leo	$11^h 22.9^m$	+16° 35'	1.5' × 1.0'
charts 6, 13			m_v 11.7	p.a. 30°

Pretty bright, pretty small, irregularly round; several bright arms; small, bright nucleus. Distance 38 million ly; Hubble class S.

NGC 3659

Gx	Leo	$11^h 23.8^m$	+17° 49'	2.3' × 1.3'
charts 6, 13			m_v 12.3	p.a. 60°

Quite faint, small, and slightly elongated. Distance 52 million ly; Hubble class S.

NGC 3660

Gx	Crt	$11^h 23.5^m$	−08° 39'	2.7' × 2.2'
chart 13			m_v 13.2	p.a. 115°

Faint, pretty large, irregularly roundish; with a bright nucleus. Distance 150 million ly; Hubble class SBb.

NGC 3663

Gx	Crt	$11^h 24.0^m$	−12° 18'	1.9' × 1.3'
chart 13			m_v 12.5	p.a. 85°

Extremely faint; fan shaped; Hubble class Sb. Stars close by.

NGC 3665

Gx	UMa	$11^h 24.7^m$	+38° 46'	3.9' × 3.2'
charts 6, 7			m_v 10.8	p.a. 30°

Quite bright, quite large, irregularly round; very bright, diffuse nucleus. Distance 88 million ly; Hubble class E2. Paired with galaxy NGC 3658.

NGC 3666

Gx	Leo	$11^h 24.4^m$	+11° 21'	4.5' × 1.3'
chart 13			m_v 12	p.a. 100°

Faint and elongated; nearly edge on; many arms; small, bright nucleus. Distance 41 million ly; Hubble class Sb.

NGC 3669

Gx	UMa	$11^h 25.5^m$	+57° 43'	2.0' × 0.5'
chart 2			m_v 12.4	p.a. 153°

High surface brightness; patchy; brightest in S part. Distance 89 million ly; Hubble class SBc.

NGC 3672

Gx	Crt	$11^h 25.0^m$	−09° 48'	3.8' × 1.9'
chart 13			m_v 11.4	p.a. 8°

Pretty bright, large, and elongated; a multi-armed spiral; small, bright nucleus. Distance 76 million ly; Hubble class Sb.

NGC 3673

Gx	Hya	$11^h 25.2^m$	−26° 44'	3.7' × 2.4'
chart 20			m_v 11.5	p.a. 70°

Faint and very large; thin arms; small, bright nucleus. Distance 74 million ly; Hubble class S(B)b.

NGC 3675

Gx	UMa	$11^h 26.1^m$	+43° 35'	5.9' × 3.2'
charts 2, 6, 7			m_v 10.2	p.a. 178°

Very bright, quite large, and elongated; two main arms; small, bright nucleus. Distance 32 million ly; Hubble class Sb.

NGC 3680

OC	Cen	$11^h 25.7^m$	−43° 15'	12'
charts 20, 21			m_v 7.6	

30 stars; detached, strong concentration of stars; moderate range in brightness; mag. of brightest star 10.1. Distance 2,600 ly; age 1.8 billion years; Trumpler class I 2 p.

NGC 3681

Gx	Leo	11^h 26.5^m	+16° 52′	2.8′
charts 6, 13			m$_v$ 11.2	

Bright, pretty small, and round; many arms; diffuse, bright nucleus. Distance 53 million ly; Hubble class S(B)b. In a group with galaxies NGC 3684, NGC 3686, and NGC 3691.

NGC 3683

Gx	UMa	11^h 27.5^m	+56° 53′	1.8′ × 0.7′
chart 2			m$_v$ 12.4	p.a. 128°

Quite bright, pretty large, and elongated. Distance 78 million ly; Hubble class S.

NGC 3683A

Gx	UMa	11^h 29.2^m	+57° 08′	2.4′ × 1.7′
chart 2			m$_v$ 11.9	p.a. 75°

Small, bright nucleus; several knotty, filamentary arms. Distance 110 million ly; Hubble class SBc.

NGC 3684

Gx	Leo	11^h 27.2^m	+17° 02′	3.5′ × 2.3′
charts 6, 13			m$_v$ 11.4	p.a. 130°

Pretty bright, pretty large, and elongated; two main arms; small, bright nucleus. Distance 58 million ly; Hubble class Sc. In a group with galaxies NGC 3681, NGC 3686, and NGC 3691.

NGC 3686

Gx	Leo	11^h 27.7^m	+17° 13′	3.1′ × 2.5′
charts 6, 13			m$_v$ 11.3	p.a. 15°

Pretty bright, large, and slightly elongated; two main arms; very small, extremely bright nucleus. Distance 41 million ly; Hubble class Sc. In a group with galaxies NGC 3681, NGC 3684, and NGC 3691.

NGC 3687

Gx	UMa	11^h 28.0^m	+29° 31′	1.9′
chart 6			m$_v$ 12	

Pretty bright, pretty small, and round. Distance 100 million ly; Hubble class Sb.

NGC 3689

Gx	Leo	11^h 28.2^m	+25° 40′	1.6′ × 1.0′
chart 6			m$_v$ 12.3	p.a. 97°

Pretty bright, pretty large, and slightly elongated; three arms; small, bright nucleus. Distance 120 million ly; Hubble class Sc.

NGC 3690

Gx	UMa	11^h 28.6^m	+58° 34′	2.1′ × 1.4′
chart 2			m$_v$ 11.5	p.a. 50°

Pretty bright, pretty small, and slightly elongated. Distance 140 million ly; Hubble class S. Interacting with galaxy IC 694.

NGC 3691

Gx	Leo	11^h 28.1^m	+16° 55′	1.3′ × 1.0′
charts 6, 13			m$_v$ 11.8	p.a. 15°

Faint, pretty small, and slightly elongated. Distance 39 million ly; Hubble class S. In a group with galaxies NGC 3681, NGC 3684, and NGC 3686.

NGC 3692

Gx	Leo	11^h 28.4^m	+09° 24′	3.2′ × 0.7′
chart 13			m$_v$ 12.1	p.a. 95°

Faint and elongated. Distance 70 million ly; Hubble class Sb.

NGC 3699

PN	Cen	11^h 28.0^m	−59° 57′	45″
chart 25			m$_v$ 11.3	

PK292+1.1. Expansion velocity greater than 18 km/sec (11 miles/sec).

NGC 3705

Gx	Leo	11^h 30.1^m	+09° 17′	4.6′ × 1.9′
chart 13			m$_v$ 11.1	p.a. 122°

Pretty faint, pretty large, and round; small, bright nucleus. Distance 39 million ly; Hubble class Sb.

NGC 3706

Gx	Cen	11ʰ 29.7ᵐ	−36° 24′	3.2′ × 1.9′	
charts 20, 21			m_v 11.2	p.a. 78°	

Pretty bright, quite small, and round. Distance 120 million ly; Hubble class S0.

NGC 3717

Gx	Hya	11ʰ 31.5ᵐ	−30° 18′	6.0′ × 1.1′	
chart 20			m_v 11.4	p.a. 33°	

Pretty bright, small, and very elongated; nearly edge on; small, bright nucleus; absorption lanes. Distance 64 million ly; Hubble class Sb. Paired with galaxy IC 2913.

NGC 3718

Gx	UMa	11ʰ 32.6ᵐ	+53° 04′	8.7′ × 4.5′	
charts 2, 6			m_v 10.8	p.a. 345°	

Pretty bright, very large, and round; very small, very bright nucleus. An unusual bar-shaped dust lane cuts across nucleus. Distance 48 million ly; Hubble class SBap. Interacting with galaxy NGC 3729. A peculiar chain of galaxies (Arp 322) lies about 7′ away.

NGC 3726

Gx	UMa	11ʰ 33.3ᵐ	+47° 02′	6.0′ × 4.1′	
charts 2, 6, 7			m_v 10.4	p.a. 10°	

Pretty bright, very large, and slightly elongated; several large arms; small, very bright nucleus. A Seyfert galaxy. Distance 36 million ly; Hubble class Sc.

NGC 3729

Gx	UMa	11ʰ 33.8ᵐ	+53° 08′	3.0′ × 2.1′	
charts 2, 6			m_v 11.4	p.a. 345°	

Pretty bright, pretty large, and slightly elongated; small, bright nucleus. Distance 49 million ly; Hubble class P. Interacting with galaxy NGC 3718.

NGC 3732

Gx	Crt	11ʰ 34.2ᵐ	−09° 51′	1.3′ × 1.2′	
chart 13			m_v 12.5	p.a. 85°	

Faint, small, and round; very bright, very small nucleus. Distance 66 million ly; Hubble class E0. Brightest in a group of galaxies that includes NGC 3723 and NGC 3763.

NGC 3733

Gx	UMa	11ʰ 35.0ᵐ	+54° 51′	4.9′ × 2.1′	
charts 2, 6			m_v 12.4	p.a. 170°	

Extremely faint, small, irregularly roundish. Distance 56 million ly; Hubble class S(B)c. A 6th-mag. star lies to the SE. Forms a line of galaxies with NGC 3738 and NGC 3756.

NGC 3735

Gx	Dra	11ʰ 36.0ᵐ	+70° 32′	4.1′ × 0.8′	
chart 2			m_v 11.8	p.a. 131°	

Pretty bright, large, and very elongated; several knotty arms; small, bright nucleus. Distance 120 million ly; Hubble class Sb.

NGC 3738

Gx	UMa	11ʰ 35.8ᵐ	+54° 31′	2.6′ × 2.1′	
charts 2, 6			m_v 11.7	p.a. 155°	

Pretty bright and pretty large; very faint nucleus. Distance 14 million ly; Hubble class P. Forms a line of galaxies with NGC 3756 and NGC 3733.

NGC 3742

Gx	Cen	11ʰ 35.6ᵐ	−37° 57′	2.4′ × 1.7′	
charts 20, 21			m_v 12	p.a. 116°	

Pretty faint, pretty large, and very elongated; bright nucleus. Distance 110 million ly; Hubble class SBa.

NGC 3756

Gx	UMa	11ʰ 36.8ᵐ	+54° 18′	4.2′ × 2.1′	
charts 2, 6			m_v 11.5	p.a. 177°	

Pretty faint, large, and elongated; two main arms; small, bright nucleus. Distance 50 million ly; Hubble class Sc. Forms a line of galaxies with NGC 3733 and NGC 3738.

NGC 3766

OC	Cen	11ʰ 36.1ᵐ	−61° 37′	12′
chart 25			m_v 5.3	

C97. 100 stars; detached, strong concentration of stars; small range in brightness; mag. of brightest star 7.2; pretty large. Distance 5,500 ly; age 22 million years; Trumpler class I 1 p.

NGC 3769

| Gx | UMa | 11ʰ 37.7ᵐ | +47° 54′ | 2.8′ × 0.9′ |
| charts 2, 6, 7 | | | mᵥ 11.8 | p.a. 152° |

Pretty bright, small, and elongated. Distance 34 million ly; Hubble class Sb. Paired with galaxy NGC 3769A.

NGC 3780

| Gx | UMa | 11ʰ 39.4ᵐ | +56° 16′ | 3.0′ × 2.4′ |
| chart 2 | | | mᵥ 11.5 | p.a. 90° |

Pretty faint and large; many arms; small, bright nucleus. Distance 120 million ly; Hubble class Sc. Paired with galaxy NGC 3804.

NGC 3783

| Gx | Cen | 11ʰ 39.0ᵐ | −37° 44′ | 1.9′ |
| charts 20, 21 | | | mᵥ 11.6 | |

Quite bright and round; a Seyfert galaxy with a very small, very bright nucleus. Distance 120 million ly; Hubble class SBa.

NGC 3800

| Gx | Leo | 11ʰ 40.2ᵐ | +15° 21′ | 1.8′ × 0.5′ |
| charts 6, 13 | | | mᵥ 12.7 | p.a. 52° |

Faint, pretty small, and elongated; bright nucleus. Distance 150 million ly; Hubble class S(B)bp.

NGC 3801

| Gx | Leo | 11ʰ 40.3ᵐ | +17° 44′ | 2.9′ × 1.7′ |
| charts 6, 13 | | | mᵥ 12 | p.a. 120° |

Pretty faint, pretty large, and roundish; bright nucleus. Distance 140 million ly; Hubble class S0p.

NGC 3810

| Gx | Leo | 11ʰ 41.0ᵐ | +11° 28′ | 4.1′ × 2.7′ |
| chart 13 | | | mᵥ 10.8 | p.a. 15° |

Bright, large, and slightly elongated; many arms; very small, bright nucleus. Distance 38 million ly; Hubble class Sc.

NGC 3811

| Gx | UMa | 11ʰ 41.3ᵐ | +47° 42′ | 2.2′ × 1.7′ |
| charts 2, 6, 7 | | | mᵥ 12.3 | p.a. 160° |

Faint, small, and slightly elongated; with a bright nucleus. Distance 140 million ly; Hubble class SBcp.

NGC 3813

| Gx | UMa | 11ʰ 41.3ᵐ | +36° 33′ | 2.3′ × 1.2′ |
| charts 6, 7 | | | mᵥ 11.7 | p.a. 87° |

Quite bright, pretty large, and quite elongated; very bright arms; very small, very bright nucleus. Distance 61 million ly; Hubble class Sc.

NGC 3818

| Gx | Vir | 11ʰ 42.0ᵐ | −06° 09′ | 2.1′ × 1.3′ |
| chart 13 | | | mᵥ 11.7 | p.a. 103° |

Faint, pretty small, and round; small, bright nucleus. Distance 57 million ly; Hubble class E2p.

NGC 3836

| Gx | Crt | 11ʰ 43.5ᵐ | −16° 48′ | 1.4′ |
| charts 13, 20 | | | mᵥ 12.7 | |

Faint and small. Distance 150 million ly; Hubble class Sb. Faint star close by to N.

NGC 3842

| Gx | Leo | 11ʰ 44.0ᵐ | +19° 57′ | 1.4′ × 1.0′ |
| charts 6, 13 | | | mᵥ 12 | p.a. 5° |

Faint, small, and roundish; bright nucleus. Distance 270 million ly; Hubble class E3. This is the brightest galaxy in the Abell 1367 galaxy group.

NGC 3865

| Gx | Crt | 11ʰ 44.9ᵐ | −09° 14′ | 2.0′ × 1.5′ |
| chart 13 | | | mᵥ 12 | p.a. 135° |

Faint, pretty large, and diffuse. Distance 240 million ly; Hubble class Sb. Numerous small galaxies within field.

NGC 3872

| Gx | Leo | 11ʰ 45.8ᵐ | +13° 46′ | 2.3′ × 1.8′ |
| charts 6, 13 | | | mᵥ 11.7 | p.a. 25° |

Bright, small, and round; small, bright nucleus. Distance 130 million ly; Hubble class E4.

NGC 3877

| Gx | UMa | 11ʰ 46.1ᵐ | +47° 30′ | 5.4′ × 1.5′ |
| charts 2, 6, 7 | | | m_v 11 | p.a. 35° |

Bright, large, and very elongated; very faint nucleus. Distance 39 million ly; Hubble class Sb. Foreground star superimposed on nucleus.

NGC 3882

| Gx | Cen | 11ʰ 46.1ᵐ | −56° 23′ | 2.5′ × 1.4′ |
| chart 25 | | | m_v 12.5 | p.a. 126° |

Very faint and slightly elongated. Distance 70 million ly; Hubble class SBb.

NGC 3885

| Gx | Hya | 11ʰ 46.8ᵐ | −27° 55′ | 2.4′ × 0.8′ |
| charts 20, 21 | | | m_v 11.8 | p.a. 123° |

Quite faint, very small, and slightly elongated; very bright nucleus; dark lanes. Distance 74 million ly; Hubble class Sb.

NGC 3887

| Gx | Crt | 11ʰ 47.1ᵐ | −16° 51′ | 3.5′ × 2.8′ |
| charts 13, 20 | | | m_v 10.6 | p.a. 20° |

Pretty bright, large, irregularly round; small, bright nucleus with dark lanes. Distance 40 million ly; Hubble class Sc.

NGC 3888

| Gx | UMa | 11ʰ 47.6ᵐ | +55° 58′ | 1.8′ × 1.4′ |
| chart 2 | | | m_v 12.1 | p.a. 120° |

Pretty bright and small; two main arms; small, very bright nucleus. Distance 110 million ly; Hubble class Sb. Paired with galaxy NGC 3898.

NGC 3892

| Gx | Crt | 11ʰ 48.0ᵐ | −10° 58′ | 3.2′ × 2.3′ |
| chart 13 | | | m_v 11.5 | p.a. 95° |

Pretty bright, pretty large, and round; very small, very bright nucleus. Distance 67 million ly; Hubble class S(B)a.

NGC 3893

| Gx | UMa | 11ʰ 48.7ᵐ | +48° 43′ | 4.5′ × 2.5′ |
| charts 2, 6, 7 | | | m_v 10.5 | p.a. 165° |

Bright, pretty large, and round; two main arms; very small, bright nucleus. Distance 45 million ly; Hubble class Sc. Paired with galaxy NGC 3896.

NGC 3894

| Gx | UMa | 11ʰ 48.9ᵐ | +59° 25′ | 2.1′ × 1.3′ |
| chart 2 | | | m_v 11.6 | p.a. 20° |

Bright and pretty large. Distance 140 million ly; Hubble class E4. Paired with galaxy NGC 3895.

NGC 3898

| Gx | UMa | 11ʰ 49.3ᵐ | +56° 05′ | 3.7′ × 2.5′ |
| chart 2 | | | m_v 10.7 | p.a. 107° |

Bright, pretty large, and slightly elongated; very faint arms; very bright, large nucleus. Distance 49 million ly; Hubble class Sb. Paired with fainter galaxy NGC 3888.

NGC 3900

| Gx | Leo | 11ʰ 49.1ᵐ | +27° 01′ | 3.5′ × 1.8′ |
| charts 6, 7 | | | m_v 11.3 | p.a. 2° |

Bright, pretty large, and slightly elongated; bright nucleus; dark lane. Distance 73 million ly; Hubble class Sb.

NGC 3904

| Gx | Hya | 11ʰ 49.2ᵐ | −29° 17′ | 2.8′ × 2.0′ |
| charts 20, 21 | | | m_v 10.8 | p.a. 8° |

Pretty bright, small, and round; bright nucleus. Distance 60 million ly; Hubble class E2. Paired with galaxy NGC 3923.

NGC 3912

| Gx | Leo | 11ʰ 50.1ᵐ | +26° 29′ | 1.5′ × 0.9′ |
| charts 6, 7 | | | m_v 12.4 | p.a. 5° |

Faint, pretty large, and round. Distance 75 million ly; Hubble class Sb.

NGC 3917

| Gx | UMa | 11ʰ 50.8ᵐ | +51° 50′ | 4.9′ × 1.4′ |
| charts 2, 6, 7 | | | m_v 11.8 | p.a. 77° |

Faint, large, and very elongated; nearly edge on; many arms; small, bright nucleus. Distance 45 million ly; Hubble class Sc. In a group with galaxy NGC 3931 and an anonymous spiral galaxy.

NGC 3918

PN	Cen	$11^h 50.3^m$	$-57°\,11'$	19″
chart 25			m_v 8.1	

Blue Planetary Nebula, PK294+04.1. Small, round, smooth disk of uniform brightness; bluish color; phot. mag. 8.4; central star mag. 13.2. Per *NGC*, a (!) remarkable object. Expansion velocity 20 km/sec (12 miles/sec); distance 2,600 ly; V-V class 2b.

NGC 3923

Gx	Hya	$11^h 51.0^m$	$-28°\,48'$	$6.0' \times 4.2'$
charts 20, 21			m_v 9.6	p.a. 50°

Bright, pretty large, and slightly elongated; small, bright, elongated nucleus. Distance 67 million ly; Hubble class E3. Paired with galaxy NGC 3904.

NGC 3936

Gx	Hya	$11^h 52.3^m$	$-26°\,54'$	$3.9' \times 0.7'$
charts 20, 21			m_v 11.8	p.a. 63°

Very faint, quite large, and very elongated; nearly edge on. Distance 78 million ly; Hubble class Sb.

NGC 3938

Gx	UMa	$11^h 52.8^m$	$+44°\,07'$	5.4′
charts 2, 6, 7			m_v 10.4	

Bright, very large, and round; a face-on spiral; several bright arms; diffuse, small, bright nucleus. Distance 36 million ly; Hubble class Sc.

NGC 3941

Gx	UMa	$11^h 52.9^m$	$+36°\,59'$	$3.5' \times 2.5'$
charts 6, 7			m_v 10.3	p.a. 10°

Very bright, pretty large, and round; small, bright nucleus. Distance 42 million ly; Hubble class E3.

NGC 3945

Gx	UMa	$11^h 53.2^m$	$+60°\,41'$	$5.5' \times 3.6'$
chart 2			m_v 10.8	p.a. 15°

Bright, pretty large, and round; very small, bright nucleus. Distance 58 million ly; Hubble class SBa.

NGC 3949

Gx	UMa	$11^h 53.7^m$	$+47°\,52'$	$2.9' \times 1.7'$
charts 2, 6, 7			m_v 11.1	p.a. 120°

Pretty large, elongated, and quite faint. Distance 32 million ly; Hubble class Sb.

NGC 3953

Gx	UMa	$11^h 53.8^m$	$+52°\,20'$	$6.6' \times 3.6'$
charts 2, 6, 7			m_v 10.1	p.a. 13°

Quite bright, large, and elongated; many arms; small, very bright nucleus. It has a peculiar appearance in the field, there being a coarse double star to the N and a vertical line of five equidistant stars to the E (Smyth). A good example of a spiral galaxy. Distance 45 million ly; Hubble class Sb.

NGC 3955

Gx	Crt	$11^h 54.0^m$	$-23°\,10'$	$3.1' \times 1.0'$
charts 20, 21			m_v 11.3	p.a. 165°

Quite faint, small, and elongated; bright center with dark lanes. Distance 49 million ly; Hubble class Sb.

NGC 3956

Gx	Crt	$11^h 54.0^m$	$-20°\,34'$	$3.5' \times 1.0'$
chart 20			m_v 12.2	p.a. 58°

Quite faint, pretty large, and elongated; has several filamentary arms. Distance 22 million ly; Hubble class S–.

NGC 3957

Gx	Crt	$11^h 54.0^m$	$-19°\,34'$	$3.1' \times 0.7'$
charts 13, 20			m_v 12	p.a. 173°

Faint, small, and elongated; small, bright nucleus with dark lane. Distance 70 million ly; Hubble class S0.

NGC 3960

OC	Cen	$11^h 50.9^m$	$-55°\,42'$	7′
chart 25			m_v 8.3	

45 stars; detached, strong concentration of stars; moderate range in brightness; mag. of brightest star 11.5; pretty large. Trumpler class I 2 m.

NGC 3962

Gx	Crt	11ʰ 54.7ᵐ	−13° 58′	3.5′ × 2.8′
charts 13, 20			m_v 10.7	p.a. 15°

Quite bright, pretty large, irregularly round. Distance 71 million ly; Hubble class E2.

NGC 3963

Gx	UMa	11ʰ 55.0ᵐ	+58° 30′	2.7′ × 2.4′
chart 2			m_v 11.9	p.a. 100°

Pretty faint, quite large, and round; two large arms; small, very bright nucleus. Distance 140 million ly; Hubble class Sc. Paired with galaxy NGC 3958.

NGC 3968

Gx	Leo	11ʰ 55.5ᵐ	+11° 58′	2.7′ × 1.9′
chart 13			m_v 11.8	p.a. 10°

Pretty bright, large, irregularly roundish; bright nucleus. Distance 270 million ly; Hubble class S(B)b.

NGC 3972

Gx	UMa	11ʰ 55.8ᵐ	+55° 19′	3.7′ × 1.0′
chart 2			m_v 12.3	p.a. 120°

Pretty bright and elongated; two main arms; small, bright nucleus. Distance 41 million ly; Hubble class Sb. In a group with galaxies NGC 3982 and NGC 3998.

NGC 3976

Gx	Vir	11ʰ 55.9ᵐ	+06° 45′	3.5′ × 1.1′
chart 13			m_v 11.5	p.a. 53°

Bright, pretty large, and very elongated; a nearly edge-on spiral; two main arms; small, bright, diffuse nucleus. Distance 100 million ly; Hubble class Sb.

NGC 3981

Gx	Crt	11ʰ 56.1ᵐ	−19° 54′	5.0′ × 2.8′
charts 13, 20			m_v 11	p.a. 15°

Very faint and pretty large; nearly edge on; faint outer arms; bright center. Distance 65 million ly; Hubble class Sb.

NGC 3982

Gx	UMa	11ʰ 56.5ᵐ	+55° 07′	2.3′ × 2.0′
chart 2			m_v 11	p.a. 0°

Bright, pretty large, and round; several knotty arms; small, very bright nucleus. Distance 53 million ly; Hubble class Sb. In a group with galaxies NGC 3972 and NGC 3998.

NGC 3992

See M109.

NGC 3995

Gx	UMa	11ʰ 57.7ᵐ	+32° 18′	2.7′ × 1.0′
charts 6, 7			m_v 12.4	p.a. 33°

Faint, pretty large, irregularly round; three arms; small, bright nucleus. Distance 150 million ly; Hubble class Sc.

NGC 3998

Gx	UMa	11ʰ 57.9ᵐ	+55° 27′	2.9′ × 2.3′
chart 2			m_v 10.7	p.a. 140°

Quite bright, pretty small, and round; very bright nucleus. Distance 54 million ly; Hubble class E2p. In a group with galaxies NGC 3972 and NGC 3982.

NGC 4008

Gx	Leo	11ʰ 58.3ᵐ	+28° 12′	2.4′ × 1.3′
charts 6, 7			m_v 12	p.a. 167°

Pretty bright, pretty small, and elongated; small, bright nucleus. Distance 160 million ly; Hubble class Sa.

NGC 4013

Gx	UMa	11ʰ 58.5ᵐ	+43° 57′	4.8′ × 1.0′
charts 6, 7			m_v 11.2	p.a. 66°

Bright, quite large, and very elongated; edge on; very faint nucleus. Distance 33 million ly; Hubble class Sb.

NGC 4017

Gx	Com	11ʰ 58.8ᵐ	+27° 27′	1.8′ × 1.4′
charts 6, 7			m_v 12.2	p.a. 25°

Faint, large, and elongated; Hubble class S(B)b.

NGC 4024

Gx	Crv	11ʰ 58.5ᵐ	−18° 21′	1.9′ × 1.7′
charts 13, 20			m_v 11.9	p.a. 125°

Faint and very small; small, bright nucleus. Distance 64 million ly; Hubble class SB0.

NGC 4026

Gx	UMa	$11^h 59.4^m$	+50° 58′	4.8′ × 1.3′
charts 2, 6, 7			m_v 10.8	p.a. 178°

Very bright, quite large, and very elongated; edge on; very small, very bright nucleus. Distance 42 million ly; Hubble class S0.

NGC 4027

Gx	Crv	$11^h 59.5^m$	−19° 16′	3.2′ × 2.5′
charts 13, 20, 21			m_v 11.2	p.a. 167°

Pretty faint, pretty large, and round; small, bright, elongated nucleus. Distance 64 million ly; Hubble class Sc. Interacting with irregular dwarf galaxy NGC 4027A.

NGC 4030

Gx	Vir	$12^h 00.4^m$	−01° 06′	4.1′ × 3.1′
charts 13, 14			m_v 10.6	p.a. 27°

Quite bright, large, and slightly elongated; a nearly face-on spiral; many arms; small, bright nucleus. Distance 55 million ly; Hubble class Sc. Bright star nearby.

NGC 4032

Gx	Com	$12^h 00.6^m$	+20° 05′	1.9′ × 1.8′
charts 6, 7, B1			m_v 12.2	p.a. 130°

Pretty faint, pretty large, and round. Distance 51 million ly; Hubble class Ir+.

NGC 4033

Gx	Crv	$12^h 00.6^m$	−17° 51′	2.5′ × 1.2′
charts 13, 14, 20, 21			m_v 11.8	p.a. 47°

Pretty bright, small, and slightly elongated; very bright, small nucleus. Distance 57 million ly; Hubble class E5.

NGC 4036

Gx	UMa	$12^h 01.5^m$	+61° 54′	4.0′ × 1.8′
chart 2			m_v 10.7	p.a. 85°

Very bright, very large, and elongated; spindle shaped; very bright nucleus. Distance 66 million ly; Hubble class E6. Paired with galaxy NGC 4041.

NGC 4037

Gx	Com	$12^h 01.4^m$	+13° 24′	2.5′ × 2.3′
charts 13, 14, B1			m_v 11.9	p.a. 15°

Extremely faint, pretty large, and round; with a very small, bright nucleus. Distance 37 million ly; Hubble class S(B).

NGC 4038

Gx	Crv	$12^h 01.9^m$	−18° 52′	6.0′ × 3.4′
charts 13, 14, 20, 21			m_v 10.5	p.a. 100°

C60, one half (the N part) of the Antennae or Ring Tail. Quite large, pretty bright, and round; bright nucleus; has a pair of long, thin, streaming arms. Distance 63 million ly; Hubble class Sc. Colliding with galaxy NGC 4039.

NGC 4039

Gx	Crv	$12^h 01.9^m$	−18° 53′	5.0′ × 2.8′
charts 13, 14, 20, 21			m_v 10.3	p.a. 50°

C61, the other half (the S part) of the Antennae or Ring Tail. Pretty faint and pretty large; Hubble class Smp. Colliding with galaxy NGC 4038.

NGC 4041

Gx	UMa	$12^h 02.2^m$	+62° 08′	2.7′
chart 2			m_v 11.3	

Bright, quite large, and round; several filamentary arms; bright nucleus. Distance 57 million ly; Hubble class Sc. Paired with galaxy NGC 4036.

NGC 4045

Gx	Vir	$12^h 02.7^m$	+01° 59′	3.0′ × 1.9′
charts 13, 14			m_v 12	p.a. 95°

Pretty faint, large, and round; very small, very bright nucleus. Distance 79 million ly; Hubble class Sp. Paired with galaxy NGC 4045A.

NGC 4047

Gx	UMa	$12^h 02.9^m$	+48° 38′	1.2′ × 1.0′
charts 2, 6, 7			m_v 12.2	p.a. 105°

Pretty bright, pretty small, and round; many arms;

very small, very bright nucleus. Distance 150 million ly; Hubble class Sb.

NGC 4051

| Gx | UMa | $12^h\,03.2^m$ | +44° 32′ | 5.0′ × 4.5′ |
| charts 2, 6, 7 | | | m_v 10.2 | p.a. 135° |

Bright and very large; two thick, spiral arms; very small, extremely bright nucleus. A Seyfert galaxy. Distance 32 million ly; Hubble class Sc.

NGC 4062

| Gx | UMa | $12^h\,04.1^m$ | +31° 54′ | 4.1′ × 1.8′ |
| charts 6, 7 | | | m_v 11.1 | p.a. 100° |

Pretty bright, very large, and very elongated; many arms; very small, bright nucleus. Distance 32 million ly; Hubble class Sb.

NGC 4064

| Gx | Com | $12^h\,04.2^m$ | +18° 27′ | 4.3′ × 1.7′ |
| charts 7, 13, 14, B1 | | | m_v 11.4 | p.a. 150° |

Bright and elongated; bright center with dark lanes. Distance 42 million ly; Hubble class Sb.

NGC 4067

| Gx | Vir | $12^h\,04.2^m$ | +10° 51′ | 1.3′ × 0.9′ |
| charts 13, 14, B1 | | | m_v 12.5 | p.a. 35° |

Faint, pretty small, and roundish; bright nucleus. Distance 100 million ly; Hubble class Sb.

NGC 4073

| Gx | Vir | $12^h\,04.4^m$ | +01° 54′ | 2.8′ × 2.0′ |
| charts 13, 14 | | | m_v 11.4 | p.a. 105° |

Faint, pretty small, and round. Distance 250 million ly; Hubble class E1. In a group with NGC 4077 and other faint galaxies.

NGC 4085

| Gx | UMa | $12^h\,05.4^m$ | +50° 21′ | 2.5′ × 0.8′ |
| charts 2, 6, 7 | | | m_v 12.4 | p.a. 78° |

Bright, pretty large, and elongated; very faint nucleus. Distance 35 million ly; Hubble class Sb. Paired with galaxy NGC 4088.

NGC 4088

| Gx | UMa | $12^h\,05.6^m$ | +50° 33′ | 5.8′ × 2.5′ |
| charts 2, 6, 7 | | | m_v 10.6 | p.a. 43° |

Bright, quite large, and elongated; large, knotty arms; very small, bright nucleus. Distance 36 million ly; Hubble class Sc. Paired with fainter galaxy NGC 4085.

NGC 4094

| Gx | Crv | $12^h\,05.9^m$ | −14° 32′ | 4.1′ × 1.4′ |
| charts 13, 14, 21 | | | m_v 11.8 | p.a. 63° |

Extremely faint, large, and elongated. Distance 54 million ly; Hubble class S–.

NGC 4096

| Gx | UMa | $12^h\,06.0^m$ | +47° 29′ | 6.0′ × 1.7′ |
| charts 2, 6, 7 | | | m_v 10.8 | p.a. 20° |

Pretty bright, very large, and very elongated; nearly edge on; small, bright nucleus. Distance 24 million ly; Hubble class Sc.

NGC 4100

| Gx | UMa | $12^h\,06.1^m$ | +49° 35′ | 5.2′ × 1.9′ |
| charts 2, 6, 7 | | | m_v 11.2 | p.a. 167° |

Pretty bright, very large, and very elongated; nearly edge on; two main knotty arms; very small, bright nucleus. Distance 50 million ly; Hubble class Sb.

NGC 4102

| Gx | UMa | $12^h\,06.4^m$ | +52° 43′ | 3.1′ × 1.7′ |
| charts 2, 6, 7 | | | m_v 11.2 | p.a. 38° |

Bright, pretty small, and round; two main arms that form a pseudo-ring; small, bright nucleus. Distance 43 million ly; Hubble class Sc.

NGC 4103

| OC | Cru | $12^h\,06.7^m$ | −61° 15′ | 7′ |
| chart 25 | | | m_p 7.4 | |

45 stars; detached, strong concentration of stars; large range in brightness; brightest star is phot. mag. 10; pretty large, irregularly round. Distance 3,900 ly; age 22 million years; Trumpler class I 3 m.

NGC 4105

Gx   Hya $12^h 06.7^m$ $-29° 46'$ $3.0' \times 2.3'$
charts 20, 21 m_v 10.4   p.a. 151°

Pretty faint, pretty small, and round. Distance 72 million ly; Hubble class E2. Interacting with galaxy NGC 4106.

NGC 4106

Gx   Hya $12^h 06.7^m$ $-29° 46'$ $2.3' \times 1.7'$
charts 20, 21 m_v 10.6   p.a. 77°

Pretty faint, pretty small, and round; very bright nucleus. Distance 84 million ly; Hubble class E0. Interacting with galaxy NGC 4105.

NGC 4111

Gx   CVn $12^h 07.0^m$ $+43° 04'$ $4.6' \times 1.0'$
charts 6, 7 m_v 10.7   p.a. 150°

Very bright, pretty small, and very elongated; very small, bright nucleus with dark lane. Distance 37 million ly; Hubble class S0. Paired with smaller galaxy NGC 4109 6' to SW.

NGC 4112

Gx   Cen $12^h 07.2^m$ $-40° 12'$ $1.6' \times 0.9'$
charts 20, 21 m_v 12   p.a. 5°

Faint, small, and slightly elongated; bright nucleus. Distance 110 million ly; Hubble class SBb. Three bright stars nearby to S.

NGC 4116

Gx   Vir $12^h 07.6^m$ $+02° 42'$ $3.6' \times 2.2'$
charts 13, 14 m_v 12   p.a. 155°

Very faint and elongated; hook shaped; several filamentary arms; bright, very small nucleus. Distance 52 million ly; Hubble class SBc. Paired with galaxy NGC 4123.

NGC 4123

Gx   Vir $12^h 08.2^m$ $+02° 53'$ $4.3' \times 3.2'$
charts 13, 14, B1 m_v 11.4   p.a. 135°

Quite faint, very large, and elongated; several filamentary arms; very small, very bright nucleus. Distance 53 million ly; Hubble class SBb. Paired with galaxy NGC 4116.

NGC 4124

Gx   Vir $12^h 08.2^m$ $+10° 23'$ $4.0' \times 1.9'$
charts 13, 14, B1 m_v 11.3   p.a. 114°

Pretty bright, pretty large, and very elongated; very small, very bright nucleus. Hubble class Sa. Member of the Virgo Galaxy Cluster.

NGC 4125

Gx   Dra $12^h 08.1^m$ $+65° 10'$ $5.1' \times 3.2'$
chart 2 m_v 9.5   p.a. 95°

Pretty bright, pretty large, and quite elongated. Distance 64 million ly; Hubble class E5p. Paired with NGC 4121.

NGC 4128

Gx   Dra $12^h 08.5^m$ $+68° 46'$ $2.4' \times 0.9'$
chart 2 m_v 12   p.a. 58°

Holmberg 337a. Quite bright and spindle shaped; small, very bright nucleus. Distance 110 million ly; Hubble class Sa. Paired with galaxy Holmberg 337b.

NGC 4136

Gx   Com $12^h 09.3^m$ $+29° 56'$ $3.9'$
charts 6, 7 m_v 11

Faint and very large; small, very bright nucleus. Distance 19 million ly; Hubble class Sc.

NGC 4138

Gx   CVn $12^h 09.5^m$ $+43° 41'$ $3.0' \times 1.8'$
charts 6, 7 m_v 11.3   p.a. 150°

Bright, pretty large, and slightly elongated; small, bright nucleus; dark lane. Distance 47 million ly; Hubble class E4.

NGC 4143

Gx   CVn $12^h 09.6^m$ $+42° 32'$ $2.9' \times 1.9'$
charts 6, 7 m_v 10.7   p.a. 144°

Quite bright and round; small, bright, diffuse nucleus. Distance 36 million ly; Hubble class E4.

NGC 4144

Gx   UMa $12^h 10.0^m$ $+46° 27'$ $5.9' \times 1.5'$
charts 2, 6, 7 m_v 11.6   p.a. 104°

Pretty faint, quite large, and very elongated; very faint nucleus. Distance 45 million ly; Hubble class Sb.

NGC 4145

Gx	CVn	12^h 10.0^m	+39° 53′	5.8′ × 4.4′
charts 6, 7			m$_v$ 11.3	p.a. 100°

Holmberg 342a. Bright and very large. Distance 45 million ly; Hubble class Sc. Paired with a faint galaxy (Holmberg 342b).

NGC 4147

GC	Com	12^h 10.1^m	+18° 33′	4.7′
charts 7, 13, 14, B1			m$_v$ 10.4	

Very bright, pretty large, and round; very well resolved; medium concentration of stars. Distance 57,000 ly; S-S class 6.

NGC 4150

Gx	Com	12^h 10.6^m	+30° 24′	2.3′ × 1.6′
charts 6, 7			m$_v$ 11.6	p.a. 147°

Bright, small, and round; very small, very bright nucleus. Distance 10 million ly; Hubble class E2.

NGC 4151

Gx	CVn	12^h 10.5^m	+39° 24′	5.9′ × 4.4′
charts 6, 7			m$_v$ 10.8	p.a. 130°

Very bright, small, and round; extremely bright, small nucleus. A Seyfert galaxy. Distance 44 million ly; Hubble class P. In a group with galaxies NGC 4156 and Holmberg 345c.

NGC 4152

Gx	Com	12^h 10.6^m	+16° 02′	2.3′ × 1.9′
charts 7, 13, 14, B1			m$_v$ 12.2	p.a. 115°

Pretty bright, pretty large, and round; several filamentary arms; small, bright nucleus. Distance 91 million ly; Hubble class Sc.

NGC 4157

Gx	UMa	12^h 11.1^m	+50° 29′	6.9′ × 1.7′
charts 2, 6, 7			m$_v$ 11.3	p.a. 66°

Pretty faint, quite large, and very elongated; edge on; bright center with dark lane; several filamentary arms; a Seyfert galaxy. Distance 39 million ly; Hubble class Sb.

NGC 4158

Gx	Com	12^h 11.2^m	+20° 11′	2.0′ × 1.7′
charts 7, B1			m$_v$ 12.1	p.a. 80°

Faint, pretty small, and slightly elongated; many filamentary arms; very bright nucleus. Distance 110 million ly; Hubble class Sa.

NGC 4162

Gx	Com	12^h 11.9^m	+24° 07′	2.3′ × 1.3′
chart 7			m$_v$ 12.2	p.a. 174°

Bright, large, and slightly elongated; very bright nucleus with star attached. Distance 110 million ly; Hubble class Sc.

NGC 4168

Gx	Vir	12^h 12.3^m	+13° 12′	2.8′ × 2.2′
charts 13, 14, B1			m$_v$ 11.2	p.a. 120°

Pretty bright and pretty large. Distance 100 million ly; Hubble class E0. Brightest in a group of galaxies; paired with galaxy NGC 4165.

NGC 4177

Gx	Crv	12^h 12.7^m	−14° 01′	1.7′ × 1.1′
charts 13, 14, 21			m$_v$ 13.2	p.a. 65°

Very faint, pretty large, and roundish; bright nucleus. Distance 170 million ly; Hubble class Sb.

NGC 4178

Gx	Vir	12^h 12.8^m	+10° 52′	5.0′ × 1.7′
charts 13, 14, B1			m$_v$ 11.4	p.a. 30°

Very faint, very large, and elongated; several faint arms. Hubble class SBc. Member of the Virgo Galaxy Cluster.

NGC 4179

Gx	Vir	12^h 12.9^m	+01° 18′	4.2′ × 1.3′
charts 13, 14			m$_v$ 11	p.a. 143°

Pretty bright, pretty small, and elongated. Distance 50 million ly; Hubble class S0.

NGC 4183

Gx CVn $12^h 13.3^m$ +43° 42' 5.0' × 0.9'
charts 6, 7 m_v 12.3 p.a. 166°

Very faint, quite large, and very elongated; filamentary arms; very faint nucleus. Distance 43 million ly; Hubble class S–.

NGC 4189

Gx Com $12^h 13.8^m$ +13° 26' 2.4' × 1.7'
charts 7, 13, 14, B1 m_v 11.7 p.a. 85°

Faint, large, and slightly elongated; several filamentary arms; small, bright nucleus. Hubble class Sc. Member of the Virgo Galaxy Cluster; in a group with galaxies NGC 4164 and NGC 4193.

NGC 4192

See M98.

NGC 4193

Gx Vir $12^h 13.9^m$ +13° 10' 2.2' × 1.1'
charts 13, 14, B1 m_v 12.3 p.a. 93°

Very faint, pretty large, and elongated; small, bright nucleus. Distance 110 million ly; Hubble class S(B)c. In a group with the galaxies NGC 4164 and NGC 4189.

NGC 4194

Gx UMa $12^h 14.2^m$ +54° 32' 1.8' × 1.1'
charts 2, 7 m_v 12.5 p.a. 150°

Pretty bright and very small; bright nucleus. Distance 110 million ly; Hubble class Irp+.

NGC 4203

Gx Com $12^h 15.1^m$ +33° 12' 3.5' × 3.2'
charts 6, 7 m_v 10.9 p.a. 10°

Very bright, small, and round; very bright, diffuse nucleus. Distance 44 million ly; Hubble class Ep.

NGC 4206

Gx Vir $12^h 15.3^m$ +13° 02' 5.0' × 0.9'
charts 13, 14, B1 m_v 12.2 p.a. 0°

Faint and very elongated; many arms; dark lanes; very small, bright nucleus. Hubble class Sb. In a group with galaxies NGC 4216, NGC 4222, and IC 771; member of the Virgo Galaxy Cluster.

NGC 4212

Gx Com $12^h 15.7^m$ +13° 54' 3.1' × 2.0'
charts 7, 13, 14, B1 m_v 11.2 p.a. 75°

Bright, large, and elongated; several knotty arms; bright nucleus. Hubble class Sc. Member of the Virgo Galaxy Cluster.

NGC 4214

Gx CVn $12^h 15.7^m$ +36° 20' 7.9' × 6.3'
charts 6, 7 m_v 9.8 p.a. 130°

Quite bright, quite large, and slightly elongated. Distance 13 million ly; Hubble class Ir+.

NGC 4215

Gx Vir $12^h 15.9^m$ +06° 24' 1.9' × 0.8'
charts 13, 14, B1 m_v 12.1 p.a. 174°

Bright, pretty small, and elongated; edge on; extremely small, very bright nucleus. Distance 86 million ly; Hubble class Sb.

NGC 4216

Gx Vir $12^h 15.9^m$ +13° 09' 8.3' × 2.2'
charts 13, 14, B1 m_v 10 p.a. 19°

Very bright, very large, and very elongated; nearly edge on; small, extremely bright nucleus with dark lane. A very curious object, in shape resembling a weaver's shuttle (Smyth). Distance 50 million ly; Hubble class Sb. In a group with galaxies NGC 4206, NGC 4222, and IC 771; member of the Virgo Galaxy Cluster.

NGC 4217

Gx CVn $12^h 15.8^m$ +47° 06' 4.9' × 1.5'
charts 2, 6, 7 m_v 11.2 p.a. 50°

Pretty faint, large, and very elongated. Distance 46 million ly; Hubble class Sb. Paired with fainter galaxy NGC 4226 to E.

NGC 4219

Gx Cen $12^h 16.5^m$ –43° 19' 4.3' × 1.2'
charts 20, 21, 25 m_v 11.9 p.a. 36°

Pretty faint, pretty large, and quite elongated; very faint arms; bright nucleus. Distance 76 million ly; Hubble class Sb.

NGC 4220

Gx	CVn	12^h 16.2^m	+47° 53′	3.8′ × 1.5′
charts 2, 6, 7			m$_v$ 11.4	p.a. 141°

Quite bright, pretty large, and elongated; small, bright nucleus. Distance 46 million ly; Hubble class Sa. Paired with galaxy NGC 4218.

NGC 4224

Gx	Vir	12^h 16.6^m	+07° 28′	2.5′ × 1.0′
charts 13, 14, B1			m$_v$ 11.8	p.a. 57°

Pretty bright, pretty small, and slightly elongated; edge on; dust lane. Distance 110 million ly; Hubble class Sa.

NGC 4233

Gx	Vir	12^h 17.1^m	+07° 37′	2.3′ × 1.1′
charts 13, 14, B1			m$_v$ 11.9	p.a. 174°

Pretty faint and round. Hubble class S0. Member of the Virgo Galaxy Cluster.

NGC 4235

Gx	Vir	12^h 17.1^m	+07° 12′	3.8′ × 0.9′
charts 13, 14, B1			m$_v$ 11.6	p.a. 48°

Pretty bright, pretty large, and very elongated; nearly edge on; smooth arms with dark lane; small, very bright, diffuse nucleus; a Seyfert galaxy. Distance 110 million ly; Hubble class Sa. In a group with galaxies NGC 4246 and NGC 4247.

NGC 4236

Gx	Dra	12^h 16.7^m	+69° 28′	22′ × 6′
chart 2			m$_v$ 9.6	p.a. 162°

C3, Holmberg 357a. Very faint, extremely large, and very elongated. Distance 7 million ly; Hubble class SB+. Paired with galaxy Holmberg 357b.

NGC 4237

Gx	Com	12^h 17.2^m	+15° 19′	2.1′ × 1.3′
charts 7, 13, 14, B1			m$_v$ 11.6	p.a. 108°

Pretty bright, pretty large, and slightly elongated; many arms; small, bright nucleus. Distance 38 million ly; Hubble class S(B)c. Member of the Virgo Galaxy Cluster.

NGC 4240

Gx	Vir	12^h 17.4^m	−09° 57′	1.3′ × 1.2′
charts 13, 14			m$_v$ 12.7	p.a. 115°

Pretty bright, small. Distance 79 million ly; Hubble class E/S0. A 12th-mag. star is located approx. 0.5′ toward SW.

NGC 4241

Gx	Vir	12^h 17.4^m	+06° 41′	2.6′ × 1.5′
charts 13, 14, B1			m$_v$ 11.9	p.a. 128°

Very faint and large; bright nucleus. Distance 94 million ly; Hubble class Sa. A 7th-mag. star to S.

NGC 4242

Gx	CVn	12^h 17.5^m	+45° 37′	4.8′ × 3.8′
charts 6, 7			m$_v$ 10.8	p.a. 25°

Very faint, quite large, irregularly round; very faint arms. Distance 33 million ly; Hubble class S–.

NGC 4244

Gx	CVn	12^h 17.5^m	+37° 48′	16.2′ × 2.5′
charts 6, 7			m$_v$ 10.4	p.a. 48°

C26. Pretty bright, very large, and extremely elongated; edge on; very small, bright nucleus. Like a somewhat fainter version of the famed NGC 4565, but the dark lane is hard to detect visually (Mullaney). Distance 12 million ly; Hubble class S–.

NGC 4245

Gx	Com	12^h 17.6^m	+29° 37′	3.2′ × 2.7′
charts 6, 7			m$_v$ 11.4	p.a. 135°

Quite bright, pretty large, and slightly elongated; very small, very bright nucleus. Distance 38 million ly; Hubble class S–. Paired with galaxy NGC 4253.

NGC 4248

Gx	CVn	12^h 17.8^m	+47° 25′	3.0′ × 1.2′
charts 2, 6, 7			m$_v$ 12.5	p.a. 108°

Very faint, small, and quite elongated; bright nucleus. Distance 24 million ly; Hubble class Ir–.

NGC 4251

Gx	Com	12^h 18.1^m	+28° 11'	3.6' × 2.2'
chart 7			m$_v$ 10.7	p.a. 100°

Very bright, small, and elongated; very bright, diffuse nucleus. Distance 43 million ly; Hubble class E7.

NGC 4254

See M99.

NGC 4256

Gx	Dra	12^h 18.7^m	+65° 54'	4.1' × 0.7'
chart 2			m$_v$ 11.9	p.a. 42°

Pretty bright, large, and quite elongated; edge on; small, bright nucleus; dark lane. Distance 120 million ly; Hubble class Sb. Located between two 7th-mag. stars.

NGC 4258

See M106.

NGC 4260

Gx	Vir	12^h 19.4^m	+06° 06'	2.5' × 1.2'
charts 13, 14, B1			m$_v$ 11.8	p.a. 58°

Pretty bright and elongated; has two arms; small, very bright nucleus. Distance 75 million ly; Hubble class SBb.

NGC 4261

Gx	Vir	12^h 19.4^m	+05° 50'	4.0' × 3.5'
charts 13, 14, B1			m$_v$ 10.4	p.a. 160°

Pretty bright, pretty small, and round. Distance 91 million ly; Hubble class E2. Paired with galaxy NGC 4264.

NGC 4262

Gx	Com	12^h 19.5^m	+14° 53'	1.9' × 1.7'
charts 7, 13, 14, B1			m$_v$ 11.6	p.a. 125°

Bright, small, and round; extremely bright nucleus. Hubble class E1. Member of the Virgo Galaxy Cluster.

NGC 4267

Gx	Vir	12^h 19.8^m	+12° 48'	3.2'
charts 13, 14, B1			m$_v$ 10.9	

Pretty bright, very small, and round; very bright, diffuse nucleus. Hubble class E2. Member of the Virgo Galaxy Cluster.

NGC 4270

Gx	Vir	12^h 19.8^m	+05° 28'	1.9' × 0.9'
charts 13, 14, B1			m$_v$ 12.2	p.a. 110°

Pretty bright, small, and elongated. Distance 97 million ly; Hubble class Sa. In a group with galaxies NGC 4273 and NGC 4281.

NGC 4273

Gx	Vir	12^h 19.9^m	+05° 21'	2.3' × 1.5'
charts 13, 14, B1			m$_v$ 11.9	p.a. 10°

Pretty bright, large, and elongated. Distance 95 million ly; Hubble class Sc. Brightest in a group of galaxies that includes NGC 4270 and NGC 4281.

NGC 4274

Gx	Com	12^h 19.8^m	+29° 37'	6.9' × 2.8'
chart 7			m$_v$ 10.4	p.a. 102°

Very bright, very large, and elongated; faint arms; very bright nucleus. Distance 31 million ly; Hubble class Sb.

NGC 4278

Gx	Com	12^h 20.1^m	+29° 17'	4.0' × 3.7'
chart 7			m$_v$ 10.2	p.a. 22°

Very bright, pretty large, and round; very bright center. Distance 28 million ly; Hubble class E1. Paired with galaxy NGC 4283.

NGC 4281

Gx	Vir	12^h 20.4^m	+05° 23'	3.0' × 1.6'
charts 13, 14, B1			m$_v$ 11.3	p.a. 88°

Bright, large, and round. Distance 110 million ly; Hubble class E5. In a group with galaxies NGC 4270 and NGC 4273.

NGC 4283

Gx	Com	12^h 20.3^m	+29° 19'	1.4'
chart 7			m$_v$ 12.1	

Bright, small, and round; small, very bright center. Distance 49 million ly; Hubble class E0.

Paired with galaxy NGC 4278.

NGC 4290

Gx	UMa	12^h 20.8^m	+58° 06′	2.3′ × 1.6′
chart 2			m$_v$ 11.8	p.a. 90°

Pretty bright, large, and round; very small, very bright nucleus. Distance 120 million ly; Hubble class SBb. Paired with galaxy NGC 4284.

NGC 4291

Gx	Dra	12^h 20.3^m	+75° 22′	2.0′ × 1.7′
charts 1, 2			m$_v$ 11.5	p.a. 110°

Pretty bright, very small, and round. Distance 86 million ly; Hubble class E1. Paired with galaxy NGC 4319.

NGC 4293

Gx	Com	12^h 21.2^m	+18° 23′	5.0′ × 2.8′
charts 7, 13, 14, B1			m$_v$ 10.4	p.a. 72°

Faint, very large, elongated; very small, bright nucleus with dark lane. Distance 36 million ly; Hubble class Sap. Member of the Virgo Galaxy Cluster.

NGC 4294

Gx	Vir	12^h 21.3^m	+11° 31′	3.0′ × 1.2′
charts 13, 14, B1			m$_v$ 12.1	p.a. 155°

Faint, large, and very elongated; very faint nucleus. Hubble class S(B)c. Paired with galaxy NGC 4299; member of the Virgo Galaxy Cluster.

NGC 4298

Gx	Com	12^h 21.5^m	+14° 36′	3.1′ × 1.8′
charts 7, 13, 14, B1			m$_v$ 11.3	p.a. 140°

Faint, large, and elongated; many arms with dark lanes; very small, bright nucleus. Distance 45 million ly; Hubble class Sc. Paired with galaxy NGC 4302; member of the Virgo Galaxy Cluster.

NGC 4299

Gx	Vir	12^h 21.7^m	+11° 30′	1.7′ × 1.6′
charts 13, 14, B1			m$_v$ 12.5	p.a. 26°

Faint, large, and slightly elongated; very faint nucleus. Hubble class S. Paired with galaxy NGC 4294; member of the Virgo Galaxy Cluster.

NGC 4302

Gx	Com	12^h 21.7^m	+14° 36′	5.2′ × 1.1′
charts 7, 13, 14, B1			m$_v$ 11.6	p.a. 178°

Large and very elongated; nearly edge on; bright center with dark lane. Distance 45 million ly; Hubble class Sc. Member of the Virgo Galaxy Cluster.

NGC 4303

See M61.

NGC 4304

Gx	Hya	12^h 22.2^m	−33° 29′	2.8′ × 2.7′
charts 20, 21			m$_v$ 11.7	p.a. 110°

Very faint, very large, and round; patchy arms; bright nucleus with dust lane. Distance 100 million ly; Hubble class SBb. An 8th-mag. double star lies 7′ to SE of this object.

NGC 4307

Gx	Vir	12^h 22.1^m	+09° 03′	3.5′ × 0.8′
charts 13, 14, B1			m$_v$ 12	p.a. 24°

Holmberg 380a. Pretty faint, large, and very elongated; edge on. Hubble class Sb. Paired with Holmberg 380b; member of the Virgo Galaxy Cluster.

NGC 4312

Gx	Com	12^h 22.5^m	+15° 32′	4.6′ × 1.1′
charts 7, 13, 14, B1			m$_v$ 11.7	p.a. 170°

Pretty bright, quite large, and elongated. Hubble class Sb. Member of the Virgo Galaxy Cluster; in a group with M100 (NGC 4321) and several faint galaxies. A faint, wide double star lies about 3′ to the east of NGC 4312.

NGC 4313

Gx	Vir	12^h 22.6^m	+11° 48′	3.8′ × 0.9′
charts 13, 14, B1			m$_v$ 11.6	p.a. 143°

Very faint, large, and elongated; edge on; small, bright nucleus. Hubble class Sb. Member of the Virgo Galaxy Cluster.

NGC 4314

Gx	Com	$12^h 22.5^m$	$+29° 54'$	$3.9' \times 3.7'$
chart 7			m_v 10.6	p.a. 65°

Quite bright, large, and elongated; extremely bright nucleus; unconnected inner and outer spiral patterns. Distance 38 million ly; Hubble class SBa.

NGC 4321

See M100.

NGC 4324

Gx	Vir	$12^h 23.1^m$	$+05° 15'$	$2.8' \times 1.2'$
charts 13, 14, B1			m_v 11.6	p.a. 53°

Pretty bright, and round; small, very bright nucleus. Distance 70 million ly; Hubble class Sb.

NGC 4339

Gx	Vir	$12^h 23.6^m$	$+06° 05'$	2.2'
charts 13, 14, B1			m_v 11.3	

Bright, pretty large, and round. Distance 51 million ly; Hubble class E0. Brightest in a group of three galaxies.

NGC 4340

Gx	Com	$12^h 23.6^m$	$+16° 43'$	$3.2' \times 2.5'$
charts 7, 13, 14, B1			m_v 11.2	p.a. 102°

Pretty bright, small, and round; small, very bright nucleus. Hubble class SBa. Paired with galaxy NGC 4350; member of the Virgo Galaxy Cluster.

NGC 4343

Gx	Vir	$12^h 23.6^m$	$+06° 57'$	$2.3' \times 0.7'$
charts 13, 14, B1			m_v 12.1	p.a. 133°

Pretty faint, small, and elongated. Distance 44 million ly; Hubble class Sa. Member of the Virgo Galaxy Cluster.

NGC 4346

Gx	CVn	$12^h 23.5^m$	$+47° 00'$	$3.2' \times 1.3'$
charts 2, 6, 7			m_v 11.1	p.a. 99°

Very faint, small, and elongated. Distance 40 million ly; Hubble class E6.

NGC 4349

OC	Cru	$12^h 24.5^m$	$-61° 54'$	15'
chart 25			m_p 7.4	

30 stars; detached, strong concentration of stars; moderate range in brightness; mag. of brightest star 10.9; very bright, very large. Distance 5,500 ly; age 220 million years; Trumpler class I 2 m.

NGC 4350

Gx	Com	$12^h 24.0^v$	$+16° 42'$	$2.9' \times 1.5'$
charts 7, 13, 14, B1			m_v 11	p.a. 28°

Quite bright, very small, and very elongated; very small, very bright nucleus. Distance 49 million ly; Hubble class E7. Member of the Virgo Galaxy Cluster; paired with galaxy NGC 4340.

NGC 4351

Gx	Vir	$12^m 24.0^m$	$+12° 12'$	$1.9' \times 1.3'$
charts 13, 14, B1			m_v 12.6	p.a. 80°

Faint, pretty large, irregularly roundish; bright nucleus. Hubble class SBbp. Member of the Virgo Galaxy Cluster.

NGC 4361

PN	Crv	$12^h 24.5^m$	$-18° 48'$	63"
charts 13, 14, 21			m_v 10.9	

PK294+43.1. Very bright, large, round disk involved in a larger, fainter disk; phot. mag. 10.3; central star mag. 13.2. Expansion velocity 38 km/sec (24 miles/sec); distance 2,600 ly; V-V class 3a + 2.

NGC 4365

Gx	Vir	$12^h 24.5^m$	$+07° 19'$	$6.2' \times 4.6'$
charts 13, 14, B1			m_v 9.6	p.a. 40°

Quite bright, pretty large, and slightly elongated; very bright center; paired with galaxy NGC 4370. Distance 47 million ly; Hubble class E2. Member of the Virgo Galaxy Cluster.

NGC 4369

Gx	CVn	$12^h 24.6^m$	$+39° 23'$	2.1'
charts 6, 7			m_v 11.7	p.a. 127°

Quite bright, small, and round; very bright center. Distance 46 million ly; Hubble class Sa.

NGC 4371

| Gx | Vir | 12ʰ 24.9ᵐ | +11° 42′ | 4.1′ × 2.1′ |
| charts 13, 14, B1 | | | m$_v$ 10.8 | p.a. 95° |

Bright, pretty small, and round; very bright nucleus. Distance 39 million ly; Hubble class SBa. Member of the Virgo Galaxy Cluster.

NGC 4372

| GC | Mus | 12ʰ 25.8ᵐ | −72° 40′ | 5′ |
| charts 25, 26 | | | m$_v$ 7.2 | |

C108. Low concentration of stars; pretty faint, large, and round. Distance 1,600 ly; S-S class 12. A 6.6-magnitude foreground star lies at the cluster's NW edge.

NGC 4373

| Gx | Cen | 12ʰ 25.3ᵐ | −39° 46′ | 4.1′ × 3.0′ |
| charts 20, 21 | | | m$_v$ 10.6 | p.a. 43° |

Pretty bright, small, and round; bright nucleus. Distance 140 million ly; Hubble class S0. Paired with galaxy IC 3290.

NGC 4373A

| Gx | Cen | 12ʰ 25.6ᵐ | −39° 19′ | 2.8′ × 1.0′ |
| charts 20, 21 | | | m$_v$ 12.1 | p.a. 149° |

Distance 120 million ly; Hubble class S0.

NGC 4374

See M84.

NGC 4377

| Gx | Com | 12ʰ 25.2ᵐ | +14° 46′ | 1.7′ × 1.3′ |
| charts 7, 13, 14, B1 | | | m$_v$ 11.9 | p.a. 177° |

Bright, small, and round; small, very bright nucleus. Hubble class E1. Member of the Virgo Galaxy Cluster; located in a group with two faint galaxies.

NGC 4378

| Gx | Vir | 12ʰ 25.3ᵐ | +04° 56′ | 2.5′ × 2.3′ |
| charts 13, 14, B1 | | | m$_v$ 11.7 | p.a. 167° |

Bright and small with a small, very bright, diffuse nucleus. Distance 110 million ly; Hubble class Sa.

NGC 4379

| Gx | Com | 12ʰ 25.2ᵐ | +15° 36′ | 1.9′ × 1.6′ |
| charts 7, 13, 14, B1 | | | m$_v$ 11.7 | p.a. 105° |

Pretty small and round. Distance 42 million ly; Hubble class E1. Member of the Virgo Galaxy Cluster.

NGC 4380

| Gx | Vir | 12ʰ 25.4ᵐ | +10° 01′ | 3.2′ × 1.8′ |
| charts 13, 14, B1 | | | m$_v$ 11.7 | p.a. 153° |

Very faint, pretty large, and round; very small, bright nucleus. Distance 38 million ly; Hubble class Sb. Member of the Virgo Galaxy Cluster.

NGC 4382

See M85.

NGC 4383

| Gx | Com | 12ʰ 25.4ᵐ | +16° 28′ | 2.0′ × 1.0′ |
| charts 7, 13, 14, B1 | | | m$_v$ 12.1 | p.a. 28° |

Extremely small; appears stellar; very small, very bright nucleus with dark lane. Hubble class Ep. Member of the Virgo Galaxy Cluster.

NGC 4386

| Gx | Dra | 12ʰ 24.5ᵐ | +75° 32′ | 2.6′ × 1.6′ |
| charts 1, 2 | | | m$_v$ 11.7 | p.a. 135° |

Pretty bright and quite large. Distance 81 million ly; Hubble class Sa.

NGC 4387

| Gx | Vir | 12ʰ 25.7ᵐ | +12° 49′ | 1.7′ × 1.1′ |
| charts 13, 14, B1 | | | m$_v$ 12.1 | p.a. 140° |

Pretty faint, very small, and roundish. Hubble class E5. Member of the Virgo Galaxy Cluster. A 13th-mag. star is about 1.5′ to NW.

NGC 4388

| Gx | Vir | 12ʰ 25.8ᵐ | +12° 40′ | 5.1′ × 1.4′ |
| charts 13, 14, B1 | | | m$_v$ 11 | p.a. 92° |

Very faint and elongated; edge on. Hubble class Sb. Member of the Virgo Galaxy Cluster.

NGC 4389

Gx	CVn	12^h 25.6^m	+45° 41′	2.5′ × 1.4′
charts 6, 7			m$_v$ 11.7	p.a. 105°

Pretty bright, pretty large, and slightly elongated; small, bright nucleus with dark lane. Distance 34 million ly; Hubble class SB–.

NGC 4394

Gx	Com	12^h 25.9^m	+18° 13′	3.4′ × 3.2′
charts 7, 13, 14, B1			m$_v$ 10.9	p.a. 140°

Pretty bright and slightly elongated; extremely bright nucleus with dark lane. Hubble class S+. Member of the Virgo Galaxy Cluster; paired with galaxy M85.

NGC 4395

Gx	CVn	12^h 25.8^m	+33° 33′	13′ × 11′
chart 7			m$_v$ 10.2	p.a. 147°

Extremely faint and very large. Distance 13 million ly; Hubble class S+. In a group with three faint galaxies (NGC 4399, NGC 4400, and NGC 4401) forming a trapezium.

NGC 4402

Gx	Vir	12^h 26.1^m	+13° 07′	3.6′ × 1.1′
charts 13, 14, B1			m$_v$ 11.8	p.a. 90°

Faint, large, and very elongated; edge on; has dust lanes. Hubble class Sb. Paired with galaxy M86; member of the Virgo Galaxy Cluster.

NGC 4405

Gx	Com	12^h 26.1^m	+16° 11′	1.7′ × 1.1′
charts 7, 13, 14, B1			m$_v$ 12	p.a. 20°

Pretty faint, small, and roundish; bright nucleus. Hubble class Sa. Member of the Virgo Galaxy Cluster.

NGC 4406

See M86.

NGC 4411B

Gx	Vir	12^h 26.8^m	+08° 53′	2.5′
charts 13, 14, B1			m$_v$ 12.3	

Hubble class S(B)d. Member of the Virgo Galaxy Cluster.

NGC 4413

Gx	Vir	12^h 26.5^m	+12° 37′	2.3′ × 1.4′
charts 13, 14, B1			m$_v$ 12.2	p.a. 60°

Quite faint and small; Hubble class SBb. Member of the Virgo Galaxy Cluster.

NGC 4414

Gx	Com	12^h 26.5^m	+31° 13′	3.8′ × 2.2′
chart 7			m$_v$ 10.1	p.a. 155°

Very bright, large, and elongated; bright, diffuse nucleus in bright bulge; multiple arms. Distance 31 million ly; Hubble class Sc.

NGC 4417

Gx	Vir	12^h 26.8^m	+09° 35′	3.3′ × 1.4′
charts 13, 14, B1			m$_v$ 11.1	p.a. 49°

Faint, pretty large, and elongated; edge on; faint, elongated nucleus. Hubble class E7. Member of the Virgo Galaxy Cluster.

NGC 4419

Gx	Com	12^h 26.9^m	+15° 03′	3.3′ × 1.1′
charts 7, 13, 14, B1			m$_v$ 11.2	p.a. 133°

Bright and elongated; strong arms with dark lanes; bright center. Hubble class Ep. Member of the Virgo Galaxy Cluster.

NGC 4420

Gx	Vir	12^h 27.0^m	+02° 30′	2.0′ × 1.0′
charts 13, 14			m$_v$ 12.1	p.a. 8°

Faint, pretty large, and slightly elongated; very faint nucleus. Distance 68 million ly; Hubble class Sc.

NGC 4421

Gx	Com	12^h 27.0^m	+15° 28′	2.8′ × 2.1′
charts 7, 13, 14, B1			m$_v$ 11.6	p.a. 20°

Pretty bright and pretty large; bright nucleus. Hubble class SBa. Bright star to NW. Member of the Virgo Galaxy Cluster.

NGC 4424

Gx	Vir	12^h 27.2^m	+09° 25′	3.5′ × 1.7′
charts 13, 14, B1			m$_v$ 11.7	p.a. 95°

Faint, pretty large, irregularly round. Hubble class Sb. Member of the Virgo Galaxy Cluster.

NGC 4425

Gx	Vir	12^h 27.2^m	+12° 44′	2.9′ × 1.0′
charts 13, 14, B1			m$_v$ 11.8	p.a. 27°

Pretty faint, small, and round. Hubble class SB–. Member of the Virgo Galaxy Cluster.

NGC 4429

Gx	Vir	12^h 27.4^m	+11° 06′	5.5′ × 2.6′
charts 13, 14, B1			m$_v$ 10	p.a. 99°

Bright, large, and quite elongated; very bright nucleus with dark crescent. Distance 45 million ly; Hubble class S0. Member of the Virgo Galaxy Cluster.

NGC 4430

Gx	Vir	12^h 27.4^m	+06° 16′	2.4′ × 2.1′
charts 13, 14, B1			m$_v$ 12	p.a. 80°

Quite faint, large, and roundish; bright nucleus. Distance 59 million ly; Hubble class SBbp.

NGC 4434

Gx	Vir	12^h 27.6^m	+08° 09′	1.4′ × 1.3′
charts 13, 14, B1			m$_v$ 12.2	p.a. 133°

Pretty faint and very small. Distance 40 million ly; Hubble class E0.

NGC 4435

Gx	Vir	12^h 27.7^m	+13° 05′	3.0′ × 2.1′
charts 13, 14, B1			m$_v$ 10.8	p.a. 13°

One of the Eyes. Very bright, quite large, and round; very bright nucleus. Hubble class E4. Member of the Virgo Galaxy Cluster; interacting with galaxy NGC 4438 (the other Eye).

NGC 4438

Gx	Vir	12^h 27.8^m	+13° 01′	9.3′ × 3.9′
charts 13, 14, B1			m$_v$ 10.2	p.a. 27°

One of the Eyes. Bright, quite large, slightly elongated; very small, very bright nucleus; dark lane. Hubble class Sap. Member of the Virgo Galaxy Cluster; interacting with galaxy NGC 4435 (the other Eye).

NGC 4439

OC	Cru	12^h 28.4^m	−60° 06′	4′
chart 25			m$_v$ 8.4	

Few stars; detached, weak concentration of stars; small range in brightness; mag. of brightest star 10.3; small. Distance 5,200 ly; age 63 million years; Trumpler class II 1 p.

NGC 4440

Gx	Vir	12^h 27.9^m	+12° 18′	1.8′ × 1.5′
charts 13, 14, B1			m$_v$ 11.7	p.a. 93°

Bright, small, and roundish; very bright nucleus. Hubble class SBa. Member of the Virgo Galaxy Cluster.

NGC 4442

Gx	Vir	12^h 28.1^m	+09° 48′	4.5′ × 1.8′
charts 13, 14, B1			m$_v$ 10.4	p.a. 87°

Very bright, pretty large, and round; very bright, diffuse nucleus. Hubble class E5p. Member of the Virgo Galaxy Cluster.

NGC 4444

Gx	Cen	12^h 28.6^m	−43° 16′	2.8′ × 2.5′
charts 20, 21, 25			m$_v$ 12.3	p.a. 120°

Extremely faint, large, and roundish; bright nucleus. Distance 120 million ly; Hubble class SBb.

NGC 4448

Gx	Com	12^h 28.3^m	+28° 37′	3.6′ × 1.3′
chart 7			m$_v$ 11.1	p.a. 94°

Bright, large, and elongated; two arms with dark lanes; bright, diffuse nucleus. Distance 30 million ly; Hubble class Sb.

NGC 4449

Gx	CVn	12^h 28.2^m	+44° 06′	6.0′ × 4.9′
charts 6, 7			m$_v$ 9.6	p.a. 45°

C21; very bright and quite large; faint nucleus. Distance 11 million ly; Hubble class Ir+.

NGC 4450

Gx	Com	$12^h 28.5^m$	+17° 05′	5.0′ × 4.0′
charts 7, 13, 14, B1			m_v 10.1	p.a. 175°

Bright, large, and round; small, very bright, diffuse nucleus; bright star nearby; circular dust lanes start outside nucleus. Hubble class Sb. Member of the Virgo Galaxy Cluster.

NGC 4452

Gx	Vir	$12^h 28.7^m$	+11° 45′	3.0′ × 0.7′
charts 13, 14, B1			m_v 12	p.a. 32°

Pretty bright, small, and very elongated. Hubble class S0. Member of the Virgo Galaxy Cluster.

NGC 4454

Gx	Vir	$12^h 28.9^m$	−01° 56′	2.0′ × 1.8′
charts 13, 14			m_v 11.9	p.a. 100°

Faint, large, and round; very small, bright nucleus. Distance 97 million ly; Hubble class Sb.

NGC 4457

Gx	Vir	$12^h 29.0^m$	+03° 34′	2.8′ × 2.4′
charts 13, 14, B1			m_v 10.9	p.a. 75°

Quite bright, pretty small, and round; very bright nucleus; one spiral arm brighter than other. Distance 27 million ly; Hubble class S(B)a.

NGC 4458

Gx	Vir	$12^h 29.0^m$	+13° 15′	1.6′ × 1.5′
charts 13, 14, B1			m_v 12.1	p.a. 45°

Pretty bright, small, and roundish; bright nucleus. Hubble class E0. Paired with galaxy NGC 4461; member of the Virgo Galaxy Cluster.

NGC 4459

Gx	Com	$12^h 29.0^m$	+13° 59′	3.8′ × 3.0′
charts 13, 14, B1			m_v 10.4	p.a. 110°

Pretty bright, pretty large, irregularly round; dark crescent; small, very bright nucleus; internal absorption ring. Distance 45 million ly; Hubble class E2. Member of the Virgo Galaxy Cluster.

NGC 4460

Gx	CVn	$12^h 28.8^m$	+44° 52′	4.2′ × 1.3′
charts 6, 7			m_v 11.3	p.a. 40°

Bright, pretty large, and elongated; faint nucleus. Distance 27 million ly; Hubble class SB0.

NGC 4461

Gx	Vir	$12^h 29.0^m$	+13° 11′	3.4′ × 1.4′
charts 13, 14, B1			m_v 11.2	p.a. 9°

Pretty faint, small, and round; very small, very bright, diffuse nucleus. Hubble class Sa. Paired with galaxy NGC 4458; member of the Virgo Galaxy Cluster.

NGC 4462

Gx	Crv	$12^h 29.4^m$	−23° 10′	3.2′ × 1.1′
chart 21			m_v 11.9	p.a. 124°

Pretty bright, pretty small, and elongated; very bright nucleus. Distance 72 million ly; Hubble class Sb.

NGC 4463

OC	Mus	$12^h 30.0^m$	−64° 48′	5′
chart 25			m_v 7.2	

30 stars; detached, strong concentration of stars; large range in brightness; mag. of brightest star 8.3; Trumpler class I 3 p.

NGC 4469

Gx	Vir	$12^h 29.5^m$	+08° 45′	3.5′ × 1.2′
charts 13, 14, B1			m_v 11.2	p.a. 89°

Pretty faint, pretty large, and very elongated; edge on; very bright nucleus. Hubble class Sp. Member of the Virgo Galaxy Cluster.

NGC 4470

Gx	Vir	$12^h 29.6^m$	+07° 49′	1.3′ × 0.9′
charts 13, 14, B1			m_v 12.1	p.a. 0°

Faint, pretty large, irregularly roundish; bright nucleus. Distance 100 million ly; Hubble class Sa. In a group with galaxies NGC 4467, M49 (NGC 4472), and a faint irregular dwarf galaxy.

NGC 4472

See M49.

NGC 4473

Gx   Com $12^h 29.8^m$ +13° 26'   4.1' × 2.5'
charts 13, 14, B1 m_v 10.2   p.a. 100°

Pretty bright. Hubble class E4. Member of the Virgo Galaxy Cluster.

NGC 4474

Gx   Com $12^h 29.9^m$ +14° 04'   2.5' × 1.6'
charts 13, 14, B1 m_v 11.5   p.a. 80°

Pretty faint and round. Hubble class E6. Member of the Virgo Galaxy Cluster.

NGC 4477

Gx   Com $12^h 30.0^m$ +13° 38'   3.6' × 3.3'
charts 13, 14, B1 m_v 10.4   p.a. 15°

Pretty bright and quite large. Hubble class S(B)a. Member of the Virgo Galaxy Cluster.

NGC 4478

Gx   Vir $12^h 30.3^m$ +12° 20'   1.7' × 1.4'
charts 13, 14, B1 m_v 11.4   p.a. 140°

Pretty bright, small, and round; extremely bright center. Hubble class E1. Paired with galaxy NGC 4476; member of the Virgo Galaxy Cluster.

NGC 4480

Gx   Vir $12^h 30.4^m$ +04° 15'   2.2' × 1.1'
charts 13, 14, B1 m_v 12.4   p.a. 175°

Pretty faint, pretty small, and elongated. Distance 95 million ly; Hubble class S(B)c.

NGC 4485

Gx   CVn $12^h 30.5^m$ +41° 42'   2.5' × 1.8'
charts 6, 7 m_v 11.9   p.a. 15°

Bright, pretty small, irregularly round. Distance 36 million ly; Hubble class Ir+. Along with galaxy NGC 4490, it forms the interacting system Arp 269.

NGC 4486

See M87.

NGC 4487

Gx   Vir $12^h 31.1^m$ −08° 03'   3.9' × 2.7'
charts 13, 14 m_v 10.9   p.a. 75°

Faint, very large; very small, very bright nucleus. Distance 38 million ly; Hubble class Sc.

NGC 4488

Gx   Vir $12^h 30.9^m$ +08° 22'   3.9' × 1.4'
charts 13, 14, B1 m_v 12.2   p.a. 160°

Very faint, very small, and slightly elongated. Distance 41 million ly; Hubble class SBap.

NGC 4489

Gx   Com $12^h 30.9^m$ +16° 46'   1.7' × 1.6'
charts 7, 13, 14, B1 m_v 12   p.a. 115°

Pretty faint, quite small, and roundish; bright nucleus. Distance 35 million ly; Hubble class E1. Member of the Virgo Galaxy Cluster.

NGC 4490

Gx   CVn $12^h 30.6^m$ +41° 38'   6.0' × 3.2'
charts 6, 7 m_v 9.8   p.a. 125°

Very bright, very large, and very elongated. Distance 27 million ly; Hubble class Sc. Along with galaxy NGC 4485, it forms the interacting system Arp 269.

NGC 4494

Gx   Com $12^h 31.4^m$ +25° 46'   4.5' × 4.3'
chart 7 m_v 9.8   p.a. 175°

Very bright, pretty large, and round; very bright center. Distance 56 million ly; Hubble class E1.

NGC 4496A

Gx   Vir $12^h 31.7^m$ +03° 56'   3.8' × 2.8'
charts 13, 14, B1 m_v 11.4   p.a. 70°

Distance 72 million ly; Hubble class S(B)c. Possibly colliding or interacting with galaxy NGC 4496B.

NGC 4496B

Gx   Vir $12^h 31.7^m$ +03° 56'   2.8' × 2.2'
charts 13, 14, B1 m_v 13.5   p.a. 100°

Distance 72 million ly; Hubble class Ir+. Possibly colliding or interacting with galaxy NGC 4496A.

NGC 4498

Gx Com $12^h 31.7^m$ +16° 51′ 3.0′ × 1.6′
charts 7, 13, 14, B1 m_v 12.2 p.a. 133°

Very faint, pretty large, and elongated. Hubble class S(B)c. Member of the Virgo Galaxy Cluster.

NGC 4501

See M88.

NGC 4503

Gx Vir $12^h 32.1^m$ +11° 11′ 3.5′ × 1.7′
charts 13, 14, B1 m_v 11.1 p.a. 12°

Pretty bright, small, and round; small, very bright nucleus. Hubble class E2. Member of the Virgo Galaxy Cluster.

NGC 4504

Gx Vir $12^h 32.3^m$ −07° 34′ 4.2′ × 2.6′
charts 13, 14 m_v 11.2 p.a. 145°

Pretty bright, quite large, and slightly elongated; two strong arms; very small, bright nucleus. Distance 37 million ly; Hubble class Sc.

NGC 4507

Gx Cen $12^h 35.6^m$ −39° 55′ 1.6′ × 1.3′
charts 20, 21, 25 m_v 12.1 p.a. 55°

Pretty bright, small, and round; bright arms; very bright, very small nucleus. A Seyfert galaxy. Distance 140 million ly; Hubble class S(B)0. Member of the Centaurus Galaxy Cluster.

NGC 4517

Gx Vir $12^h 32.8^m$ +00° 07′ 10.0′ × 1.7′
charts 13, 14 m_v 10.4 p.a. 83°

Quite bright, very large, and extremely elongated; very faint nucleus. Distance 44 million ly; Hubble class Sc. Paired with galaxy NGC 4517A; pretty bright star in contact.

NGC 4517A

Gx Vir $12^h 32.5^m$ +00° 23′ 4.1′ × 2.6′
charts 13, 14 m_v 12.5 p.a. 30°

Reinmuth 80. Distance 61 million ly; Hubble class SBd. Paired with galaxy NGC 4517.

NGC 4519

Gx Vir $12^h 33.5^m$ +08° 39′ 3.0′ × 2.4′
charts 13, 14, B1 m_v 11.8 p.a. 145°

Holmberg 418a. Faint, pretty large, round galaxy; three arms; bright, diffuse. Distance 47 million ly; Hubble class Sc. Member of the Virgo Galaxy Cluster; paired with galaxy Holmberg 418b.

NGC 4522

Gx Vir $12^h 33.7^m$ +09° 10′ 3.5′ × 1.0′
charts 13, 14, B1 m_v 12.3 p.a. 33°

Extremely faint, pretty large, and slightly elongated; edge-on spiral. Hubble class Scp. Member of the Virgo Galaxy Cluster.

NGC 4525

Gx Com $12^h 33.9^m$ +30° 17′ 2.6′ × 1.3′
chart 7 m_v 12.2 p.a. 53°

Faint, pretty large, irregularly roundish; bright nucleus. Distance 49 million ly; Hubble class Sc.

NGC 4526

Gx Vir $12^h 34.0^m$ +07° 42′ 7.0′ × 2.6′
charts 13, 14, B1 m_v 9.7 p.a. 112°

Lost Galaxy. Very bright, very large, and very elongated; small, bright diffuse nucleus; internal absorption ring. Hubble class E7. Member of the Virgo Galaxy Cluster.

NGC 4527

Gx Vir $12^h 34.1^m$ +02° 39′ 6.3′ × 2.3′
charts 13, 14 m_v 10.5 p.a. 67°

Pretty bright, large, and very elongated; two main arms; very small, extremely bright nucleus. Distance 70 million ly; Hubble class Sb.

NGC 4528

Gx Vir $12^h 34.1^m$ +11° 19′ 1.6′ × 1.0′
charts 13, 14, B1 m_v 12.1 p.a. 5°

Pretty faint, quite small, and roundish; bright nucleus. Hubble class S0. Member of the Virgo Galaxy Cluster.

NGC 4531

Gx Vir 12^h 34.3^m +13° 05' 3.5' × 2.3'
charts 13, 14, B1 m$_v$ 11.4 p.a. 155°

Faint, pretty large, and roundish; bright nucleus. Hubble class Sa. Member of the Virgo Galaxy Cluster.

NGC 4532

Gx Vir 12^h 34.3^m +06° 28' 2.8' × 1.2'
charts 13, 14, B1 m$_v$ 11.9 p.a. 160°

Pretty bright, pretty large, and very elongated. Distance 90 million ly; Hubble class Ir+.

NGC 4534

Gx CVn 12^h 34.1^m +35° 31' 2.9' × 2.3'
chart 7 m$_v$ 12.3 p.a. 125°

Quite faint, large, and slightly elongated; very gradually brighter toward the nucleus. Distance 36 million ly; Hubble class Sd.

NGC 4535

Gx Vir 12^h 34.3^m +08° 12' 6.8' × 5.0'
charts 13, 14, B1 m$_v$ 10 p.a. 0°

Holmberg 420a. Pretty faint and very large; an S-shaped spiral; two main arms; very small, extremely bright nucleus. Hubble class S(B)c. Member of the Virgo Galaxy Cluster; paired with galaxy Holmberg 420b.

NGC 4536

Gx Vir 12^h 34.4^m +02° 11' 7.0' × 3.0'
charts 13, 14 m$_v$ 10.6 p.a. 130°

Bright, very large, and very elongated; two main arms; very small, extremely bright nucleus. Distance 79 million ly; Hubble class Sc. Paired with galaxy NGC 4533.

NGC 4539

Gx Com 12^h 34.6^m +18° 12' 3.2' × 1.4'
charts 7, 13, 14, B1 m$_v$ 12 p.a. 95°

Pretty bright and quite elongated. Distance 55 million ly; Hubble class SBa.

NGC 4540

Gx Com 12^h 34.8^m +15° 33' 2.1' × 1.7'
charts 7, 13, 14, B1 m$_v$ 11.7 p.a. 40°

Pretty small and faint; very faint nucleus; many arms. Hubble class Ir. Paired with galaxy IC 3528; member of the Virgo Galaxy Cluster.

NGC 4546

Gx Vir 12^h 35.5^m −03° 48' 3.3' × 1.7'
charts 13, 14 m$_v$ 10.3 p.a. 80°

Very bright, quite large, and elongated; small, bright, diffuse nucleus. Distance 38 million ly; Hubble class E6.

NGC 4548

See M91.

NGC 4550

Gx Vir 12^h 35.5^m +12° 13' 3.2' × 0.9'
charts 13, 14, B1 m$_v$ 11.7 p.a. 178°

Pretty bright, small, and slightly elongated. Hubble class E7. Paired with galaxy NGC 4551; member of the Virgo Galaxy Cluster.

NGC 4551

Gx Vir 12^h 35.6^m +12° 16' 1.7' × 1.4'
charts 13, 14, B1 m$_v$ 12 p.a. 70°

Pretty bright, small, and roundish; bright nucleus. Distance 39 million ly; Hubble class E3. Paired with galaxy NGC 4550; member of the Virgo Galaxy Cluster.

NGC 4552

See M89.

NGC 4559

Gx Com 12^h 36.0^m +27° 58' 10.5' × 4.9'
chart 7 m$_v$ 10 p.a. 150°

C36. Very bright, very large, and very elongated; a multi-arm spiral; very small, faint nucleus. Distance 35 million ly; Hubble class Sc.

NGC 4561

Gx Com $12^h 36.1^m$ +19° 19' 1.3' × 0.9'
charts 7, 13, 14, B1 m_v 12.5   p.a. 30°

Pretty bright, pretty large, and slightly elongated; very faint nucleus. Distance 62 million ly; Hubble class Sc.

NGC 4564

Gx Vir $12^h 36.5^m$ +11° 26' 3.2' × 1.8'
charts 13, 14, B1 m_v 11.1   p.a. 47°

Pretty bright, small, and slightly elongated. Distance 41 million ly; Hubble class E6. Member of the Virgo Galaxy Cluster.

NGC 4565

Gx Com $12^h 36.3^m$ +25° 59' 16.2' × 2.8'
chart 7 m_v 9.6   p.a. 136°

C38. Bright, extremely large, and extremely elongated; edge on. A curious, long, and streaky object (Smyth). Like a ray of bright nebulosity (Denning). Even through a small telescope it looks like a pencil-thin shaft of grayish light (Harrington). Distance 49 million ly; Hubble class Sb.

NGC 4567

Gx Vir $12^h 36.5^m$ +11° 16' 3.0' × 2.1'
charts 13, 14, B1 m_v 11.3   p.a. 85°

One of the Siamese Twins. Very faint and large; very bright, diffuse nucleus. Hubble class Sc. Member of the Virgo Galaxy Cluster; interacting with galaxy NGC 4568 (the other Twin).

NGC 4568

Gx Vir $12^h 36.6^m$ +11° 14' 4.5' × 2.0'
charts 13, 14, B1 m_v 10.8   p.a. 23°

One of the Siamese Twins; very faint and large; bright arms; very small, bright nucleus. Hubble class Sc. Member of the Virgo Galaxy Cluster; interacting with galaxy NGC 4567 (the other Twin).

NGC 4569

See M90.

NGC 4570

Gx Vir $12^h 36.9^m$ +07° 15' 3.8' × 1.2'
charts 13, 14, B1 m_v 10.9   p.a. 159°

Quite bright, pretty small, and very elongated; very bright nucleus. Hubble class S0. Member of the Virgo Galaxy Cluster.

NGC 4571

Gx Com $12^h 36.9^m$ +14° 13' 3.7' × 3.4'
charts 13, 14, B1 m_v 11.3   p.a. 55°

Very faint, large, and elongated; very small, faint nucleus. Hubble class Sc. Member of the Virgo Galaxy Cluster; paired with galaxy M91 (NGC 4548).

NGC 4578

Gx Vir $12^h 37.5^m$ +09° 33' 3.1' × 2.3'
charts 13, 14, B1 m_v 11.5   p.a. 35°

Holmberg 429a. Pretty faint, pretty small, and round. Hubble class Sa. Paired with galaxy Holmberg 429b; member of the Virgo Galaxy Cluster.

NGC 4579

See M58.

NGC 4580

Gx Vir $12^h 37.8^m$ +05° 22' 2.0' × 1.5'
charts 13, 14, B1 m_v 11.8   p.a. 165°

Pretty bright and large; very small, bright nucleus. Distance 52 million ly; Hubble class Sb.

NGC 4586

Gx Vir $12^h 38.5^m$ +04° 19' 3.8' × 1.2'
charts 13, 14, B1 m_v 11.7   p.a. 115°

Pretty bright, large, and elongated; small, bright nucleus; dark lane. Distance 31 million ly; Hubble class Sb.

NGC 4589

Gx Dra $12^h 37.4^m$ +74° 12' 3.3' × 2.7'
chart 2 m_v 10.7   p.a. 90°

Quite bright and large; bright, diffuse nucleus.

Distance 87 million ly; Hubble class Sa. In a group with galaxies NGC 4572 and NGC 4648.

NGC 4590

See M68.

NGC 4592

Gx	Vir	$12^h 39.3^m$	−00° 32′	4.6′ × 1.5′
charts 13, 14			m_v 11.7	p.a. 97°

Faint, large, and elongated; very faint nucleus. Distance 42 million ly; Hubble class Sb.

NGC 4593

Gx	Vir	$12^h 39.7^m$	−05° 21′	3.5′ × 2.5′
charts 13, 14			m_v 10.9	p.a. 55°

Pretty bright, quite large, and elongated; two asymmetrical arms; very bright nucleus; has a broken internal ring with bar. Distance 110 million ly; Hubble class SBb.

NGC 4594

See Sombrero Galaxy.

NGC 4595

Gx	Com	$12^h 39.9^m$	+15° 18′	1.7′ × 1.0′
charts 7, 13, 14, B1			m_v 12.1	p.a. 110°

Pretty faint, pretty large, and round. Hubble class Sc. Member of the Virgo Galaxy Cluster.

NGC 4596

Gx	Vir	$12^h 39.9^m$	+10° 11′	4.1′ × 3.5′
charts 13, 14, B1			m_v 10.4	p.a. 135°

Bright, pretty small, and round; very bright, diffuse nucleus. Hubble class SBc. Paired with galaxy NGC 4608; member of the Virgo Galaxy Cluster.

NGC 4597

Gx	Vir	$12^h 40.2^m$	−05° 48′	3.7′ × 1.5′
charts 13, 14			m_v 12.1	p.a. 40°

Faint and pretty small; asymmetrical arms. Distance 39 million ly; Hubble class SBc.

NGC 4602

Gx	Vir	$12^h 40.6^m$	−05° 08′	2.9′ × 1.0′
charts 13, 14			m_v 11.5	p.a. 100°

Faint, large, and elongated; several arms and dark lanes; small, bright nucleus. Distance 100 million ly; Hubble class Sc.

NGC 4605

Gx	UMa	$12^h 40.0^m$	+61° 37′	6.0′ × 2.5′
chart 2			m_v 10.3	p.a. 125°

Bright, large, and very elongated. Distance 12 million ly; Hubble class SBcp.

NGC 4606

Gx	Vir	$12^h 41.0^m$	+11° 55′	3.0′ × 1.5′
charts 13, 14, B1			m_v 11.8	p.a. 33°

Very faint, pretty small, and elongated; bright nucleus. Hubble class SBa. Member of the Virgo Galaxy Cluster.

NGC 4608

Gx	Vir	$12^h 41.2^m$	+10° 09′	3.2′ × 3.0′
charts 13, 14, B1			m_v 11	p.a. 25°

Pretty bright, pretty large, and round; very bright, diffuse nucleus. Hubble class SBa. Member of the Virgo Galaxy Cluster; paired with brighter galaxy NGC 4596.

NGC 4609

OC	Cru	$12^h 42.3^m$	−62° 58′	5′
chart 25			m_v 6.9	

C98. 40 stars; detached, weak concentration of stars; small range in brightness; mag. of brightest star 9.0; pretty large, elongated, and compressed. Distance 4,900 ly; age 36 million years; Trumpler class II 1 p.

NGC 4612

Gx	Vir	$12^h 41.5^m$	+07° 19′	2.3′ × 1.8′
charts 13, 14, B1			m_v 10.9	p.a. 145°

Pretty bright, small, and round; small, diffuse, very bright nucleus. Hubble class Ep. Member of the Virgo Galaxy Cluster.

NGC 4618

Gx	CVn	$12^h 41.6^m$	$+41° 09'$	$4.4' \times 3.6'$
chart 7			m_v 10.8	p.a. 25°

Bright, large, and elongated; curving, branching arm. Distance 27 million ly; Hubble class Sc. Possibly interacting with galaxy NGC 4625.

NGC 4621

See M59.

NGC 4625

Gx	CVn	$12^h 41.9^m$	$+41° 16'$	$2.3' \times 2.0'$
chart 7			m_v 12.3	p.a. 125°

Pretty faint, small, and roundish; bright nucleus. Distance 29 million ly; Hubble class S(B)mp. Possibly interacting with galaxy NGC 4618.

NGC 4627

Gx	CVn	$12^h 42.0^m$	$+32° 34'$	$2.2' \times 1.7'$
chart 7			m_v 12.4	p.a. 10°

Faint, small, and roundish; has a diffuse countertail. Distance 30 million ly; Hubble class E4p. Interacting with galaxy NGC 4631.

NGC 4631

Gx	CVn	$12^h 42.2^m$	$+32° 33'$	$14.0' \times 2.6'$
chart 7			m_v 9.2	p.a. 86°

C32. Very bright, very large, and extremely elongated; edge on; star attached. Per NGC, a (!) remarkable object. Distance 28 million ly; Hubble class Sc. Interacting with galaxy NGC 4627.

NGC 4632

Gx	Vir	$12^h 42.5^m$	$-00° 05'$	$3.0' \times 1.1'$
charts 13, 14			m_v 11.7	p.a. 63°

Pretty bright, large, and elongated; several arms and a dark lane; very faint nucleus. Distance 68 million ly; Hubble class Sc.

NGC 4636

Gx	Vir	$12^h 42.8^m$	$+02° 41'$	$6.2' \times 5.0'$
charts 13, 14			m_v 9.5	p.a. 150°

Bright, large, irregularly round; very bright nucleus. Distance 38 million ly; Hubble class E1.

NGC 4638

Gx	Vir	$12^h 42.8^m$	$+11° 27'$	$2.5' \times 1.7'$
charts 13, 14, B1			m_v 11.2	p.a. 125°

Faint and round; very bright center. Distance 44 million ly; Hubble class E5. Paired with galaxy NGC 4637; member of the Virgo Galaxy Cluster.

NGC 4639

Gx	Vir	$12^h 42.9^m$	$+13° 15'$	$3.0' \times 2.0'$
charts 13, 14, B1			m_v 11.5	p.a. 123°

Pretty bright, small, and elongated. Distance 39 million ly; Hubble class S(B)b. Member of the Virgo Galaxy Cluster.

NGC 4643

Gx	Vir	$12^h 43.3^m$	$+01° 59'$	$3.2' \times 2.8'$
charts 13, 14			m_v 10.8	p.a. 130°

Quite bright, pretty small, and slightly elongated; very bright, diffuse nucleus. Distance 54 million ly; Hubble class SBa.

NGC 4645

Gx	Cen	$12^h 44.2^m$	$-41° 45'$	$2.3' \times 1.4'$
charts 21, 25			m_v 11.8	p.a. 52°

Pretty bright and small; small, bright nucleus. Distance 100 million ly; Hubble class E3. In a group with galaxies NGC 4645A and NGC 4645B; member of the Centaurus Galaxy Cluster.

NGC 4647

Gx	Vir	$12^h 43.5^m$	$+11° 35'$	$2.9' \times 2.3'$
charts 13, 14, B1			m_v 11.3	p.a. 125°

Pretty faint, pretty large, and slightly elongated; dark lane; very small, bright nucleus. Hubble class Sc. Member of the Virgo Galaxy Cluster. Galaxy M60 is about 2.5' away.

NGC 4648

Gx	Dra	$12^h 41.8^m$	$+74° 25'$	$1.7' \times 1.3'$
chart 2			m_v 12	p.a. 70°

Pretty bright, quite small, and roundish; bright nucleus; Hubble class E3. In a group with galaxies NGC 4572 and NGC 4589.

NGC 4649

See M60.

NGC 4650

| Gx | Cen | 12^h 44.3^m | −40° 44′ | 3.4′ × 2.9′ |
| charts 21, 25 | | | m$_v$ 11.8 | p.a. 83° |

Very faint and roundish; bright nucleus. Distance 100 million ly; Hubble class SBa.

NGC 4651

| Gx | Com | 12^h 43.7^m | +16° 24′ | 4.1′ × 2.7′ |
| charts 7, 13, 14, B1 | | | m$_v$ 10.8 | p.a. 80° |

Quite bright, large, and elongated; filamentary arms; very bright, diffuse nucleus. Hubble class Scp. Member of the Virgo Galaxy Cluster.

NGC 4653

| Gx | Vir | 12^h 43.8^m | −00° 34′ | 3.1′ × 2.8′ |
| charts 13, 14 | | | m$_v$ 12.2 | p.a. 30° |

Very faint and pretty large. Distance 110 million ly; Hubble class S−. In a group with galaxies NGC 4666 and NGC 4668.

NGC 4654

| Gx | Vir | 12^h 43.9^m | +13° 08′ | 4.7′ × 3.0′ |
| charts 13, 14, B1 | | | m$_v$ 10.6 | p.a. 128° |

Faint, very large, and elongated. Distance 42 million ly; Hubble class Sc. Member of the Virgo Galaxy Cluster.

NGC 4656

| Gx | CVn | 12^h 44.0^m | +32° 10′ | 13.8′ × 3.3′ |
| chart 7 | | | m$_v$ 10.5 | p.a. 33° |

Pretty bright, large, and extremely elongated. Distance 29 million ly; Hubble class Sc. A companion galaxy (NGC 4657) is located at NE end. Some astronomers consider them to be a single galaxy, misshapen after a close encounter with NGC 4631 lying 0.5° away.

NGC 4658

| Gx | Vir | 12^h 44.6^m | −10° 05′ | 2.0′ × 0.9′ |
| charts 13, 14 | | | m$_v$ 12.5 | p.a. 0° |

Very faint, large, and elongated; two main arms; bright center with several dark lanes. Distance 98 million ly; Hubble class Sc. Pair with galaxy NGC 4663.

NGC 4660

| Gx | Vir | 12^h 44.5^m | +11° 11′ | 2.1′ × 1.7′ |
| charts 13, 14, B1 | | | m$_v$ 11.2 | p.a. 100° |

Very bright and small. Distance 41 million ly; Hubble class E5. Member of the Virgo Galaxy Cluster.

NGC 4665

| Gx | Vir | 12^h 45.1^m | +03° 03′ | 4.3′ × 3.6′ |
| charts 13, 14, B1 | | | m$_v$ 10.5 | p.a. 5° |

Bright, pretty large, irregularly round; bright, diffuse nucleus. Distance 30 million ly; Hubble class S(B)a.

NGC 4666

| Gx | Vir | 12^h 45.1^m | −00° 28′ | 4.5′ × 1.5′ |
| charts 13, 14 | | | m$_v$ 10.7 | p.a. 42° |

Bright, very large, and very elongated; many arms; dark lanes; bright nucleus. Distance 61 million ly; Hubble class Sc. In a group with galaxies NGC 4653 and NGC 4668.

NGC 4684

| Gx | Vir | 12^h 47.3^m | −02° 44′ | 2.8′ × 1.0′ |
| charts 13, 14 | | | m$_v$ 11.4 | p.a. 23° |

Bright, pretty large, and elongated; almost edge on; very bright nucleus. Distance 64 million ly; Hubble class Sa.

NGC 4689

| Gx | Com | 12^h 47.8^m | +13° 46′ | 4.6′ × 3.9′ |
| charts 13, 14, B1 | | | m$_v$ 10.9 | p.a. 170° |

Pretty bright, very large, and elongated; very small, very bright nucleus. Hubble class Sb. Member of the Virgo Galaxy Cluster.

NGC 4691

| Gx | Vir | 12^h 48.2^m | −03° 20′ | 3.0′ × 2.6′ |
| charts 13, 14 | | | m$_v$ 11.1 | p.a. 15° |

Pretty bright, pretty large, and elongated; bright center. Distance 43 million ly; Hubble class SBbp.

NGC 4694

Gx	Vir	12ʰ 48.3ᵐ	+10° 59′	3.2′ × 1.7′
charts 13, 14, B1			m_v 11.4	p.a. 140°

Pretty faint, small, and elongated; very bright nucleus with dark lanes. Distance 49 million ly; Hubble class E5. Member of the Virgo Galaxy Cluster.

NGC 4696

Gx	Cen	12ʰ 48.8ᵐ	−41° 19′	4.5′ × 3.2′
charts 21, 25			m_v 10.2	p.a. 95°

Pretty bright, large, and round; small, very bright nucleus. Distance 130 million ly; Hubble class E1p. Brightest galaxy in the Centaurus Galaxy Cluster.

NGC 4697

Gx	Vir	12ʰ 48.6ᵐ	−05° 48′	6.0′ × 4.1′
charts 13, 14			m_v 9.2	p.a. 70°

C52. Very bright, large, and slightly elongated; very bright center. Distance 51 million ly; Hubble class E4. Interacting with an anonymous, barred-spiral galaxy.

NGC 4698

Gx	Vir	12ʰ 48.4ᵐ	+08° 29′	3.9′ × 2.6′
charts 13, 14, B1			m_v 10.6	p.a. 170°

Quite bright, pretty large, irregularly round; two main arms; very bright, diffuse nucleus. Distance 38 million ly; Hubble class Sb. Member of the Virgo Galaxy Cluster.

NGC 4699

Gx	Vir	12ʰ 49.0ᵐ	−08° 40′	3.9′ × 2.9′
charts 13, 14			m_v 9.5	p.a. 45°

Very bright and round; many arms; very bright, very small nucleus. Distance 59 million ly; Hubble class Sa.

NGC 4700

Gx	Vir	12ʰ 49.1ᵐ	−11° 25′	2.9′ × 0.5′
charts 13, 14			m_v 11.9	p.a. 50°

Faint, large, and very elongated; an edge-on spiral; bright nucleus with dark areas. Distance 54 million ly; Hubble class S. Bright star nearby; paired with an anonymous spiral galaxy.

NGC 4701

Gx	Vir	12ʰ 49.2ᵐ	+03° 23′	3.1′ × 2.6′
charts 13, 14, B1			m_v 12.4	p.a. 45°

Faint and small; many arms; very small, bright nucleus. Distance 27 million ly; Hubble class Sc.

NGC 4709

Gx	Cen	12ʰ 50.1ᵐ	−41° 23′	2.3′ × 2.0′
charts 21, 25			m_v 11.5	p.a. 90°

Pretty bright, quite small, and roundish; very bright nucleus. Distance 190 million ly; Hubble class E1.

NGC 4710

Gx	Com	12ʰ 49.7ᵐ	+15° 10′	4.9′ × 1.6′
charts 13, 14, B1			m_v 11	p.a. 27°

Quite bright, pretty large, and very elongated; edge on with broken dust lane; small, bright nucleus. Distance 47 million ly; Hubble class S0. Member of the Virgo Galaxy Cluster.

NGC 4713

Gx	Vir	12ʰ 50.0ᵐ	+05° 19′	2.9′ × 1.8′
charts 13, 14, B1			m_v 11.7	p.a. 100°

Pretty bright, large, and slightly elongated; has a small, bright nucleus. Distance 25 million ly; Hubble class Sc.

NGC 4725

Gx	Com	12ʰ 50.4ᵐ	+25° 30′	11.0′ × 7.9′
chart 7			m_v 9.4	p.a. 35°

Very bright, very large, and elongated; internal ring; very small, extremely bright nucleus. Distance 49 million ly; Hubble class S(B)b. Paired with galaxy NGC 4712.

NGC 4727

Gx	Crv	12ʰ 51.0ᵐ	−14° 20′	1.7′ × 1.2′
charts 13, 14			m_v 13	p.a. 130°

Faint, pretty large, and roundish; fairly bright nucleus. Distance 320 million ly; Hubble class SBb.

NGC 4731

Gx	Vir	12^h 51.0^m	–06° 23′	6.0′ × 2.5′
charts 13, 14			m$_v$ 11.5	p.a. 85°

Very faint, pretty large, and elongated; two main arms; bright center. Distance 59 million ly; Hubble class SBcp. Paired with irregular dwarf galaxy.

NGC 4733

Gx	Vir	12^h 51.1^m	+10° 55′	1.9′
charts 13, 14, B1			m$_v$ 11.8	

Quite faint, pretty large, and slightly elongated. Distance 42 million ly; Hubble class S0. Member of the Virgo Galaxy Cluster; a 12th-mag. star to W.

NGC 4736

See M94.

NGC 4739

Gx	Vir	12^h 51.6^m	–08° 25′	1.5′
charts 13, 14			m$_v$ 12.6	p.a. 171°

Faint, pretty large, and slightly elongated; moderately bright nucleus. Distance 160 million ly; Hubble class Sb.

NGC 4742

Gx	Vir	12^h 51.8^m	–10° 27′	2.2′ × 1.4′
charts 13, 14			m$_v$ 11.3	p.a. 75°

Pretty bright and very small. Distance 51 million ly; Hubble class E3.

NGC 4747

Gx	Com	12^h 51.8^m	+25° 46′	3.3′ × 1.3′
chart 7			m$_v$ 12.3	p.a. 30°

Faint, pretty large, and slightly elongated; faint extensions to NE. Distance 52 million ly; Hubble class P.

NGC 4750

Gx	Dra	12^h 50.1^m	+72° 53′	2.1′ × 1.9′
chart 2			m$_v$ 11.2	p.a. 125°

Pretty bright, large, and round; very small, very bright nucleus; dark ring around nucleus. Distance 74 million ly; Hubble class Sap.

NGC 4751

Gx	Cen	12^h 52.9^m	–42° 40′	3.1′ × 1.3′
charts 21, 25			m$_v$ 12	p.a. 175°

Bright, pretty small, and roundish; bright nucleus. Distance 78 million ly; Hubble class E.

NGC 4753

Gx	Vir	12^h 52.4^m	–01° 12′	5.4′ × 2.9′
charts 13, 14			m$_v$ 9.9	p.a. 80°

Quite bright, large, and slightly elongated; extremely small, bright nucleus with dark lanes. Distance 49 million ly; Hubble class P.

NGC 4754

Gx	Vir	12^h 52.3^m	+11° 19′	4.4′ × 2.5′
charts 13, 14, B1			m$_v$ 10.6	p.a. 23°

Bright, pretty large, and round; diffuse, very bright nucleus. Hubble class SB0. Member of the Virgo Galaxy Cluster; paired with galaxy NGC 4762.

NGC 4755

See Jewel Box Cluster.

NGC 4760

Gx	Vir	12^h 53.1^m	–10° 30′	2.0′ × 1.8′
charts 13, 14			m$_v$ 11.4	p.a. 10°

Pretty bright and round. Distance 180 million ly; Hubble class E2. Paired with galaxy NGC 4757.

NGC 4762

Gx	Vir	12^h 52.9^m	+11° 14′	8.7′ × 1.6′
charts 13, 14, B1			m$_v$ 10.3	p.a. 32°

Pretty bright and extremely elongated; one of the flattest edge-on galaxies known; very small, very bright nucleus. Like a paper kite, beautifully grouped with three stars (Webb). Distance 38 million ly; Hubble class SB0. Paired with galaxy NGC 4754; member of the Virgo Galaxy Cluster.

NGC 4767

Gx Cen 12ʰ 53.9ᵐ −39° 43′ 2.6′ × 1.2′
charts 21, 25 m_v 11.6 p.a. 0°

Bright, pretty small, and slightly elongated. Distance 130 million ly; Hubble class E5. Member of the Centaurus Galaxy Cluster.

NGC 4771

Gx Vir 12ʰ 53.4ᵐ +01° 16′ 3.8′ × 0.9′
charts 13, 14 m_v 12.3 p.a. 133°

Faint, pretty large, and very elongated; edge on; very faint nucleus; many arms. Distance 48 million ly; Hubble class Sb.

NGC 4772

Gx Vir 12ʰ 53.5ᵐ +02° 10′ 3.2′ × 1.9′
charts 13, 14 m_v 11 p.a. 147°

Pretty faint, pretty small, and round; bright, diffuse nucleus; dark lane. Distance 43 million ly; Hubble class Sa.

NGC 4775

Gx Vir 12ʰ 53.8ᵐ −06° 37′ 2.2′ × 2.0′
charts 13, 14 m_v 11.1 p.a. 80°

Faint, quite large, and round; face on; very small, bright nucleus. Distance 61 million ly; Hubble class Sc.

NGC 4781

Gx Vir 12ʰ 54.4ᵐ −10° 32′ 3.3′ × 1.3′
charts 13, 14 m_v 11.1 p.a. 120°

Quite bright, very large, and elongated. Distance 32 million ly; Hubble class Sc. In a group with galaxies NGC 4784 and NGC 4790.

NGC 4782

Gx Crv 12ʰ 54.6ᵐ −12° 34′ 2.6′ × 1.7′
charts 13, 14 m_v 11.7 p.a. 155°

Pretty faint, pretty small, and round. Distance 170 million ly; Hubble class E0. Interacting with galaxy NGC 4783.

NGC 4783

Gx Crv 12ʰ 54.6ᵐ −12° 33′ 2.3′ × 1.3′
charts 13, 14 m_v 11.5 p.a. 105°

Pretty faint, pretty small, and round. Distance 200 million ly; Hubble class E0. Interacting with galaxy NGC 4782.

NGC 4786

Gx Vir 12ʰ 54.5ᵐ −06° 51′ 1.6′ × 1.3′
charts 13, 14 m_v 11.7 p.a. 15°

Pretty bright and pretty small. Distance 200 million ly; Hubble class E2p.

NGC 4790

Gx Vir 12ʰ 54.9ᵐ −10° 15′ 1.5′ × 0.9′
charts 13, 14 m_v 12.1 p.a. 85°

Pretty faint, pretty small, irregularly round; bright nucleus. Distance 52 million ly; Hubble class Sc. In a group with galaxies NGC 4781 and NGC 4784.

NGC 4793

Gx Com 12ʰ 54.7ᵐ +28° 56′ 2.9′ × 1.5′
chart 7 m_v 11.6 p.a. 50°

Pretty bright, pretty small, and slightly elongated; very small, very bright nucleus. Distance 110 million ly; Hubble class Sc.

NGC 4800

Gx CVn 12ʰ 54.6ᵐ +46° 32′ 1.6′ × 1.2′
charts 6, 7 m_v 11.5 p.a. 25°

Pretty bright, quite small, and round; very bright nucleus; many arms. Distance 36 million ly; Hubble class Sb.

NGC 4802

Gx Crv 12ʰ 55.8ᵐ −12° 03′ 2.5′ × 1.9′
charts 13, 14 m_v 11.8 p.a. 20°

Very faint and small. Distance 37 million ly; Hubble class S0. A 10th-mag. star is attached.

NGC 4808

Gx Vir 12ʰ 55.8ᵐ +04° 18′ 2.4′ × 1.0′
charts 13, 14, B1 m_v 11.7 p.a. 127°

Pretty bright, quite large, and elongated; many arms; very small, bright nucleus. Distance 30 mil-

NGC 4814

| Gx | UMa | 12^h 55.4^m | +58° 21' | 3.3' × 2.3' |
| chart 2 | | | m$_v$ 12 | p.a. 135° |

Bright and pretty small; two main arms; dark lanes; small, very bright nucleus. Distance 120 million ly; Hubble class Sbp.

NGC 4815

| OC | Mus | 12^h 58.0^m | −64° 57' | 3' |
| chart 25 | | | m$_v$ 8.6 | |

100 stars; detached, strong concentration of stars; large range in brightness; mag. of brightest star 9.6; Trumpler class I 3 m.

NGC 4818

| Gx | Vir | 12^h 56.8^m | −08° 32' | 3.3' × 1.2' |
| charts 13, 14 | | | m$_v$ 11.1 | p.a. 3° |

Pretty bright, large, and elongated. Distance 44 million ly; Hubble class S(B).

NGC 4825

| Gx | Vir | 12^h 57.2^m | −13° 40' | 2.1' × 1.2' |
| charts 13, 14 | | | m$_v$ 11.7 | p.a. 135° |

Pretty bright; bright, diffuse nucleus. Distance 190 million ly; Hubble class E2p. In a group with galaxies NGC 4820, NGC 4823, NGC 4829, and others.

NGC 4826

See M64.

NGC 4833

| GC | Mus | 12^h 59.6^m | −70° 53' | 13' |
| charts 25, 26 | | | m$_v$ 8.4 | |

C105. Medium concentration of stars; bright, large, and round. Distance 18,000 ly; S-S class 8.

NGC 4835

| Gx | Cen | 12^h 58.1^m | −46° 16' | 4.5' × 0.8' |
| charts 21, 25 | | | m$_v$ 11.5 | p.a. 150° |

Faint, pretty large, and very elongated; very bright nucleus. Distance 85 million ly; Hubble class S(B)b. Paired with an irregularly shaped dwarf galaxy.

NGC 4845

| Gx | Vir | 12^h 58.0^m | +01° 35' | 5.0' × 1.6' |
| charts 13, 14 | | | m$_v$ 11.2 | p.a. 89° |

Pretty faint, pretty large, and very elongated; edge on; a small, bright, diffuse nucleus in a bright bulge with dark lane. Distance 49 million ly; Hubble class Sb.

NGC 4856

| Gx | Vir | 12^h 59.4^m | −15° 03' | 4.2' × 1.5' |
| charts 13, 14 | | | m$_v$ 10.5 | p.a. 37° |

Bright and round; small, bright, diffuse nucleus. Distance 47 million ly; Hubble class S(B)a.

NGC 4866

| Gx | Vir | 12^h 59.5^m | +14° 10' | 6.0' × 1.4' |
| charts 13, 14 | | | m$_v$ 11.2 | p.a. 87° |

Bright, pretty large, and very elongated; very small, extremely bright nucleus with dark lanes. Distance 81 million ly; Hubble class SB−.

NGC 4868

| Gx | CVn | 12^h 59.2^m | +37° 19' | 1.5' × 1.4' |
| chart 7 | | | m$_v$ 12.2 | p.a. 90° |

Pretty bright, small, and round. Distance 210 million ly; Hubble class Sb.

NGC 4874

| Gx | Com | 12^h 59.6^m | +27° 58' | 2.8' × 2.5' |
| chart 7 | | | m$_v$ 11.7 | p.a. 39° |

A faint galaxy. Distance 310 million ly; Hubble class E0. Second brightest member of the Coma Galaxy Cluster; paired with galaxy NGC 4889.

NGC 4877

| Gx | Vir | 13^h 00.4^m | −15° 17' | 2.2' × 0.9' |
| chart 14 | | | m$_v$ 12.4 | p.a. 6° |

Pretty bright and pretty large; bright nucleus. Distance 210 million ly; Hubble class Sb.

NGC 4880

Gx	Vir	13^h 00.2^m	+12° 29′	3.4′ × 2.7′
chart 14			m$_v$ 11.4	p.a. 165°

Quite faint, pretty large, and round. Distance 65 million ly; Hubble class Sa.

NGC 4889

Gx	Com	13^h 00.1^m	+27° 59′	3.2′ × 2.2′
chart 7			m$_v$ 11.5	p.a. 80°

C35. Pretty bright and elongated. Distance 280 million ly; Hubble class E4. Brightest member of the Coma Galaxy Cluster; paired with galaxy NGC 4874.

NGC 4899

Gx	Vir	13^h 00.9^m	−13° 57′	2.2′ × 1.3′
chart 14			m$_v$ 11.9	p.a. 15°

Pretty faint and elongated; several arms; small, bright nucleus. Distance 110 million ly; Hubble class Sp.

NGC 4900

Gx	Vir	13^h 00.7^m	+02° 30′	2.2′ × 1.7′
chart 14			m$_v$ 11.4	p.a. 90°

Quite bright and quite elongated; small, very bright nucleus. Distance 41 million ly; Hubble class Sc. A 10th-mag. star is attached.

NGC 4902

Gx	Vir	13^h 01.0^m	−14° 31′	2.8′ × 2.5′
chart 14			m$_v$ 10.9	p.a. 70°

Pretty bright, pretty large, irregularly round; very small, bright nucleus; several arms. Distance 110 million ly; Hubble class SBb.

NGC 4904

Gx	Vir	13^h 01.0^m	−00° 02′	2.2′ × 1.5′
chart 14			m$_v$ 12	p.a. 25°

Pretty bright, pretty small, and round; bright, narrow bar. Distance 51 million ly; Hubble class S(B)c.

NGC 4914

Gx	CVn	13^h 00.7^m	+37° 19′	3.5′ × 2.0′
chart 7			m$_v$ 11.6	p.a. 155°

Pretty bright, quite small, and round. Distance 210 million ly; Hubble class E2p.

NGC 4915

Gx	Vir	13^h 01.3^m	−04° 34′	1.5′ × 1.2′
chart 14			m$_v$ 12.1	p.a. 55°

Pretty bright, small, and round. Distance 130 million ly; Hubble class Sa.

NGC 4928

Gx	Vir	13^h 03.0^m	−08° 05′	1.1′ × 0.9′
chart 14			m$_v$ 12.5	p.a. 125°

Faint, pretty small, and slightly elongated; two bright arms; small, very bright nucleus. Distance 67 million ly; Hubble class S+.

NGC 4930

Gx	Cen	13^h 04.1^m	−41° 25′	4.7′ × 3.8′
charts 21, 25			m$_v$ 11.7	p.a. 40°

Very faint and roundish; bright nucleus. Distance 100 million ly; Hubble class SBb.

NGC 4933A

Gx	Vir	13^h 04.0^m	−11° 30′	2.2′ × 1.3′
chart 14			m$_v$ 11.7	p.a. 45°

Bright nucleus. Distance 130 million ly; Hubble class Ep. Interacting with galaxy NGC 4933B.

NGC 4936

Gx	Cen	13^h 04.3^m	−30° 31′	3.2′
chart 21			m$_v$ 10.8	

Pretty bright, small, and round; bright, diffuse nucleus. Distance 130 million ly; Hubble class E3. Paired with galaxy IC 844.

NGC 4939

Gx	Vir	13^h 04.2^m	−10° 20′	5.8′ × 3.2′
chart 14			m$_v$ 11.3	p.a. 5°

Pretty bright, large, and round; a well-defined spiral; very small, very bright nucleus. Distance 130 million ly; Hubble class Sb.

NGC 4941

Gx   Vir $13^h\,04.2^m$ $-05°\,33'$ $3.5' \times 2.2'$
chart 14 $m_v\,11.1$ p.a. 15°

Pretty faint, large, and elongated; small, very bright nucleus. Distance 26 million ly; Hubble class Sbp.

NGC 4945

Gx   Cen $13^h\,05.4^m$ $-49°\,28'$ $20.0' \times 4.4'$
charts 21, 25 $m_v\,8.8$ p.a. 43°

C83. Bright, very large, and extremely elongated; nearly edge on. Distance 16 million ly; Hubble class SBc. Small galaxy nearby.

NGC 4947

Gx   Cen $13^h\,05.3^m$ $-35°\,20'$ $2.5' \times 1.3'$
charts 21, 25 $m_v\,11.9$ p.a. 10°

Faint, pretty large, and round; small, bright nucleus; patchy arms. Distance 95 million ly; Hubble class SBb.

NGC 4948

Gx   Vir $13^h\,04.9^m$ $-07°\,57'$ $2.0' \times 0.7'$
chart 14 $m_v\,14.4$ p.a. 145°

Extremely faint, pretty small, and slightly elongated. Distance 50 million ly; Hubble class SBd. Double star approx. 1' to SE. In a group with galaxies NGC 4948A and NGC 4958.

NGC 4951

Gx   Vir $13^h\,05.1^m$ $-06°\,30'$ $3.2' \times 1.1'$
chart 14 $m_v\,11.9$ p.a. 90°

Faint, pretty large, and slightly elongated; nearly edge on; small, bright nucleus with dark lanes. Distance 46 million ly; Hubble class S(B)c.

NGC 4958

Gx   Vir $13^h\,05.8^m$ $-08°\,01'$ $4.2' \times 1.3'$
chart 14 $m_v\,10.7$ p.a. 15°

Very bright, pretty small, and elongated; very bright nucleus. Distance 60 million ly; Hubble class E6. In a group with galaxies NGC 4948 and NGC 4948A.

NGC 4965

Gx   Hya $13^h\,07.2^m$ $-28°\,14'$ $2.4' \times 2.0'$
chart 21 $m_v\,12.1$ p.a. 130°

Very faint, very large, and quite elongated; moderately bright nucleus; broad, patchy arms. Distance 280 million ly; Hubble class Sd.

NGC 4976

Gx   Cen $13^h\,08.6^m$ $-49°\,30'$ $5.0' \times 2.9'$
charts 21, 25 $m_v\,10.1$ p.a. 161°

Bright, pretty large, and round; bright nucleus. Distance 49 million ly; Hubble class E4p.

NGC 4981

Gx   Vir $13^h\,08.8^m$ $-06°\,47'$ $2.7' \times 1.9'$
chart 14 $m_v\,11.3$ p.a. 150°

Bright, pretty large, and round; small, very bright nucleus. Distance 67 million ly; Hubble class Sc.

NGC 4984

Gx   Vir $13^h\,09.0^m$ $-15°\,31'$ $3.1' \times 2.3'$
chart 14 $m_v\,11.3$ p.a. 15°

Bright, pretty large, and round; extremely bright nucleus. Distance 48 million ly; Hubble class S(B)a.

NGC 4995

Gx   Vir $13^h\,09.7^m$ $-07°\,50'$ $2.5' \times 1.7'$
chart 14 $m_v\,11.1$ p.a. 92°

Pretty bright, pretty large, and round; very bright nucleus; knotty arms. Distance 69 million ly; Hubble class Sb.

NGC 4999

Gx   Vir $13^h\,09.6^m$ $+01°\,40'$ $2.3' \times 1.9'$
chart 14 $m_v\,11.8$ p.a. 35°

Quite faint, pretty large, and round; small, very bright nucleus. Distance 130 million ly; Hubble class SBb.

NGC 5005

Gx   CVn $13^h\,10.9^m$ $+37°\,03'$ $5.4' \times 2.7'$
chart 7 $m_v\,9.8$ p.a. 65°

C29. Very bright, very large, and very elongated; extremely bright nucleus with circular dust lanes. Distance 47 million ly; Hubble class Sb.

NGC 5011

Gx Cen $13^h 12.9^m$ −43° 06′ 2.5′ × 2.1′
charts 21, 25 m_v 11.1 p.a. 154°

Pretty bright, quite small, and round; very bright nucleus. Distance 120 million ly; Hubble class E1. Located among four stars.

NGC 5012

Gx Com $13^h 11.6^m$ +22° 55′ 2.9′ × 1.7′
chart 7 m_v 12.2 p.a. 10°

Pretty faint, quite large, and elongated; several filamentary arms; small, very bright nucleus. Distance 110 million ly; Hubble class Sb. Paired with an anonymous, peculiar galaxy.

NGC 5015

Gx Vir $13^h 12.4^m$ −04° 20′ 1.8′ × 1.5′
chart 14 m_v 12.1 p.a. 40°

Extremely faint, extremely small, and quite elongated. Distance 130 million ly; Hubble class SBb.

NGC 5018

Gx Vir $13^h 13.0^m$ −19° 31′ 3.5′ × 2.6′
charts 14, 21 m_v 10.7 p.a. 112°

Quite bright, small, and round; very bright nucleus. Distance 120 million ly; Hubble class Sa. Paired with galaxy NGC 5022.

NGC 5020

Gx Vir $13^h 12.7^m$ +12° 36′ 3.2′ × 2.7′
chart 14 m_v 11.7 p.a. 85°

Quite faint, quite large, and slightly elongated; moderately bright nucleus. Distance 140 million ly; Hubble class SBa.

NGC 5023

Gx CVn $13^h 12.2^m$ +44° 02′ 6.0′ × 0.8′
chart 7 m_v 12.3 p.a. 28°

Pretty faint, large, and very elongated. Distance 21 million ly; Hubble class Sc.

NGC 5024

See M53.

NGC 5026

Gx Cen $13^h 14.2^m$ −42° 58′ 3.4′ × 2.1′
charts 21, 25 m_v 11.5 p.a. 52°

Pretty bright, pretty large, and roundish; small, bright nucleus. Distance 160 million ly; Hubble class SBa.

NGC 5028

Gx Vir $13^h 13.8^m$ −13° 03′ 1.8′ × 1.1′
chart 14 m_v 12.7 p.a. 130°

Very faint and small, distance 280 million ly; Hubble class E6.

NGC 5033

Gx CVn $13^h 13.5^m$ +36° 36′ 10.5′ × 5.6′
chart 7 m_v 10.2 p.a. 170°

Very bright and pretty large; elongated; several filamentary arms; small, very bright, diffuse nucleus. Distance 42 million ly; Hubble class Sb. Holmberg VIII is a companion galaxy.

NGC 5037

Gx Vir $13^h 15.0^m$ −16° 35′ 2.1′ × 0.6′
charts 14, 21 m_v 12.2 p.a. 50°

Quite faint, pretty small, and elongated; almost edge on; bright nucleus. Distance 75 million ly; Hubble class Sb. Part of NGC 5044 galaxy group.

NGC 5042

Gx Hya $13^h 15.5^m$ −23° 59′ 4.3′ × 2.2′
chart 21 m_v 11.8 p.a. 22°

Faint, large, and roundish; moderately bright nucleus. Distance 55 million ly; Hubble class S(B)c.

NGC 5044

Gx Vir $13^h 15.4^m$ −16° 23′ 2.9′
chart 14 m_v 10.8

Pretty bright, pretty large, and round. Distance 110 million ly; Hubble class E0. Brightest in NGC 5044 galaxy group.

NGC 5053

GC	Com	13^h 16.5^m	+17° 42'	8'
charts 7, 14			m$_v$ 9	

Low concentration of stars; pretty large, very faint, irregularly round. A little gem of woven fairy fire (Houston). Distance 50,000 ly; S-S class 11.

NGC 5054

Gx	Vir	13^h 17.0^m	−16° 38'	4.8' × 3.0'
charts 14, 21			m$_v$ 10.9	p.a. 155°

Faint, pretty small, irregularly round; three bright arms; small, very bright nucleus. Distance 69 million ly; Hubble class Sb. Paired with an anonymous barred-spiral galaxy.

NGC 5055

See M63.

NGC 5061

Gx	Hya	13^h 18.1^m	−26° 50'	4.0' × 3.2'
chart 21			m$_v$ 10.2	p.a. 110°

Very bright, small, and round; very bright, diffuse nucleus. Distance 77 million ly; Hubble class E2. Bright star about 2.3' to E.

NGC 5064

Gx	Cen	13^h 19.0^m	−47° 55'	2.4' × 1.2'
charts 21, 25			m$_v$ 12	p.a. 38°

Bright, small, and round; bright nucleus. Distance 120 million ly; Hubble class Sb.

NGC 5068

Gx	Vir	13^h 18.9^m	−21° 02'	6.9' × 6.3'
chart 21			m$_v$ 9.6	p.a. 110°

Faint, large, irregularly round; two main arms; very small, faint nucleus. Distance 18 million ly; Hubble class S(B)c.

NGC 5073

Gx	Vir	13^h 19.3^m	−14° 51'	3.4' × 0.6'
chart 14			m$_v$ 12.3	p.a. 151°

Very faint, pretty large, and quite elongated. Distance 110 million ly; Hubble class SBc.

NGC 5077

Gx	Vir	13^h 19.5^m	−12° 39'	2.0' × 1.6'
chart 14			m$_v$ 11.3	p.a. 10°

Pretty bright, small, and slightly elongated; very bright, diffuse nucleus. Distance 120 million ly; Hubble class E3. Paired with galaxy NGC 5079; brightest in a group of galaxies.

NGC 5078

Gx	Hya	13^h 19.8^m	−27° 25'	19.0' × 5.0'
chart 21			m$_v$ 10.6	p.a. 148°

Pretty bright, pretty small, and quite elongated; nearly edge on; very bright, diffuse nucleus with dark lane; Hubble class Sa.

NGC 5079

Gx	Vir	13^h 19.6^m	−12° 42'	1.5' × 0.8'
chart 14			m$_v$ 13	p.a. 35°

Quite faint, pretty small, and slightly elongated. Distance 120 million ly; Hubble class SBbp. Paired with galaxy NGC 5077.

NGC 5084

Gx	Vir	13^h 20.3^m	−21° 50'	10.0' × 2.1'
chart 21			m$_v$ 10.4	p.a. 80°

Quite bright, quite small, and very elongated; small, very bright nucleus. Distance 68 million ly; Hubble class S0. Several other galaxies are nearby.

NGC 5087

Gx	Vir	13^h 20.4^m	−20° 37'	2.5' × 1.9'
chart 21			m$_v$ 11	p.a. 10°

Quite faint and very small; very small, very bright nucleus. Distance 72 million ly; Hubble class E4.

NGC 5090

Gx	Cen	13^h 21.2^m	−43° 42'	2.5' × 1.9'
charts 21, 25			m$_v$ 11.6	p.a. 85°

Pretty bright, pretty large, and round. Distance 130 million ly; Hubble class E2. Paired with galaxy NGC 5091; several other galaxies nearby.

NGC 5101

Gx	Hya	$13^h 21.8^m$	$-27° 26'$	$5.0' \times 4.7'$
chart 21			m_v 10.4	p.a. 125°

Quite bright, pretty small, and slightly elongated; very bright, diffuse nucleus. Distance 73 million ly; Hubble class SBa.

NGC 5102

Gx	Cen	$13^h 22.0^m$	$-36° 38'$	$7.0' \times 2.6'$
charts 21, 25			m_v 8.8	p.a. 48°

Bright galaxy; very bright nucleus. Distance 11 million ly; Hubble class S0.

NGC 5105

Gx	Vir	$13^h 21.8^m$	$-13° 12'$	$1.9' \times 1.4'$
chart 14			m_v 11.8	p.a. 140°

Extremely faint, pretty small, and slightly elongated. Distance 120 million ly; Hubble class SBc.

NGC 5112

Gx	CVn	$13^h 21.9^m$	$+38° 44'$	$4.0' \times 3.0'$
chart 7			m_v 12.1	p.a. 130°

Faint, large, irregularly round. Distance 45 million ly; Hubble class Sc. Paired with galaxy NGC 5107.

NGC 5121

Gx	Cen	$13^h 24.8^m$	$-37° 41'$	$2.0' \times 1.5'$
charts 21, 25			m_v 11.6	p.a. 36°

Quite bright, small, and round; bright nucleus. Distance 58 million ly; Hubble class Sa.

NGC 5128

Gx	Cen	$13^h 25.5^m$	$-43° 01'$	$27' \times 20'$
charts 21, 25			m_v 6.7	p.a. 35°

C77, the Centaurus A radio galaxy. Very bright, very large, and very elongated; dark central band; might be a pair of colliding galaxies; a radio source. From southern latitudes, giant binoculars are all that is needed to detect both the galaxy and the dark lane (Harrington). Per *NGC,* a (!!) very remarkable object. Distance 14 million ly; Hubble class S0p.

NGC 5129

Gx	Vir	$13^h 24.2^m$	$+13° 59'$	$1.7' \times 1.4'$
chart 14			m_v 12.1	p.a. 10°

Pretty bright, very small, and roundish; bright nucleus. Distance 300 million ly; Hubble class E.

NGC 5134

Gx	Vir	$13^h 25.3^m$	$-21° 08'$	$3.0' \times 2.0'$
chart 21			m_v 11.2	p.a. 155°

Faint, pretty small, slightly elongated or oval; hazy arms; very small, very bright nucleus. Distance 67 million ly; Hubble class Sb.

NGC 5135

Gx	Hya	$13^h 25.7^m$	$-29° 50'$	$2.5' \times 2.3'$
chart 21			m_v 11.8	p.a. 10°

Pretty bright, small, and elongated; patchy arms; very bright nucleus. Distance 170 million ly; Hubble class SBb. Paired with galaxy IC 4248.

NGC 5138

OC	Cen	$13^h 27.3^m$	$-59° 01'$	$7'$
chart 25			m_v 7.6	

40 stars; detached, weak concentration of stars; moderate range in brightness; mag. of brightest star 10.3. Distance 4,600 ly; age 320 million years; Trumpler class II 2 p.

NGC 5139

See Omega Centauri.

NGC 5140

Gx	Cen	$13^h 26.3^m$	$-33° 52'$	$2.0' \times 1.7'$
chart 21			m_v 11.8	p.a. 52°

Very faint, small, and roundish; very bright nucleus. Distance 150 million ly; Hubble class S(B)0.

NGC 5147

Gx	Vir	$13^h 26.3^m$	$+02° 06'$	$1.9' \times 1.6'$
chart 14			m_v 11.8	p.a. 120°

Pretty bright, pretty large, and slightly elongated; two strong arms; very faint nucleus. Distance 45 million ly; Hubble class Sc. Bright star nearby.

NGC 5153

Gx	Hya	13ʰ 27.9ᵐ	−29° 37′	2.0′ × 1.4′
chart 21			m_v 12.3	p.a. 175°

Pretty faint and small. Distance 180 million ly; Hubble class E.

NGC 5156

Gx	Cen	13ʰ 28.7ᵐ	−48° 55′	2.5′ × 2.1′
charts 21, 25			m_v 11.7	p.a. 90°

Pretty bright, quite small, and slightly elongated; small, bright nucleus. Distance 120 million ly; Hubble class SBb.

NGC 5161

Gx	Cen	13ʰ 29.2ᵐ	−33° 10′	5.4′ × 2.3′
chart 21			m_v 11.4	p.a. 77°

Pretty faint, large, and very elongated; similar in structure to M81. Distance 88 million ly; Hubble class S. Located between two pretty bright stars.

NGC 5170

Gx	Vir	13ʰ 29.8ᵐ	−17° 58′	8.1′ × 1.3′
charts 14, 21			m_v 11.3	p.a. 127°

Quite faint, large, and very elongated; an edge-on spiral; dark lanes; very small, very bright nucleus. Distance 59 million ly; Hubble class Sb.

NGC 5172

Gx	Com	13ʰ 29.3ᵐ	+17° 03′	3.4′ × 1.8′
charts 7, 14			m_v 11.9	p.a. 103°

Faint, pretty large, and round; three main arms; small, bright nucleus. Distance 180 million ly; Hubble class Sb.

NGC 5188

Gx	Cen	13ʰ 31.5ᵐ	−34° 48′	3.0′ × 1.3′
chart 21			m_v 11.8	p.a. 104°

Faint, pretty large, slightly elongated; very bright center. Distance 93 million ly; Hubble class SBbp.

NGC 5189

PN	Mus	13ʰ 33.6ᵐ	−65° 59′	2.3′
chart 25			m_v 9.9	

PK307−3.1. Bright, pretty large, slightly elongated; curved or irregular shape; phot. mag. 10.3; central star mag. 14.0. Per *NGC*, a (!) remarkable object. Distance 2,600 ly; expansion velocity 31 km/sec (19 miles/sec); V-V class 5.

NGC 5193

Gx	Cen	13ʰ 31.9ᵐ	−33° 14′	1.7′
chart 21			m_v 11.7	

Pretty bright, small, and round. Distance 150 million ly; Hubble class E2. Paired with galaxy NGC 5193A.

NGC 5194

See Whirlpool Galaxy.

NGC 5195

Gx	CVn	13ʰ 30.0ᵐ	+47° 16′	5.0′ × 4.7′
chart 7			m_v 9.6	p.a. 79°

Part of the Whirlpool Galaxy (M51). Bright, pretty small, slightly elongated. Distance 37 million ly; Hubble class P.

NGC 5198

Gx	CVn	13ʰ 30.2ᵐ	+46° 40′	2.0′ × 1.7′
chart 7			m_v 11.8	p.a. 15°

Pretty faint, pretty small, and round; very small, very bright, diffuse nucleus. Distance 110 million ly; Hubble class E2.

NGC 5204

Gx	UMa	13ʰ 29.6ᵐ	+58° 25′	4.8′ × 3.0′
chart 2			m_v 11.3	p.a. 5°

Pretty bright, quite large, irregularly round; irregular arms; very faint nucleus. Distance 15 million ly; Hubble class Ir. In M101 galaxy group.

NGC 5206

Gx	Cen	13ʰ 33.7ᵐ	−48° 09′	3.5′ × 3.0′
charts 21, 25			m_v 10.6	p.a. 16°

Faint, pretty large, and roundish; moderately bright nucleus. Distance 18 million ly; Hubble class S0.

NGC 5230

Gx	Vir	13ʰ 35.5ᵐ	+13° 40′	2.1′
chart 14			m_p 12.1	

Faint, large, and elongated; bright, very small nucleus. Distance 300 million ly; Hubble class Sc. Brightest in a group of three galaxies.

NGC 5236

See M83.

NGC 5247

Gx	Vir	13ʰ 38.1ᵐ	−17° 53′	5.4′ × 4.7′
charts 14, 21			m_v 10.1	p.a. 20°

Quite faint and very large; fine example of an S-shaped spiral; large, bright nucleus. Distance 66 million ly; Hubble class Sb.

NGC 5248

Gx	Boo	13ᵐ 37.5ᵐ	+08° 53′	6.5′ × 4.9′
chart 14			m_v 10.3	p.a. 110°

C45. Bright, large, and elongated; extremely bright nucleus with many dark lanes. Distance 48 million ly; Hubble class Sc.

NGC 5253

Gx	Cen	13ʰ 39.9ᵐ	−31° 39′	4.8′ × 1.8′
chart 21			m_v 10.2	p.a. 45°

Bright, pretty large, and elongated; very bright center. Distance 9 million ly; Hubble class E5. Similar to M82.

NGC 5254

Gx	Vir	13ʰ 39.6ᵐ	−11° 30′	2.9′ × 1.5′
chart 14			m_v 12.3	p.a. 125°

Pretty bright, large, and quite elongated; moderately bright nucleus. Distance 97 million ly; Hubble class Sc.

NGC 5264

Gx	Hya	13ʰ 41.6ᵐ	−29° 55′	2.6′ × 1.7′
chart 21			m_v 12.1	p.a. 70°

Very faint, pretty large, and roundish; faint nucleus. Distance 13 million ly; Hubble class Ir+.

NGC 5266

Gx	Cen	13ʰ 43.0ᵐ	−48° 10′	3.3′ × 2.3′
charts 21, 25			m_v 10.9	p.a. 19°

Bright, pretty large, and slightly elongated; small, bright nucleus with dark lane. Distance 120 million ly; Hubble class S0. Three stars nearby.

NGC 5266A

Gx	Cen	13ʰ 40.6ᵐ	−48° 21′	3.2′ × 2.5′
charts 21, 25			m_v 12.1	p.a. 150°

Bright bar; patchy arms with knots. Distance 120 million ly; Hubble class Sc.

NGC 5272

See M3.

NGC 5273

Gx	CVn	13ʰ 42.1ᵐ	+35° 39′	2.9′ × 2.5′
chart 7			m_v 11.6	p.a. 10°

Quite bright, pretty large, and round; small, bright, diffuse nucleus with two faint, dark patches. Distance 47 million ly; Hubble class E1p.

NGC 5281

OC	Cen	13ʰ 46.6ᵐ	−62° 54′	5′
chart 25			m_v 5.9	

40 stars; detached, strong concentration of stars; small range in brightness; mag. of brightest star 6.6; irregularly round shape; small and bright. Distance 4,200 ly; age 51 million years; Trumpler class I 1 p.

NGC 5286

GC	Cen	13ʰ 46.5ᵐ	−51° 22′	10′
charts 21, 25			m_v 7.4	

C84. Medium concentration of stars; pretty large, very bright, round, very well resolved. Distance 29,000 ly; S-S class 5.

NGC 5292

Gx	Cen	13ʰ 47.7ᵐ	−30° 56′	2.0′ × 1.7′
chart 21			m_v 11.9	p.a. 55°

Pretty faint, small, and roundish. Distance 180 million ly; Hubble class Sa. Two stars nearby.

NGC 5297

Gx	CVn	$13^h 46.4^m$	$+43°\,52'$	$5.6' \times 1.4'$
chart 7			$m_v\,11.8$	p.a. 148°

Quite bright, large, and very elongated; very small, bright nucleus with dark lane. Distance 110 million ly; Hubble class Sb. Possibly interacting with galaxy NGC 5296.

NGC 5300

Gx	Vir	$13^h 48.3^m$	$+03°\,57'$	$3.8' \times 2.5'$
chart 14			$m_v\,11.4$	p.a. 150°

Very faint, very large, and slightly elongated; almost face-on spiral. Distance 49 million ly; Hubble class Sc.

NGC 5307

PN	Cen	$13^h 51.1^m$	$-51°\,12'$	13"
charts 21, 25			$m_v\,11.2$	

PK312+10.1. Very faint, extremely small, and irregular; phot. mag. 12.1; central star mag. 14.7. Expansion velocity 15 km/sec (9 miles/sec); V-V class 3.

NGC 5308

Gx	UMa	$13^h 47.0^m$	$+60°\,58'$	$3.6' \times 0.7'$
chart 2			$m_v\,11.4$	p.a. 60°

Bright, pretty large, and very elongated; very bright, diamond-shaped nucleus. Distance 93 million ly; Hubble class S0.

NGC 5313

Gx	CVn	$13^h 49.7^m$	$+39°\,59'$	$1.7' \times 1.0'$
chart 7			$m_v\,12$	p.a. 40°

Pretty bright, pretty small, and slightly elongated. Distance 120 million ly; Hubble class Sb.

NGC 5315

PN	Cir	$13^h 54.0^m$	$-66°\,31'$	6"
chart 25			$m_v\,9.8$	

PK309–4.2. Appears stellar; phot. mag. 13.0; central star mag. 14.4. Distance 9,000 ly; expansion velocity 40 km/sec (25 miles/sec); V-V class 2.

NGC 5316

OC	Cen	$13^h 53.9^m$	$-61°\,52'$	14'
charts 25, A3			$m_v\,6$	

80 stars; detached, no concentration of stars; small range in brightness; mag. of brightest star 7.8; pretty large. Distance 3,700 ly; age 190 million years; Trumpler class III 1 p.

NGC 5320

Gx	CVn	$13^h 50.3^m$	$+41°\,22'$	$3.6' \times 1.8'$
chart 7			$m_v\,12.1$	p.a. 18°

Quite faint, pretty large, and roundish; moderately bright nucleus. Distance 120 million ly; Hubble class S(B)c.

NGC 5322

Gx	UMa	$13^h 49.3^m$	$+60°\,11'$	$6.0' \times 4.1'$
chart 2			$m_v\,10.2$	p.a. 95°

Very bright and pretty large; bright, diffuse nucleus. Distance 90 million ly; Hubble class E2.

NGC 5324

Gx	Vir	$13^h 52.1^m$	$-06°\,03'$	$2.2' \times 2.0'$
chart 14			$m_v\,11.7$	p.a. 170°

Quite faint, large, irregularly round; many filamentary arms; bright, diffuse nucleus. Distance 130 million ly; Hubble class Sc.

NGC 5326

Gx	CVn	$13^h 50.8^m$	$+39°\,34'$	$2.2' \times 1.2'$
chart 7			$m_v\,11.9$	p.a. 137°

Quite faint, small, and slightly elongated. Distance 120 million ly; Hubble class Sb.

NGC 5328

Gx	Hya	$13^h 52.9^m$	$-28°\,29'$	$1.8' \times 1.3'$
chart 21			$m_v\,11.7$	p.a. 87°

Pretty bright, small, and round. Distance 200 million ly; Hubble class E2. Paired with much fainter galaxy NGC 5330.

NGC 5333

Gx	Cen	$13^h 54.4^m$	$-48°\,31'$	$1.9' \times 1.0'$
charts 21, 25			$m_p\,12.8$	p.a. 52°

Very faint, very small, and roundish. Distance 110 million ly; Hubble class S0.

NGC 5334

Gx	Vir	13ʰ 52.9ᵐ	−01° 07′	4.1′ × 3.2′
chart 14			m_v 11.3	p.a. 15°

Quite faint, very large, and round. Distance 57 million ly; Hubble class S–.

NGC 5350

Gx	CVn	13ʰ 53.4ᵐ	+40° 22′	2.7′ × 2.1′
chart 7			m_v 11.3	p.a. 40°

Quite faint and pretty large; very small, very bright nucleus; two filamentary arms. Distance 100 million ly; Hubble class Sb. In a group with galaxies NGC 5353, NGC 5354, and NGC 5355.

NGC 5351

Gx	CVn	13ʰ 53.5ᵐ	+37° 55′	2.9′ × 1.6′
chart 7			m_v 12.1	p.a. 100°

Quite faint, large, and slightly elongated; small, bright nucleus. Distance 100 million ly; Hubble class Sb. In a group with galaxies NGC 5341 and NGC 5349.

NGC 5353

Gx	CVn	13ʰ 53.4ᵐ	+40° 17′	1.9′ × 1.3′
chart 7			m_v 11	p.a. 145°

Pretty bright, small, and round; very small, bright nucleus. Distance 92 million ly; Hubble class E5. In a group with galaxies NGC 5350, NGC 5354, and NGC 5355.

NGC 5354

Gx	CVn	13ʰ 53.4ᵐ	+40° 18′	1.7′ × 1.5′
chart 7			m_v 11.4	p.a. 95°

Pretty faint, small, and round. Distance 140 million ly; Hubble class S0. In a group with galaxies NGC 5350, NGC 5353, and NGC 5355.

NGC 5357

Gx	Cen	13ʰ 56.0ᵐ	−30° 20′	1.7′ × 1.4′
chart 21			m_v 12.1	p.a. 23°

Pretty faint, small, and round. Distance 210 million ly; Hubble class E0. Located between two 10th-mag. stars; in the IC 4329 galaxy group.

NGC 5363

Gx	Vir	13ʰ 56.1ᵐ	+05° 15′	4.9′ × 3.2′
chart 14			m_v 10.1	p.a. 135°

Bright, pretty large, and round; bright nucleus. Distance 47 million ly; Hubble class Ep. In a group with galaxies NGC 5360 and NGC 5364.

NGC 5364

Gx	Vir	13ʰ 56.2ᵐ	+05° 01′	6.0′ × 4.4′
chart 14			m_v 10.5	p.a. 30°

Quite faint, large, and round; small, bright nucleus; internal ring with dark matter; two main arms. Distance 59 million ly; Hubble class Sbp. In a group with galaxies NGC 5360 and NGC 5363.

NGC 5365

Gx	Cen	13ʰ 57.8ᵐ	−43° 56′	3.5′ × 2.1′
charts 21, 25			m_v 11.2	p.a. 4°

Pretty bright, quite small, and round; small, very bright nucleus. Distance 92 million ly; Hubble class SB0. Located among stars.

NGC 5367

BNr	Cen	13ʰ 57.7ᵐ	−39° 59′	4.0′ × 3.0′
chart 21				

Stars of mag. 10 and 11 involved in a very bright, very large, and round nebula. Per *NGC*, a (!) remarkable object.

NGC 5371

Gx	CVn	13ʰ 55.7ᵐ	+40° 28′	4.2′ × 3.4′
chart 7			m_v 10.6	p.a. 8°

Pretty bright, large, and round; filamentary arms; very small, bright, diffuse nucleus. Distance 120 million ly; Hubble class Sb.

NGC 5375

Gx	CVn	13ʰ 56.9ᵐ	+29° 10′	3.2′ × 2.8′
chart 7			m_v 11.5	p.a. 0°

Pretty bright, pretty large, and roundish. Distance 110 million ly; Hubble class SBb.

NGC 5376

Gx	UMa	$13^h 55.3^m$	+59° 31′	2.1′ × 1.3′
chart 2			m_v 12.1	p.a. 70°

Quite bright and pretty large; small, bright nucleus. Distance 97 million ly; Hubble class Sa. In a group with galaxies NGC 5379 and NGC 5389.

NGC 5377

Gx	CVn	$13^h 56.3^m$	+47° 14′	3.5′ × 2.0′
chart 7			m_v 11.3	p.a. 20°

Bright, large, and elongated; small, extremely bright nucleus. Distance 85 million ly; Hubble class Sap.

NGC 5383

Gx	CVn	$13^h 57.1^m$	+41° 51′	2.7′ × 2.2′
chart 7			m_v 11.4	p.a. 85°

Quite bright, quite large, and round; two main arms; extremely bright nucleus with dark lanes. Distance 100 million ly; Hubble class SBb.

NGC 5389

Gx	UMa	$13^h 56.1^m$	+59° 45′	4.0′ × 1.1′
chart 2			m_v 12.1	p.a. 3°

Pretty bright, pretty large, and elongated; very bright nucleus with dark lane. Distance 87 million ly; Hubble class S(B)a. In a group with galaxies NGC 5376 and NGC 5379.

NGC 5395

Gx	CVn	$13^h 58.6^m$	+37° 26′	2.6′ × 1.3′
chart 7			m_v 11.4	p.a. 167°

Quite faint, quite large, and elongated; single arm; small, bright nucleus. Distance 160 million ly; Hubble class Sb. In contact with fainter galaxy NGC 5394.

NGC 5408

Gx	Cen	$14^h 03.4^m$	−41° 23′	2.0′ × 1.2′
chart 21			m_v 12.1	p.a. 62°

Extremely faint and elongated. Distance 16 million ly; Hubble class SBm. Located between two very faint stars.

NGC 5419

Gx	Cen	$14^h 03.6^m$	−33° 59′	3.9′ × 3.1′
chart 21			m_v 10.8	p.a. 75°

Pretty bright, pretty large, and round. Distance 180 million ly; Hubble class E4. Several small galaxies nearby.

NGC 5422

Gx	UMa	$14^h 00.7^m$	+55° 10′	3.5′ × 0.6′
charts 2, 7			m_v 11.8	p.a. 152°

Pretty bright, small, and elongated; small, very bright nucleus. Distance 86 million ly; Hubble class S0. Galaxies M101, NGC 5473, NGC 5474, and NGC 5485 all lie within an area of about 1°.

NGC 5426

Gx	Vir	$14^h 03.4^m$	−06° 04′	2.9′ × 1.6′
chart 14			m_v 12.1	p.a. 10°

Pretty faint, quite large, and round; bright, diffuse nucleus. Distance 100 million ly; Hubble class Sc. Interacting with galaxy NGC 5427.

NGC 5427

Gx	Vir	$14^h 03.4^m$	−06° 02′	2.8′ × 2.2′
chart 14			m_v 11.4	p.a. 170°

Pretty faint, quite large, and round; a face-on spiral; very bright nucleus. Distance 110 million ly; Hubble class Sc. Interacting with galaxy NGC 5426.

NGC 5430

Gx	UMa	$14^h 00.8^m$	+59° 20′	2.2′ × 1.4′
chart 2			m_v 11.9	p.a. 0°

Pretty bright, small, and slightly elongated; two main arms; small, very bright nucleus. Distance 160 million ly; Hubble class Sb.

NGC 5444

Gx	CVn	$14^h 03.4^m$	+35° 08′	2.5′ × 2.0′
chart 7			m_v 11.8	p.a. 90°

Pretty bright, pretty large, and slightly elongated. Distance 180 million ly; Hubble class E1. Brightest in a group of galaxies.

NGC 5448

Gx	UMa	14^h 02.8^m	+49° 10′	3.9′ × 2.0′
chart 7			m$_v$ 11	p.a. 115°

Pretty bright, quite large, and elongated; small, very bright nucleus. Distance 91 million ly; Hubble class Sb.

NGC 5457

See M101.

NGC 5460

OC	Cen	14^h 07.6^m	−48° 19′	25′
charts 21, 25			m$_v$ 5.6	

40 stars; detached, weak concentration of stars; large range in brightness; mag. of brightest star 8.0; very large. Distance 1,600 ly; age 110 million years; Trumpler class II 3 m. Galaxy E 221-26 located approx. 0.5° to N.

NGC 5466

GC	Boo	14^h 05.5^m	+28° 32′	8′
chart 7			m$_v$ 9.2	

A large cluster with low concentration of stars. Distance 47,000 ly; S-S class 12.

NGC 5468

Gx	Vir	14^h 06.6^m	−05° 27′	2.4′ × 2.2′
chart 14			m$_v$ 12.5	p.a. 105°

Faint, large, and round; two filamentary arms; small, very bright nucleus. Distance 120 million ly; Hubble class Sc.

NGC 5473

Gx	UMa	14^h 04.7^m	+54° 54′	2.2′ × 1.7′
charts 2, 7			m$_v$ 11.4	p.a. 160°

Pretty bright, small, and round; very bright, small nucleus. Distance 94 million ly; Hubble class E2. Galaxies M101, NGC 5422, NGC 5474, and NGC 5485 all lie within an area of about 1°.

NGC 5474

Gx	UMa	14^h 05.0^m	+53° 40′	5′
charts 2, 7			m$_v$ 10.8	

Pretty bright and large; faint, small nucleus. Distance 18 million ly; Hubble class Sc. In M101 galaxy group; Galaxies M101, NGC 5422, NGC 5473, and NGC 5485 all lie within an area of about 1°.

NGC 5480

Gx	UMa	14^h 06.4^m	+50° 44′	1.6′ × 1.0′
charts 2, 7			m$_v$ 12.1	p.a. 0°

Faint and pretty small; several filamentary, branching arms; small, very bright nucleus. Distance 84 million ly; Hubble class Sc. Paired with slightly smaller galaxy NGC 5481.

NGC 5483

Gx	Cen	14^h 10.4^m	−43° 20′	4.3′ × 4.0′
chart 21			m$_v$ 11.2	p.a. 25°

Pretty faint, very large, and round; two patchy arms; very faint nucleus. Distance 71 million ly; Hubble class Sc.

NGC 5485

Gx	UMa	14^h 07.2^m	+55° 00′	2.8′ × 2.1′
charts 2, 7			m$_v$ 11.4	p.a. 170°

Quite bright and round; small, bright nucleus; dark lane. Distance 93 million ly; Hubble class Sa. In a group with galaxies NGC 5484 and NGC 5486; in addition, galaxies M101, NGC 5422, NGC 5473, and NGC 5474 all lie within an area of about 1°.

NGC 5488

Gx	Cen	14^h 08.1^m	−33° 19′	3.0′ × 1.0′
chart 21			m$_v$ 11.9	p.a. 22°

Faint and round. Distance 180 million ly; Hubble class SBb. An 8th-mag. star is nearby.

NGC 5493

Gx	Vir	14^h 11.5^m	−05° 03′	2.0′ × 1.7′
chart 14			m$_v$ 11.4	p.a. 124°

Pretty bright, very small, and round; very bright, diffuse nucleus. Distance 110 million ly; Hubble class S0p.

NGC 5496

Gx	Vir	14^h 11.6^m	−01° 09′	4.7′ × 0.9′
chart 14			m$_v$ 12.1	p.a. 172°

Pretty bright, very large, and elongated; edge on; dust lanes. Distance 64 million ly; Hubble class S–.

NGC 5506

Gx	Vir	14^h 13.2^m	–03° 12′	2.6′ × 0.7′
chart 14			m$_v$ 11.9	p.a. 91°

Pretty bright, large, and elongated; has a moderately bright nucleus. Distance 73 million ly; Hubble class Sap.

NGC 5516

Gx	Cen	14^h 15.9^m	–48° 07′	1.7′ × 1.1′
charts 21, 25			m$_v$ 12	p.a. 169°

Pretty faint, small, and roundish; bright nucleus. Distance 190 million ly; Hubble class E. Many faint superimposed stars.

NGC 5523

Gx	Boo	14^h 14.9^m	+25° 19′	4.7′ × 1.3′
chart 7			m$_v$ 12.1	p.a. 99°

Faint, pretty large, and very elongated; many filamentary arms; very faint nucleus. Distance 48 million ly; Hubble class Sb.

NGC 5529

Gx	Boo	14^h 15.6^m	+36° 14′	6.0′ × 0.7′
chart 7			m$_v$ 11.9	p.a. 115°

Quite faint and pretty large; bright nucleus; almost exactly edgewise. Distance 130 million ly; Hubble class Sc.

NGC 5530

Gx	Lup	14^h 18.5^m	–43° 23′	4.3′ × 1.8′
chart 21			m$_v$ 11.1	p.a. 127°

Very faint and very elongated; very patchy arms. Per *NGC*, a (!) remarkable object. Distance 8 million ly; Hubble class Sb.

NGC 5532

Gx	Boo	14^h 16.9^m	+10° 48′	1.6′
chart 14			m$_v$ 11.9	

Very faint, very small, and roundish; moderately bright nucleus. Distance 310 million ly; Hubble class S0.

NGC 5533

Gx	Boo	14^h 16.1^m	+35° 21′	3.2′ × 1.9′
chart 7			m$_v$ 11.8	p.a. 30°

Pretty bright and round; very bright nucleus. Distance 170 million ly; Hubble class S.

NGC 5534

Gx	Vir	14^h 17.7^m	–07° 25′	1.3′ × 1.0′
chart 14			m$_v$ 12.3	p.a. 55°

Pretty faint and very small. Distance 110 million ly; Hubble class S.

NGC 5556

Gx	Hya	14^h 20.6^m	–29° 15′	4.2′ × 3.3′
chart 21			m$_v$ 11.8	p.a. 148°

Extremely faint and large; very patchy arms; very faint nucleus. Distance 54 million ly; Hubble class S(B)–. Paired with an anonymous spiral galaxy.

NGC 5557

Gx	Boo	14^h 18.4^m	+36° 30′	2.3′ × 1.9′
chart 7			m$_v$ 11	p.a. 105°

Quite bright, small, and round; bright nucleus. Distance 140 million ly; Hubble class E1.

NGC 5566

Gx	Vir	14^h 20.3^m	+03° 56′	6.0′ × 2.3′
chart 14			m$_v$ 10.6	p.a. 35°

Bright, pretty large, and round; smooth outer arms with dark lane; extremely bright nucleus; internal ring with bar. Distance 65 million ly; Hubble class Sb.

NGC 5576

Gx	Vir	14^h 21.1^m	+03° 16′	3.0′ × 2.3′
chart 14			m$_v$ 11	p.a. 95°

Bright, small, and round; bright, diffuse nucleus. Distance 65 million ly; Hubble class E2. Brightest in a group that includes galaxies NGC 5574 and NGC 5577.

NGC 5582

Gx	Boo	14^h 20.7^m	+39° 42′	2.8′ × 1.7′
chart 7			m$_v$ 11.6	p.a. 25°

Pretty bright, pretty small, and roundish; faint nucleus. Distance 66 million ly; Hubble class E.

NGC 5584

Gx	Vir	$14^h\,22.4^m$	$-00°\,23'$	$3.3' \times 2.7'$
chart 14			$m_v\,11.4$	p.a. 140°

Faint, large, and elongated; filamentary arms; short, bright bar. Distance 69 million ly; Hubble class Sc. Similar to M101 (but much smaller in angular extent).

NGC 5585

Gx	UMa	$14^h\,19.8^m$	$+56°\,44'$	$5.5' \times 3.7'$
charts 2, 7			$m_v\,10.7$	p.a. 30°

Pretty faint, large, and irregularly round; very faint nucleus. Distance 20 million ly; Hubble class S. In M101 galaxy group.

NGC 5595

Gx	Lib	$14^h\,24.2^m$	$-16°\,43'$	$2.3' \times 1.3'$
chart 14			$m_v\,12$	p.a. 70°

Faint, pretty large, and round; two main arms; small, bright nucleus. Distance 110 million ly; Hubble class Sc. Paired with galaxy NGC 5597.

NGC 5597

Gx	Lib	$14^h\,24.5^m$	$-16°\,46'$	$2.0' \times 1.7'$
chart 14			$m_v\,12$	p.a. 60°

Very faint, large, and slightly elongated; several filamentary arms; very small, extremely bright nucleus. Distance 110 million ly; Hubble class Sb. Paired with galaxy NGC 5595.

NGC 5600

Gx	Boo	$14^h\,23.8^m$	$+14°\,38'$	$1.4'$
chart 14			$m_v\,12.1$	

Pretty bright and pretty small; very bright nucleus. Distance 100 million ly; Hubble class Sc.

NGC 5605

Gx	Lib	$14^h\,25.1^m$	$-13°\,10'$	$1.5' \times 1.2'$
chart 14			$m_v\,12.3$	p.a. 70°

Very faint, pretty large, and round. Distance 140 million ly; Hubble class Sc.

NGC 5606

OC	Cen	$14^h\,27.8^m$	$-59°\,38'$	$3'$
charts 25, A3			$m_v\,7.7$	

15 stars; detached, strong concentration; small brightness range; mag. of brightest star 7.9. Distance 5,500 ly; age 13 million years; Trumpler class I 1 p.

NGC 5612

Gx	Aps	$14^h\,34.0^m$	$-78°\,23'$	$1.9' \times 1.1'$
charts 25, 26			$m_v\,12.1$	p.a. 63°

Very faint and elongated; pretty faint arms. Distance 110 million ly; Hubble class S0.

NGC 5614

Gx	Boo	$14^h\,24.1^m$	$+34°\,51'$	$2.5' \times 2.0'$
chart 7			$m_v\,11.7$	p.a. 130°

Pretty bright, small, and round; small, very bright nucleus that is noticeably off center in the glow of the spiral arms. Distance 170 million ly; Hubble class S. Interacting with two much fainter objects not shown in the atlas: NGC 5615 (a bright knot on the rim of NGC 5614, NW of nucleus) and the galaxy NGC 5613 lying 2′ N.

NGC 5617

OC	Cen	$14^h\,29.8^m$	$-60°\,43'$	$10'$
charts 25, A3			$m_v\,6.3$	

80 stars; detached, strong concentration of stars; large range in brightness; mag. of brightest star 8.8; large. Distance 3,900 ly; age 46 million years; Trumpler class I 3 m.

NGC 5631

Gx	UMa	$14^h\,26.6^m$	$+56°\,35'$	$2.0' \times 1.6'$
charts 2, 7			$m_v\,11.5$	p.a. 110°

Bright, small, and round; very bright, diffuse nucleus. Distance 93 million ly; Hubble class Sa.

NGC 5633

Gx	Boo	$14^h\,27.5^m$	$+46°\,09'$	$2.1' \times 1.1'$
chart 7			$m_v\,12.4$	p.a. 10°

Quite bright, pretty small, and round; small, bright nucleus. Distance 110 million ly; Hubble class Sb.

NGC 5634

GC	Vir	$14^h 29.6^m$	$-05°\,59'$	$5'$
chart 14			$m_v\,9.5$	

High concentration of stars; very bright, quite large, round, and well resolved. Distance 70,000 ly; S-S class 4. An 11th-mag. star is on one edge.

NGC 5638

Gx	Vir	$14^h 29.7^m$	$+03°\,14'$	$2.3' \times 2.1'$
chart 14			$m_v\,11.2$	p.a. 150°

Quite bright, pretty large, and round. Distance 72 million ly; Hubble class E1. Paired with galaxy NGC 5636.

NGC 5643

Gx	Lup	$14^h 32.7^m$	$-44°\,10'$	$4.8'$
chart 21			$m_v\,10.4$	

Pretty bright, large, and round; very small, very bright nucleus. Distance 42 million ly; Hubble class S(B)c.

NGC 5645

Gx	Vir	$14^h 30.7^m$	$+07°\,17'$	$2.8' \times 1.6'$
chart 14			$m_v\,12.5$	p.a. 80°

Quite faint, pretty large, irregularly round. Distance 60 million ly; Hubble class Sc.

NGC 5653

Gx	Boo	$14^h 30.2^m$	$+31°\,13'$	$1.7' \times 1.3'$
chart 7			$m_v\,12.2$	p.a. 125°

Faint, pretty small, and round; several bright arms; very bright nucleus. Distance 160 million ly; Hubble class S.

NGC 5656

Gx	Boo	$14^h 30.4^m$	$+35°\,19'$	$1.9' \times 1.5'$
chart 7			$m_v\,11.8$	p.a. 50°

Pretty faint, pretty large, and round; bright nucleus. Distance 140 million ly; Hubble class Sb.

NGC 5660

Gx	Boo	$14^h 29.8^m$	$+49°\,37'$	$2.8' \times 2.6'$
chart 7			$m_v\,11.9$	p.a. 90°

Pretty bright, large, irregularly round; several branching arms; small, very bright nucleus. Distance 110 million ly; Hubble class Sc. Paired with a faint, anonymous, irregular galaxy.

NGC 5662

OC	Cen	$14^h 35.2^m$	$-56°\,33'$	$12'$
charts 21, 25			$m_v\,5.5$	

70 stars; detached, weak concentration of stars; large range in brightness; mag. of brightest star 7.0; Trumpler class II 3 m.

NGC 5665

Gx	Boo	$14^h 32.4^m$	$+08°\,05'$	$2.1' \times 1.3'$
chart 14			$m_v\,12$	p.a. 145°

Pretty bright, pretty large, and round; small, bright nucleus. Distance 98 million ly; Hubble class Sc.

NGC 5668

Gx	Vir	$14^h 33.4^m$	$+04°\,27'$	$3.2' \times 3.0'$
chart 14			$m_v\,11.5$	p.a. 120°

Faint, pretty small, and slightly elongated; many arms; very small, faint nucleus. Distance 68 million ly; Hubble class Sc.

NGC 5669

Gx	Boo	$14^h 32.7^m$	$+09°\,53'$	$4.5' \times 3.6'$
chart 14			$m_v\,11.3$	p.a. 50°

Faint, large, and round. Distance 60 million ly; Hubble class Sc.

NGC 5670

Gx	Lup	$14^h 35.6^m$	$-45°\,58'$	$2.2' \times 0.9'$
charts 21, 25			$m_p\,13.0$	p.a. 74°

Very faint, small, and quite elongated. Distance 120 million ly; Hubble class S. Located between two stars.

NGC 5673

Gx	Boo	$14^h 31.5^m$	$+49°\,58'$	$2.3' \times 0.5'$
charts 2, 7			$m_v\,12.1$	p.a. 136°

Faint, small, and quite elongated. Distance 98 million ly; Hubble class SBc.

NGC 5676

| Gx | Boo | 14^h 32.8^m | +49° 27' | 3.9' × 1.8' |
| chart 7 | | | m$_v$ 11.2 | p.a. 47° |

Bright, large, and elongated; many filamentary arms; small, bright nucleus. Distance 100 million ly; Hubble class Sc.

NGC 5678

| Gx | Dra | 14^h 32.1^m | +57° 55' | 3.3' × 1.5' |
| chart 2 | | | m$_v$ 11.3 | p.a. 5° |

Bright and large; very small, bright nucleus. Distance 100 million ly; Hubble class Scp. Paired with an anonymous class E3 galaxy.

NGC 5687

| Gx | Boo | 14^h 34.9^m | +54° 29' | 2.6' × 1.9' |
| charts 2, 7 | | | m$_v$ 11.8 | p.a. 105° |

Pretty faint and small with a faint halo; has a bright, diffuse nucleus. Distance 99 million ly; Hubble class Sa.

NGC 5688

| Gx | Lup | 14^h 39.6^m | −45° 01' | 4.2' × 2.5' |
| chart 21 | | | m$_v$ 11.9 | p.a. 85° |

Faint and small; moderately bright nucleus. Distance 120 million ly; Hubble class SBb. Located among stars.

NGC 5689

| Gx | Boo | 14^h 35.5^m | +48° 45' | 3.3' × 1.0' |
| chart 7 | | | m$_v$ 11.9 | p.a. 85° |

Quite bright, pretty large, and elongated; very bright, diffuse nuclear bulge with dark lane. Distance 100 million ly; Hubble class SBa. Brightest in a group with galaxies NGC 5682, NGC 5683, NGC 5693 (and others).

NGC 5690

| Gx | Vir | 14^h 37.7^m | +02° 17' | 3.4' × 1.0' |
| chart 14 | | | m$_v$ 11.8 | p.a. 143° |

Very faint and very elongated; several arms with dark lanes; bright middle; very faint nucleus. Distance 75 million ly; Hubble class Sb.

NGC 5691

| Gx | Vir | 14^h 37.9^m | −00° 24' | 1.9' × 1.6' |
| chart 14 | | | m$_v$ 12.3 | p.a. 110° |

Pretty bright, pretty small, and slightly elongated; a rather featureless galaxy; very bright center. Distance 80 million ly; Hubble class S(B)ap.

NGC 5694

| GC | Hya | 14^h 39.6^m | −26° 32' | 4.2' |
| chart 21 | | | m$_v$ 10.2 | |

C66. Medium concentration of stars; quite bright, quite small, and round. Distance 105,000 ly; S–S class 7.

NGC 5701

| Gx | Vir | 14^h 39.2^m | +05° 22' | 4.7' |
| chart 14 | | | m$_v$ 10.9 | |

Quite bright, pretty small, and round; very small, very bright nucleus. Distance 65 million ly; Hubble class S(B) I-II. A small, anonymous spiral galaxy is visible near the outer ring.

NGC 5713

| Gx | Vir | 14^h 40.2^m | −00° 17' | 2.8' × 2.5' |
| chart 14 | | | m$_v$ 11.2 | p.a. 10° |

Quite bright, pretty large, and round; extremely bright nucleus. Distance 80 million ly; Hubble class Sc. Interacting with galaxy NGC 5719.

NGC 5716

| Gx | Lib | 14^h 41.1^m | −17° 29' | 1.9' × 1.3' |
| charts 14, 21 | | | m$_v$ 12.9 | p.a. 80° |

Very faint, pretty large, and roundish. Distance 180 million ly; Hubble class SBc. In a group with galaxy NGC 5728 and a small, anonymous, barred spiral; two stars located approx. 1' to NE.

NGC 5728

| Gx | Lib | 14^h 42.4^m | −17° 15' | 3.2' × 1.9' |
| charts 14, 21 | | | m$_v$ 11.5 | p.a. 33° |

Pretty faint, pretty large, and elongated; very bright nucleus with spiraling dark lanes. Distance 120 million ly; Hubble class S(B)b. In a group with galaxy NGC 5716 and a small, anonymous, barred spiral.

NGC 5729

Gx	Lib	14^h 42.1^m	−09° 01′	2.6′ × 0.7′
chart 14			m$_v$ 12.6	p.a. 170°

Faint, pretty large, and elongated. Distance 78 million ly; Hubble class Sb.

NGC 5740

Gx	Vir	14^h 44.4^m	+01° 41′	2.8′ × 1.5′
chart 14			m$_v$ 11.9	p.a. 160°

Pretty bright, large, irregularly round; several branching arms; very bright nucleus. Distance 68 million ly; Hubble class Sb. Paired with galaxy NGC 5746.

NGC 5746

Gx	Vir	14^h 44.9^m	+01° 57′	6.0′ × 1.2′
chart 14			m$_v$ 10.3	p.a. 170°

Bright, large, and very elongated; edge on; very small, bright nucleus. Distance 78 million ly; Hubble class Sb. Paired with galaxy NGC 5740.

NGC 5750

Gx	Vir	14^h 46.2^m	−00° 13′	3.0′ × 1.5′
chart 14			m$_v$ 11.6	p.a. 65°

Pretty faint, pretty small, and slightly elongated; small, extremely bright nucleus. Distance 87 million ly; Hubble class Sb.

NGC 5774

Gx	Vir	14^h 53.7^m	+03° 35′	3.1′ × 2.5′
chart 14			m$_v$ 12.1	p.a. 145°

Pretty faint, pretty large, and roundish. Distance 69 million ly; Hubble class S(B)d.

NGC 5775

Gx	Vir	14^h 54.0^m	+03° 33′	3.9′ × 1.1′
chart 14			m$_v$ 11.4	p.a. 146°

Faint, pretty small, and very elongated; edge on; several arms with dark lanes; bright central bulge. Distance 69 million ly; Hubble class Sb.

NGC 5786

Gx	Cen	14^h 58.9^m	−42° 01′	2.4′ × 1.3′
chart 21			m$_v$ 11.2	p.a. 63°

Faint and moderately elongated; pretty bright nucleus. Distance 120 million ly; Hubble class SBb. Extremely bright star located about 6′ to SE.

NGC 5791

Gx	Lib	14^h 58.8^m	−19° 16′	2.5′ × 1.3′
charts 14, 21			m$_v$ 11.6	p.a. 163°

Pretty faint, small, and round; bright, diffuse nucleus. Distance 140 million ly; Hubble class Sa. Paired with galaxy IC 1077.

NGC 5792

Gx	Lib	14^h 58.4^m	−01° 05′	6.0′ × 1.6′
chart 14			m$_v$ 11.2	p.a. 84°

Pretty bright, pretty large, and round; bright, very small nucleus with dark lane. Distance 86 million ly; Hubble class Sbp.

NGC 5796

Gx	Lib	14^h 59.4^m	−16° 37′	2.4′ × 1.8′
charts 14, 21			m$_v$ 11.6	p.a. 95°

A faint galaxy; bright, diffuse nucleus; small star in center. Distance 120 million ly; Hubble class E0. Paired with galaxy NGC 5793.

NGC 5806

Gx	Vir	15^h 00.0^m	+01° 53′	3.0′ × 1.5′
charts 14, 15			m$_v$ 11.7	p.a. 170°

Quite bright, quite large, and elongated; smooth outer arms; extremely bright nucleus. Distance 56 million ly; Hubble class Sb. In NGC 5846 galaxy group.

NGC 5812

Gx	Lib	15^h 00.9^m	−07° 27′	2.3′ × 2.0′
charts 14, 15			m$_v$ 11.2	p.a. 130°

Quite bright, small, and round; paired with galaxy IC 1084. Distance 88 million ly; Hubble class E1. In NGC 5846 galaxy group.

NGC 5813

Gx	Vir	15^h 01.2^m	+01° 42′	4.2′ × 3.2′
charts 14, 15			m$_v$ 10.5	p.a. 145°

Bright, pretty small, and round. Distance 82 million ly; Hubble class E1. In NGC 5846 galaxy group.

NGC 5822

OC	Lup	$15^h 05.2^m$	$-54°21'$	39'
charts 21, 25			m_p 6.5	

150 stars; detached, weak concentration of stars; small range in brightness; brightest star is phot. mag. 10; very large. Distance 1,800 ly; age 890 million years; Trumpler class II 1 r.

NGC 5823

OC	Cir	$15^h 05.7^m$	$-55°36'$	10'
charts 21, 25			m_v 7.9	

C88. 100 stars; detached, no concentration of stars; moderate range in brightness; mag. of brightest star 9.7. Distance 2,300 ly; age 200 million years; Trumpler class III 2 m.

NGC 5824

GC	Lup	$15^h 04.0^m$	$-33°04'$	7'
chart 21			m_v 9.1	

High concentration of stars; a pretty bright and small cluster; stellar nucleus. Distance 77,000 ly; S-S class 1.

NGC 5831

Gx	Vir	$15^h 04.1^m$	$+01°13'$	$2.2' \times 1.9'$
charts 14, 15			m_v 11.5	p.a. 55°

Pretty bright and small. Distance 73 million ly; Hubble class Ep.

NGC 5833

Gx	Aps	$15^h 11.9^m$	$-72°52'$	$3.0' \times 1.8'$
charts 25, 26			m_v 12	p.a. 128°

Faint, quite small, and slightly elongated; moderately bright nucleus. Distance 120 million ly; Hubble class Sb. Located among stars.

NGC 5838

Gx	Vir	$15^h 05.4^m$	$+02°06'$	$3.8' \times 1.5'$
charts 14, 15			m_v 10.9	p.a. 43°

Pretty bright and pretty small; bright, diffuse nucleus. Distance 62 million ly; Hubble class Sa. In NGC 5846 galaxy group.

NGC 5843

Gx	Lup	$15^h 07.5^m$	$-36°20'$	$1.9' \times 1.2'$
chart 21			m_v 12.3	p.a. 70°

Very faint, small, and slightly elongated; moderately bright nucleus. Distance 180 million ly; Hubble class SBa.

NGC 5846

Gx	Vir	$15^h 06.5^m$	$+01°36'$	3.7'
charts 14, 15			m_v 10	

Very bright, pretty large, and round; bright, diffuse nucleus. Distance 75 million ly; Hubble class E0. Paired with galaxy NGC 5846A. Brightest in a group of galaxies.

NGC 5850

Gx	Vir	$15^h 07.1^m$	$+01°33'$	$4.4' \times 3.5'$
charts 14, 15			m_v 10.8	p.a. 140°

Quite faint, small, and slightly elongated; very bright nucleus. Distance 100 million ly; Hubble class SBb. In NGC 5846 galaxy group.

NGC 5854

Gx	Vir	$15^h 07.8^m$	$+02°34'$	$2.7' \times 0.8'$
charts 14, 15			m_v 11.9	p.a. 55°

Pretty bright, small, and slightly elongated; very small, faint, diffuse nucleus. Distance 71 million ly; Hubble class Sa. Located among stars.

NGC 5861

Gx	Lib	$15^h 09.3^m$	$-11°19'$	$3.0' \times 1.7'$
charts 14, 15			m_v 11.6	p.a. 150°

Faint, large, and elongated; two filamentary arms; very faint nucleus. Distance 79 million ly; Hubble class Scp. Paired with galaxy NGC 5858.

NGC 5864

Gx	Vir	$15^h 09.6^m$	$+03°03'$	$2.5' \times 0.9'$
charts 14, 15			m_v 11.8	p.a. 68°

Pretty faint, quite small, and slightly elongated; small, bright diffuse nucleus. Distance 71 million ly; Hubble class E7p.

NGC 5866

Gx	Dra	15^h 06.5^m	+55° 46′	6.0′ × 3.1′
charts 2, 7			m$_v$ 9.9	p.a. 128°

Very bright and quite large; elongated, thin, edge on; very bright nucleus with dark lane. A Seyfert galaxy. Distance 38 million ly; Hubble class E6p. Galaxy NGC 5867 is nearby.

NGC 5869

Gx	Vir	15^h 09.8^m	+00° 28′	2.4′ × 1.7′
charts 14, 15			m$_p$ 12.9	p.a. 125°

Pretty faint, small, and elongated; bright nucleus. Distance 93 million ly; Hubble class S0.

NGC 5873

PN	Lup	15^h 12.9^m	−38° 07′	7″
chart 21			m$_v$ 11	

PK331+16.1. Appears stellar; phot. mag. 13.3; central star mag. 15.5. Distance 16,000 ly; expansion velocity 40 km/sec (25 miles/sec); V-V class 2.

NGC 5878

Gx	Lib	15^h 13.8^m	−14° 16′	3.2′ × 1.4′
charts 14, 15			m$_v$ 11.5	p.a. 5°

Pretty bright, pretty large, and very elongated; bright, diffuse nucleus. Distance 90 million ly; Hubble class Sb.

NGC 5879

Gx	Dra	15^h 09.8^m	+57° 00′	4.0′ × 1.4′
chart 2			m$_v$ 11.6	p.a. 0°

Quite bright, small, and elongated; very small, bright nucleus; many filamentary arms. Distance 44 million ly; Hubble class Sb.

NGC 5882

PN	Lup	15^h 16.8^m	−45° 39′	14″
charts 21, 22			m$_v$ 9.4	

PK327+10.1. Very small, round; phot. mag. 10.5; central star mag. 13.4. Expansion velocity 40 km/sec (25 miles/sec); distance 6,200 ly.

NGC 5885

Gx	Lib	15^h 15.1^m	−10° 05′	3.7′ × 3.0′
charts 14, 15			m$_v$ 11.8	p.a. 65°

Faint, quite large, and round; several filamentary arms; small, bright nucleus. Distance 86 million ly; Hubble class S−.

NGC 5892

Gx	Lib	15^h 13.8^m	−15° 28′	3.5′ × 2.9′
charts 14, 15, 21			m$_v$ 11.7	p.a. 105°

Extremely faint and large; bright nucleus. Distance 98 million ly; Hubble class SBc.

NGC 5897

GC	Lib	15^h 17.4^m	−21° 01′	10′
chart 21			m$_v$ 8.4	

Low concentration of stars; large, pretty faint, but very well resolved. Distance 39,000 ly; S-S class 11.

NGC 5898

Gx	Lib	15^h 18.2^m	−24° 06′	2.5′
chart 21			m$_v$ 11.4	

Faint, small, and round; elongated nucleus. Distance 93 million ly; Hubble class E1p. In a group with galaxy NGC 5903 and an anonymous galaxy.

NGC 5899

Gx	Boo	15^h 15.1^m	+42° 03′	3.3′ × 1.4′
chart 7			m$_v$ 11.7	p.a. 18°

Quite bright, pretty large, and elongated; several arms. Distance 120 million ly; Hubble class Sb. Paired with galaxy NGC 5900.

NGC 5903

Gx	Lib	15^h 18.6^m	−24° 04′	2.9′ × 2.1′
chart 21			m$_v$ 11.1	p.a. 168°

Quite faint, small, and round; elongated nucleus. Distance 100 million ly; Hubble class E1. In a group with galaxy NGC 5898 and an anonymous galaxy.

NGC 5904

See M5.

NGC 5907

Gx Dra 15^h 15.9^m +56° 20′ 11′ × 1.3′
chart 2   m$_v$ 10.3   p.a. 155°

Quite bright, very large, and very elongated; nearly edge on; very small nucleus with dark lane; a Seyfert galaxy. A faint streak, requiring the eye to "settle" before its outline is well seen (Smyth). Distance 34 million ly; Hubble class Sb. Galaxy NGC 5906 is attached.

NGC 5908

Gx Dra 15^h 16.7^m +55° 25′ 3.2′ × 1.4′
charts 2, 7   m$_v$ 11.8   p.a. 154°

Pretty faint, pretty small, and round; very small, bright nucleus with dark lane. Distance 150 million ly; Hubble class Sb. Paired with galaxy NGC 5905.

NGC 5915

Gx Lib 15^h 21.6^m −13° 06′ 1.5′ × 1.1′
charts 14, 15   m$_v$ 12.3   p.a. 35°

Bright, small, and round; two bright arms; bright center. Distance 97 million ly; Hubble class Sb. Interacting with galaxies NGC 5916 and NGC 5916A.

NGC 5916

Gx Lib 15^h 21.6^m −13° 10′ 2.2′ × 0.7′
charts 14, 15   m$_v$ 13.4   p.a. 15°

Faint, small, and slightly elongated; very bright nucleus. Distance 97 million ly; Hubble class SBap. Interacting with galaxies NGC 5915 and NGC 5916A.

NGC 5921

Gx Ser 15^h 21.9^m +05° 04′ 4.8′ × 4.0′
charts 14, 15   m$_v$ 10.8   p.a. 130°

Quite bright, quite large, and irregularly round; filamentary arms; extremely bright nucleus with dark lanes. Distance 65 million ly; Hubble class S(B)b. Located in Serpens Caput.

NGC 5925

OC Nor 15^h 27.7^m −54° 31′ 15′
charts 21, 25   m$_p$ 8.4

120 stars; detached, no concentration of stars; small range in brightness; very large; Trumpler class III 1 m.

NGC 5927

GC Lup 15^h 28.0^m −50° 40′ 5′
charts 21, 22, 25   m$_v$ 8

Medium concentration of stars; quite bright, large, round, and well resolved. Distance 24,000 ly; S-S class 8.

NGC 5937

Gx Ser 15^h 30.8^m −02° 50′ 1.9′ × 1.1′
charts 14, 15   m$_v$ 12.3   p.a. 150°

Pretty bright, pretty small, and roundish; brighter toward middle; Hubble class Sbp. Three stars located to E.

NGC 5946

GC Nor 15^h 35.5^m −50° 40′ 3.0′
charts 21, 22, 25   m$_v$ 8.4

Low concentration of stars; quite bright, pretty large, round, and very well resolved. Distance 31,000 ly; S-S class 9.

NGC 5949

Gx Dra 15^h 28.0^m +64° 46′ 2.2′ × 1.0′
chart 2   m$_v$ 12   p.a. 147°

Faint and small; many bright arms; has a bright, very small nucleus. Distance 26 million ly; Hubble class Sc.

NGC 5957

Gx Ser 15^h 35.4^m +12° 03′ 3.2′ × 3.0′
charts 14, 15   m$_v$ 11.7   p.a. 100°

Pretty bright and pretty large; cometary shape; moderately bright nucleus. Distance 82 million ly; Hubble class S(B)b.

NGC 5962

Gx Ser 15^h 36.5^m +16° 37′ 3.0′ × 2.2′
charts 7, 14, 15   m$_v$ 11.3   p.a. 110°

Pretty faint, pretty large, and slightly elongated; filamentary arms; small, very bright nucleus. Distance 90 million ly; Hubble class Sc. Located in Serpens Caput.

NGC 5963

| Gx | Dra | 15^h 33.5^m | +56° 34′ | 3.3′ × 2.3′ |
| chart 2 | | | m$_v$ 12.5 | p.a. 55° |

Pretty faint, pretty small, irregular shape. Distance 37 million ly; Hubble class Sp.

NGC 5965

| Gx | Dra | 15^h 34.0^m | +56° 41′ | 4.7′ × 0.7′ |
| chart 2 | | | m$_v$ 11.7 | p.a. 53° |

Quite faint, quite large, and elongated. Distance 160 million ly; Hubble class Sb.

NGC 5967

| Gx | Aps | 15^h 48.3^m | −75° 40′ | 2.5′ × 1.7′ |
| charts 25, 26 | | | m$_v$ 12 | p.a. 90° |

Faint, pretty large, and round; several arms; bright nucleus. Distance 120 million ly; Hubble class S(B)c. Paired with galaxy NGC 5967A.

NGC 5970

| Gx | Ser | 15^h 38.5^m | +12° 11′ | 3.0′ × 1.9′ |
| charts 14, 15 | | | m$_v$ 11.5 | p.a. 88° |

Pretty faint, pretty large, and round; very small, bright nucleus. Distance 90 million ly; Hubble class Sc. Paired with galaxy IC 1131; located in Serpens Caput.

NGC 5979

| PN | TrA | 15^h 47.7^m | −61° 13′ | 8″ |
| chart 25 | | | m$_v$ 11.5 | |

PK322−5.1. The central star is mag. 15.3. Distance 11,000 ly.

NGC 5982

| Gx | Dra | 15^h 38.7^m | +59° 21′ | 3.0′ × 2.1′ |
| chart 2 | | | m$_v$ 11.1 | p.a. 110° |

Quite bright, small, and round; very bright nucleus. Several stars in the field, of which one is pretty close W (Smyth). Distance 130 million ly; Hubble class E3p. In a group with galaxies NGC 5981 and NGC 5985.

NGC 5984

| Gx | Ser | 15^h 42.9^m | +14° 14′ | 3.0′ × 0.7′ |
| charts 7, 14, 15 | | | m$_v$ 12.5 | p.a. 144° |

Pretty bright, small, and elongated; bright bar. Distance 52 million ly; Hubble class Sb. Located in Serpens Caput.

NGC 5985

| Gx | Dra | 15^h 39.6^m | +59° 20′ | 5.0′ × 2.7′ |
| chart 2 | | | m$_v$ 11.1 | p.a. 13° |

Pretty bright and quite large; filamentary arms; small, bright nucleus. Distance 120 million ly; Hubble class Sb. In a group with galaxies NGC 5981 and NGC 5982.

NGC 5986

| GC | Lup | 15^h 46.1^m | −37° 47′ | 8′ |
| charts 21, 22 | | | m$_v$ 7.6 | |

Medium concentration of stars; very bright, large, and round. Per *NGC*, a (!) remarkable object. Distance 33,000 ly; S-S class 7.

NGC 5987

| Gx | Dra | 15^h 40.0^m | +58° 05′ | 4.2′ × 1.3′ |
| chart 2 | | | m$_v$ 11.7 | p.a. 165° |

Pretty faint and quite small. Distance 140 million ly; Hubble class Sb.

NGC 5990

| Gx | Ser | 15^h 46.3^m | +02° 25′ | 1.5′ × 0.9′ |
| charts 14, 15 | | | m$_v$ 12.4 | p.a. 115° |

Very faint, very small, and roundish; pretty bright nucleus. Distance 170 million ly; Hubble class Sa (?).

NGC 5996

| Gx | Ser | 15^h 47.0^m | +17° 53′ | 1.7′ × 1.0′ |
| charts 7, 14, 15 | | | m$_v$ 12.8 | p.a. 30° |

Pretty faint, quite small, and roundish. Distance 150 million ly; Hubble class Sc. Located between two double stars.

NGC 6010

| Gx | Ser | 15^h 54.3^m | +00° 33′ | 1.9′ × 0.5′ |
| charts 14, 15 | | | m$_v$ 12.6 | p.a. 105° |

Pretty faint, small, and elongated; has a moderately bright nucleus. Distance 84 million ly; Hubble class Sa.

NGC 6012

Gx Ser $15^h 54.2^m$ +14° 36' 2.1' × 1.5'
charts 7, 14, 15 m_v 12   p.a. 168°

Faint. Distance 88 million ly; Hubble class SBb. Located between two bright stars.

NGC 6015

Gx Dra $15^h 51.4^m$ +62° 19' 5.4' × 2.3'
charts 2, 3 m_v 11.1   p.a. 28°

Bright and elongated; dark lanes, very small, bright nucleus; a Seyfert galaxy. Distance 46 million ly; Hubble class Sc.

NGC 6025

OC TrA $16^h 03.7^m$ −60° 30' 12'
charts 25, 26 m_v 5.1

C95. 60 stars; detached, weak concentration of stars; moderate range in brightness; mag. of brightest star 7.3; bright, a very large cluster. Distance 2,700 ly; age 110 million years; Trumpler class I 2 p.

NGC 6031

OC Nor $16^h 07.6^m$ −54° 04' 2'
charts 22, 25, 26 m_v 8.5

20 stars; detached, strong concentration of stars; moderate range in brightness; mag. of brightest star 10.9. Distance 10,400 ly; age 22 million years; Trumpler class I 2 p.

NGC 6058

PN Her $16^h 04.4^m$ +40° 41' 23"
charts 7, 8 m_v 12.9

PK64+48.1. Pretty faint, very small, and round; phot. mag. 13.3; central star mag. 13.9. Distance 8,500 ly; expansion velocity 33 km/sec (20 miles/sec).

NGC 6067

OC Nor $16^h 13.2^m$ −54° 13' 12'
charts 22, 25, 26 m_v 5.6

100 stars; detached, strong concentration of stars; moderate range in brightness; mag. of brightest star 8.3; very bright and very large. Distance 6,900 ly; age 78 million years; Trumpler class I 2 r.

NGC 6068

Gx UMi $15^h 55.4^m$ +79° 00' 1.2' × 0.8'
charts 2, 3 m_v 12.8   p.a. 155°

Very faint, very small, and slightly elongated. Distance 180 million ly; Hubble class SBb.

NGC 6070

Gx Ser $16^h 10.0^m$ +00° 43' 3.5' × 1.8'
chart 15 m_v 11.8   p.a. 62°

Holmberg 729a. Faint, large, and slightly elongated. Distance 90 million ly; Hubble class Sc. In a group with galaxies Holmberg 729b and Holmberg 729c; located in Serpens Caput.

NGC 6087

OC Nor $16^h 18.9^m$ −57° 54' 12'
charts 25, 26 m_v 5.4

C89. 40 stars; detached, strong concentration of stars; moderate range in brightness; mag. of brightest star 7.9; bright, large. Distance 2,900 ly; age 55 million years; Trumpler class I 2 p.

NGC 6093

See M80.

NGC 6101

GC Aps $16^h 25.8^m$ −72° 12' 5'
chart 26 m_v 9.2

C107. Low concentration of stars; large, pretty faint, irregularly round. Distance 40,000 ly; S-S class 10.

NGC 6106

Gx Her $16^h 18.8^m$ +07° 25' 2.5' × 1.3'
chart 15 m_v 12.2   p.a. 140°

Faint, pretty large, and slightly elongated; knotty arms; very small, bright, diffuse nucleus. Distance 68 million ly; Hubble class Sb. Looks like M101 (but much smaller). Paired with a small, anonymous spiral galaxy.

NGC 6118

Gx	Ser	16ʰ 21.8ᵐ	−02° 17′	4.7′ × 1.9′
chart 15			m_v 11.7	p.a. 58°

Very faint, quite large, and quite elongated; three filamentary arms; faint nucleus. Distance 70 million ly; Hubble class Sb. Located in Serpens Caput.

NGC 6121

See M4.

NGC 6124

OC	Sco	16ʰ 25.6ᵐ	−40° 40′	28′
charts 21, 22			m_v 5.8:	

C75. 100 stars; detached, weak concentration of stars; large range in brightness; mag. of brightest star 8.7; bright, large. Distance 1,600 ly; age 51 million years; Trumpler class II 3 m.

NGC 6127

Gx	Dra	16ʰ 19.2ᵐ	+57° 59′	1.4′
charts 2, 3			m_v 12	

Pretty faint, very small, and roundish. Distance 210 million ly; Hubble class E.

NGC 6134

OC	Nor	16ʰ 27.7ᵐ	−49° 09′	7′
charts 21, 22, 26			m_v 7.2	

Detached, weak concentration of stars; large range in brightness; mag. of brightest star 9.3; quite large. Distance 2,600 ly; age 630 million years; Trumpler class II 3 m.

NGC 6139

GC	Sco	16ʰ 27.7ᵐ	−38° 51′	7′
charts 21, 22			m_v 9.1	

High concentration of stars; bright, pretty large, round, and partially resolved. Distance 29,000 ly; S-S class 2.

NGC 6140

Gx	Dra	16ʰ 21.0ᵐ	+65° 24′	6.0′ × 4.6′
charts 2, 3			m_v 11.3	p.a. 95°

Quite faint, pretty large, irregularly roundish. Distance 50 million ly; Hubble class SBcp.

NGC 6144

GC	Sco	16ʰ 27.2ᵐ	−26° 02′	16′
chart 22			m_v 9	

Low concentration of stars; quite large and very well resolved. Distance 26,000 ly; S-S class 11.

NGC 6152

OC	Nor	16ʰ 32.7ᵐ	−52° 37′	30′
charts 22, 26			m_p 8.1	

70 stars; detached, weak concentration of stars; moderate range in brightness; brightest star is about phot. mag. 11; large. Distance 3,400 ly; Trumpler class II 2 m.

NGC 6153

PN	Sco	16ʰ 31.5ᵐ	−40° 15′	24″
charts 21, 22			m_v 10.9	

PK341+05.1. Ring structure; phot. mag. 11.5; central star mag. 15.4; V-V class 4.

NGC 6156

Gx	TrA	16ʰ 34.9ᵐ	−60° 37′	1.6′ × 1.4′
chart 26			m_v 11.5	p.a. 0°

Pretty faint, pretty large, and slightly elongated; moderately bright nucleus; Hubble class SBb.

NGC 6166

Gx	Her	16ʰ 28.6ᵐ	+39° 33′	2.2′ × 1.7′
charts 7, 8			m_v 11.8	p.a. 35°

Pretty faint, small, and slightly elongated; bright nucleus. Distance 390 million ly; Hubble class E2p. Large number of faint galaxies nearby.

NGC 6167

OC	Nor	16ʰ 34.4ᵐ	−49° 36′	7′
charts 21, 22, 26			m_v 6.7	

Detached, weak concentration of stars; large range in brightness; mag. of brightest star 7.4; large and moderately rich. Perhaps not a true cluster.

NGC 6169

OC	Nor	16ʰ 34.1ᵐ	−44° 03′	7′
charts 21, 22			m_p 6.6	

40 stars; detached, no concentration; small brightness range. Distance 3,600 ly; Trumpler class III 1 m.

NGC 6171

See M107.

NGC 6178

OC	Sco	16^h 35.7^m	–45° 38′	4′
charts 21, 22			m$_v$ 7.2	

12 stars; detached, strong concentration of stars; large range in brightness; mag. of brightest star 8.4; small, bright; Trumpler class I 3 p.

NGC 6181

Gx	Her	16^h 32.3^m	+19° 50′	2.5′ × 1.1′
charts 8, 15			m$_v$ 11.9	p.a. 175°

Pretty bright, pretty large, and slightly elongated; three knotty arms; very small, bright nucleus. Distance 110 million ly; Hubble class Sc.

NGC 6188

BNer	Ara	16^h 40.5^m	–48° 47′	19′ × 12′
charts 21, 22, 26				

A faint, very large, and slightly elongated patch surrounding open cluster NGC 6193. Per *NGC*, a (!) remarkable object. Dark and bright nebulae nearby.

NGC 6192

OC	Sco	16^h 40.3^m	–43° 22′	7′
charts 21, 22			m$_p$ 8.5	

60 stars; detached, strong concentration of stars; moderate range in brightness; pretty large, irregularly round; brightest star is phot. mag. 11; Trumpler class I 2 p.

NGC 6193

OC	Ara	16^h 41.3^m	–48° 46′	15′
charts 21, 22, 26			m$_v$ 5.2	

C82. 15 stars; detached, weak concentration of stars; large range in brightness; mag. of brightest star 5.7; very large. Distance 4,400 ly; age 1 million years; involved in a faint, very large (19′ × 12′) nebula, NGC 6188.

NGC 6200

OC	Ara	16^h 44.2^m	–47° 29′	12′
charts 21, 22			m$_v$ 7.4	

40 stars; detached, no concentration; moderate brightness range; mag. of brightest star 9.2. Distance 7,800 ly; Trumpler class III 2 m.

NGC 6204

OC	Ara	16^h 46.5^m	–47° 01′	5′
charts 21, 22			m$_v$ 8.2	

45 stars; detached, strong concentration of stars; moderate range in brightness; mag. of brightest star 9.8. Distance 8,500 ly; age 13 million years; Trumpler class I 2 p. Adjacent to open cluster Ho 22.

NGC 6205

See M13.

NGC 6207

Gx	Her	16^h 43.1^m	+36° 50′	3.0′ × 1.2′
chart 8			m$_v$ 11.6	p.a. 15°

Pretty bright, pretty large, and elongated; very faint nucleus. Distance 46 million ly; Hubble class Sc. Bright star near center. Globular cluster M13 lies approx. 0.5° to SW.

NGC 6208

OC	Ara	16^h 49.5^m	–53° 49′	15′
charts 22, 26			m$_v$ 7.2	

60 stars; detached, weak concentration of stars; small range in brightness; mag. of brightest star 10.0; large. Distance 3,300 ly; over 1 billion years old; Trumpler class II 1 m.

NGC 6210

PN	Her	16^h 44.5^m	+23° 49′	16″
chart 8			m$_v$ 8.8	

PK43+37.1. Very bright, very small, smooth disk, involved in larger, fainter disk; traces of ring structure; phot. mag. 9.3; central star mag. 12.7. Exactly like a star out of focus (Webb). There are four stars in the field, of which that in the SE quadrant is reddish, affording a fair test of comparison with the pale blue nebula (Smyth). Distance 3,600 ly; expansion velocity 20 km/sec (12 miles/sec); V-V class 2 + 3b.

NGC 6215

| Gx | Ara | $16^h 51.1^m$ | $-59°\,00'$ | $2.1' \times 1.9'$ |
| chart 26 | | | m_v 10.9 | p.a. 78° |

Pretty faint and round; thick arms; very small, bright nucleus. Distance 60 million ly; Hubble class Sc. Paired with galaxy NGC 6221.

NGC 6217

| Gx | UMi | $16^h 32.7^m$ | $+78°\,12'$ | $3.3' \times 3.2'$ |
| charts 2, 3 | | | m_v 11.2 | p.a. 160° |

Bright and quite large; inner arms curved more than outer arms; small, very bright nucleus. Distance 73 million ly; Hubble class Sc.

NGC 6218

See M12.

NGC 6221

| Gx | Ara | $16^h 52.8^m$ | $-59°\,13'$ | $4.1' \times 2.7'$ |
| chart 26 | | | m_v 10.1 | p.a. 5° |

Pretty bright, quite large, and round; small, very bright nucleus with dark lanes. Distance 55 million ly; Hubble class SBc. Paired with galaxy NGC 6215.

NGC 6223

| Gx | Dra | $16^h 43.1^m$ | $+61°\,35'$ | $3.5' \times 2.6'$ |
| charts 2, 3 | | | m_v 11.8 | p.a. 85° |

Faint, small, and roundish; bright nucleus. Distance 260 million ly; Hubble class P.

NGC 6229

| GC | Her | $16^h 47.0^m$ | $+47°\,32'$ | 4.2' |
| charts 7, 8 | | | m_v 9.4 | |

High concentration of stars; very bright, large, and round. Faint with 3.7-inch but beautifully grouped in a triangle with two stars (Webb). Distance 102,000 ly; S-S class 4.

NGC 6231

| OC | Sco | $16^h 54.0^m$ | $-41°\,48'$ | 15' |
| chart 22 | | | m_v 2.6 | |

C76, Table of Scorpius. 120 stars; detached, strong concentration of stars; large range in brightness; mag. of brightest star 4.7; bright, quite large. Distance 5,900 ly; age 3.2 million years; Trumpler class I 3 p. This cluster is involved in very large (about 6°) emission nebula that is best seen with a nebula filter.

NGC 6235

| GC | Oph | $16^h 53.4^m$ | $-22°\,11'$ | 5' |
| chart 22 | | | m_v 8.9 | |

Low concentration of stars; pretty bright, quite large, irregularly round cluster. Distance 32,600 ly; S-S class 10.

NGC 6239

| Gx | Her | $16^h 50.1^m$ | $+42°\,44'$ | $2.3' \times 1.0'$ |
| chart 8 | | | m_v 12.4 | p.a. 118° |

Very faint and elongated; double nucleus. Distance 50 million ly; Hubble class Sb.

NGC 6242

| OC | Sco | $16^h 55.6^m$ | $-39°\,30'$ | 8' |
| chart 22 | | | m_v 6.4 | |

40 stars; detached, strong concentration of stars; large range in brightness; mag. of brightest star 7.3; bright and large. Distance 3,900 ly; age 51 million years; Trumpler class I 3 m.

NGC 6249

| OC | Sco | $16^h 57.6^m$ | $-44°\,47'$ | 6' |
| chart 22 | | | m_v 8.2 | |

30 stars; detached, weak concentration; small brightness range; mag. of brightest star 9.8. Distance 3,300 ly; Trumpler class II 1 p.

NGC 6250

| OC | Ara | $16^h 58.0^m$ | $-45°\,48'$ | 7' |
| chart 22 | | | m_v 5.9 | |

60 stars; not well detached from surrounding star field; large range in brightness; mag. of brightest star 7.6; large cluster. Distance 3,300 ly; age 14 million years; Trumpler class IV 3 p.

NGC 6254

See M10.

NGC 6259

OC	Sco	$17^h 00.7^m$	$-44° 40'$	10'
chart 22			$m_v 8$	

120 stars; detached, no concentration of stars; moderate range in brightness; mag. of brightest star 11.6; bright, very large. Per *NGC*, a (!) remarkable object. Distance 2,500 ly; age 220 million years; Trumpler class III 2 m.

NGC 6266

See M62.

NGC 6273

See M19.

NGC 6281

OC	Sco	$17^h 04.8^m$	$-37° 54'$	7'
chart 22			$m_v 5.4$	

70 stars; detached, weak concentration of stars; moderate range in brightness; mag. of brightest star 7.9; large cluster; involved in a large, faint emission nebula. Distance 2,000 ly; age 220 million years; Trumpler class II 2 p n.

NGC 6284

GC	Oph	$17^h 04.5^m$	$-24° 46'$	3.3'
chart 22			$m_v 8.9$	

Low concentration of stars; bright, large, round, and very well resolved. Distance 33,000 ly; S-S class 9.

NGC 6287

GC	Oph	$17^h 05.2^m$	$-22° 42'$	5'
chart 22			$m_v 9.3$	

Medium concentration of stars; quite bright, large, round, and very well resolved. Distance 29,000 ly; S-S class 7.

NGC 6293

GC	Oph	$17^h 10.2^m$	$-26° 35'$	5'
chart 22			$m_v 8.3$	

High concentration of stars; very bright, large, round, and very well resolved. Distance 24,000 ly; S-S class 4.

NGC 6300

Gx	Ara	$17^h 17.0^m$	$-62° 49'$	$4.9' \times 3.0'$
chart 26			$m_v 10.1$	p.a. 118°

Faint, very large, and slightly elongated; faint nucleus with dark lane; located among stars. Distance 43 million ly; Hubble class SBb.

NGC 6302

See Bug Nebula.

NGC 6304

GC	Oph	$17^h 14.5^m$	$-29° 28'$	7'
chart 22			$m_v 8.3$	

Medium concentration of stars; bright, quite large, round, and very well resolved. Distance 18,000 ly; S-S class 6.

NGC 6309

PN	Oph	$17^h 14.1^m$	$-12° 55'$	16"
chart 15			$m_v 11.5$	

PK9+14.1, the Box Nebula. A bright, small, irregular disk with traces of ring structure; phot. mag. 10.8; central star mag. 13.0. Expansion velocity 35 km/sec (22 miles/sec); distance 8,500 ly. Located between two stars.

NGC 6316

GC	Oph	$17^h 16.6^m$	$-28° 08'$	6'
chart 22			$m_v 8.1$	

High concentration of stars; bright, pretty small, and round. Distance 39,000 ly; S-S class 3.

NGC 6322

OC	Sco	$17^h 18.5^m$	$-42° 57'$	10'
chart 22			$m_v 6$	

30 stars; detached, strong concentration of stars; moderate range in brightness; mag. of brightest star 7.5; very large cluster. Distance 3,900 ly; age 10 million years; Trumpler class I 2 p.

NGC 6325

GC	Oph	17^h 18.0^m	−23° 46′	3.9′
chart 22			m$_v$ 10.2	

High concentration of stars; pretty faint, large, round, and partially resolved. Distance 63,000 ly; S-S class 4.

NGC 6326

PN	Ara	17^h 20.8^m	−51° 45′	12″
charts 22, 26			m$_v$ 11	

PK338−8.1. Pretty bright, very small, irregular disk with traces of ring structure; phot. mag. 12.2; central star mag. 15.7. Per *NGC*, a (!!!) most remarkable object. Distance 7,800 ly; V-V class 3b.

NGC 6333

See M9.

NGC 6334

BNe	Sco	17^h 20.5^m	−35° 43′	39′ × 30′
chart 22				

Numerous stars involved in several faint, very large patches of nebulosity.

NGC 6340

Gx	Dra	17^h 10.4^m	+72° 18′	3.0′ × 2.8′
charts 2, 3			m$_v$ 11	p.a. 120°

Quite faint, pretty large, and round; small, bright nucleus. Distance 93 million ly; Hubble class Sap.

NGC 6341

See M92.

NGC 6342

GC	Oph	17^h 21.2^m	−19° 35′	5′
charts 15, 22			m$_v$ 9.5	

High concentration of stars; quite bright, pretty small, slightly elongated, and extremely rich. Distance 49,000 ly; S-S class 4.

NGC 6352

GC	Ara	17^h 25.5^m	−48° 25′	8′
chart 22			m$_v$ 7.8	

C81. Low concentration of stars; pretty faint and large; might actually be an open cluster. Distance 17,600 ly; S-S class 11.

NGC 6355

GC	Oph	17^h 24.0^m	−26° 21′	5′
chart 22			m$_v$ 8.6	

Quite faint, large, round, and very well resolved. Distance 22,000 ly.

NGC 6356

GC	Oph	17^h 23.6^m	−17° 49′	7′
charts 15, 22			m$_v$ 8.2	

High concentration of stars; very bright, quite large, and very well resolved. Distance 56,000 ly; S-S class 2.

NGC 6357

BNe	Sco	17^h 24.6^m	−34° 10′	50′ × 39′
chart 22				

A faint, large, elongated, mottled, and irregularly shaped nebula. On photographs taken with a red filter it looks something like a faint version of M42.

NGC 6362

GC	Ara	17^h 31.9^m	−67° 03′	14′
chart 26			m$_v$ 8.1	

Low concentration of stars; large, bright, and very well resolved. Distance 23,000 ly; S-S class 10.

NGC 6366

GC	Oph	17^h 27.7^m	−05° 05′	12′
chart 15			m$_v$ 9.5:	

Low concentration of stars; faint and large. Distance 13,000 ly; S-S class 11.

NGC 6369

PN	Oph	17^h 29.3^m	−23° 46′	38″
chart 22			m$_v$ 11.4	

PK2+05.1, the Little Ghost Nebula. A round ring and smooth disk; phot. mag. 12.9; central star mag. 15.9. Pretty bright, small ring nebula (Denning). Best seen in 8-inch and larger telescopes on a dark, transparent night (Mullaney). A faint, ghostly halo measuring fully 1′ surrounds the planetary's brighter, inner ring on photographs, but this feature is probably too subtle to be seen visually (Harrington). Per *NGC*, a (!!) very remarkable object. Distance 3,900 ly; V-V class 4 + 2.

NGC 6383

| OC | Sco | 17^h 34.8^m | −32° 34′ | 5′ |
| chart 22 | | | m$_v$ 5.5 | |

40 stars; not well detached from surrounding star field; large range in brightness; mag. of brightest star 5.6. Distance 4,500 ly; age 4.5 million years; involved in a faint, large (80′ × 30′) emission nebula; Trumpler class IV 3 p n.

NGC 6384

| Gx | Oph | 17^h 32.4^m | +07° 04′ | 5.0′ × 3.9′ |
| chart 15 | | | m$_v$ 10.4 | p.a. 30° |

Pretty bright, small, and slightly elongated; four filamentary arms; small, bright nucleus. Distance 78 million ly; Hubble class Sb.

NGC 6388

| GC | Sco | 17^h 36.3^m | −44° 44′ | 6′ |
| chart 22 | | | m$_v$ 6.8 | |

High concentration of stars; very bright, large, and round. Distance 47,000 ly; S-S class 3.

NGC 6389

| Gx | Her | 17^h 32.7^m | +16° 24′ | 2.8′ × 1.9′ |
| chart 15 | | | m$_v$ 12.1 | p.a. 130° |

Faint, small, irregularly shaped. Distance 140 million ly; Hubble class Sb.

NGC 6392

| Gx | Aps | 17^h 43.5^m | −69° 47′ | 1.4′ |
| chart 26 | | | m$_p$ 12.5 | |

Quite faint, small, and roundish; moderately bright nucleus. Distance 160 million ly; Hubble class Sa. A 13th-mag. star to SW.

NGC 6395

| Gx | Dra | 17^h 26.5^m | +71° 06′ | 2.5′ × 0.8′ |
| charts 2, 3 | | | m$_v$ 12.3 | p.a. 15° |

Very faint, pretty large, and elongated. Distance 62 million ly; Hubble class Sc. Double star to N.

NGC 6396

| OC | Sco | 17^h 38.1^m | −35° 00′ | 3.0′ |
| chart 22 | | | m$_v$ 8.5 | |

30 stars; detached, weak concentration; large brightness range; mag. of brightest star 9.8; Trumpler class II 3 p.

NGC 6397

| GC | Ara | 17^h 40.7^m | −53° 40′ | 30′ |
| charts 22, 26 | | | m$_v$ 5.3 | |

C86. Low concentration of stars; very large and bright. Distance 7,200 ly; S-S class 9.

NGC 6401

| GC | Oph | 17^h 38.6^m | −23° 55′ | 4.4′ |
| chart 22 | | | m$_v$ 7.4 | |

Medium concentration of stars; pretty bright, pretty large, and round. Distance 22,000 ly; S-S class 8.

NGC 6402

See M14.

NGC 6405

See M6.

NGC 6411

| Gx | Dra | 17^h 35.5^m | +60° 49′ | 2.3′ × 1.8′ |
| chart 3 | | | m$_v$ 11.8 | p.a. 70° |

Very small, with a bright nucleus. Distance 170 million ly; Hubble class E.

NGC 6412

| Gx | Dra | 17^h 29.6^m | +75° 42′ | 2.5′ × 2.3′ |
| charts 2, 3 | | | m$_v$ 11.8 | p.a. 125° |

Quite large and round; small, bright nucleus; two main arms. Distance 72 million ly; Hubble class Sc. Similar to M33 (but much smaller).

NGC 6416

OC Sco $17^h\,44.4^m$ $-32°\,21'$ 18'
chart 22 $m_v\,5.7$

40 stars; not well detached from surrounding star field; small range in brightness; mag. of brightest star 8.4; very large cluster. Distance 2,600 ly; Trumpler class IV 1 p.

NGC 6425

OC Sco $17^h\,46.9^m$ $-31°\,32'$ 7'
chart 22 $m_v\,7.2$

35 stars; detached, strong concentration of stars; small range in brightness; mag. of brightest star is 10.2; pretty small. Distance 2,600 ly; Trumpler class I 1 p.

NGC 6426

GC Oph $17^h\,44.9^m$ $+03°\,10'$ 4.0'
charts 15, A1 $m_v\,10.9$

Low concentration of stars; very faint, quite large, and elongated. Distance 52,000 ly; S-S class 9.

NGC 6438

Gx Oct $18^h\,22.3^m$ $-85°\,24'$ $1.5'\times1.3'$
charts 26, A5 $m_v\,12.4$ p.a. 140°

Pretty bright and roundish; bright nucleus. Distance 97 million ly; Hubble class S0. In collision with galaxy NGC 6438A, the separation being about 0.5'.

NGC 6438A

Gx Oct $18^h\,22.7^m$ $-85°\,24'$ $2.7'\times1.1'$
charts 26, A5 $m_v\,12.1$ p.a. 30°

In collision with galaxy NGC 6438 about 0.5' away. Distance 100 million ly; Hubble class Ir.

NGC 6439

PN Sgr $17^h\,48.3^m$ $-16°\,28'$ 5"
chart 15 $m_v\,12.6$

PK11+05.1. Appears stellar; phot. mag. 13.8; central star mag. 16.1. Distance 16,000 ly; expansion velocity 24 km/sec (15 miles/sec); V-V class 2a.

NGC 6440

GC Sgr $17^h\,48.9^m$ $-20°\,22'$ 5'
chart 22 $m_v\,9.3$

Medium concentration of stars; pretty bright, pretty large, and round. Distance 12,000 ly; S-S class 5.

NGC 6441

GC Sco $17^h\,50.2^m$ $-37°\,03'$ 10'
chart 22 $m_v\,7.2$

High concentration of stars; very bright, pretty large, and round. Distance 34,000 ly; S-S class 3.

NGC 6445

PN Sgr $17^h\,49.3^m$ $-20°\,01'$ 33"
charts 15, 22 $m_v\,11.2$

PK8+03.1. Pretty bright, small, round; irregularly shaped disk with traces of ring structure; phot. mag. 13.2; central star mag. 19.0. Distance 4,600 ly; expansion velocity 38 km/sec (24 miles/sec); V-V class 3b + 3.

NGC 6451

OC Sco $17^h\,50.7^m$ $-30°\,13'$ 7'
chart 22 $m_p\,8.2$

Tom Thumb Cluster. 80 stars; detached, weak concentration of stars; small range in brightness; brightest star is about phot. mag. 12. Pretty large, with a central dark lane. Distance 1,900 ly; Trumpler class II 1 p.

NGC 6453

GC Sco $17^h\,50.9^m$ $-34°\,36'$ 3.5'
chart 22 $m_v\,10.2$

High concentration of stars; irregularly round. Distance 23,000 ly; S-S class 4.

NGC 6469

OC Sgr $17^h\,52.9^m$ $-22°\,21'$ 12'
chart 22 $m_p\,8.2$

50 stars; detached, no concentration of stars; moderate range in brightness. Distance 5,200 ly; Trumpler class III 2 p.

NGC 6475

See M7.

NGC 6482

| Gx | Her | 17^h 51.8^m | +23° 04′ | 2.1′ × 1.8′ |
| chart 8 | | | m$_v$ 11.4 | p.a. 70° |

Very faint, small, and round. Per *NGC*, a (!) remarkable object. Distance 180 million ly; Hubble class E3p. Superimposed star.

NGC 6483

| Gx | Pav | 17^h 59.5^m | −63° 40′ | 1.5′ × 0.9′ |
| chart 26 | | | m$_v$ 12.2 | p.a. 122° |

Faint, small, and elongated; bright nucleus. Distance 210 million ly; Hubble class E. Located between two stars.

NGC 6487

| Gx | Her | 17^h 52.7^m | +29° 50′ | 1.9′ |
| chart 8 | | | m$_v$ 11.9 | |

Faint, small, and roundish; bright nucleus. Distance 340 million ly; Hubble class E.

NGC 6492

| Gx | Pav | 18^h 02.8^m | −66° 26′ | 2.5′ × 1.3′ |
| chart 26 | | | m$_v$ 12.3 | p.a. 75° |

Pretty faint, small, and quite elongated. Distance 180 million ly; Hubble class S.

NGC 6494

See M23.

NGC 6496

| GC | Sco | 17^h 59.0^m | −44° 16′ | 3.3′ |
| chart 22 | | | m$_v$ 8.6 | |

Low concentration of stars; pretty large and elongated. Distance 29,000 ly; S-S class 12.

NGC 6500

| Gx | Her | 17^h 56.0^m | +18° 20′ | 2.2′ × 1.4′ |
| charts 8, 15 | | | m$_v$ 12.2 | p.a. 50° |

Very faint and very small. Distance 140 million ly; Hubble class Sb.

NGC 6503

| Gx | Dra | 17^h 49.5^m | +70° 09′ | 7.0′ × 2.5′ |
| charts 2, 3 | | | m$_v$ 10.2 | p.a. 123° |

Pretty faint, large, and very elongated; bright arms; very small, very bright nucleus. Distance 14 million ly; Hubble class Sb.

NGC 6514

See Trifid Nebula.

NGC 6517

| GC | Oph | 18^h 01.9^m | −08° 57′ | 3.3′ |
| charts 15, 16 | | | m$_v$ 10.1 | |

High concentration of stars; pretty bright, pretty large, round, and partially resolved. Distance 25,000 ly; S-S class 4.

NGC 6520

| OC | Sgr | 18^h 03.4^m | −27° 54′ | 6′ |
| chart 22 | | | m$_p$ 7.6 | |

60 stars; detached, strong concentration of stars; moderate range in brightness; brightest star is phot. mag. 9; pretty small. Distance 5,400 ly; age 54 million years; Trumpler class I 2 m. Dark nebula B86 is on one edge.

NGC 6522

| GC | Sgr | 18^h 03.6^m | −30° 02′ | 7′ |
| chart 22 | | | m$_v$ 9.9 | |

Medium concentration of stars; bright, pretty large, round, and very well resolved. Distance 21,000 ly; S-S class 6.

NGC 6523

See Lagoon Nebula.

NGC 6526

| BNe | Sgr | 18^h 02.6^m | −23° 35′ | 40′ × 30′ |
| chart 22 | | | | |

Faint, large, and quite elongated.

NGC 6528

GC	Sgr	18^h 04.8^m	−30° 03′	5′
chart 22			m$_v$ 9.6	

Medium concentration of stars; pretty faint and quite small. Distance 24,000 ly; S-S class 5.

NGC 6530

OC	Sgr	18^h 04.8^m	−24° 20′	15′
chart 22			m$_v$ 4.6	

100 stars; detached, weak concentration of stars; moderate range in brightness; mag. of brightest star 6.9. Distance 5,200 ly; age 2 million years; Trumpler class II 2 m n. Involved in the nebulosity of the Lagoon Nebula (M8).

NGC 6531

See M21.

NGC 6535

GC	Ser	18^h 03.9^m	−00° 18′	3.3′
charts 15, 16			m$_v$ 9.3	

Low concentration of stars; pretty faint and very small. Distance 36,000 ly; S-S class 11. Located in Serpens Cauda.

NGC 6537

PN	Sgr	18^h 05.2^m	−19° 51′	10″
charts 15, 16, 22			m$_v$ 11.6	

PK10+00.1. Bright and small; appears stellar; has a large, faint halo; phot. mag. 12.5; central star mag. 18.8. Expansion velocity 10 km/sec (6 miles/sec); distance 3,900 ly; V-V class 2a + 6.

NGC 6539

GC	Ser	18^h 04.8^m	−07° 35′	7′
charts 15, 16			m$_v$ 8.9	

Low concentration of stars; very small and faint. Distance 7,500 ly; S-S class 10. Located in Serpens Cauda.

NGC 6541

GC	CrA	18^h 08.0^m	−43° 42′	14′
chart 22			m$_v$ 6.3	

C78. High concentration of stars; very well resolved, bright, and round. Distance 23,000 ly; S-S class 3.

NGC 6543

PN	Dra	17^h 58.6^m	+66° 38′	20″
chart 3			m$_v$ 8.1	

PK96+29.1, C6, the Cat's Eye Nebula. Very bright, irregular, oblong disk; phot. mag. 8.8; central star mag. 11.1. With a low power it looks like a star out of focus (Denning). A blue-green egg with nuclear sun (Mullaney). Blue spheroid, like a snail in photographs (Copeland). The central star looks blazing yellow due to its strong contrast with the blue nebula (Houston). Distance 3,600 ly; expansion velocity 19 km/sec (12 miles/sec).

NGC 6544

GC	Sgr	18^h 07.4^m	−25° 00′	7′
chart 22			m$_v$ 7.5:	

Quite faint, pretty large, irregularly round. Distance 15,000 ly.

NGC 6546

OC	Sgr	18^h 07.2^m	−23° 20′	12′
chart 22			m$_v$ 8	

150 stars; detached, no concentration of stars; moderate range in brightness; mag. of brightest star 10.6; very large. Distance 2,700 ly; age 250 million years; Trumpler class III 2 m.

NGC 6548

Gx	Her	18^h 06.0^m	+18° 35′	2.9′ × 2.6′
charts 8, 15, 16			m$_v$ 11.7	p.a. 75°

Quite faint, small, and slightly elongated. Distance 100 million ly; Hubble class S0.

NGC 6553

GC	Sgr	18^h 09.3^m	−25° 54′	6′
chart 22			m$_v$ 8.3	

Low concentration of stars; faint, large, slightly elongated, partially resolved. Distance 19,000 ly; S-S class 11.

NGC 6555

Gx	Her	$18^h 07.8^m$	$+17°\,36'$	$2.0' \times 1.7'$
charts 8, 15, 16			$m_v\,12.4$	p.a. 110°

Faint, large, and roundish; bright nucleus. Distance 100 million ly; Hubble class S(B)c.

NGC 6558

GC	Sgr	$18^h 10.3^m$	$-31°\,46'$	$4.2'$
chart 22			$m_v\,8.6$	

Pretty bright, pretty large, round, very well resolved. Distance 30,000 ly.

NGC 6559

BNe	Sgr	$18^h 10.0^m$	$-24°\,06'$	$7' \times 5'$
chart 22				

Very faint, large, and slightly elongated; double star involved in nebulosity. Brightest in a group that includes nebulae IC 1274, IC 1275, and IC 4685.

NGC 6563

PN	Sgr	$18^h 12.0^m$	$-33°\,52'$	$48''$
chart 22			$m_v\,11$	

PK358–7.1. A faint, elongated disk with very irregular brightness distribution; has a hazy edge; phot. mag. 13.8; central star mag. 17.3. Distance 4,900 ly; expansion velocity 11 km/sec (7 miles/sec).

NGC 6565

PN	Sgr	$18^h 11.9^m$	$-28°\,11'$	$14''$
chart 22			$m_v\,13.2$	

PK3–4.5. Ring structure; phot. mag. 13.2; central star mag. 18.5. Distance 9,800 ly; V-V class 3a.

NGC 6567

PN	Sgr	$18^h 13.8^m$	$-19°\,05'$	$8''$
charts 15, 16, 22			$m_v\,11$	

PK11–0.2. Smooth, irregular disk; central star mag. 14.4. Distance 3,900 ly; V-V class 2a + 3. Located in the Sagittarius Star Cloud.

NGC 6568

OC	Sgr	$18^h 12.8^m$	$-21°\,36'$	$13'$
chart 22			$m_p\,8.6$	

50 stars; detached, no concentration of stars; small range in brightness. A large and coarse cluster of minute stars (Smyth). Trumpler class III 1 m.

NGC 6569

GC	Sgr	$18^h 13.7^m$	$-31°\,50'$	$7'$
chart 22			$m_v\,8.4$	

Medium concentration of stars; quite bright, large, round, very well resolved. Distance 25,000 ly; S-S class 8.

NGC 6572

PN	Oph	$18^h 12.1^m$	$+06°\,51'$	$11''$
charts 15, 16			$m_v\,9.1$	

PK34+11.1. Very bright, very small, round, and hazy; brighter toward center; phot. mag. 9.0; central star mag. 12.9. A fine planetary nebula, with many telescopic stars in the field (Smyth). Distance 2,000 ly; expansion velocity 16 km/sec (10 miles/sec); V-V class 2a.

NGC 6574

Gx	Her	$18^h 11.8^m$	$+14°\,59'$	$1.5' \times 1.1'$
charts 15, 16			$m_v\,12$	p.a. 160°

Pretty bright, small, and round; several massive arms; small, bright, diffuse nucleus. Distance 110 million ly; Hubble class S.

NGC 6578

PN	Sgr	$18^h 16.3^m$	$-20°\,27'$	$9''$
chart 22			$m_v\,12.9$	

PK10–1.1. Appears stellar; phot. mag. 13.1; central star mag. 15.8. Distance 9,500 ly; V-V class 2a.

NGC 6584

GC	Tel	$18^h 18.6^m$	$-52°\,13'$	$4.2'$
charts 22, 26			$m_v\,7.9$	

Medium concentration of stars; quite bright, quite large, round, and very well resolved. Distance 49,000 ly; S-S class 8.

NGC 6589

BNr	Sgr	$18^h 16.3^m$	$-19°\,48'$	$6' \times 4'$
charts 15, 16, 22				

A pair of 10th-mag. stars involved in an extremely faint, pretty large patch of nebulosity.

NGC 6590

BNr   Sgr $18^h\ 17.0^m$ $-19°\ 53'$ $5' \times 4'$
charts 15, 16, 22

Small, roundish nebula.

NGC 6595

OC   Sgr $18^h\ 17.0^m$ $-19°\ 53'$ $10'$
charts 15, 16, 22 $m_p\ 7.0$

30 stars, possibly an asterism; involved in nebulosity. Distance 6,400 ly.

NGC 6604

OC   Ser $18^h\ 18.1^m$ $-12°\ 14'$ $2'$
charts 15, 16 $m_v\ 6.5$

30 stars; detached, strong concentration; large brightness range; mag. of brightest star 7.5; involved in nebulosity. Distance 2,300 ly; age 4 million years; Trumpler class I 3 p n.

NGC 6611

See Eagle Nebula.

NGC 6613

See M18.

NGC 6618

See Omega Nebula.

NGC 6624

GC   Sgr $18^h\ 23.7^m$ $-30°\ 22'$ $4.3'$
chart 22 $m_v\ 7.6$

Medium concentration of stars; very bright, pretty large, round, and very well resolved. Distance 28,000 ly; S-S class 6.

NGC 6626

See M28.

NGC 6629

PN   Sgr $18^h\ 25.7^m$ $-23°\ 12'$ $16''$
chart 22 $m_v\ 11.3$

PK9–5.1. Pretty bright, very small, and round; phot. mag. 11.6; central star mag. 12.9. Distance 6,200 ly; expansion velocity 6 km/sec (4 miles/sec); V-V class 2a.

NGC 6632

Gx   Her $18^h\ 25.1^m$ $+27°\ 32'$ $3.0' \times 1.4'$
chart 8 $m_v\ 12.1$ p.a. 155°

Faint, small, and roundish; moderately bright nucleus. Distance 210 million ly; Hubble class Sb.

NGC 6633

OC   Oph $18^h\ 27.7^m$ $+06°\ 34'$ $26'$
charts 15, 16 $m_v\ 4.6$

30 stars; detached, no concentration of stars; moderate range in brightness; mag. of brightest star 7.6; large. Distance 1,000 ly; age 660 million years; Trumpler class III 2 m.

NGC 6637

See M69.

NGC 6638

GC   Sgr $18^h\ 30.9^m$ $-25°\ 30'$ $7'$
chart 22 $m_v\ 9.2$

Medium concentration of stars; bright, small, round, and partially resolved. Distance 26,000 ly; S-S class 6.

NGC 6642

GC   Sgr $18^h\ 31.9^m$ $-23°\ 28'$ $8'$
chart 22 $m_v\ 8.9$

Medium concentration of stars; pretty bright, pretty large, irregularly round. Distance 20,000 ly; S-S class 5.

NGC 6643

Gx   Dra $18^h\ 19.8^m$ $+74°\ 34'$ $3.7' \times 1.8'$
charts 2, 3 $m_v\ 11.1$ p.a. 38°

Pretty bright, pretty large, and elongated; very small, bright nucleus. Distance 76 million ly; Hubble class Sc.

NGC 6644

PN   Sgr $18^h\ 32.6^m$ $-25°\ 08'$ $3''$
chart 22 $m_v\ 10.7$

PK8–7.2. Appears stellar; phot. mag. 12.2; central star mag. 15.6. Distance 9,000 ly.

NGC 6645

OC	Sgr	18^h 32.6^m	–16° 54′	10′
charts 15, 16, 22			m$_p$ 8.5	

40 stars; detached, no concentration of stars; small range in brightness; brightest star is phot. mag. 12; pretty large; Trumpler class III 1 m.

NGC 6652

GC	Sgr	18^h 35.8^m	–32° 59′	5′
chart 22			m$_v$ 8.5	

Medium concentration of stars; bright, small, slightly elongated, and very well resolved. Distance 49,000 ly; S-S class 6.

NGC 6654

Gx	Dra	18^h 24.1^m	+73° 11′	2.8′ × 2.2′
charts 2, 3			m$_v$ 12	p.a. 0°

Pretty bright and pretty large; nucleus looks like a 12th- or 13th-mag. star involved in nebulosity. Distance 90 million ly; Hubble class SBa.

NGC 6656

See M22.

NGC 6664

OC	Sct	18^h 36.7^m	–08° 13′	15′
charts 15, 16			m$_v$ 7.8	

50 stars; detached, no concentration of stars; moderate range in brightness; mag. of brightest star 10.2; large. Distance 4,200 ly; age 140 million years; Trumpler class III 2 m.

NGC 6673

Gx	Pav	18^h 45.1^m	–62° 18′	2.3′ × 1.0′
chart 26			m$_v$ 11.7	p.a. 26°

Pretty faint, small, and roundish; bright nucleus. Distance 45 million ly; Hubble class E.

NGC 6674

Gx	Her	18^h 38.6^m	+25° 23′	4.2′ × 2.0′
chart 8			m$_v$ 12.2	p.a. 143°

Faint, pretty small, irregularly roundish; very bright nucleus. Distance 160 million ly; Hubble class SBb.

NGC 6681

See M70.

NGC 6684

Gx	Pav	18^h 49.0^m	–65° 10′	4.1′ × 2.6′
chart 26			m$_v$ 10.4	p.a. 35°

Very bright, pretty large, and round; bright nucleus. Distance 30 million ly; Hubble class SB0.

NGC 6694

See M26.

NGC 6699

Gx	Pav	18^h 52.0^m	–57° 19′	1.4′
chart 26			m$_v$ 12.1	

Pretty faint, pretty small, and slightly elongated; small, bright nucleus. Distance 150 million ly; Hubble class S(B)b.

NGC 6701

Gx	Dra	18^h 43.2^m	+60° 39′	1.5′ × 1.3′
chart 3			m$_v$ 12.1	p.a. 25°

Pretty bright, pretty small, and slightly elongated. Distance 180 million ly; Hubble class SBa. Faint star close by to E.

NGC 6703

Gx	Lyr	18^h 47.3^m	+45° 33′	2.6′
charts 3, 8			m$_v$ 11.3	

Bright, small, and round; bright nucleus. Distance 110 million ly; Hubble class S0.

NGC 6705

See M11.

NGC 6709

OC	Aql	18^h 51.5^m	+10° 21′	12′
charts 15, 16			m$_v$ 6.7	

40 stars; detached, no concentration of stars; moderate range in brightness; mag. of brightest star 9.1. Distance 3,000 ly; age 78 million years; Trumpler class III 2 m.

NGC 6712

GC	Sct	$18^h 53.1^m$	$-08° 42'$	8'
charts 15, 16			m_v 8.1	

Low concentration of stars; pretty bright, very large, very well resolved but irregularly shaped. Distance 25,000 ly; S-S class 9.

NGC 6715

See M54.

NGC 6716

OC	Sgr	$18^h 54.6^m$	$-19° 53'$	7'
charts 15, 16, 22			m_v 6.9	

20 stars; not well detached from surrounding star field; small range in brightness; mag. of brightest star 8.3. Distance 2,000 ly; age 160 million years; Trumpler class IV 1 p.

NGC 6717

GC	Sgr	$18^h 55.1^m$	$-22° 42'$	10'
chart 22			m_v 8.4	

Palomar 9. Low concentration of stars toward center. Distance 52,000 ly; S-S class 8.

NGC 6720

See Ring Nebula.

NGC 6723

GC	Sgr	$18^h 59.6^m$	$-36° 38'$	12'
chart 22			m_v 6.8	

Medium concentration of stars; very large and well resolved. Distance 28,000 ly; S-S class 7.

NGC 6726, 7

BNr	CrA	$19^h 01.7^m$	$-36° 53'$	9' × 7'
chart 22				

Two fairly bright stars involved in a high-surface-brightness nebula; illuminated by the variable star TY CrA. Dark nebula SL 42 is located about 1.8° to E.

NGC 6729

BNer	CrA	$19^h 01.9^m$	$-36° 57'$	1'
chart 22				

C68. Comet shaped. The variable star R CrA (which can be as bright as mag. 9.7) is located at the apex.

NGC 6730

Gx	Pav	$19^h 07.6^m$	$-68° 55'$	1.6' × 1.4'
chart 26			m_p 13.0	p.a. 35°

Very faint, small, and roundish; bright nucleus. Distance 180 million ly; Hubble class E. A star of mag. 7 or 8 is located to NE.

NGC 6738

OC	Aql	$19^h 01.4^m$	$+11° 36'$	15'
chart 16			m_p 8.3	

Not well detached; moderate brightness range; a poor cluster; Trumpler class IV 2 p.

NGC 6741

PN	Aql	$19^h 02.6^m$	$-00° 27'$	8"
chart 16			m_v 11.4	

PK33–2.1, the Phantom Streak. Nearly stellar; phot. mag. 10.8; central star mag. 17.6. Expansion velocity 21 km/sec (17 miles/sec); distance 5,200 ly; V-V class 4.

NGC 6744

Gx	Pav	$19^h 09.8^m$	$-63° 51'$	20' × 13'
chart 26			m_v 8.6	p.a. 15°

C101. Quite bright, quite large, and round; filamentary, knotty arms; bright, diffuse nucleus. Distance 23 million ly; Hubble class S(B)b.

NGC 6751

PN	Aql	$19^h 05.9^m$	$-06° 00'$	20"
chart 16			m_v 11.9	

PK29–5.1. Pretty bright, small, irregular disk; central star mag. 15.5. Expansion velocity 24 km/sec (15 miles/sec); distance 6,500 ly; V-V class 3.

NGC 6752

GC	Pav	19^h 10.9^m	−59° 59′	28′
chart 26			m$_v$ 5.3	

C93. Medium concentration of stars; bright, very large, irregularly roundish. Distance 13,700 ly; S-S class 6.

NGC 6753

Gx	Pav	19^h 11.4^m	−57° 03′	2.5′ × 2.1′
chart 26			m$_v$ 11.1	p.a. 30°

Pretty bright, pretty large, and round; very bright, very small nucleus. Distance 130 million ly; Hubble class Sb.

NGC 6754

Gx	Tel	19^h 11.4^m	−50° 39′	1.8′ × 0.9′
charts 22, 23, 26			m$_v$ 12.1	p.a. 80°

Pretty faint, pretty large, and very elongated; massive arms; small, faint nucleus. Distance 140 million ly; Hubble class SBb.

NGC 6755

OC	Aql	19^h 07.8^m	+04° 14′	15′
chart 16			m$_v$ 7.5	

100 stars; not well detached from surrounding star field; moderate range in brightness; mag. of brightest star 10.2; very large. Distance 4,900 ly; age 35 million years; Trumpler class IV 2 m.

NGC 6758

Gx	Tel	19^h 13.9^m	−56° 19′	2.3′ × 1.7′
chart 26			m$_v$ 11.5	p.a. 121°

Pretty bright, small, and round. Distance 140 million ly; Hubble class E1. In a group with many faint galaxies.

NGC 6760

GC	Aql	19^h 11.2^m	+01° 02′	8′
chart 16			m$_v$ 9	

Low concentration of stars; pretty bright and pretty large. Distance 13,000 ly; S-S class 9.

NGC 6764

Gx	Cyg	19^h 08.3^m	+50° 56′	2.2′ × 1.2′
charts 3, 8, 9			m$_v$ 11.8	p.a. 62°

Pretty faint, pretty large, and elongated. Distance 120 million ly; Hubble class SBb. Several very faint superimposed stars.

NGC 6769

Gx	Pav	19^h 18.4^m	−60° 30′	2.2′ × 1.5′
chart 26			m$_v$ 11.5	p.a. 123°

Very faint, small, and round; filamentary arms; dark lane; very bright, diffuse nucleus. Distance 160 million ly; Hubble class S(B)bp. Brightest in a group of galaxies; interacting with galaxy NGC 6770.

NGC 6770

Gx	Pav	19^h 18.6^m	−60° 30′	2.1′ × 1.6′
chart 26			m$_v$ 11.6	p.a. 20°

Extremely faint and very small. Distance 160 million ly; Hubble class S(B)bp. Interacting with galaxy NGC 6769.

NGC 6776

Gx	Pav	19^h 25.3^m	−63° 52′	1.7′ × 1.5′
chart 26			m$_v$ 12.2	p.a. 15°

Pretty bright, small, and round; very bright nucleus. Distance 240 million ly; Hubble class E2.

NGC 6778

PN	Aql	19^h 18.4^m	−01° 36′	16″
chart 16			m$_v$ 12.3	

PK34−6.1. Small and elongated; an ill-defined disk with a larger and fainter outer disk; phot. mag. 13.3; central star mag. 14.8. Expansion velocity 23 km/sec (14 miles/sec); V-V class 3 + 3. Dark nebula B139 lies just to NW.

NGC 6779

See M56.

NGC 6781

PN	Aql	19^h 18.4^m	+06° 33′	1.8′
chart 16			m$_v$ 11.4	

PK41−2.1. A faint, large, irregularly round disk with a larger, fainter outer disk; phot. mag. 11.8; central star mag. 16.8. Distance 2,600 ly; expansion velocity 12 km/sec (7 miles/sec); V-V class 3b + 3.

NGC 6782

Gx	Pav	$19^h 24.0^m$	$-59°55'$	$2.1' \times 1.6'$
chart 26			m_v 11.8	p.a. 45°

Quite faint, quite small, and round; faint, ringed outer arms; very bright nucleus. Distance 160 million ly; Hubble class SB0.

NGC 6788

Gx	Tel	$19^h 26.8^m$	$-54°57'$	$3.0' \times 0.9'$
charts 22, 26			m_p 12.8	p.a. 71°

Pretty bright, small, and very elongated; moderately bright nucleus; Hubble class Sa.

NGC 6790

PN	Aql	$19^h 23.0^m$	$+01°31'$	7"
chart 16			m_v 10.5	

PK37–6.1. Bright and extremely small; has an almost stellar appearance; phot. mag. 10.2; central star mag. 15.5. Distance 5,200 ly; expansion velocity 22 km/sec (14 miles/sec); V-V class 2.

NGC 6803

PN	Aql	$19^h 31.3^m$	$+10°03'$	6"
chart 16			m_v 11.4	

PK46–4.1. Appears stellar; phot. mag. 11.3; central star mag. 15.2. Distance 5,200 ly; expansion velocity 14 km/sec (10 miles/sec); V-V class 2a.

NGC 6804

PN	Aql	$19^h 31.6^m$	$+09°13'$	$1.1' \times 0.5'$
chart 16			m_v 12	

PK45–4.1. Quite bright, irregularly round; phot. mag. 12.2; central star mag. 14.4. Distance 4,200 ly; V-V class 4 + 5. Star visible on one edge.

NGC 6807

PN	Aql	$19^h 34.6^m$	$+05°41'$	2"
chart 16			m_v 12	

PK42–6.1. Appears stellar; phot. mag. 13.8; central star's phot. mag. is 16.3. Distance 17,000 ly; expansion velocity 17 km/sec (11 miles/sec); V-V class 2.

NGC 6809

See M55.

NGC 6810

Gx	Pav	$19^h 43.6^m$	$-58°39'$	$3.2' \times 0.9'$
chart 26			m_v 11.4	p.a. 176°

Pretty small and round; dark lane; very bright nucleus. Distance 75 million ly; Hubble class Sb. Paired with large, very faint galaxy.

NGC 6811

OC	Cyg	$19^h 38.2^m$	$+46°34'$	12'
charts 3, 8, 9			m_v 6.8	

Hole in a Cluster. About 70 stars; not well detached from surrounding star field; large brightness range; mag. of brightest star 9.9; a large cluster. Distance 2,900 ly; age 540 million years; Trumpler class IV 3 p.

NGC 6813

BNe	Vul	$19^h 40.4^m$	$+27°18'$	3'
chart 8				

Very faint and small; located in a very rich star field.

NGC 6814

Gx	Aql	$19^h 42.7^m$	$-10°19'$	$3.4' \times 3.3'$
chart 16			m_v 11.2	p.a. 50°

Pretty faint and pretty large; knotty arms; very small, very bright nucleus. A Seyfert galaxy. Distance 69 million ly; Hubble class Sb.

NGC 6818

PN	Sgr	$19^h 44.0^m$	$-14°09'$	20"
charts 16, 22			m_v 9.3	

PK25–17.1, the Little Gem Nebula. Bright and very small; ring structure; phot. mag. 9.9; central star mag. 15.0. Blue, like a star out of focus (Webb). Accompanied by several small stars, four of which form a square about it (Smyth). Distance 5,000 ly; expansion velocity 30 km/sec (19 miles/sec); V-V class 4.

NGC 6819

OC	Cyg	$19^h 41.3^m$	$+40°11'$	5'
charts 8, 9			m_v 7.3	

150 stars; detached, strong concentration of stars; small range in brightness; mag. of brightest star

11.5; very large. Distance 7,200 ly; age 3.5 billion years; Trumpler class I 1 r.

NGC 6820

BNe	Vul	19^h 43.1^m	+23° 17′	39′ × 30′
chart 8				

A faint, large, and irregularly shaped nebula with a bright rim and dark clouds. Contains an open cluster (NGC 6823).

NGC 6822

Gx	Sgr	19^h 45.0^m	−14° 48′	15′ × 14′
charts 16, 22			m$_v$ 8.8	p.a. 5°

C57, Barnard's Galaxy. Very faint, large, and elongated. A barely discernible smudge whose huge area makes a wide-field eyepiece a must (Harrington). Extends 12,000 ly; distance 2.8 million ly; Hubble class Ir+. A member of the Local Group.

NGC 6823

OC	Vul	19^h 43.1^m	+23° 18′	12′
chart 8			m$_v$ 7.1	

30 stars; detached, strong concentration of stars; large range in brightness; mag. of brightest star 8.8; a slightly elongated cluster. Distance 8,800 ly; age 2 million years; involved in a large (39′ × 30′) emission nebula (NGC 6820); Trumpler class I 3 p n.

NGC 6824

Gx	Cyg	19^h 43.7^m	+56° 07′	2.1′ × 1.5′
chart 3			m$_v$ 12.2	p.a. 60°

Pretty bright and irregularly shaped; very bright nucleus. Distance 160 million ly; Hubble class Sb .

NGC 6826

PN	Cyg	19^h 44.8^m	+50° 31′	25″
charts 3, 8, 9			m$_v$ 8.8	

PK83+12.1, C15, the Blinking Planetary. A bright, pretty large, blue-green disk with a bright center, within a larger, fainter outer disk; phot. mag. 9.8; central star mag. 10.4. Stare right at the star and the nebula disappears, but switch to averted vision and it suddenly reappears and swamps the star with its bluish radiance (Mullaney). Distance 3,300 ly; expansion velocity 13 km/sec (8 miles/sec); V-V class 3a.

NGC 6830

OC	Vul	19^h 51.0^m	+23° 04′	12′
chart 8			m$_v$ 7.9	

20 stars; detached, weak concentration of stars; moderate range in brightness; mag. of brightest star 9.9; a large cluster. Distance 4,800 ly; age 100 million years; Trumpler class II 2 p.

NGC 6833

PN	Cyg	19^h 49.7^m	+48° 58′	2″
charts 3, 8, 9			m$_v$ 12.1	

PK82+11.1. Appears stellar; phot. mag. 13.8; central star mag. 14.5. Distance 5,500 ly; expansion velocity 14 km/sec (9 miles/sec); V-V class 2.

NGC 6834

OC	Cyg	19^h 52.2^m	+29° 25′	5′
charts 8, 9			m$_v$ 7.8	

50 stars; detached, weak concentration of stars; moderate range in brightness; mag. of brightest star 9.7. Distance 7,500 ly; age 79 million years; Trumpler class II 2 m.

NGC 6838

See M71.

NGC 6842

PN	Vul	19^h 55.0^m	+29° 17′	57″
charts 8, 9			m$_v$ 13.1	

PK65+00.1. Faint, pretty large, and slightly elongated; irregular disk with traces of ring structure; phot. mag. 13.6; central star mag. 16.0. Distance 4,000 ly; V-V class 3b.

NGC 6849

Gx	Sgr	20^h 06.3^m	−40° 12′	1.8′ × 1.1′
charts 22, 23			m$_v$ 11.9	p.a. 18°

Pretty bright, small, and roundish. Distance 260 million ly; Hubble class E. Star superimposed at 0.5′ toward SW.

NGC 6851

Gx	Tel	20ʰ 03.6ᵐ	−48° 17′	1.9′ × 1.5′
charts 22, 23, 26			m_v 11.8	p.a. 160°

Pretty faint, small, and slightly elongated; very bright nucleus. Distance 130 million ly; Hubble class E4.

NGC 6853

See Dumbbell Nebula.

NGC 6861

Gx	Tel	20ʰ 07.3ᵐ	−48° 22′	2.9′ × 1.9′
charts 22, 23, 26			m_v 11.1	p.a. 140°

Bright, small, and quite elongated; very bright nucleus. Distance 120 million ly; Hubble class S0. In a group along with galaxies NGC 6868 and NGC 6870.

NGC 6864

See M75.

NGC 6866

OC	Cyg	20ʰ 03.7ᵐ	+44° 00′	7′
charts 8, 9			m_v 7.6	

80 stars; detached, weak concentration of stars; moderate range in brightness; mag. of brightest star 10.7. In a 10-inch at 150×, the field is filled with sparkling starry excitement (Houston). Distance 3,900 ly; age 230 million years; Trumpler class II 2 m.

NGC 6868

Gx	Tel	20ʰ 09.9ᵐ	−48° 23′	3.6′ × 2.8′
charts 22, 23, 26			m_v 10.6	p.a. 86°

Very bright, small, and round; bright nucleus. Distance 120 million ly; Hubble class E2. In a group with galaxies NGC 6861 and somewhat fainter NGC 6870.

NGC 6871

OC	Cyg	20ʰ 05.9ᵐ	+35° 47′	19′
charts 8, 9			m_v 5.2	

15 stars; not well detached from surrounding star field; large range in brightness; mag. of brightest star 6.8. Distance 5,400 ly; age 10 million years; Trumpler class IV 3 p.

NGC 6872

Gx	Pav	20ʰ 16.9ᵐ	−70° 46′	6.0′ × 1.4′
chart 26			m_v 11.2	p.a. 66°

Faint, pretty small, and quite elongated; has a very bright nucleus. Distance 200 million ly; Hubble class SBbp.

NGC 6875

Gx	Tel	20ʰ 13.2ᵐ	−46° 10′	2.3′ × 1.3′
charts 22, 23, 26			m_v 11.9	p.a. 22°

Faint, very small, and elongated; very bright nucleus. Distance 130 million ly; Hubble class E6.

NGC 6876

Gx	Pav	20ʰ 18.3ᵐ	−70° 51′	2.8′ × 2.3′
chart 26			m_v 10.8	p.a. 80°

Pretty bright, small, and round; very bright center. Distance 160 million ly; Hubble class E3. In a group with galaxies NGC 6877 and IC 4972.

NGC 6879

PN	Sge	20ʰ 10.5ᵐ	+16° 55′	5″
charts 9, 16			m_v 12.5	

PK57−8.1. Appears stellar; phot. mag. 13.0; central star mag. 14.8. Distance 17,000 ly; expansion velocity 23 km/sec (14 miles/sec); V-V class 2a.

NGC 6882

OC	Vul	20ʰ 11.7ᵐ	+26° 33′	18′
charts 8, 9			m_v 8.1:	

Detached, weak concentration; moderate brightness range; mag. of brightest star 9.9; a rather poor cluster. Distance 1,900 ly; Trumpler class II 2 p. Located immediately NW of the open cluster NGC 6885.

NGC 6883

OC	Cyg	20ʰ 11.3ᵐ	+35° 51′	15′
charts 8, 9			m_p 8.0	

30 stars; detached, strong concentration of stars;

large range in brightness. Distance 4,500 ly; age 15 million years; Trumpler class I 3 p. Double star involved.

NGC 6884

PN	Cyg	20^h 10.4^m	+46° 28′	6″
charts 3, 8, 9			m$_v$ 10.9	

PK82+07.1. Appears stellar; phot. mag. 12.6; central star mag. 15.8. Distance 9,000 ly; expansion velocity 23 km/sec (14 miles/sec); V-V class 2b.

NGC 6885

OC	Vul	20^h 12.0^m	+26° 29′	7′
charts 8, 9			m$_v$ 8.1	

C37. 30 stars; detached, no concentration of stars; moderate range in brightness; brightest star is phot. mag. 6; very bright, a very large cluster. Distance 1,900 ly; Trumpler class III 2 p. Open cluster NGC 6882 is located immediately to NW.

NGC 6886

PN	Sge	20^h 12.7^m	+19° 59′	6″
charts 9, 16			m$_v$ 11.4	

PK60–7.2. Appears stellar; phot. mag. 12.2; central star mag. 18.0. Distance 10,000 ly; expansion velocity 19 km/sec (12 miles/sec); V-V class 2 + 3.

NGC 6887

Gx	Tel	20^h 17.3^m	–52° 48′	3.2′ × 1.3′
charts 22, 23, 26			m$_v$ 12	p.a. 102°

Pretty faint, quite large, and very elongated; has a very faint nucleus. Distance 120 million ly; Hubble class Sb.

NGC 6888

BNe	Cyg	20^h 12.0^m	+38° 21′	19′ × 9′
charts 8, 9				

C27, Crescent Nebula. Faint, very large, very elongated; crescent shaped; illuminated by a 7.4-mag. star. Double star attached.

NGC 6890

Gx	Sgr	20^h 18.3^m	–44° 48′	1.5′ × 1.3′
charts 22, 23, 26			m$_v$ 12.1	p.a. 152°

Pretty faint, small, and round; two main arms; very small, bright nucleus. Distance 100 million ly; Hubble class Sb. In a group of galaxies.

NGC 6891

PN	Del	20^h 15.2^m	+12° 42′	15″
chart 16			m$_v$ 10.5	

PK54–12.1. Bright, pretty small, and somewhat elongated; phot. mag. 11.7; central star mag. 12.4. Expansion velocity 7 km/sec (4 miles/sec), distance 7,200 ly; V-V class 2a + 2b.

NGC 6893

Gx	Tel	20^h 20.8^m	–48° 14′	2.7′ × 1.9′
charts 22, 23, 26			m$_v$ 11.7	p.a. 10°

Pretty faint, small, and round; extremely bright nucleus. Distance 140 million ly; Hubble class S(B)0.

NGC 6902

Gx	Sgr	20^h 24.5^m	–43° 39′	2.2′ × 1.8′
charts 22, 23, 26			m$_v$ 11	p.a. 153°

Faint, quite small, and round; very bright nucleus. Distance 120 million ly; Hubble class SBa.

NGC 6903

Gx	Cap	20^h 23.7^m	–19° 19′	2.6′
charts 16, 23			m$_v$ 11.9	

Bright nucleus. Distance 150 million ly; Hubble class S0. Pretty bright star superimposed about 0.4′ to NE.

NGC 6905

PN	Del	20^h 22.4^m	+20° 06′	40″
chart 9			m$_v$ 11.1	

PK61–9.1, the Blue Flash Nebula. Bright, pretty small, and round; phot. mag. 11.9; central star mag. 15.7. A fine though small planetary, in a coarse cluster (Smyth). Per NGC, a (!!) very remarkable object. Distance 4,200 ly; expansion velocity 27 km/sec (17 miles/sec); V-V class 3 + 3. Four stars nearby.

NGC 6907

Gx	Cap	20^h 25.1^m	–24° 48′	3.6′ × 2.7′
chart 23			m$_v$ 11.1	p.a. 46°

Quite faint, quite large, and slightly elongated; S shaped; two main arms; very small, bright nucleus with dark lanes. Distance 140 million ly; Hubble class S(B)b.

NGC 6909

Gx	Tel	20^h 27.6^m	–47° 02′	2.3′ × 1.1′
charts 22, 23, 26			m$_v$ 11.8	p.a. 68°

Pretty bright and pretty large; bright nucleus. Distance 110 million ly; Hubble class E6. Two 10th-mag. stars nearby.

NGC 6910

OC	Cyg	20^h 23.1^m	+40° 47′	7′
charts 8, 9			m$_v$ 7.4	

50 stars; detached, strong concentration of stars; moderate range in brightness; mag. of brightest star 9.6; pretty bright and pretty small. Distance 5,400 ly; age 10 million years; Trumpler class I 2 p.

NGC 6913

See M29.

NGC 6914

BNr	Cyg	20^h 24.7^m	+42° 29′	12′ × 12′
charts 8, 9				

Very faint, very large, diffuse, irregularly round; two stars attached. Connected to nebulae NGC 6914A and NGC 6914B.

NGC 6923

Gx	Mic	20^h 31.6^m	–30° 50′	2.9′ × 1.4′
chart 23			m$_v$ 12	p.a. 78°

Pretty faint, quite small, and round. Distance 120 million ly; Hubble class SB+. Located between two stars.

NGC 6925

Gx	Mic	20^h 34.3^m	–31° 59′	4.4′ × 1.2′
chart 23			m$_v$ 11.3	p.a. 5°

Quite bright, large, and very elongated; bright, diffuse nucleus; thin, filamentary arms. Distance 110 million ly; Hubble class Sb.

NGC 6934

GC	Del	20^h 34.2^m	+07° 24′	7′
chart 16			m$_v$ 8.9	

C47. Medium concentration of stars; bright, large, round, and very well resolved. A mass of very small stars (Smyth). Distance 48,000 ly; S-S class 8.

NGC 6935

Gx	Ind	20^h 38.3^m	–52° 07′	2.2′
charts 23, 26			m$_v$ 11.9	p.a. 8°

Pretty bright, quite large, and round; faint arms; small, bright nucleus. Distance 210 million ly; Hubble class S(B)a.

NGC 6939

OC	Cep	20^h 31.4^m	+60° 38′	7′
chart 3			m$_v$ 7.8	

80 stars; detached, strong concentration of stars; small range in brightness; mag. of brightest star 11.9; pretty large. Distance 4,000 ly; age 1.8 billion years; Trumpler class I 1 m. Makes a unique cluster-galaxy combo with NGC 6946 in the same eyepiece field (Mullaney).

NGC 6940

OC	Vul	20^h 34.6^m	+28° 18′	30′
chart 9			m$_v$ 6.3	

60 stars; detached, no concentration of stars; moderate range in brightness; mag. of brightest star 9.3; very bright, very large. Distance 2,600 ly; over 1 billion years old; Trumpler class III 2 m.

NGC 6941

Gx	Aql	20^h 36.4^m	–04° 37′	1.9′ × 1.3′
chart 16			m$_v$ 12.7	p.a. 50°

Extremely faint and slightly elongated; moderately bright nucleus. Distance 280 million ly; Hubble class Sb.

NGC 6942

Gx	Ind	20^h 40.6^m	–54° 18′	2.2′ × 1.6′
charts 23, 26			m$_v$ 11.9	p.a. 150°

Pretty bright, pretty large, and round; very bright center. Distance 170 million ly; Hubble class SB0.

NGC 6943

Gx　Pav　20ʰ 44.6ᵐ　−68° 45′　3.9′ × 2.2′
chart 26　　　　　m_v 11.2　p.a. 130°

Pretty faint, large, and very elongated; patchy arms; small, bright nucleus. Distance 130 million ly; Hubble class S(B)c.

NGC 6946

Gx　Cyg　20ʰ 34.9ᵐ　+60° 09′　11.0′ × 9.8′
chart 3　　　　　m_v 8.8　p.a. 85°

C12. Very faint and very large; several massive arms; extremely small, bright nucleus. Distance 15 million ly; Hubble class Sc. Open cluster NGC 6939 lies just 0.6° NW.

NGC 6951

Gx　Cep　20ʰ 37.3ᵐ　+66° 06′　3.9′ × 3.5′
chart 3　　　　　m_v 10.7　p.a. 170°

Pretty bright, pretty large, and slightly elongated; extremely bright, extremely small nucleus. Distance 71 million ly; Hubble class Sbp.

NGC 6958

Gx　Mic　20ʰ 48.7ᵐ　−38° 00′　2.2′ × 1.7′
charts 23, 26　　m_v 11.3　p.a. 107°

Bright, quite small, and round; very bright nucleus. Distance 120 million ly; Hubble class E1.

NGC 6960

BNe　Cyg　20ʰ 45.7ᵐ　+30° 43′　70′ × 6′
chart 9

C34, the W segment of the Veil Nebula. This segment, by itself, is also known as the Filamentary Nebula. Pretty bright, quite large, and crescent shaped; contains the 4.2-mag. star 52 Cyg; use an O III filter for best view. Per *NGC*, this segment is a (!!) very remarkable object.

NGC 6962

Gx　Aqr　20ʰ 47.3ᵐ　+00° 19′　3.0′ × 2.3′
chart 16　　　　m_v 12.1　p.a. 75°

Quite faint, small, and roundish; bright nucleus. Distance 190 million ly; Hubble class S(B)b. Brightest and largest of a dense group of small, faint galaxies.

NGC 6981

See M72.

NGC 6992

BNe　Cyg　20ʰ 56.4ᵐ　+31° 43′　60′ × 7′
chart 9

Northern part of C33, the E segment of the Veil Nebula. This part, by itself, is also known as the Network Nebula. Very faint, large, and crescent shaped. Distance 1,300 ly; best seen with an O III filter. Per *NGC*, this segment is a (!!) very remarkable object.

NGC 6994

See M73.

NGC 6995

BNe　Cyg　20ʰ 57.1ᵐ　+31° 13′　12′
chart 9

Southern part of C33, the E segment the Veil Nebula. Faint and small; best seen with the help of an O III filter.

NGC 7000

See North America Nebula.

NGC 7006

GC　Del　21ʰ 01.5ᵐ　+16° 11′　3.3′
charts 9, 16, 17　m_v 10.6

C42. High concentration of stars; faint, small, and round. Distance 11,300 ly; S-S class 1.

NGC 7007

Gx　Ind　21ʰ 05.5ᵐ　−52° 33′　2.2′ × 1.3′
charts 23, 26　　m_v 11.9　p.a. 2°

Pretty bright, small, and round; dust lane. Distance 130 million ly; Hubble class S0. Located among stars.

NGC 7008

PN　Cyg　21ʰ 00.6ᵐ　+54° 33′　1.4′
charts 3, 9　　　m_v 10.7

PK93+05.2. Quite bright, large, and oval shaped; phot. mag. 13.3; central star mag. 13.2. Distance 2,600 ly; expansion velocity 11 km/sec (7 miles/sec); V-V class 3. Star at one end.

NGC 7009

See Saturn Nebula.

NGC 7013

Gx	Cyg	21^h 03.6^m	+29° 54′	4.5′ × 1.4′
chart 9			m$_v$ 11.3	p.a. 157°

Pretty bright, quite small, and roundish; bright nucleus. Distance 38 million ly; Hubble class Sa. Pretty bright star located to NW.

NGC 7020

Gx	Pav	21^h 11.3^m	−64° 02′	3.3′ × 1.6′
chart 26			m$_v$ 11.8	p.a. 165°

Pretty bright, quite small, and slightly elongated; patchy outer ring; very bright nucleus. Distance 130 million ly; Hubble class S0.

NGC 7023

BNr+C	Cep	21^h 00.5^m	+68° 10′	18′
chart 3				

C4. One of the brightest of the reflection nebulae; has bright and dark filaments; contains a 5′ open cluster and one bright 7th-mag. star. The star has such a foggy, veiled appearance that the observer thinks the telescope's lenses are dewed, but neighboring stars are sharp and clear on a dark sky, and the contrast is very pronounced (Denning).

NGC 7026

PN	Cyg	21^h 06.3^m	+47° 51′	20″
chart 9			m$_v$ 10.9	

PK89−0.1. Pretty bright, irregular disk; double nucleus; phot. mag. 12.7; central star mag. 14.2. Distance 5,500 ly; expansion velocity 41 km/sec (25 miles/sec); V-V class 3a.

NGC 7027

PN	Cyg	21^h 07.1^m	+42° 14′	14″
chart 9			m$_v$ 8.5	

PK84−3.1. Extremely bright and small; phot. mag. 10.4; central star mag. 16.3. Distance 3,600 ly; expansion velocity 22 km/sec (14 miles/sec); V-V class 3a.

NGC 7029

Gx	Ind	21^h 11.9^m	−49° 17′	2.6′ × 1.5′
charts 23, 26			m$_v$ 11.5	p.a. 71°

Bright, quite small, and round; extremely bright nucleus. Distance 120 million ly; Hubble class E6. In a group with galaxies NGC 7041 and NGC 7049.

NGC 7038

Gx	Ind	21^h 15.1^m	−47° 13′	3.4′ × 1.5′
charts 23, 26			m$_v$ 12	p.a. 127°

Pretty bright, pretty large, and slightly elongated; knotty arms; very small, bright nucleus. Distance 210 million ly; Hubble class Sc. In a group with galaxy NGC 7014 and other small galaxies.

NGC 7039

OC	Cyg	21^h 11.2^m	+45° 39′	25′
chart 9			m$_v$ 7.6	

50 stars; detached, no concentration; moderate brightness range; mag. of brightest star 11.3. Distance 2,300 ly; age 1 billion years; Trumpler class III 2 m.

NGC 7041

Gx	Ind	21^h 16.5^m	−48° 22′	3.5′ × 1.4′
charts 23, 26			m$_v$ 11.2	p.a. 85°

Bright, quite small, and quite elongated. Distance 80 million ly; Hubble class S(B)0. In a group with galaxies NGC 7029 and NGC 7049.

NGC 7042

Gx	Peg	21^h 13.8^m	+13° 35′	2.2′ × 1.9′
charts 16, 17			m$_v$ 12	p.a. 140°

Very faint, small, and roundish. Distance 240 million ly; Hubble class Sb.

NGC 7048

PN	Cyg	21^h 14.2^m	+46° 16′	1.0′
chart 9			m$_v$ 12.1	

PK88–1.1. Pretty large, pretty faint, diffuse, irregularly round; phot. mag. 11.3; central star mag. 19.1. Distance 3,900 ly; expansion velocity 11 km/sec (7 miles/sec); V-V class 3b.

NGC 7049

Gx	Ind	$21^h 19.0^m$	$-48° 34'$	$4.5' \times 2.8'$
charts 23, 26			m_v 10.3	p.a. 57°

Very bright, pretty small, and elongated; has a dark crescent; very bright nucleus. Distance 92 million ly; Hubble class S0. Brightest in a group with galaxies NGC 7029 and NGC 7041.

NGC 7059

Gx	Pav	$21^h 27.4^m$	$-60° 01'$	$3.3' \times 1.7'$
chart 26			m_v 11.9	p.a. 98°

Bright and pretty large; bright nucleus with dark lanes. Distance 74 million ly; Hubble class S(B)c.

NGC 7062

OC	Cyg	$21^h 23.2^m$	$+46° 23'$	7'
chart 9			m_v 8.3	

30 stars; detached, no concentration of stars; small range in brightness; mag. of brightest star 10.1; pretty small. Distance 6,200 ly; age 100 million years; Trumpler class III 1 p.

NGC 7063

OC	Cyg	$21^h 24.4^m$	$+36° 30'$	7'
chart 9			m_v 7	

12 stars; detached, no concentration; moderate brightness range; mag. of brightest star 8.9. Distance 2,000 ly; age 140 million years; Trumpler class III 2 p.

NGC 7064

Gx	Ind	$21^h 29.0^m$	$-52° 46'$	$3.8' \times 0.7'$
charts 23, 26			m_v 12.2	p.a. 91°

Extremely faint, pretty large, and very elongated; two main arms; bright center. Distance 32 million ly; Hubble class SBc.

NGC 7070

Gx	Gru	$21^h 30.4^m$	$-43° 05'$	$2.3' \times 1.9'$
charts 23, 26			m_v 12	p.a. 30°

Faint and quite elongated; two main, branching arms; small, bright nucleus. Distance 89 million ly; Hubble class Sc. In a group with galaxies NGC 7070A, NGC 7072, and NGC 7072A.

NGC 7078

See M15.

NGC 7079

Gx	Gru	$21^h 32.6^m$	$-44° 04'$	$2.0' \times 1.4'$
charts 23, 26			m_v 11.6	p.a. 82°

Bright, round, and quite small; very bright, elongated nucleus. Distance 120 million ly; Hubble class SB0.

NGC 7082

OC	Cyg	$21^h 29.4^m$	$+47° 05'$	25'
chart 9			m_v 7.2	

Not well detached from surrounding star field; moderate range in brightness; mag. of brightest star 9.9; large, poor cluster. Distance 4,600 ly; age 1.6 billion years.

NGC 7083

Gx	Ind	$21^h 35.8^m$	$-63° 54'$	$3.5' \times 2.2'$
chart 26			m_v 11.2	p.a. 5°

Pretty faint, quite large, and slightly elongated; several filamentary arms; small, bright, diffuse nucleus. Distance 130 million ly; Hubble class SBc.

NGC 7086

OC	Cyg	$21^h 30.5^m$	$+51° 35'$	8'
charts 3, 9			m_v 8.4	

50 stars; detached, weak concentration of stars; moderate range in brightness; mag. of brightest star 10.2; quite large. Distance 3,900 ly; age 85 million years; Trumpler class III 2 m.

NGC 7089

See M2.

NGC 7090

Gx	Ind	$21^h 36.5^m$	$-54° 33'$	$7.1' \times 1.4'$
charts 23, 26			m_v 10.7	p.a. 127°

Pretty bright, pretty large, and very elongated; bright center with dark lane. Distance 31 million ly; Hubble class SBm.

NGC 7092

See M39.

NGC 7096

| Gx | Ind | $21^h 41.3^m$ | $-63°\,55'$ | $1.8' \times 1.6'$ |
| chart 26 | | | $m_v\,12$ | p.a. 130° |

Very faint, small, and round; thin arms; very bright center. Distance 120 million ly; Hubble class Sa.

NGC 7097

| Gx | Gru | $21^h 40.2^m$ | $-42°\,32'$ | $1.9' \times 1.3'$ |
| charts 23, 26 | | | $m_v\,11.6$ | p.a. 20° |

Bright, small, and slightly elongated. Distance 100 million ly; Hubble class E5. Paired with galaxy NGC 7097A.

NGC 7098

| Gx | Oct | $21^h 44.3^m$ | $-75°\,07'$ | $4.2' \times 2.6'$ |
| charts 24, 26 | | | $m_v\,11.4$ | p.a. 74° |

Pretty faint and roundish; very bright nucleus. Distance 98 million ly; Hubble class SBa. Located among stars.

NGC 7099

See M30.

NGC 7125

| Gx | Ind | $21^h 49.3^m$ | $-60°\,43'$ | $3.0' \times 2.1'$ |
| chart 26 | | | $m_v\,12$ | p.a. 110° |

Extremely faint, pretty large, and round; two main, filamentary arms; bright center. Distance 130 million ly; Hubble class S(B)c. Paired with galaxy NGC 7126 to N.

NGC 7126

| Gx | Ind | $21^h 49.3^m$ | $-60°\,37'$ | $2.8' \times 1.4'$ |
| chart 26 | | | $m_v\,12.2$ | p.a. 80° |

Pretty bright, pretty small, and slightly elongated; very small, very bright nucleus. Distance 130 million ly; Hubble class Sc. Paired with galaxy NGC 7125 to S.

NGC 7129

| BNr | Cep | $21^h 42.5^m$ | $+66°\,10'$ | $7' \times 6'$ |
| chart 3 | | | | |

Small and quite faint; contains a quite faint, pretty large, loose cluster. Per *NGC*, a (!) remarkable object. Adjacent to reflection nebula NGC 7133.

NGC 7130

| Gx | PsA | $21^h 48.3^m$ | $-34°\,57'$ | $1.6' \times 1.5'$ |
| chart 23 | | | $m_v\,12$ | p.a. 70° |

Pretty bright, small, and roundish; very bright nucleus. Distance 210 million ly; Hubble class Sa.

NGC 7133

| BNr | Cep | $21^h 44.4^m$ | $+66°\,12'$ | $7' \times 6'$ |
| chart 3 | | | | |

Faint star involved in a very faint, pretty large patch of nebulosity. Adjacent to the reflection nebula NGC 7129.

NGC 7135

| Gx | PsA | $21^h 49.7^m$ | $-34°\,53'$ | $3.0' \times 1.9'$ |
| chart 23 | | | $m_v\,11.3$ | p.a. 47° |

Pretty bright, pretty large, and round; bright, diffuse nucleus; long, thin, cometlike extension. Distance 120 million ly; Hubble class S0p.

NGC 7139

| PN | Cep | $21^h 46.1^m$ | $+63°\,48'$ | $1.3'$ |
| chart 3 | | | $m_v\,13.3$ | |

PK104+07.1. Very faint, quite large, and round; phot. mag. 13.0; central star mag. 18.7. Distance 3,900 ly; V-V class 3b.

NGC 7141

| Gx | Ind | $21^h 52.3^m$ | $-55°\,34'$ | $4.5' \times 3.2'$ |
| charts 23, 26 | | | $m_v\,11.9$ | p.a. 18° |

Faint, large, and roundish; bright nucleus. Distance 130 million ly; Hubble class SBa.

NGC 7144

Gx	Gru	21ʰ 52.7ᵐ	−48° 15′	3.5′
charts 23, 26			m_v 10.9	

Very bright, pretty small, and round; very bright center with star. Distance 90 million ly; Hubble class E0. Paired with galaxy NGC 7145.

NGC 7145

Gx	Gru	21ʰ 53.3ᵐ	−47° 53′	2.5′
charts 23, 26			m_v 11.1	p.a. 173°

Bright, small, and round; located within a triangle of stars; very bright center. Distance 80 million ly; Hubble class E0. Paired with galaxy NGC 7144.

NGC 7154

Gx	PsA	21ʰ 55.4ᵐ	−34° 49′	2.1′ × 1.7′
chart 23			m_v 12.4	p.a. 102°

Bright, pretty large, irregularly round; patchy arms; Hubble class SBd.

NGC 7160

OC	Cep	21ʰ 53.7ᵐ	+62° 36′	7′
chart 3			m_v 6.1	

12 stars; detached, weak concentration of stars; large range in brightness; mag. of brightest star 7.1. Distance 2,900 ly; age 10 million years; Trumpler class II 3 p.

NGC 7166

Gx	Gru	22ʰ 00.5ᵐ	−43° 23′	2.4′ × 1.0′
charts 23, 26			m_v 11.6	p.a. 14°

Quite bright, small, and slightly elongated; extremely bright nucleus. Distance 100 million ly; Hubble class S0. In a group with galaxies NGC 7162 and NGC 7162A.

NGC 7167

Gx	Aqr	22ʰ 00.5ᵐ	−24° 38′	1.7′ × 1.3′
chart 23			m_v 12.5	p.a. 110°

Faint, pretty small, and roundish; bright nucleus. Distance 120 million ly; Hubble class SBc. A 10th-mag. star is located to E.

NGC 7168

Gx	Ind	22ʰ 02.1ᵐ	−51° 45′	2.1′ × 1.5′
charts 23, 26			m_v 11.9	p.a. 68°

Pretty bright, small, and round; bright, diffuse nucleus. Distance 120 million ly; Hubble class E3. Paired with an anonymous elliptical galaxy.

NGC 7171

Gx	Aqr	22ʰ 01.0ᵐ	−13° 16′	2.2′ × 1.3′
chart 17			m_v 12.2	p.a. 120°

Very faint, quite large, and elongated; two main arms; small, bright nucleus. Distance 120 million ly; Hubble class Sb. Paired with galaxy IC 1417.

NGC 7172

Gx	PsA	22ʰ 02.0ᵐ	−31° 52′	2.1′ × 1.2′
chart 23			m_v 11.8	p.a. 100°

Pretty bright, pretty large, and elongated; nearly edge on; dust lane. Distance 120 million ly; Hubble class S. In a group with four other galaxies.

NGC 7173

Gx	PsA	22ʰ 02.1ᵐ	−31° 58′	1.2′ × 0.9′
chart 23			m_v 11.1	p.a. 143°

Quite bright, quite small, and round. Distance 110 million ly; Hubble class E2. In a group with four other galaxies.

NGC 7176

Gx	PsA	22ʰ 02.1ᵐ	−31° 59′	1.4′ × 1.1′
chart 23			m_v 11.1	p.a. 80°

Bright, pretty large, and round. Distance 110 million ly; Hubble class E0p. In a group with four other galaxies.

NGC 7177

Gx	Peg	22ʰ 00.7ᵐ	+17° 44′	3.2′ × 2.1′
charts 9, 17			m_v 11.2	p.a. 90°

Pretty bright, pretty small, and round; many filamentary arms; extremely bright nucleus. Distance 63 million ly; Hubble class Sb.

NGC 7183

Gx　Aqr　22^h 02.4^m　−18° 55′　4.2′ × 1.1′
charts 17, 23　　m$_v$ 11.9　p.a. 77°

Very faint, pretty large, and elongated; moderately bright nucleus. Distance 120 million ly; Hubble class Sa. Four galaxies close by to N, another very close to S.

NGC 7184

Gx　Aqr　22^h 02.6^m　−20° 49′　6.0′ × 1.3′
chart 23　　m$_v$ 11.2　p.a. 62°

Pretty bright, very large, and very elongated; small, very bright nucleus. Distance 120 million ly; Hubble class Sb. Located among three stars.

NGC 7192

Gx　Ind　22^h 06.8^m　−64° 19′　2.1′
chart 26　　m$_v$ 11.2

Pretty bright, small, and round; bright nucleus. Distance 120 million ly; Hubble class SBa.

NGC 7196

Gx　Ind　22^h 05.9^m　−50° 07′　2.5′ × 1.9′
charts 23, 26　　m$_v$ 11.4　p.a. 53°

Quite bright, small, and round; very bright nucleus. Distance 130 million ly; Hubble class E3. Brightest in a group of small galaxies; superimposed star.

NGC 7205

Gx　Tuc　22^h 08.6^m　−57° 27′　3.8′ × 2.2′
chart 26　　m$_v$ 11.1　p.a. 73°

Pretty bright, large, and quite elongated; two main, bright arms; small, bright, diffuse nucleus. Distance 60 million ly; Hubble class Sb. Paired with nearby galaxy NGC 7205A.

NGC 7209

OC　Lac　22^h 05.2^m　+46° 30′　25′
chart 9　　m$_v$ 6.7

25 stars; detached, no concentration of stars; small range in brightness; mag. of brightest star 9.0; a large cluster. Distance 2,900 ly; age 300 million years; Trumpler class III 1 p.

NGC 7213

Gx　Gru　22^h 09.3^m　−47° 10′　3.3′ × 3.0′
charts 23, 26　　m$_v$ 10　p.a. 80°

Very bright, pretty small, and round; dust lane; extremely bright nucleus. Distance 76 million ly; Hubble class Sa.

NGC 7217

Gx　Peg　22^h 07.9^m　+31° 22′　4.1′ × 3.4′
chart 9　　m$_v$ 10.1　p.a. 95°

Bright and pretty large; knotty arms; very bright, diffuse nucleus; intermediate ring of low surface brightness. Distance 53 million ly; Hubble class Sb.

NGC 7218

Gx　Aqr　22^h 10.2^m　−16° 40′　2.5′ × 1.1′
chart 17　　m$_v$ 12　p.a. 25°

Pretty bright and slightly elongated; patchy arms; faint nucleus. Distance 78 million ly; Hubble class Sc.

NGC 7221

Gx　PsA　22^h 11.2^m　−30° 34′　2.0′ × 1.5′
chart 23　　m$_v$ 12.2　p.a. 10°

Faint, small, and roundish; bright nucleus. Distance 190 million ly; Hubble class SBbp.

NGC 7232

Gx　Gru　22^h 15.6^m　−45° 51′　2.5′ × 0.9′
charts 23, 26　　m$_v$ 11.6　p.a. 99°

Pretty bright, small, and very elongated; dark lane. Distance 86 million ly; Hubble class SBa. In a group with galaxies NGC 7232B and NGC 7233.

NGC 7235

OC　Cep　22^h 12.6^m　+57° 17′　4′
chart 3　　m$_v$ 7.7

30 stars; detached, no concentration of stars; moderate range in brightness; mag. of brightest star 8.8. Distance 12,400 ly; age 2 million years; Trumpler class III 2 p. Contains a ruby-colored 10th-mag. star.

NGC 7243

OC　Lac　22^h 15.3^m　+49° 53′　20′
chart 9　　m$_v$ 6.4

C16. 40 stars; not well detached from surrounding star field; moderate range in brightness; mag. of brightest star 8.5. Fine cluster, quickly followed by a beautiful field with three pairs (Webb). A neat double star forms the vertex of a telescopic triangle near the middle of the group (Smyth). Clustering of many bright stars (Denning). A splashy coarse star grouping (Houston). Distance 2,900 ly; age 110 million ly; Trumpler class IV 2 p.

NGC 7252

Gx	Aqr	$22^h 20.7^m$	$-24° 41'$	$2.2' \times 1.8'$
chart 23			m_v 11.4	p.a. 115°

Faint, small, and round; extremely bright nucleus; faint filaments and loops around main body. Distance 210 million ly; Hubble class E.

NGC 7261

OC	Cep	$22^h 20.4^m$	$+58° 05'$	6'
chart 3			m_v 8.4	

30 stars; detached, no concentration of stars; small range in brightness; mag. of brightest star 9.6; a large cluster. Distance 2,900 ly; age 40 million years; Trumpler class III 1 p.

NGC 7267

Gx	PsA	$22^h 24.4^m$	$-33° 42'$	$1.6' \times 1.4'$
chart 23			m_v 12.1	p.a. 6°

Quite bright, pretty small, and very slightly elongated; moderately bright nucleus. Distance 150 million ly; Hubble class SBa. Bright triple star to SW.

NGC 7280

Gx	Peg	$22^h 26.5^m$	$+16° 09'$	$2.1' \times 1.5'$
chart 17			m_v 12.1	p.a. 78°

Faint, quite small, and roundish; bright nucleus. Distance 90 million ly; Hubble class S(B)0. Three stars located to NE.

NGC 7285

Gx	Aqr	$22^h 28.6^m$	$-24° 50'$	$2.2' \times 1.4'$
chart 23			m_v 11.9	p.a. 80°

Distance 200 million ly; Hubble class SBa.

NGC 7292

Gx	Peg	$22^h 28.4^m$	$+30° 18'$	$2.1' \times 1.7'$
chart 9			m_v 12.5	p.a. 117°

Extremely faint, small, and oval shaped. Distance 54 million ly; Hubble class Ir+. Faint star involved.

NGC 7293

See Helix Nebula.

NGC 7307

Gx	Gru	$22^h 33.9^m$	$-40° 56'$	$3.5' \times 1.0'$
chart 23			m_v 12.2	p.a. 9°

Faint, pretty large, and very elongated; strong dark lane; faint nucleus. Distance 82 million ly; Hubble class S(B)cp.

NGC 7309

Gx	Aqr	$22^h 34.4^m$	$-10° 21'$	$1.9' \times 1.8'$
chart 17			m_v 12.5	p.a. 20°

Very faint, pretty large, and round; three main arms; very bright nucleus. Distance 180 million ly; Hubble class Sc.

NGC 7314

Gx	PsA	$22^h 35.8^m$	$-26° 03'$	$4.5' \times 1.9'$
chart 23			m_v 10.9	p.a. 3°

Quite faint, large, and very elongated; many patchy arms; extremely small, bright nucleus. Distance 74 million ly; Hubble class Sc. Paired with galaxy NGC 7313 (which is located about 4.3' away).

NGC 7329

Gx	Tuc	$22^h 40.4^m$	$-66° 29'$	$4.1' \times 2.9'$
charts 24, 26			m_v 11.8	p.a. 107°

Pretty bright, pretty small, and very elongated; several knotty arms; very bright nucleus. Distance 130 million ly; Hubble class SBb.

NGC 7331

Gx	Peg	$22^h 37.1^m$	$+34° 25'$	$10.7' \times 4.0'$
chart 9			m_v 9.5	p.a. 171°

C30. Bright, pretty large, very elongated; almost edge on; dust lanes. Distance 48 million ly; Hubble class Sb. Brightest, and in the foreground, of a group of galaxies. For large telescopes, Stephan's Quintet lies 0.5° to SSW.

NGC 7332

Gx	Peg	$22^h 37.4^m$	+23° 48′	3.6′ × 1.0′
chart 9			m_v 11.1	p.a. 155°

Quite bright, small, and very elongated; very bright, box-shaped nucleus. Distance 63 million ly; Hubble class E7. Paired with galaxy NGC 7339.

NGC 7354

PN	Cep	$22^h 40.3^m$	+61° 17′	23″
chart 3			m_v 12.2	

PK107+02.1. A bright, small, round ring; phot. mag. 12.9. Central star is fainter than mag. 16.2. Distance 5,200 ly; expansion velocity 26 km/sec (16 miles/sec); V-V class 4 + 3b.

NGC 7361

Gx	PsA	$22^h 42.3^m$	−30° 03′	4.0′ × 1.0′
chart 23			m_v 12.2	p.a. 4°

Faint, pretty large, and very elongated; nearly edge on; faint arms. Distance 55 million ly; Hubble class S.

NGC 7371

Gx	Aqr	$22^h 46.1^m$	−11° 00′	1.9′ × 1.8′
chart 17			m_v 11.5	p.a. 90°

Very faint, pretty large, and round; small, bright nucleus. Distance 110 million ly; Hubble class S(B)d.

NGC 7377

Gx	Aqr	$22^h 47.8^m$	−22° 19′	3.8′ × 3.1′
chart 23			m_v 10.4	p.a. 101°

Pretty bright, small, and slightly elongated; dark lanes; bright, diffuse nucleus. Distance 150 million ly; Hubble class E1.

NGC 7380

C+BNe	Cep	$22^h 47.0^m$	+58° 06′	30′ × 25′
chart 3				

40 stars in a 12′ region; detached, no concentration of stars; large range in brightness; total mag. of cluster 7.2; mag. of brightest star 8.6. The cluster is involved in a large emission nebula. Distance 11,700 ly; age 3.8 million years; Trumpler class III 3 p n.

NGC 7392

Gx	Aqr	$22^h 51.8^m$	−20° 36′	2.1′ × 1.3′
chart 23			m_v 11.8	p.a. 123°

Pretty bright, pretty small, and slightly elongated; twin arms; dark lanes; very small, very bright nucleus. Distance 130 million ly; Hubble class Sb.

NGC 7410

Gx	Gru	$22^h 55.0^m$	−39° 40′	5.0′ × 1.6′
chart 23			m_v 10.5	p.a. 45°

Quite bright, large, and very elongated; nearly edge on; bright, diffuse nucleus. Distance 71 million ly; Hubble class SBm.

NGC 7412

Gx	Gru	$22^h 55.8^m$	−42° 38′	4.0′ × 3.1′
chart 23			m_v 11.1	p.a. 65°

Extremely faint and very large; small, bright nucleus. Distance 73 million ly; Hubble class SBb. A 7th-mag. star nearby to NE.

NGC 7418

Gx	Gru	$22^h 56.6^m$	−37° 02′	3.7′ × 2.6′
chart 23			m_v 11	p.a. 139°

Quite bright, very large, and slightly elongated; two main, filamentary arms; small, very bright nucleus. Distance 66 million ly; Hubble class S(B)c. Paired with galaxy NGC 7421.

NGC 7421

Gx	Gru	$22^h 56.9^m$	−37° 21′	2.0′ × 1.9′
chart 23			m_v 11.9	p.a. 85°

Quite bright, large, and slightly elongated; three main arms. Distance 80 million ly; Hubble class SBb. Paired with galaxy NGC 7418.

NGC 7424

Gx	Gru	$22^h 57.3^m$	−41° 04′	9′
chart 23			m_v 10.2	p.a. 87°

Faint, quite large, and slightly elongated; small,

bright, diffuse nucleus. Distance 37 million ly; Hubble class S(B)c.

NGC 7448

Gx	Peg	$23^h 00.1^m$	+15° 59′	2.6′ × 1.2′
charts 9, 17			m_v 11.7	p.a. 170°

Pretty bright, large, and elongated; several bright, spiral arms; small, very bright nucleus. Distance 110 million ly; Hubble class Sc. Brightest in a group of galaxies.

NGC 7454

Gx	Peg	$23^h 01.1^m$	+16° 23′	2.1′ × 1.6′
charts 9, 17			m_v 11.8	p.a. 150°

Faint, quite small, and slightly elongated; small, diffuse, very bright nucleus. Distance 97 million ly; Hubble class E4. Paired with anonymous barred-spiral galaxy.

NGC 7456

Gx	Gru	$23^h 02.2^m$	−39° 34′	5.9′ × 1.8′
chart 23			m_v 11.6	p.a. 23°

Very faint, large, and very elongated; patchy arms; has a faint nucleus. Distance 52 million ly; Hubble class Sc.

NGC 7457

Gx	Peg	$23^h 01.0^m$	+30° 09′	4.0′ × 2.3′
chart 9			m_v 11.2	p.a. 130°

Quite bright, quite large, and slightly elongated; very small, bright nucleus. Distance 34 million ly; Hubble class Ep. Paired with a small, anonymous, spindle-shaped galaxy.

NGC 7462

Gx	Gru	$23^h 02.8^m$	−40° 50′	4.6′ × 0.9′
chart 23			m_v 11.3	p.a. 75°

Quite faint, pretty small, and very elongated; an almost spindle-shaped galaxy; dust lane. Distance 46 million ly; Hubble class SBc. Superimposed star.

NGC 7469

Gx	Peg	$23^h 03.3^m$	+08° 52′	1.5′ × 1.0′
chart 17			m_v 12.3	p.a. 125°

Very faint and very small; extremely bright nucleus. A Seyfert galaxy. Distance 220 million ly; Hubble class S. Paired with galaxy IC 5283.

NGC 7479

Gx	Peg	$23^h 05.0^m$	+12° 19′	4.3′ × 3.3′
chart 17			m_v 10.8	p.a. 25°

C44. Pretty bright, quite large, and very elongated; two main arms; dark lanes; very small, bright nucleus. Distance 110 million ly; Hubble class SBb.

NGC 7484

Gx	Scl	$23^h 07.1^m$	−36° 16′	1.7′
chart 23			m_v 11.8	p.a. 24°

Pretty bright, small, and roundish; moderately bright nucleus; Hubble class E0.

NGC 7496

Gx	Gru	$23^h 09.8^m$	−43° 26′	3.3′ × 3.2′
chart 23			m_v 11.1	p.a. 10°

Pretty bright, quite large, and slightly elongated; patchy arms; small, very bright nucleus. Distance 63 million ly; Hubble class SBb.

NGC 7497

Gx	Peg	$23^h 09.1^m$	+18° 11′	4.3′ × 1.5′
charts 9, 17			m_v 12.2	p.a. 48°

Very faint, large, and quite elongated; with a moderately bright nucleus. Distance 85 million ly; Hubble class SBc.

NGC 7507

Gx	Scl	$23^h 12.1^m$	−28° 32′	2.9′
chart 23			m_v 10.6	p.a. 109°

Pretty bright, quite small, and round. Distance 73 million ly; Hubble class E0. Paired with fainter galaxy NGC 7513.

NGC 7510

OC	Cep	$23^h 11.5^m$	+60° 34′	4′
chart 3			m_v 7.9	

60 stars; detached, weak concentration of stars; moderate range in brightness; mag. of brightest star

9.7; fan shaped; involved in a large, faint nebula. Distance 10,300 ly; age 10 million years; Trumpler class II 2 m n.

NGC 7513

Gx	Scl	23^h 13.2^m	−28° 22′	3.2′ × 2.1′
chart 23			m$_v$ 11.9	p.a. 108°

Very faint, pretty large, and elongated; moderately bright nucleus. Distance 70 million ly; Hubble class SBbp. Paired with galaxy NGC 7507.

NGC 7531

Gx	Gru	23^h 14.8^m	−43° 36′	4.6′ × 1.7′
charts 18, 23			m$_v$ 11.2	p.a. 15°

Pretty bright, small, and slightly elongated; bright, diffuse nucleus. Distance 67 million ly; Hubble class Sb.

NGC 7538

BNe	Cep	23^h 13.5^m	+61° 31′	10′ × 5′
chart 3				

A pair of 11th-mag. stars involved in a large, very faint nebula.

NGC 7541

Gx	Psc	23^h 14.7^m	+04° 32′	3.3′ × 1.1′
chart 17			m$_v$ 11.7	p.a. 102°

Bright, large, and very elongated. Only perceptible under settled gazing, when it faintly gleams among telescopic stars (Smyth). Distance 120 million ly; Hubble class Sc.

NGC 7552

Gx	Gru	23^h 16.2^m	−42° 35′	3.5′ × 3.1′
charts 18, 23			m$_v$ 10.4	p.a. 1°

Bright, small, and very elongated; dark lane; extremely bright nucleus. Distance 71 million ly; Hubble class SBb. Member of the Grus Quartet with galaxies NGC 7582, NGC 7590, and NGC 7599.

NGC 7556

Gx	Psc	23^h 15.7^m	−02° 23′	2.6′ × 1.6′
chart 17			m$_v$ 11.7	p.a. 110°

Quite faint, pretty large, and roundish. Distance 330 million ly; Hubble class E. Bright double star to E.

NGC 7562

Gx	Psc	23^h 16.0^m	+06° 41′	2.2′ × 1.7′
chart 17			m$_v$ 11.6	p.a. 83°

Quite bright, pretty small, irregularly roundish; bright nucleus. Distance 170 million ly; Hubble class E2.

NGC 7582

Gx	Gru	23^h 18.4^m	−42° 22′	5.0′ × 2.4′
charts 18, 23			m$_v$ 10.1	p.a. 157°

Pretty bright, large, and very elongated; very small, bright nucleus. Distance 62 million ly; Hubble class SBb. Member of the Grus Quartet with galaxies NGC 7552, NGC 7590, and NGC 7599.

NGC 7585

Gx	Aqr	23^h 18.0^m	−04° 39′	2.8′ × 2.3′
chart 17			m$_v$ 11.4	p.a. 60°

Pretty bright, pretty small, irregularly round; bright, diffuse nucleus; has a faint, extended envelope. Distance 150 million ly; Hubble class S. Paired with galaxy NGC 7576.

NGC 7590

Gx	Gru	23^h 18.9^m	−42° 14′	2.6′ × 1.0′
charts 18, 23			m$_v$ 11.3	p.a. 36°

Pretty bright, pretty large, and very elongated; several bright arms; very bright center. Distance 61 million ly; Hubble class Sb. Member of the Grus Quartet with galaxies NGC 7552, NGC 7582, and NGC 7599.

NGC 7599

Gx	Gru	23^h 19.3^m	−42° 15′	4.3′ × 1.3′
charts 18, 23			m$_v$ 11.1	p.a. 57°

Faint, pretty large, and very elongated; many knotty arms; very small, faint nucleus, Distance 72 million ly; Hubble class Sc. Member of the Grus Quartet with galaxies NGC 7552, NGC 7582, and NGC 7590.

NGC 7600

Gx	Aqr	23^h 18.9^m	−07° 35′	3.2′ × 1.4′
chart 17			m$_v$ 11.9	p.a. 70°

Quite faint, small, and round; small, very bright nucleus. Distance 150 million ly; Hubble class E6.

NGC 7606

Gx	Aqr	23^h 19.1^m	−08° 29′	4.3′ × 2.3′
chart 17			m$_v$ 10.8	p.a. 145°

Pretty faint, quite large, and very elongated; two filamentary arms; bright, diffuse nucleus. Distance 100 million ly; Hubble class Sb.

NGC 7619

Gx	Peg	23^h 20.2^m	+08° 12′	2.8′ × 2.5′
chart 17			m$_v$ 11.1	p.a. 30°

Quite bright, pretty small, and round. Distance 170 million ly; Hubble class E1. Brightest member of Pegasus I Galaxy Cluster.

NGC 7625

Gx	Peg	23^h 20.5^m	+17° 14′	1.5′
charts 9, 17			m$_v$ 12.1	p.a. 60°

Pretty bright, quite small, and round; narrow absorption lanes across one end; small, bright nucleus. Distance 81 million ly; Hubble class Ep.

NGC 7626

Gx	Peg	23^h 20.7^m	+08° 13′	2.8′ × 2.4′
chart 17			m$_v$ 11.1	p.a. 9°

Quite bright, pretty small, and round. Distance 160 million ly; Hubble class E2p. After galaxy NGC 7619, this is the second brightest member of the Pegasus I Galaxy Cluster.

NGC 7632

Gx	Gru	23^h 22.0^m	−42° 29′	2.5′ × 1.2′
charts 18, 23			m$_v$ 12.1	p.a. 92°

Faint, small, and roundish; moderately bright nucleus. Distance 170 million ly; Hubble class SB0.

NGC 7635

BNe	Cas	23^h 20.7^m	+61° 12′	15′ × 8′
chart 3				

C11, Bubble Nebula, A very faint, large, and luminous shell of nebulosity; illuminated by a 6.9-mag. star.

NGC 7640

Gx	And	23^h 22.1^m	+40° 51′	10.7′ × 2.5′
charts 4, 9			m$_v$ 11.3	p.a. 167°

Quite faint, large, and extremely elongated; nearly edge on; very faint nucleus. Distance 28 million ly; Hubble class S(B)b.

NGC 7654

See M52.

NGC 7662

PN	And	23^h 25.9^m	+42° 33′	17″
charts 4, 9			m$_v$ 8.3	

PK106−17.1, C22, the Blue Snowball Nebula. Very bright, pretty small, round; phot. mag. 9.2; central star mag. 13.2. Bluish disk with a woolly border and suspicion of dark center (Webb). On photographs it bears a fanciful resemblance to a lily (Copeland). Per *NGC*, a (!!!) most remarkable object. Distance 3,900 ly; extends 0.3 ly; expansion velocity 26 km/sec (16 miles/sec); V-V class 4 + 3.

NGC 7678

Gx	Peg	23^h 28.5^m	+22° 25′	2.3′ × 1.6′
chart 9			m$_v$ 11.8	p.a. 5°

Very faint, pretty large, and slightly elongated; two main, filamentary arms; one arm is very massive and brighter than the other; small, very bright nucleus. Distance 160 million ly; Hubble class Sc. Located among four stars.

NGC 7686

OC	And	23^h 30.2^m	+49° 08′	15′
charts 3, 4, 9			m$_v$ 5.6	

20 stars; not well detached from surrounding star field; small range in brightness; mag. of brightest star 6.2; possibly an asterism. Distance 3,300 ly; Trumpler class IV 1 p.

NGC 7689

Gx	Phe	23^h 33.3^m	−54° 06′	2.9′ × 1.9′
charts 23, 26			m$_v$ 11.4	p.a. 162°

Pretty faint, large, and round; knotty arms; very small, bright nucleus. Distance 82 million ly; Hubble class S(B)c.

NGC 7690

| Gx | Phe | 23ʰ 33.0ᵐ | −51° 42′ | 2.1′ × 0.9′ |
| charts 18, 23, 26 | | | m_v 12.1 | p.a. 132° |

Quite bright, small, and slightly elongated; very small, very bright nucleus. Distance 57 million ly; Hubble class Sb.

NGC 7713

| Gx | Scl | 23ʰ 36.2ᵐ | −37° 56′ | 4.5′ × 1.9′ |
| charts 18, 23 | | | m_v 11.1 | p.a. 168° |

Pretty bright, large, and elongated; faint arms; bright middle; faint nucleus. Distance 29 million ly; Hubble class SBd. Paired with galaxy NGC 7713A.

NGC 7714

| Gx | Psc | 23ʰ 36.2ᵐ | +02° 09′ | 1.9′ × 1.4′ |
| chart 17 | | | m_v 12.5 | p.a. 4° |

Pretty bright, small, and roundish; bright nucleus. Distance 130 million ly; Hubble class SBbp. A 12th-mag. star is located to SW and a 6th-mag. star to SE.

NGC 7716

| Gx | Psc | 23ʰ 36.5ᵐ | +00° 18′ | 2.2′ × 1.8′ |
| chart 17 | | | m_v 12.1 | p.a. 35° |

Faint, pretty large, and slightly elongated; very bright nucleus. Distance 120 million ly; Hubble class Sb.

NGC 7721

| Gx | Aqr | 23ʰ 38.8ᵐ | −06° 31′ | 3.0′ × 1.2′ |
| chart 17 | | | m_v 11.6 | p.a. 20° |

Pretty faint, quite large, and elongated; several filamentary arms; very small, bright nucleus. Distance 95 million ly; Hubble class Sc.

NGC 7723

| Gx | Aqr | 23ʰ 39.0ᵐ | −12° 58′ | 3.3′ × 2.2′ |
| chart 17 | | | m_v 11.2 | p.a. 35° |

Quite bright, quite large, and elongated; many knotty arms; small, very bright nucleus. Distance 85 million ly; Hubble class Sb.

NGC 7727

| Gx | Aqr | 23ʰ 39.9ᵐ | −12° 18′ | 4.3′ × 3.5′ |
| chart 17 | | | m_v 10.6 | p.a. 35° |

Pretty bright, pretty large, irregularly round; very faint outer arms; very small, bright, diffuse nucleus. Distance 84 million ly; Hubble class S(B)ap. Paired with galaxy NGC 7724.

NGC 7741

| Gx | Peg | 23ʰ 43.9ᵐ | +26° 05′ | 4.5′ × 3.1′ |
| chart 9 | | | m_v 11.3 | p.a. 170° |

Quite faint, quite large, irregularly round; very faint nucleus. Distance 44 million ly; Hubble class SBc.

NGC 7742

| Gx | Peg | 23ʰ 44.3ᵐ | +10° 46′ | 1.8′ |
| chart 17 | | | m_v 11.6 | |

Quite bright and quite small; with an extremely bright, small nucleus. Distance 79 million ly; Hubble class E0p.

NGC 7743

| Gx | Peg | 23ʰ 44.4ᵐ | +09° 56′ | 2.8′ × 2.4′ |
| chart 17 | | | m_v 11.5 | p.a. 80° |

Pretty faint, small, and round; very bright nucleus. Distance 87 million ly; Hubble class Sa.

NGC 7744

| Gx | Phe | 23ʰ 45.0ᵐ | −42° 55′ | 2.4′ × 1.9′ |
| charts 18, 23 | | | m_v 11.5 | p.a. 105° |

Quite bright, small, and slightly elongated; small, bright nucleus. Distance 130 million ly; Hubble class S(B)0.

NGC 7753

| Gx | Peg | 23ʰ 47.1ᵐ | +29° 29′ | 2.8′ × 2.0′ |
| charts 4, 9 | | | m_v 12 | p.a. 50° |

Quite faint, quite large, and slightly elongated; moderately bright nucleus. Distance 230 million ly; Hubble class S(B)b. Paired with galaxy NGC 7752.

NGC 7755

| Gx | Scl | 23ʰ 47.9ᵐ | −30° 31′ | 3.6′ × 2.8′ |
| charts 18, 23 | | | m_v 11.4 | p.a. 20° |

Bright, quite large, and round. Distance 130 million ly; Hubble class Sc. Companion galaxy at end of one arm.

NGC 7764

Gx   Phe   23ʰ 50.9ᵐ   −40° 44′   2.0′ × 1.5′
charts 18, 23   m_v 12.5   p.a. 148°

Bright, pretty large, and round; knotty arms; bright center. Distance 73 million ly; Hubble class Ir+.

NGC 7769

Gx   Peg   23ʰ 51.1ᵐ   +20° 09′   1.9′ × 1.7′
chart 9   m_v 12   p.a. 90°

Pretty faint, pretty small, and round; extremely bright nucleus. Distance 200 million ly; Hubble class Sc.

NGC 7785

Gx   Psc   23ʰ 55.3ᵐ   +05° 55′   2.0′ × 1.3′
chart 17   m_v 11.6   p.a. 143°

Pretty bright, pretty small, irregularly round; faint nucleus. Distance 180 million ly; Hubble class E5.

NGC 7789

OC   Cas   23ʰ 57.0ᵐ   +56° 44′   15′
charts 1, 3   m_v 6.7

Magnificent Cluster. 300 stars; detached, weak concentration of stars; small range in brightness; mag. of brightest star 10.7; a beautiful example of a rather large open cluster. A very glorious assemblage, both in extent and richness, having spangly rays of stars which give it a remote resemblance to a crab (Smyth). One of those rare objects that is impressive in any size instrument (Houston). Distance 6,200 ly; age 1.6 billion years; Trumpler class II 1 r.

NGC 7790

OC   Cas   23ʰ 58.4ᵐ   +61° 13′   16′
charts 1, 3   m_v 8.5

40 stars; detached, no concentration of stars; moderate range in brightness; mag. of brightest star 10.9. Distance 10,400 ly; age 78 million years; Trumpler class III 2 p.

NGC 7793

Gx   Scl   23ʰ 57.8ᵐ   −32° 35′   9.1′ × 6.6′
charts 18, 23   m_v 9.3   p.a. 98°

Cometlike; very small, very bright nucleus. Distance 9 million ly; Hubble class Scm. Chain of galaxies in background; located in Sculptor galaxy group.

NGC 7796

Gx   Phe   23ʰ 59.0ᵐ   −55° 27′   2.2′ × 1.9′
charts 24, 26   m_v 11.5   p.a. 168°

Pretty bright, quite small, and round; small, very bright nucleus. Distance 150 million ly; Hubble class E2.

NGC 7798

Gx   Peg   23ʰ 59.4ᵐ   +20° 45′   1.4′ × 1.3′
charts 4, 9   m_v 12.4   p.a. 80°

Pretty faint, small, and roundish; bright nucleus. Distance 130 million ly; Hubble class S. A 10th-mag. star is located to SW.

NGC 7814

Gx   Peg   00ʰ 03.3ᵐ   +16° 09′   5.0′ × 2.8′
charts 4, 10, 17   m_v 10.6   p.a. 135°

C43. Quite bright, quite large, and elongated; edgewise dark lane; very bright bulge in center. Distance 54 million ly; Hubble class Sb.

NGC 7817

Gx   Peg   00ʰ 04.0ᵐ   +20° 45′   3.5′ × 1.0′
charts 4, 9   m_v 11.8   p.a. 45°

Pretty faint and quite large; spindle shaped; moderately bright nucleus. Distance 110 million ly; Hubble class Sb.

NGC 7822

BNe Cep   00ʰ 03.6ᵐ   +68° 37′   60′ × 30′
charts 1, 3

An extremely faint and large arc of nebulosity; possibly a supernova remnant. Per *NGC*, this is a (!) remarkable object.

North America Nebula

BNe Cyg   20ʰ 58.8ᵐ   +44° 20′   2.0° × 1.7°
chart 9

NGC 7000, C20. Faint and extremely large; shaped

just like the continent of North America; best seen in a rich-field scope with a low-power, wide-field eyepiece and a nebula filter. Extends 60 ly; distance 1,600 ly. Illuminated by a 6th-mag. star.

Omega Centauri

GC   Cen $13^h\,26.8^m$ $-47°\,29'$ $50'$
charts 21, 25 $m_v\,3.9$

NGC 5139, C80. Extremely large, bright, and rich. This is the best globular cluster — much more impressive than M13 and easily visible to the naked eye. Per *NGC*, a (!!!) most remarkable object. Contains over a million stars. Distance 16,000 ly, extends 230 ly; S-S class 8.

Omega Nebula

BNe+C   Sgr $18^h\,20.8^m$ $-16°\,11'$ $45' \times 37'$
charts 15, 16 $m_v\,7$

M17, NGC 6618, also the Swan Nebula. Large and bright with a large (9′) and bright open cluster of about 40 stars. Mag. of brightest star 9.3. Per *NGC*, a (!!!) most remarkable object.

Omicron Velorum Cluster

See IC 2391.

Opened Box Cluster

See NGC 2360.

Orion Nebula

See Great Orion Nebula.

Owl Cluster

See NGC 457.

Owl Nebula

PN   UMa $11^h\,14.8^m$ $+55°\,01'$ $2.8'$
charts 2, 6 $m_v\,9.9$

M97, NGC 3587, PK148+57.1. Very bright, very large, and round; resembles the face of an owl with two dark "eyes" in large telescopes. Phot. mag. 12.0; central star mag. 16.0. Per *NGC*, a (!!) very remarkable object. Distance 1,300 ly; extends 1 ly; expansion velocity 41 km/sec (25 miles/sec); V-V class 3a. Galaxy M108 lies about 50′ to NW.

Pal 8

GC   Sgr $18^h\,41.5^m$ $-19°\,50'$ $4.7'$
charts 15, 16, 22 $m_v\,10.9$

Low concentration of stars. Distance 100,000 ly; S-S class 10.

Pal 11

GC   Aql $19^h\,45.2^m$ $-08°\,00'$ $3.2'$
chart 16 $m_v\,9.8$:

Low concentration of stars. Distance 36,000 ly; S-S class 11.

Parrot's Head

See B87.

Pazmino's Cluster

See St 23.

Pelican Nebula

BN   Cyg $20^h\,47.8^m$ $+44°\,22'$ $80' \times 70'$
charts 8, 9

Consists of two parts, IC 5067 (the N portion) and IC 5070 (the S portion). Very large and faint. Distance 4,000 ly.

Perseus A

See NGC 1275.

Phantom Streak

See NGC 6741.

Pi 4

OC   Vel $08^h\,34.5^m$ $-44°\,16'$ $18'$
charts 19, 20 $m_v\,5.9$

45 stars; not well detached; small brightness range; mag. of brightest star 7.3; involved in nebulosity. Distance 1,900 ly; Trumpler class IV 1 p n.

Pi 6

| OC | Vel | 08^h 39.3^m | −46° 13′ | 1.5′ |
| charts 19, 20 | | | m$_v$ 7.2 | |

15 stars; detached, weak concentration; moderate brightness range; mag. of brightest star 8.9. Distance 1,200 ly; age 32 million years; Trumpler class II 2 p.

Pi 16

| OC | Vel | 09^h 51.1^m | −53° 11′ | 1.5′ |
| charts 20, 25 | | | m$_v$ 8 | |

12 stars; detached, strong concentration; large brightness range; mag. of brightest star 8.7; Trumpler class I 3 p.

Pi 20

| OC | Cir | 15^h 15.4^m | −59° 04′ | 4.5′ |
| chart 25 | | | m$_v$ 7.8 | |

Detached, strong concentration; large brightness range; mag. of brightest star 8.2; a poor cluster. Distance 14,000 ly; Trumpler class I 3 p.

Pinwheel Galaxy

See M33.

Pipe Nebula

| DN | Oph | 17^h 30.0^m | −26° 00′ | 7° |
| chart 22 | | | | |

Made up of two sections. The Bowl of the Pipe consists of dark nebulae LDN 1773, B77, and B78, while the Stem of the Pipe is B65, B66, and B67.

PK1−6.1

| PN | Sgr | 18^h 15.4^m | −30° 32′ | <10″ |
| chart 22 | | | m$_p$ 14.0 | |

Central star mag. 12.5.

PK1−6.2

| PN | Sgr | 18^h 16.2^m | −30° 52′ | 5″ |
| chart 22 | | | m$_v$ 11.8 | |

SwSt 1. Stellar appearance; phot. mag. 12.0; central star mag. 11.8. Expansion velocity 13 km/sec (8 miles/sec); V-V class 1.

PK2−9.1

| PN | Sgr | 18^h 29.2^m | −31° 30′ | 7″ |
| chart 22 | | | m$_v$ 11.9 | |

Cn 1−5. Central star mag. 16.6; phot. mag. 15.2. Expansion velocity 18 km/sec (11 miles/sec).

PK3−6.1

| PN | Sgr | 18^h 17.7^m | −29° 08′ | 8″ |
| chart 22 | | | m$_p$ 13.0 | |

M 2−36. Smooth disk; central star mag. 15.7; V-V class 2.

PK3−14.1

| PN | Sgr | 18^h 55.6^m | −32° 16′ | 4″ |
| chart 22 | | | m$_v$ 12.8 | |

Hb 7. Smooth disk; phot. mag. 10.9; central star mag. 14.0. Distance 5,200 ly; V-V class 2.

PK3−17.1

| PN | Sgr | 19^h 05.6^m | −33° 12′ | 5″ |
| chart 22 | | | m$_v$ 12.5 | |

Hb 8. Smooth disk; phot. mag. 13.4. Distance 18,000 ly; V-V class 2.

PK4−11.1

| PN | Sgr | 18^h 39.4^m | −30° 41′ | 8″ |
| chart 22 | | | m$_v$ 13.3 | |

M 3−29. Smooth disk; phot. mag. 13.1; central star mag. 15.5. Expansion velocity 7 km/sec (4 miles/sec); V-V class 2.

PK7+7.1

| PN | Oph | 17^h 35.2^m | −18° 34′ | 9″ |
| charts 15, 22 | | | m$_v$ 13.5 | |

M 1−22. Ring structure; phot. mag. 13.3; V-V class 4.

PK7−6.2

| PN | Sgr | 18^h 28.0^m | −26° 07′ | 7″ |
| chart 22 | | | m$_v$ 14 | |

M 1−24. Stellar image; phot. mag. 13.3; central star mag. about 17.6; V-V class 1.

PK18+20.1

PN	Oph	$17^h\,12.9^m$	$-03°\,16'$	5″
chart 15			$m_v\,13.4$	

Na 1. Phot. mag. 13.4; central star mag. 16.6.

PK26−11.1

PN	Aql	$19^h\,18.3^m$	$-11°\,06'$	16″
chart 16			$m_v\,14$	

Na 2. Phot. mag. 13.3; central star mag. 14.

PK27−3.2

PN	Sct	$18^h\,54.0^m$	$-06°\,26'$	15″
charts 15, 16			$m_p\,13.4$	

Vy 1–4. Stellar image; central star mag. 15.6; V-V class 1.

PK38+12.1

PN	Oph	$18^h\,17.6^m$	$+10°\,09'$	5″
charts 15, 16			$m_v\,13.6$	

Cn 3–1. Smooth disk; phot. mag. 12.4; central star mag. 12.5. Distance 6,500 ly; V-V class 2.

PK45−2.1

PN	Aql	$19^h\,24.4^m$	$+09°\,54'$	14″
chart 16			$m_v\,12.7$	

Vy 2–2. Stellar image; phot. mag. 12.7; central star mag. 14.6; V-V class 1.

PK47+42.1

PN	Her	$16^h\,27.5^m$	$+27°\,54'$	2.9′
chart 8			$m_v\,13$	

Abell 39. Smooth disk with traces of ring structure; phot. mag. 13.7; central star mag. 15.7; V-V class 2c.

PK51+9.1

PN	Her	$18^h\,49.7^m$	$+20°\,51'$	2.6″
chart 8			$m_v\,11.4$	

Hu 2–1. Stellar appearance; phot. mag. 12.2; central star mag. 13.3. Distance 4,200 ly; V-V class I.

PK52−2.2

PN	Aql	$19^h\,39.2^m$	$+15°\,57'$	8″
charts 8, 16			$m_v\,11.8$	

Me 1–1. Ring structure; phot. mag. 12.6; central star mag. 14.1; V-V class 4.

PK64+5.1

PN	Cyg	$19^h\,34.8^m$	$+30°\,31'$	7.5″
chart 8			$m_v\,11.3$	

Campbell's Star, He 2–438. Small and very faint; ring structure; phot. mag. 9.6; central star mag. 12.5. Expansion velocity 26 km/sec (16 miles/sec); V-V class 4.

PK72−17.1

PN	Vul	$21^h\,16.8^m$	$+24°\,10'$	13.8′
chart 9			$m_p\,>12.0$	

Abell 74. Smooth disk; central star mag. 17.1; V-V class 2.

PK80−6.1

PN	Cyg	$21^h\,02.3^m$	$+36°\,42'$	6″
chart 9			$m_p\,12.0$	

The Egg Nebula. Central star mag. 12.3. Expansion velocity 13 km/sec (8 miles/sec). Possibly a proto-planetary object.

PK86−8.1

PN	Cyg	$21^h\,33.1^m$	$+39°\,38'$	8″
chart 9			$m_v\,12$	

Hu 1–2. Smooth disk; phot. mag. 12.7; central star mag. 17.3. Distance 13,000 ly; V-V class 2.

PK107−2.1

PN	Cep	$22^h\,56.3^m$	$+57°\,09'$	8″
chart 3			$m_v\,14.4$	

M 1–80. Smooth disk; phot. mag. 14.0. Distance 8,200 ly; V-V class 2.

PK107−13.1

PN	And	$23^h\,23.0^m$	$+46°\,54'$	4″
charts 4, 9			$m_v\,13.6$	

Vy 2–3. Smooth disk; phot. mag. 13.9; central star mag. 14.7. Distance 23,000 ly; expansion velocity 16 km/sec (10 miles/sec); V-V class 2.

PK111–2.1

PN	Cas	23^h 26.3^m	+58° 11'	1"
chart 3			m$_v$ 11.9	

Hb 12. Stellar appearance; possibly a protoplanetary object; phot. mag. 14.0; central star mag. 13.8. Distance 7,500 ly; expansion velocity 26 km/sec (16 miles/sec); V-V class 1.

PK119–6.1

PN	Cas	00^h 28.3^m	+55° 58'	5"
chart 1			m$_v$ 12.3	

Hu 1–1. Smooth disk; phot. mag. 13.3; central star mag. 19.1. Distance 2,900 ly; expansion velocity 15 km/sec (9 miles/sec); V-V class 2.

PK158+17.1

PN	Lyn	06^h 19.6^m	+55° 37'	20'
charts 1, 5			m$_p$ 11.2	

PuWe 1. Inner core diameter 20" and outer halo 20'; central star about mag. 15; V-V class 3.

PK164+31.1

PN	Lyn	07^h 57.8^m	+53° 25'	6.3'
charts 1, 2, 5, 6			m$_v$ 12.1	

JE 1. Large and faint with a double-lobed ring structure; phot. mag. 14.0; central star mag. 16.8. Distance 1,600 ly; expansion velocity 35 km/sec (22 miles/sec); V-V class 4.

PK171–25.1

PN	Tau	03^h 53.5^m	+19° 28'	38"
charts 4, 10, 11			m$_v$ 15.1	

Baade 1. Ring structure; phot. mag. 13.9; central star mag. 17.2. Distance 7,200 ly; V-V class 4.

PK198–6.1

PN	Ori	06^h 02.4^m	+09° 39'	37"
charts 11, 12			m$_v$ 12	

Abell 12. Phot. mag. 13.9; central star phot. mag. 19.7. Distance 7,800 ly.

PK205+14.1

PN	Gem	07^h 29.0^m	+13° 15'	10.2'
chart 12			m$_v$ 10.3	

Medusa Nebula, Abell 21. Central star mag. 16.0. Very close, only 650 ly away; expansion velocity 32 km/sec (10 miles/sec).

PK215–30.1

PN	Lep	05^h 03.2^m	–15° 36'	12.7'
chart 11			m$_p$ 13.2	

Abell 7. Irregular disk and brightness distribution; central star mag. 15.4. Distance 1,000 ly; V-V class 3a.

PK217+14.1

PN	CMi	07^h 51.7^m	+03° 00'	5.9'
chart 12			m$_v$ 13.5	

Abell 24. Ring structure involved in a larger, fainter disk, irregular in form; phot. mag. 13.6; central star mag. 17.2. Distance 1,300 ly; expansion velocity 14 km/sec (9 miles/sec); V-V class 4 + 3.

PK219+31.1

PN	Cnc	08^h 54.2^m	+08° 55'	16.2'
chart 12			m$_v$ 12	

Abell 31. Irregular disk with very irregular brightness distribution; phot. mag. 12.2; central star mag. 15.5; V-V class 3a.

PK238+34.1

PN	Hya	09^h 39.1^m	–02° 48'	4.5'
charts 12, 13			m$_v$ 12.4	

Abell 33. Smooth disk of uniform brightness; phot. mag. 13.4; central star mag. 15.5. Distance 2,000 ly; expansion velocity 32 km/sec (20 miles/sec); V-V class 2b.

PK264–8.1

PN	Vel	08^h 11.5^m	–48° 43'	45"
charts 19, 20, 25			m$_v$ 12.4	

He 2–7. Smooth disk; central star mag. 16.9; V-V class 2.

PK303+40.1

PN	Hya	12^h 53.6^m	–22° 52'	12.8'
chart 21			m$_v$ 12.7	

Abell 35. Irregular disk with very irregular bright-

ness distribution; phot. mag. 12.0; central star mag. 9.6; V-V class 3a.

PK307–4.1

| PN | Mus | $13^h 39.6^m$ | $-67° 23'$ | 4' |
| chart 25 | | | m_v 12.9 | |

He 2-95. Phot. mag. 12.2; central star mag. about 13.9. Distance 7,800 ly; expansion velocity 10 km/sec (6 miles/sec).

PK318+41.1

| PN | Vir | $13^h 40.6^m$ | $-19° 53'$ | 6.2' |
| charts 14, 21 | | | m_v 11.8 | |

Abell 36. Irregular disk with traces of ring structure, involved in a larger and fainter disk of irregular form; very uneven brightness distribution; phot. mag. 13.0; central star mag. 11.5. Distance 1,300 ly; expansion velocity 28 km/sec (18 miles/sec); V-V class 3b + 3a.

PK322–2.1

| PN | Nor | $15^h 34.3^m$ | $-59° 09'$ | 26" |
| charts 25, 26 | | | m_v 12.1 | |

Mz 1. Ring structure involved in a larger, fainter halo. Distance about 7,800 ly; phot. mag. 12.5; V-V class 4 + 6.

PK329+2.1

| PN | Nor | $15^h 51.7^m$ | $-51° 31'$ | 1.2' |
| charts 21, 22, 25 | | | m_v 12.6 | |

Sp 1. Ring structure; phot. mag. 13.6; central star mag. 14.0. Distance 3,600 ly; V-V class 4.

PK329–2.2

| PN | Nor | $16^h 14.5^m$ | $-54° 57'$ | 23" |
| charts 25, 26 | | | m_v 11.9 | |

Mz 2. Ring structure, involved in larger and fainter disk; irregular in form; expansion velocity 19 km/sec (12 miles/sec); V-V class 4 + 3.

PK342–14.1

| PN | Ara | $18^h 07.3^m$ | $-51° 02'$ | 36" |
| charts 22, 26 | | | m_p 11.9 | |

Sp 3. Central star mag. 12.6. Distance 5,400 ly; expansion velocity 22 km/sec (14 miles/sec).

PK352–7.1

| PN | CrA | $18^h 00.2^m$ | $-38° 50'$ | 25" |
| chart 22 | | | m_p 11.4 | |

Fg 3. Stellar image; central star mag. 14.3; V-V class 1.

PK353+8.1

| PN | Oph | $16^h 55.8^m$ | $-29° 50'$ | <5" |
| chart 22 | | | m_p 12.8 | |

He 2–184. Stellar image; V-V class 1.

PK356–4.1

| PN | Sco | $17^h 54.6^m$ | $-34° 22'$ | 2.4" |
| chart 22 | | | m_v 12.2 | |

Cn 2–1. Smooth disk; phot. mag. 13.9; V-V class 2.

PK359–0.1

| PN | Sgr | $17^h 47.9^m$ | $-30° 00'$ | 15" |
| chart 22 | | | m_v 11.8 | |

Hb 5. Smooth disk involved in a larger and fainter halo of nebulosity, anomalous in form; phot. mag. 13.6; central star phot. mag. 18.6. Distance 6,200 ly; V-V class 2 + 6.

Pleiades

| C+BNr | Tau | $03^h 47.0^m$ | $+24° 07'$ | 1.8° |
| charts 4, A2 | | | m_v 1.2 | |

M45, also the Seven Sisters. Mentioned in the Bible (*Job* 38:31). Most people see no more than six stars, but those with exceptional vision can count from 11 to 18. With an opera glass they look like the glimmering candles on a Christmas tree (Serviss). All told, the cluster contains at least 100 stars with a large range, the brightest being 2.9-mag. Alcyone (η Tau). Hipparcos studies make the cluster's distance 380 ly, almost 10 percent nearer than earlier determinations. It extends more than 12 ly in space. Age 78 million years; Trumpler class I 3 r. Surrounded by the extremely large and faint Merope Nebula (IC 349), which is visible telescopically under very dark skies but more easily photographed on blue-sensitive film.

Praesepe

OC	Cnc	08ʰ 40.1ᵐ	+19° 59′	1.7°
charts 6, 12			m_v 3.1	

M44, also the Beehive Cluster or the Manger, NGC 2632. 50 stars; detached, weak concentration of stars; moderate range in brightness; mag. of brightest star 6.3; very large, very bright. Usually like a patch of nebulosity to the naked eye, though on a very clear, dark night, stars may be glimpsed sparkling about the spot; a very small glass will show it as a nest of stars (Denning). I have noticed 36 stars (Galileo). Two triangles will be noted (Webb). Extends 17 ly. Distance 580 ly according to Hipparcos measurements; age 660 million years; Trumpler class II 2 m.

R47

BNe	Car	10ʰ 05.0ᵐ	−58° 55′	25′ × 20′
chart 25				

Irregular shape; mottled; located in a field very rich in faint stars; W border has a 9th-mag. star involved in a nova shell.

R50

BNe	Car	10ʰ 27.0ᵐ	−57° 10′	10′ × 7′
chart 25				

Slightly elongated to an irregular oval; low surface brightness; involved in a field that is very rich in faint stars.

R58

BNe	Car	11ʰ 06.0ᵐ	−65° 35′	8′ × 5′
chart 25				

A Wolf-Rayet shell, shaped like a fragmented oval ring; SW segments are brightest. An 8th-mag. star is located at the center.

R126

BNe	Sco	17ʰ 17.0ᵐ	−36° 20′	16′ × 4′
chart 22				

Very faint, diffuse; extends ENE–WSW; irregular surface brightness.

Rho Ophiuchi Nebula

See IC 4604.

Ring Nebula

PN	Lyr	18ʰ 53.6ᵐ	+33° 02′	76″
chart 8			m_v 8.8	

M57, NGC 6720, PK63+13.1, also the Smoke Ring Nebula or Donut Nebula. Bright, pretty large; irregular ring structure; a very impressive and stunning object; phot. mag. 9.7. Small telescopes usually show it as a featureless patch or disk, but a 4-inch brings out the annular shape. The very blue central star (mag. 15.2) is difficult to see visually and has even been suspected of being variable; very good seeing and high power are essential. Much easier to notice is the 13.2-mag. star very close outside the ring's E edge. Per *NGC*, a (!!!) most remarkable object. The nebula's expansion velocity is 30 km/sec (19 miles/sec). Distance variously estimated as 1,400 to 2,000 ly; V-V class 4 + 3.

Ring Tail

See NGC 4038 and NGC 4039.

Rosette Nebula

BNe	Mon	06ʰ 32.3ᵐ	+05° 03′	80′ × 60′
charts 11, 12				

C49, consisting of emission nebulae NGC 2237, NGC 2238, and NGC 2239 along with open cluster NGC 2244. Large, bright, roundish nebula involved in a sparse cluster of 100 stars. One of the few deep-sky objects better seen with a telescope's finder than with the main instrument (Houston). Distance 5,000 ly.

Ru 32

OC	Pup	07ʰ 45.0ᵐ	−25° 31′	6′
chart 19			m_v 8.4	

30 stars; detached, no concentration; moderate brightness range; mag. of brightest star 9.6; involved in nebulosity. Distance 13,000 ly; Trumpler class III 2 p n.

Ru 44

OC	Pup	07ʰ 59.0ᵐ	−28° 35′	5′
charts 19, 20			m_v 7.2	

40 stars; detached, no concentration; small brightness range; mag. of brightest star 9.4. Distance 22,000 ly; Trumpler class III 1 p.

Ru 55

OC	Pup	08^h 12.3^m	−32° 36′	16′
charts 19, 20			m$_v$ 7.8	

12 stars; not well detached; moderate brightness range; mag. of brightest star 8.6; possibly an asterism. Distance 14,000 ly; Trumpler class IV 2 p.

Ru 82

OC	Vel	09^h 45.6^m	−53° 59′	3.6′
charts 20, 25			m$_v$ 8.1	

20 stars; detached, weak concentration; moderate brightness range; mag. of brightest star 10.8; Trumpler class II 3 p.

Ru 92

OC	Car	10^h 53.9^m	−61° 44′	2.2′
chart 25			m$_v$ 8.6	

15 stars; detached, strong concentration of stars; large range in brightness; mag. of brightest star 10.9. Distance 9,000 ly; Trumpler class I 3 p.

Ru 93

OC	Car	11^h 04.4^m	−61° 22′	4′
chart 25			m$_v$ 7.7	

30 stars; detached, no concentration; moderate brightness range; mag. of brightest star 11.2; Trumpler class III 2 p.

Ru 98

OC	Cru	11^h 58.0^m	−64° 29′	10′
chart 25			m$_v$ 7	

50 stars; detached, weak concentration; moderate brightness range; mag. of brightest star 8.9. Distance 1,300 ly; Trumpler class II 2 m.

Ru 106

GC	Cen	12^h 38.7^m	−51° 09′	
charts 20, 21, 25			m$_v$ 10.9:	

With a 9th-mag. star to SSE.

Ru 108

OC	Cen	13^h 32.2^m	−58° 29′	12′
chart 25			m$_v$ 7.5	

15 stars; detached, no concentration; moderate brightness range; mag. of brightest star 8.5. Distance 2,300 ly; Trumpler class III 2 p.

Running Chicken Nebula

See IC 2944, 48.

S Nebula

See Snake Nebula.

Saturn Nebula

PN	Aqr	21^h 04.2^m	−11° 22′	28″
charts 16, 17			m$_p$ 8.3	

NGC 7009, PK37–34.1, C55. Small and very bright; phot. mag. 8.3; central star mag. 12.8. An oblong, blue-green disk, somewhat resembling Saturn. The featureless disk has an eerie radiance especially striking in larger apertures (Mullaney). Per *NGC*, a (!!!) most remarkable object. Distance 2,900 ly; expansion velocity 21 km/sec (13 miles/sec); V-V class 4 + 6.

Sculptor Galaxy

See NGC 253.

Sculptor Dwarf Galaxy

Gx	Scl	01^h 00.2^m	−33° 43′	35′ × 28′
chart 18			m$_v$ 8.8	p.a. 110°

E 351-30. Distance 5 million ly; Hubble class E.

Seven Sisters

See Pleiades.

Sh2-1

BNer	Sco	15^h 58.9^m	−26° 09′	90′ × 10′
charts 21, 22				

Brightest part is S of star π Sco.

Sh2-3

BNe	Sco	17^h 12.3^m	−38° 29′	12′
chart 22				

Egg shaped; extends NNW–SSE. An 8th-mag. star is in the N part.

Sh2-9

BNer Sco $16^h 21.1^m$ $-25° 35'$ $60' \times 50'$
chart 22

Very diffuse.

Sh2-13

BNe Sco $17^h 29.1^m$ $-31° 33'$ $39'$
chart 22

The surrounding region is littered with faint nebulosities and is extremely rich in faint stars.

Sh2-16

BNe Sgr $17^h 46.6^m$ $-29° 18'$ $12'$
chart 22

Located in a rich star field; small patches of nebulosity are off the NE edge.

Sh2-35

BNe Sgr $18^h 15.9^m$ $-20° 15'$ $10' \times 7'$
chart 22

Very irregular shape; fairly diffuse. Set against a background that is extremely rich in very faint stars.

Sh2-46

BNe Ser $18^h 06.1^m$ $-14° 10'$ $30' \times 19'$
charts 15, 16

Elliptical shape. This region is rich in stars, especially to E.

Sh2-53

BNe Sct $18^h 25.2^m$ $-13° 13'$ $15'$
charts 15, 16

Irregular shape; consists of several parts. Set in a starry field.

Sh2-55

BNe Sct $18^h 32.2^m$ $-11° 46'$ $19'$
charts 15, 16

Irregular shape; very faint and diffuse nebulosity set in a rich star field. W edge has two bright stars separated by approx. 0.5'.

Sh2-64

BNe Ser $18^h 31.6^m$ $-01° 55'$ $19' \times 7'$
charts 15, 16

Irregular shape; divided by absorption matter into three main components.

Sh2-82

BNer Sge $19^h 30.3^m$ $+18° 16'$ $7'$
charts 8, 16

Irregular shape; large, roundish, and faint; has a faint central star. A smaller, partly detached nebulosity lies due N.

Sh2-84

BNe Sge $19^h 49.0^m$ $+18° 24'$ $15' \times 3'$
charts 8, 16

Broad and V shaped; located in a very rich star field.

Sh2-88

BNe Vul $19^h 46.0^m$ $+25° 20'$ $18' \times 6'$
chart 8

Irregular shape. With two bright knots SE of center, aligned ENE–WSW and spaced about 2.4' apart.

Sh2-90

BNr Vul $19^h 49.3^m$ $+26° 52'$ $7' \times 3'$
charts 8, 9

An 8.3-mag. star involved in an irregularly shaped nebulosity.

Sh2-101

BNe Cyg $20^h 00.0^m$ $+35° 17'$ $15' \times 8'$
charts 8, 9

Crown shaped. Extends NNE–SSW; the SE and E sections are brightest.

Sh2-104

BNe Cyg $20^h 17.8^m$ $+36° 44'$ $7'$
charts 8, 9

Circular shape.

Sh2-108

BNe   Cyg $20^h\,19.1^m$ +39° 21'   60' × 30'
charts 8, 9

Irregular shape. This is the SW portion of the nebula complex surrounding the star γ Cyg.

Sh2-112

BNe   Cyg $20^h\,33.9^m$ +45° 39'   8' × 6'
charts 8, 9

Irregular shape.

Sh2-115

BNe   Cyg $20^h\,34.5^m$ +46° 52'   30' × 19'
charts 8, 9

Irregular and filamentary.

Sh2-129

BNe   Cep $21^h\,11.8^m$ +59° 57'   1.7° × 1.0°
chart 3

Very faint; an incomplete filamentary ring of nebulosity.

Sh2-132

BNe   Cep $22^h\,18.7^m$ +56° 08'   30' × 19'
charts 3, 9

Irregular and filamentary; located in a very rich star field; NE part has higher surface brightness.

Sh2-155

BNe   Cep $22^h\,56.8^m$ +62° 37'   50' × 30'
chart 3

C9, also the Cave Nebula. Crescent shaped, with the N part brightest.

Sh2-157

BNe   Cas $23^h\,16.1^m$ +60° 02'   60' × 50'
chart 3

Irregular and filamentary.

Sh2-188

BNe   Cas $01^h\,30.6^m$ +58° 22'   10' × 3'
chart 1

Filamentary and crescent shaped; SE portion is most sharply defined.

Sh2-205

BNe   Cam $03^h\,56.1^m$ +53° 12'   1.7° × 1.0°
charts 1, 4, 5

Very faint and diffuse; irregular shape. Brightest part is centered on an 8th-mag. star.

Sh2-224

BNe   Aur $05^h\,27.3^m$ +42° 59'   19' × 3'
chart 5

Incomplete oval ring; N part is brightest. A supernova remnant.

Sh2-231

BNe   Aur $05^h\,39.4^m$ +35° 56'   10' × 5'
chart 5

Irregular shape, extended N–S; very diffuse and faint.

Sh2-235

BNe   Aur $05^h\,41.1^m$ +35° 52'   10'
chart 5

Irregular shape.

Sh2-240

See Simeis 147.

Sh2-241

BNer   Aur $06^h\,04.1^m$ +30° 15'   10'
chart 5

Fan shaped; very faint and very diffuse.

Sh2-247

BNe   Gem $06^h\,08.5^m$ +21° 37'   10'
chart 5

Circular shape.

Sh2-261

BNe   Ori $06^h\,08.9^m$ +15° 49'   30' × 15'
charts 11, 12

Also called Lower's Nebula. Roundish; very mottled; S part is brightest.

Sh2-276

See Barnard's Loop.

Sh2-282

BNe   Mon $06^h 38.0^m$ $+01° 31'$ $39' \times 14'$
charts 11, 12

Irregular shape containing several stars involved in nebulosity.

Sh2-294

BNe   Mon $07^h 16.6^m$ $-09° 26'$ $7' \times 6'$
chart 12

Irregular shape.

Sh2-301

BNe   CMa $07^h 09.8^m$ $-18° 29'$ $7' \times 6'$
charts 12, 19

Elliptical shape.

Sh2-302

BNe   Pup $07^h 31.6^m$ $-16° 58'$ $19'$
charts 12, 19

The E section is split by a narrow N–S dark lane; NE end is brightest.

Sh2-307

BNe   Pup $07^h 35.5^m$ $-18° 46'$ $4'$
charts 12, 19

Irregular shape.

Siamese Twins

See NGC 4567 and NGC 4568.

Simeis 147

BNe   Tau $05^h 39.1^m$ $+28° 00'$ $3.3° \times 3.0°$
chart 5

Also Sh2-240. A supernova remnant; extremely large and extremely faint; very wispy and filamentary structure. This is a very difficult object visually, normally detectable only on red-sensitive film or with a CCD camera.

SL 4

DN   Vel $08^h 53.6^m$ $-42° 13'$ $60' \times 9'$
chart 20

High opacity. Located in E portion of emission nebula Gum 17.

SL 7

DN   Lup $16^h 01.8^m$ $-41° 52'$ $2.7° \times 0.2°$
charts 21, 22

High opacity. Consists of several small clouds strung together.

SL 8

DN   Nor $16^h 14.2^m$ $-44° 04'$ $25' \times 5'$
charts 21, 22

High opacity. Irregular, curved shape.

SL 11

DN   Lup $15^h 57.0^m$ $-37° 48'$ $39' \times 5'$
charts 21, 22

High opacity. Irregular shape.

SL 15

DN   Sco $16^h 46.6^m$ $-44° 30'$ $7' \times 3'$
charts 21, 22

High opacity. Elliptical shape; elongated NE–SW; darkest at SW end.

SL 17

DN   Sco $16^h 53.0^m$ $-43° 35'$ $15' \times 7'$
chart 22

High opacity. Irregular shape.

SL 18

DN   Sco $16^h 44.8^m$ $-40° 23'$ $5'$
chart 22

High opacity. A very dark, irregularly shaped object; sharpest on SE side. A 9th-mag. star is close by to N.

SL 26

DN	Sco	17ʰ 34.2ᵐ	−40° 25′	10′ × 5′
chart 22				

High opacity. Elliptical shape with faint extensions.

SL 28

DN	Sco	17ʰ 35.3ᵐ	−39° 14′	30′ × 15′
chart 22				

High opacity. Irregular shape.

SL 42

DN	CrA	19ʰ 10.3ᵐ	−37° 08′	12′ × 7′
chart 22				

High opacity. Elliptical shape.

Small Magellanic Cloud

Gx	Tuc	00ʰ 52.7ᵐ	−72° 50′	4.6° × 2.7°
chart 24			m_v 2.3	

Irregular galaxy of the Local Group and a possible satellite of the Milky Way. Prominently visible to the naked eye. Has been assigned to Hubble class SBmp, but the resemblance to a barred spiral is less obvious than that of its famous companion, the Large Magellanic Cloud. Distance about 190,000 ly.

Smoke Ring Nebula

See Ring Nebula.

Snake Nebula

DN	Oph	17ʰ 23.0ᵐ	−23° 32′	10′
chart 22				

B72, also the S Nebula. High opacity; S shaped; includes globule. Giant binoculars reveal its slender form, while telescopes add depth to the overall effect (Harrington). Located approx. 1.5° NNE of star θ Oph.

Sombrero Galaxy

Gx	Vir	12ʰ 40.0ᵐ	−11° 37′	8.9′ × 4.4′
charts 13, 14			m_v 8.3	p.a. 89°

M104, NGC 4594. Very bright, very large, and elongated; with a prominent dark lane and large nuclear bulge. Per *NGC*, a (!) remarkable object. Extends 100,000 ly; distance 41 million ly; Hubble class Sb.

Southern Pleiades

OC	Car	10ʰ 43.2ᵐ	−64° 24′	50′
chart 25			m_v 1.9	

IC 2602, C102. 60 stars; detached, weak concentration of stars; large range in brightness; very large, moderately bright; mag. of brightest star 2.8. Distance is 480 ly according to Hipparcos measurements; age 36 million years; Trumpler class II 3 m.

Spindle Galaxy

See NGC 3115.

St 1

OC	Vul	19ʰ 35.8ᵐ	+25° 13′	60′
chart 8			m_v 5.3:	

40 stars; not well detached from surrounding star field; moderate range in brightness; mag. of brightest star 7.0; Trumpler class IV 2 p. Located midway between open clusters NGC 6815 and NGC 6800.

St 2

OC	Cas	02ʰ 15.0ᵐ	+59° 16′	60′
chart 1			m_v 4.4	

50 stars; detached, strong concentration; moderate brightness range; mag. of brightest star 8.2. Distance 1,000 ly; age 100 million years; Trumpler class I 2 m.

St 13

OC	Car	11ʰ 13.1ᵐ	−58° 55′	3′
chart 25			m_v 7	

15 stars; detached, strong concentration; moderate brightness range; mag. of brightest star 8.5; involved in nebulosity. Distance 8,800 ly; Trumpler class I 2 p n.

St 14

OC	Cen	11ʰ 44.0ᵐ	−62° 30′	4′
chart 25			m_v 6.3	

10 stars; detached, weak concentration; small brightness range; mag. of brightest star 8.4. Distance 8,500 ly; Trumpler class II 1 p.

St 23

OC　Cam　03ʰ 16.3ᵐ　+60° 02'　　14'
chart 1

Pazmino's Cluster. 25 stars; detached, no concentration; large brightness range; involved in nebulosity; Trumpler class III 3 p n.

Starfish Cluster

See NGC 2482.

Star Queen Nebula

See Eagle Nebula.

Stem of Pipe Nebula

See Pipe Nebula.

Steph 1

OC　Lyr　18ʰ 54.2ᵐ　+36° 55'　　19'
charts 3, 8　　　　　m_v 3.8

Also the Delta Lyrae Cluster. 15 stars; detached, no concentration; large brightness range; mag. of brightest star 4.3.

Stephan's Quintet

—　Peg　22ʰ 36.0ᵐ　+33° 58'　　0.4'
chart 9　　　　　m_v 13.1

A cluster of five very faint galaxies (NGC 7317, NGC 7318, NGC 7318A, NGC 7319, and NGC 7320). Not charted in *Sky Atlas 2000.0*. Two have been seen in a 6-inch reflector and all five by averted vision in a 10-inch. Sizable telescopes bring them out better. Small, fuzzy patches nestled among a collection of stars (Harrington). The much brighter galaxy NGC 7331 lies 0.5° to NNE.

Sunflower Galaxy

See M63.

Swan Nebula

See Omega Nebula.

Table of Scorpius

See NGC 6231.

Tank Trap Nebula

See NGC 2024.

Tarantula Nebula

BNe　Dor　05ʰ 38.6ᵐ　−69° 05'　　5'
charts 24, 25　　　m_v 8.3

NGC 2070, C103. Very bright, very large nebula. Contains a small (5') open cluster, 30 Doradus, whose brightest star is mag. 8.3. Per *NGC*, a (!!!) most remarkable object. Located in E portion of Large Magellanic Cloud.

Taurus Moving Cluster

See Hyades.

Thor's Helmet

See NGC 2359.

Tom 5

OC　Cam　03ʰ 47.8ᵐ　+59° 03'　　16'
chart 1　　　　　m_v 8.4

60 stars; detached, no concentration; moderate brightness range; mag. of brightest star 11.6. Distance 5,900 ly; Trumpler class III 2 m.

Tom Thumb Cluster

See NGC 6451.

Tr 1

OC　Cas　01ʰ 35.7ᵐ　+61° 17'　　4.5'
chart 1　　　　　m_v 8.1

20 stars; detached, strong concentration; large brightness range; mag. of brightest star 9.6. Distance 7,200 ly; age 26 million years; Trumpler class I 3 p.

Tr 2

OC　Per　02ʰ 37.3ᵐ　+55° 59'　　19'
charts 1, 4　　　　m_v 5.9

20 stars; detached, no concentration; moderate brightness range; mag. of brightest star 7.4.

Tr 3

Distance 2,000 ly; age 78 million years; Trumpler class III 2 p.

Tr 3

OC	Cas	$03^h 11.8^m$	$+63° 09'$	22'
chart 1			m_p 7.0	

30 stars; detached, no concentration; large brightness range; Trumpler class III 3 p.

Tr 7

OC	Pup	$07^h 27.3^m$	$-23° 58'$	5'
chart 19			m_v 7.9	

30 stars; detached, weak concentration; large brightness range; mag. of brightest star 9.1; involved in nebulosity. Distance 5,200 ly; Trumpler class II 3 p n.

Tr 10

OC	Vel	$08^h 47.8^m$	$-42° 29'$	15'
chart 20			m_v 4.6	

40 stars; detached, weak concentration; moderate brightness range; mag. of brightest star 6.4. Distance 1,400 ly; age 47 million years; Trumpler class II 2 p.

Tr 15

OC	Car	$10^h 44.8^m$	$-59° 22'$	3'
chart 25			m_v 7	

20 stars; detached, no concentration; moderate brightness range; mag. of brightest star 8.4; involved in nebulosity. Distance 5,000 ly; Trumpler class III 2 p n. Located in bright Eta Carinae Nebula (NGC 3372).

Tr 17

OC	Car	$10^h 56.2^m$	$-59° 13'$	5'
chart 25			m_v 8.4	

30 stars; detached, weak concentration; moderate brightness range; mag. of brightest star 10.3. Distance 4,600 ly; Trumpler class II 2 p.

Tr 18

OC	Car	$11^h 11.4^m$	$-60° 40'$	12'
chart 25			m_v 6.9	

30 stars; detached, no concentration; moderate brightness range. Distance 8,200 ly; age 250 million years; Trumpler class III 2 p.

Tr 21

OC	Cen	$13^h 32.2^m$	$-62° 47'$	4'
chart 25			m_v 7.7	

20 stars; detached, strong concentration; large brightness range; mag. of brightest star 8.9. Distance 3,600 ly; Trumpler class I 3 p.

Tr 22

OC	Cen	$14^h 31.2^m$	$-61° 10'$	7'
charts 25, A3			m_v 7.9	

50 stars; detached, weak concentration; moderate brightness range; mag. of brightest star 10.1. Probably not a true cluster.

Tr 24

OC	Sco	$16^h 57.0^m$	$-40° 40'$	60'
chart 22			m_p 8.6	

A poor cluster, not well detached; moderate range in brightness. Distance 5,000 ly; Trumpler class IV 2 p n. Involved with nebula IC 4628.

Tr 27

OC	Sco	$17^h 36.2^m$	$-33° 29'$	7'
chart 22			m_v 6.7	

35 stars; detached, strong concentration; moderate brightness range; mag. of brightest star 8.4. Distance 6,900 ly; Trumpler class I 2 p.

Tr 28

OC	Sco	$17^h 36.8^m$	$-32° 29'$	7'
chart 22			m_v 7.7	

30 stars; detached, weak concentration; moderate brightness range; mag. of brightest star 9.8; involved in nebulosity. Distance 500 ly; Trumpler class II 2 p n.

Tr 29

OC	Sco	$17^h 41.6^m$	$-40° 06'$	8'
chart 22			m_p 7.5	

30 stars; detached, weak concentration; large

brightness range; Trumpler class II 3 p.

Trapezium

C+BNer	Ori	05ʰ 35.3ᵐ	−05° 23′	0.3′
charts 11, B2			m_v 4.7	

Four hot, young stars (age 100,000 years) that lie near the center of the Orion Nebula (M42) and illuminate it. Like diamonds on green velvet (Mullaney). Also known as θ^1 Ori, the Trapezium is the core of the Orion Nebula Cluster, which includes about 130 much fainter stars in a 15′ patch of sky. The cluster and nebula lie about 1,500 ly from the Sun.

Triangulum Galaxy

See M33.

Trifid Nebula

BNer+C Sgr 18ʰ 02.6ᵐ −23° 02′ 28′
chart 22 m_v 6.3

M20, NGC 6514. Very bright, very large nebula; trisected by prominent dark lanes at NE, S, and W. Contains a very bright, very large open cluster of 60 stars, the brightest star being mag. 7.3. Per NGC, a (!!!) most remarkable object. Located about 1.5° NNW of the Lagoon Nebula (M8).

Tuft in the Tail of the Dog

See Cr 140.

UGC 1281

Gx Tri 01ʰ 49.5ᵐ +32° 36′ 4.5′ × 0.8′
chart 4 m_v 12.4 p.a. 38°

Distance 15 million ly; Hubble class Sc.

UGC 1886

Gx And 02ʰ 26.0ᵐ +39° 28′ 3.7′ × 2.0′
charts 1, 4 m_p 12.9 p.a. 35°

Distance 220 million ly; Hubble class SBb.

UGC 2296

Gx Ari 02ʰ 49.2ᵐ +18° 20′ 0.7′ × 0.6′
charts 4, 10 m_p 13.0 p.a. 90°

Distance 440 million ly.

UGC 3580

Gx Cam 06ʰ 55.5ᵐ +69° 34′ 3.5′ × 1.8′
chart 1 m_v 11.8 p.a. 3°

Distance 59 million ly; Hubble class SBa.

UGC 3714

Gx Cam 07ʰ 12.6ᵐ +71° 45′ 1.5′ × 1.3′
chart 1 m_p 12.9 p.a. 35°

Distance 140 million ly; Hubble class Sc.

UGC 3828

Gx Lyn 07ʰ 24.6ᵐ +57° 58′ 1.7′ × 0.9′
chart 1 m_p 12.8 p.a. 0°

Distance 140 million ly; Hubble class SBb.

UGC 4305

Gx UMa 08ʰ 19.1ᵐ +70° 43′ 7.6′ × 6.2′
charts 1, 2 m_v 10.7 p.a. 15°

Holmberg II. Large and irregular; compact core with many knots. Distance 13 million ly; Hubble class Sm.

UGC 4375

Gx Cnc 08ʰ 23.2ᵐ +22° 40′ 2.5′ × 1.6′
chart 6 m_p 12.8 p.a. 0°

Distance 86 million ly; Hubble class Sc.

UGC 4841

Gx Cam 09ʰ 14.8ᵐ +74° 14′ 3.1′ × 2.5′
charts 1, 2 m_v 12.4 p.a. 150°

Holmberg III. Faint, with a very small nucleus. Distance 56 million ly; Hubble class SBc.

UGC 4883

Gx Cam 09ʰ 18.3ᵐ +74° 20′ 1.3′ × 0.6′
charts 1, 2 m_p 13.0 p.a. 27°

Distance 130 million ly; Hubble class Sb.

UGC 5364

Gx Leo 09ʰ 59.4ᵐ +30° 45′ 4.9′ × 3.2′
chart 6 m_p 13.0 p.a. 90°

Distance 1 million ly; Hubble class SBm.

UGC 5373

Gx   Sex $10^h 00.0^m$ +05° 20′   5.0′ × 3.5′
charts 12, 13 m_p 11.8   p.a. 110°

Distance 7 million ly; Hubble class Ir.

UGC 5470

See Leo I.

UGC 5612

Gx   UMa $10^h 24.1^m$ +70° 53′   3.4′ × 2.2′
charts 1, 2 m_p 12.7   p.a. 165°

Distance 49 million ly; Hubble class SBc.

UGC 5720

Gx   UMa $10^h 32.5^m$ +54° 24′   1.0′ × 0.9′
charts 2, 6 m_p 13.0   p.a. 125°

Distance 66 million ly; Hubble class S.

UGC 6253

Gx   Leo $11^h 13.5^m$ +22° 09′   12′ × 10′
chart 6 m_v 11.5   p.a. 0°

Leo II. Dwarf galaxy of very low surface brightness. Extends 2,600 ly; distance 750,000 ly; Hubble class dE0. A member of the Local Group.

UGC 6887

Gx   Leo $11^h 55.2^m$ +22° 42′   1.1′ × 0.5′
charts 6, 7 m_p 10.6   p.a. 157°

Distance 300 million ly; Hubble class SBc.

UGC 6930

Gx   UMa $11^h 57.3^m$ +49° 17′   4.3′ × 3.2′
charts 2, 6, 7 m_p 12.7   p.a. 40°

Distance 36 million ly; Hubble class SBc.

UGC 7577

Gx   CVn $12^h 27.7^m$ +43° 30′   4.2′ × 2.6′
charts 6, 7 m_p 12.9   p.a. 130°

Distance 11 million ly; Hubble class Sm.

UGC 8201

Gx   Dra $13^h 06.4^m$ +67° 42′   3.5′ × 2.1′
chart 2 m_p 12.9   p.a. 90°

Distance 7 million ly; Hubble class Ir.

UGC 8320

Gx   CVn $13^h 14.5^m$ +45° 55′   3.6′ × 1.4′
chart 7 m_p 12.7   p.a. 150°

Distance 12 million ly; Hubble class Ir.

UGC 10822

See Draco Dwarf.

UGC 11453

Gx   Cyg $19^h 31.1^m$ +54° 06′   1.8′ × 1.3′
charts 3, 8 m_p 12.9   p.a. 80°

Distance 180 million ly; Hubble class Sb.

UGC 11920

Gx   Lac $22^h 08.5^m$ +48° 26′   2.7′ × 1.7′
chart 9 m_p 12.9   p.a. 45°

Distance 60 million ly; Hubble class SBa.

UGC 12613

Gx   Peg $23^h 28.6^m$ +14° 45′   4.6′ × 3.0′
charts 9, 17 m_p 12.8   p.a. 120°

Distance 1 million ly; Hubble class Sm.

vdB 1

BNr   Cas $00^h 11.0^m$ +58° 46′   5′
charts 1, 3

Three stars involved in nebulosity.

vdB 8

BNr   Cas $02^h 51.6^m$ +67° 52′   4′ × 1′
chart 1

An 8.6-mag. star involved in nebulosity extending NW–SE.

vdB 14

BNr   Cam $03^h 29.2^m$ +59° 57′   20′ × 8′
chart 1

Bluish nebula.

vdB 15

BNr Cam 03^h 30.1^m +58° 54′ 25′ × 10′
chart 1

Bluish; slightly brighter than vdB 14.

vdB 16

BNr Ari 03^h 28.3^m +29° 43′ 4.5′
chart 4

A 9.1-mag. star involved in nebulosity.

vdB 24

BNr Per 03^h 49.6^m +38° 59′ 5′ × 3′
charts 4, 5

Nebula whose brightest part is comet shaped and fans away from an 8.8-mag. star located immediately to N.

vdB 26

BNr Tau 04^h 13.6^m +10° 13′ 6′
chart 11

A 7.2-mag. star involved in nebulosity.

vdB 29

BNr Tau 04^h 48.4^m +29° 47′ 7′ × 5′
chart 5

A 6.5-mag. star involved in the E side of a diffuse nebulosity.

vdB 31

BNr Aur 04^h 55.7^m +30° 33′ 8′ × 5′
chart 5

A 6.8-mag. star involved in nebulosity.

vdB 37

BNr Ori 05^h 18.2^m +13° 24′ 7′ × 4′
chart 11

An 8.2-mag. star involved in the SE portion of nebulosity.

vdB 38

BNr Ori 05^h 21.7^m +08° 27′ 30′ × 25′
chart 11

A 5.8-mag. star involved in the NE edge of an irregular nebulosity.

vdB 49

BNr Ori 05^h 39.2^m +04° 10′ 6′
chart 11

A fairly bright nebula on red-sensitive emulsions, even more so on blue. Illuminated by ω Ori.

vdB 96

BNr CMa 07^h 19.6^m −23° 58′ 10′ × 5′
chart 19

Nebulosity involving a chain of three 9th-mag. stars; brightest around the star to NW.

vdB 97

BNr Pup 07^h 32.6^m −16° 54′ 2′
charts 12, 19

A 9.9-mag. star involved in the E edge of a small, round nebulous patch.

vdB 98

BNr Pup 07^h 36.4^m −25° 20′ 10′
chart 19

A 7.3-mag. star involved in nebulosity.

vdB 107

BNr Sco 16^h 29.2^m −26° 27′ 90′ × 80′
chart 22

Ced 132. Very red nebulosity shining by the reflected light of Antares (α Sco).

vdB 111

BNr Oph 17^h 19.0^m +06° 05′ 12′
chart 15

Predominantly blue nebulosity.

vdB 123

BNr Ser 18^h 30.5^m +01° 11′ 3′ × 2′
charts 15, 16

Nebulosity involving a 9.1-mag. star; brightest part is WNW of the star.

vdB 126

BNr Vul $19^h 27.1^m$ +22° 43′ 7′ × 5′
chart 8

An 8.3-mag. star involved in nebulosity.

vdB 128

BNr Cyg $20^h 04.6^m$ +32° 15′ 7′
charts 8, 9

A 5.6-mag. star involved in nebulosity.

vdB 133

BNr Cyg $20^h 30.7^m$ +36° 56′ 10′
charts 8, 9

A 6.2-mag. star involved in the SE portion of nebulosity.

vdB 140

BNr Cep $21^h 17.5^m$ +58° 36′ 12′ × 10′
chart 3

A 6.4-mag. star involved in the SW portion of nebulosity.

vdB 142

BNr Cep $21^h 37.1^m$ +57° 29′ 1′
chart 3

An 8.8-mag. star involved in nebulosity.

vdB 143

BNr Cep $21^h 37.1^m$ +68° 12′ 7′
chart 3

An 8.3-mag. star involved in nebulosity that is more or less diamond shaped.

vdB 145

BNr Cyg $21^h 43.7^m$ +48° 55′ 2′
chart 9

A 7.4-mag. star involved in nebulosity.

vdB 152

BNr Cep $22^h 13.6^m$ +70° 18′ 4′ × 3′
chart 3

An 8.8-mag. star involved in nebulosity; a dust ball at the edge of a small, very dense cloud.

vdBH 63

BNr Cir $14^h 49.4^m$ −65° 14′ 1.5′
charts 25, 26

A star involved in the W side of a small, roundish, well-defined nebulosity.

vdBH 65a

BNr Cir $15^h 01.1^m$ −63° 17′ 0.3′
charts 25, 26

Lies 0.7° SE of star θ Cir (mag. 5.1).

vdBH 81

BNr Ara $17^h 04.0^m$ −51° 05′ 6′
charts 22, 26

Comet-shaped nebula. Its brightest part forms an equilateral triangle with a brighter star about 2′ SW and a fainter star about 2′ NW.

Veil Nebula

BNe Cyg $20^h 50.0^m$ +31° 00′ 3.0° × 2.5°
chart 9

Also the Cirrus Nebula. Very faint and extremely large; best seen with an O III filter and an eyepiece of very low power and wide field. Irregular and extensive streams of faint nebulosity, like a telescopic Milky Way (Denning). Distance 1,300 ly. Parts of the Veil have individual designations and names: NGC 6960 (Filamentary Nebula), NGC 6979, NGC 6992 (Network Nebula), NGC 6995, and IC 1340. A truly incredible object!

Vela SNR

BN Vel $08^h 35.0^m$ −44° 00′ 20′ × 12′
charts 19, 20

Supernova remnant; extremely large, very filamentary structure.

Virgo A

See M87.

Wa 6

OC	Vel	08^h 40.4^m	−46° 09′	2.2′
charts 19, 20			m$_v$ 8.4	

8 stars; detached, weak concentration; large brightness range; mag. of brightest star 9.2. Distance 6,200 ly; age 32 million years; Trumpler class II 2 p.

Whirlpool Galaxy

Gx	CVn	13^h 29.9^m	+47° 12′	11.0′ × 7.8′
chart 7			m$_p$ 8.1	p.a. 163°

M51, NGC 5194–5. Two clouds that touch, one being fainter than the other (Messier). This is the first galaxy in which spiral structure was observed, by Lord Rosse in 1845. Several of the spiral arms can be traced visually with a 10- or 12-inch reflector under good conditions. Per *NGC*, a (!!!) most remarkable object. Distance 37 million ly. The main galaxy, NGC 5194, is of Hubble class Sc, and it is physically connected to the small neighboring galaxy, NGC 5195. A 12th-mag. foreground star, lying only 1.5′ SW of NGC 5194's nucleus, is often mistaken for a supernova.

White-Eyed Pea

See IC 4593.

Wild Duck Cluster

See M11.

Winnecke 4

See M40.

Witch's Head Nebula

See IC 2118.

Zeta Sculptoris Cluster

See Blanco 1.

Chart Key

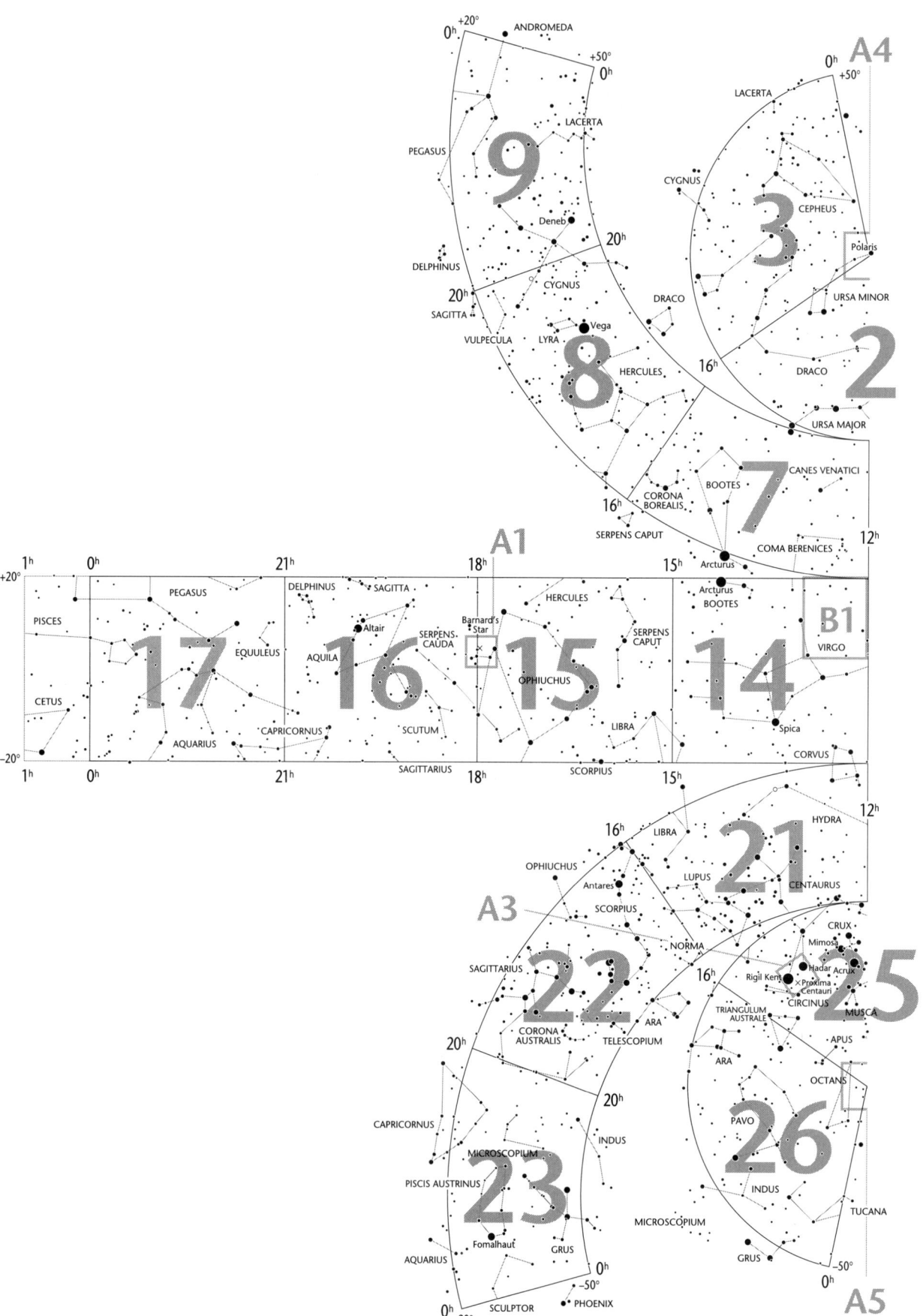

Chart Key

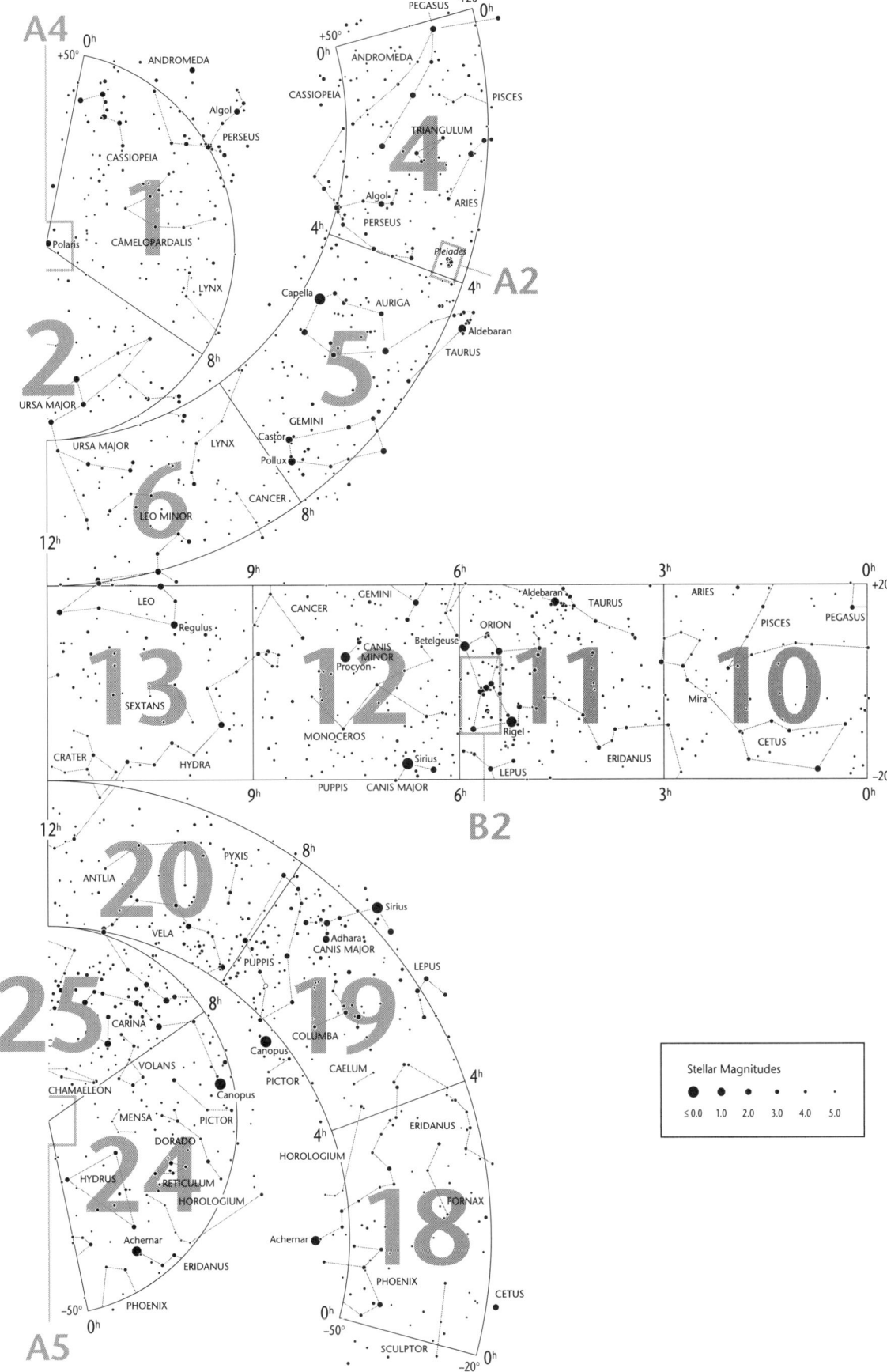

Appendix:
Objects Listed by Chart Number

Object	Type	R.A. h m	Dec. ° ′	Mag.	Const.	Object	Type	R.A. h m	Dec. ° ′	Mag.	Const.
Chart 1 B8, 9, 11, 13	DN	04 19.0	+55 03		Cam	NGC 188	OC	00 44.4	+85 20	8.1	Cep
B12	DN	04 30.0	+54 17		Cam	NGC 225	OC	00 43.4	+61 47	7	Cas
Barbell Nebula (*see* M76).						NGC 281	BNe+C	00 52.8	+56 37		Cas
Bode's Nebulae (*see* M81 and M82).						NGC 457	OC	01 19.1	+58 20	6.4	Cas
C1 (*see* NGC 188).						NGC 559	OC	01 29.5	+63 18	9.5	Cas
C2 (*see* NGC 40).						NGC 581 (*see* M103).					
C7 (*see* NGC 2403).						NGC 637	OC	01 42.9	+64 00	8.2	Cas
C8 (*see* NGC 559).						NGC 650-1 (*see* M76).					
C10 (*see* NGC 663).						NGC 654	OC	01 44.1	+61 53	6.5	Cas
C13 (*see* NGC 457).						NGC 659	OC	01 44.2	+60 42	7.9	Cas
C14 (*see* Double Cluster).						NGC 663	OC	01 46.0	+61 15	7.1	Cas
C23 (*see* NGC 891).						NGC 744	OC	01 58.4	+55 29	7.9	Per
C24 (*see* NGC 1275).						NGC 812	Gx	02 06.9	+44 34	11.2	And
Cassiopeia Nebula (*see* IC 59, IC 63).						NGC 869	OC	02 19.0	+57 09	5.3	Per
Ced 214	BNer	00 04.7	+67 10		Cep	NGC 884	OC	02 22.4	+57 07	6.1	Per
Cork Nebula (*see* M76).						NGC 891	Gx	02 22.6	+42 21	9.9	And
Cr 463	OC	01 47.7	+71 46	5.7	Cas	NGC 896	BNe	02 24.8	+61 54		Cas
Double Cluster	OC	02 20.5	+57 08	5	Per	NGC 957	OC	02 33.6	+57 32	7.6	Per
Embryo Nebula (*see* IC 1848).						NGC 1003	Gx	02 39.3	+40 52	11.4	Per
ET Cluster (*see* NGC 457).						NGC 1023	Gx	02 40.4	+39 04	9.3	Per
Heart Nebula (*see* IC 1805).						NGC 1027	OC	02 42.7	+61 33	6.7	Cas
IC 10	Gx	00 20.4	+59 18	11.3	Cas	NGC 1039 (*see* M34).					
IC 59	BNer	00 56.7	+61 04		Cas	NGC 1058	Gx	02 43.5	+37 21	11.2	Per
IC 63	BNer	00 59.5	+60 49		Cas	NGC 1122	Gx	02 52.9	+42 12	12.1	Per
IC 239	Gx	02 36.5	+38 58	11.1	And	NGC 1161	Gx	03 01.2	+44 54	11	Per
IC 284	Gx	03 06.2	+42 22	11.5	Per	NGC 1169	Gx	03 03.6	+46 23	11.2	Per
IC 289	PN	03 10.3	+61 19	13.2	Cas	NGC 1186	Gx	03 05.5	+42 50	11.4	Per
IC 334	Gx	03 45.3	+76 38	11.3	Cam	NGC 1245	OC	03 14.7	+47 15	8.4	Per
IC 342	Gx	03 46.8	+68 06	8.4	Cam	NGC 1272	Gx	03 19.4	+41 30	11.8	Per
IC 356	Gx	04 07.8	+69 49	10.5	Cam	NGC 1275	Gx	03 19.8	+41 31	11.9	Per
IC 391	Gx	04 57.4	+78 11	12.7	Cam	NGC 1444	OC	03 49.4	+52 40	6.6	Per
IC 520	Gx	08 53.7	+73 29	11.7	Cam	NGC 1491	BNe	04 03.4	+51 19		Per
IC 529	Gx	09 18.5	+73 46	11.9	Cam	NGC 1501	PN	04 07.0	+60 55	11.5	Cam
IC 1747	PN	01 57.6	+63 19	12.1	Cas	NGC 1502	OC	04 07.7	+62 20	5.7	Cam
IC 1805	BNe+C	02 33.4	+61 26		Cas	NGC 1513	OC	04 10.0	+49 31	8.4	Per
IC 1848	BNe+C	02 51.3	+60 25		Cas	NGC 1528	OC	04 15.4	+51 14	6.4	Per
IC 1871	BNe	03 03.2	+60 29		Cas	NGC 1530	Gx	04 23.5	+75 18	11.4	Cam
K14	OC	00 31.9	+63 10	8.5	Cas	NGC 1545	OC	04 20.9	+50 15	6.2	Per
Kemble's Cascade		03 57.0	+63 00		Cam	NGC 1560	Gx	04 32.8	+71 53	11.4	Cam
Letter S Cluster (*see* NGC 663).						NGC 1569	Gx	04 30.8	+64 51	11	Cam
Little Dumbbell Nebula (*see* M76).						NGC 1573	Gx	04 35.0	+73 16	11.7	Cam
M34	OC	02 42.0	+42 47	5.2	Per	NGC 1624	BNe	04 40.5	+50 27		Per
M76	PN	01 42.4	+51 34	10.1	Per	NGC 1961	Gx	05 42.1	+69 23	11	Cam
M81	Gx	09 55.6	+69 04	6.8	UMa	NGC 2146	Gx	06 18.7	+78 21	10.6	Cam
M82	Gx	09 55.8	+69 41	8.4	UMa	NGC 2258	Gx	06 47.8	+74 29	11.9	Cam
M103	OC	01 33.2	+60 42	7.4:	Cas	NGC 2268	Gx	07 14.3	+84 23	11.5	Cam
Magnificent Cluster (*see* NGC 7789).						NGC 2273	Gx	06 50.1	+60 51	11.7	Lyn
Mrk 6	OC	02 29.6	+60 39	7.1	Cas	NGC 2276	Gx	07 27.2	+85 45	11.4	Cep
NGC 40	PN	00 13.0	+72 31	12.4	Cep	NGC 2300	Gx	07 32.3	+85 43	11	Cep
NGC 129	OC	00 29.9	+60 14	6.5	Cas	NGC 2320	Gx	07 05.7	+50 35	11.9	Lyn

Charts 1 – 2

Object	Type	R.A. h m	Dec. ° ′	Mag.	Const.
NGC 2336	Gx	07 27.1	+80 11	10.4	Cam
NGC 2340	Gx	07 11.2	+50 10	11.7	Lyn
NGC 2366	Gx	07 28.9	+69 13	10.8	Cam
NGC 2403	Gx	07 36.9	+65 36	8.5	Cam
NGC 2441	Gx	07 51.9	+73 01	12.2	Cam
NGC 2460	Gx	07 56.9	+60 21	11.8	Cam
NGC 2500	Gx	08 01.9	+50 44	11.6	Lyn
NGC 2523	Gx	08 15.0	+73 35	11.9	Cam
NGC 2549	Gx	08 19.0	+57 48	11.2	Lyn
NGC 2591	Gx	08 37.4	+78 02	12.2	Cam
NGC 2633	Gx	08 48.1	+74 06	12.2	Cam
NGC 2634	Gx	08 48.4	+73 58	12	Cam
NGC 2655	Gx	08 55.6	+78 13	10.1	Cam
NGC 2715	Gx	09 08.1	+78 05	11.2	Cam
NGC 2732	Gx	09 13.4	+79 11	11.9	Cam
NGC 2748	Gx	09 13.7	+76 29	11.7	Cam
NGC 2787	Gx	09 19.3	+69 12	10.8	UMa
NGC 2976	Gx	09 47.3	+67 55	10.2	UMa
NGC 2985	Gx	09 50.3	+72 17	10.4	UMa
NGC 3027	Gx	09 55.7	+72 12	11.8	UMa
NGC 3031 (see M81).					
NGC 3034 (see M82).					
NGC 3077	Gx	10 03.4	+68 44	9.8	UMa
NGC 3147	Gx	10 16.9	+73 24	10.6	Dra
NGC 3183	Gx	10 21.8	+74 11	12.7p	Dra
NGC 3329	Gx	10 44.7	+76 49	12.2	Dra
NGC 3348	Gx	10 47.2	+72 50	11.2	UMa
NGC 3516	Gx	11 06.8	+72 34	11.7	UMa
NGC 4291	Gx	12 20.3	+75 22	11.5	Dra
NGC 4386	Gx	12 24.5	+75 32	11.7	Dra
NGC 7789	OC	23 57.0	+56 44	6.7	Cas
NGC 7790	OC	23 58.4	+61 13	8.5	Cas
NGC 7822	BNe	00 03.6	+68 37		Cep
Owl Cluster (see NGC 457).					
Pazmino's Cluster (see St 23).					
Perseus A (see NGC 1275).					
PK119-6.1	PN	00 28.3	+55 58	12.3	Cas
PK158+17.1	PN	06 19.6	+55 37	11.2p	Lyn
PK164+31.1	PN	07 57.8	+53 25	12.1	Lyn
Sh2-188	BNe	01 30.6	+58 22		Cas
Sh2-205	BNe	03 56.1	+53 12		Cam
St 2	OC	02 15.0	+59 16	4.4	Cas
St 23	OC	03 16.3	+60 02		Cam
Tom 5	OC	03 47.8	+59 03	8.4	Cam
Tr 1	OC	01 35.7	+61 17	8.1	Cas
Tr 2	OC	02 37.3	+55 59	5.9	Per
Tr 3	OC	03 11.8	+63 09	7.0p	Cas
UGC 1886	Gx	02 26.0	+39 28	12.9p	And
UGC 3580	Gx	06 55.5	+69 34	11.8	Cam
UGC 3714	Gx	07 12.6	+71 45	12.9p	Cam

Object	Type	R.A. h m	Dec. ° ′	Mag.	Const.
UGC 3828	Gx	07 24.6	+57 58	12.8p	Lyn
UGC 4305	Gx	08 19.1	+70 43	10.7	UMa
UGC 4841	Gx	09 14.8	+74 14	12.4	Cam
UGC 4883	Gx	09 18.3	+74 20	13.0p	Cam
UGC 5612	Gx	10 24.1	+70 53	12.7p	UMa
vdB 1	BNr	00 11.0	+58 46		Cas
vdB 8	BNr	02 51.6	+67 52		Cas
vdB 14	BNr	03 29.2	+59 57		Cam
vdB 15	BNr	03 30.1	+58 54		Cam

Chart 2

Bode's Nebulae (see M81 and M82).
C3 (see NGC 4236).
Coddington's Nebula (see IC 2574).
Helix Galaxy (see NGC 2685).

Object	Type	R.A. h m	Dec. ° ′	Mag.	Const.
IC 520	Gx	08 53.7	+73 29	11.7	Cam
IC 529	Gx	09 18.5	+73 46	11.9	Cam
IC 694	Gx	11 28.5	+58 33	12.0p	UMa
IC 1029	Gx	14 32.5	+49 54	12.2p	Boo
IC 2574	Gx	10 28.4	+68 25	10.4	UMa
IC 3568	PN	12 32.9	+82 33	10.6	Cam
M40	—	12 22.2	+58 05	9.1	UMa
M81	Gx	09 55.6	+69 04	6.8	UMa
M82	Gx	09 55.8	+69 41	8.4	UMa
M97 (see Owl Nebula).					
M101	Gx	14 03.2	+54 21	7.7	UMa
M102 (same object as M101).					
M106	Gx	12 19.0	+47 18	8.3	CVn
M108	Gx	11 11.5	+55 40	10	UMa
M109	Gx	11 57.6	+53 23	9.8	UMa
NGC 2441	Gx	07 51.9	+73 01	12.2	Cam
NGC 2460	Gx	07 56.9	+60 21	11.8	Cam
NGC 2500	Gx	08 01.9	+50 44	11.6	Lyn
NGC 2523	Gx	08 15.0	+73 35	11.9	Cam
NGC 2549	Gx	08 19.0	+57 48	11.2	Lyn
NGC 2552	Gx	08 19.3	+50 00	12.1	Lyn
NGC 2591	Gx	08 37.4	+78 02	12.2	Cam
NGC 2633	Gx	08 48.1	+74 06	12.2	Cam
NGC 2634	Gx	08 48.4	+73 58	12	Cam
NGC 2639	Gx	08 43.6	+50 12	11.7	UMa
NGC 2654	Gx	08 49.2	+60 13	11.8	UMa
NGC 2655	Gx	08 55.6	+78 13	10.1	Cam
NGC 2681	Gx	08 53.6	+51 19	10.3	UMa
NGC 2685	Gx	08 55.6	+58 44	11.3	UMa
NGC 2693	Gx	08 57.0	+51 21	11.9	UMa
NGC 2701	Gx	08 59.1	+53 46	12.3	UMa
NGC 2715	Gx	09 08.1	+78 05	11.2	Cam
NGC 2732	Gx	09 13.4	+79 11	11.9	Cam
NGC 2742	Gx	09 07.6	+60 29	11.4	UMa
NGC 2748	Gx	09 13.7	+76 29	11.7	Cam
NGC 2768	Gx	09 11.6	+60 02	9.9	UMa

Object	Type	R.A. h m	Dec. ° '	Mag.	Const.
NGC 2787	Gx	09 19.3	+69 12	10.8	UMa
NGC 2805	Gx	09 20.3	+64 06	11	UMa
NGC 2841	Gx	09 22.0	+50 59	9.2	UMa
NGC 2857	Gx	09 24.6	+49 21	12.3	UMa
NGC 2880	Gx	09 29.6	+62 29	11.5	UMa
NGC 2950	Gx	09 42.6	+58 51	10.9	UMa
NGC 2976	Gx	09 47.3	+67 55	10.2	UMa
NGC 2985	Gx	09 50.3	+72 17	10.4	UMa
NGC 3027	Gx	09 55.7	+72 12	11.8	UMa
NGC 3031 (see M81).					
NGC 3034 (see M82).					
NGC 3077	Gx	10 03.4	+68 44	9.8	UMa
NGC 3079	Gx	10 02.0	+55 41	10.9	UMa
NGC 3147	Gx	10 16.9	+73 24	10.6	Dra
NGC 3182	Gx	10 19.6	+58 12	12.1	UMa
NGC 3183	Gx	10 21.8	+74 11	12.7p	Dra
NGC 3184	Gx	10 18.3	+41 25	9.8	UMa
NGC 3198	Gx	10 19.9	+45 33	10.3	UMa
NGC 3206	Gx	10 21.8	+56 56	11.9	UMa
NGC 3259	Gx	10 32.6	+65 02	12.1	UMa
NGC 3264	Gx	10 32.3	+56 05	12	UMa
NGC 3294	Gx	10 36.3	+37 19	11.8	LMi
NGC 3310	Gx	10 38.8	+53 30	10.8	UMa
NGC 3319	Gx	10 39.2	+41 41	11.1	UMa
NGC 3320	Gx	10 39.6	+47 24	12.3	UMa
NGC 3329	Gx	10 44.7	+76 49	12.2	Dra
NGC 3348	Gx	10 47.2	+72 50	11.2	UMa
NGC 3359	Gx	10 46.6	+63 13	10.6	UMa
NGC 3381	Gx	10 48.4	+34 43	11.7	LMi
NGC 3432	Gx	10 52.5	+36 37	11.2	LMi
NGC 3445	Gx	10 54.6	+56 59	12.6	UMa
NGC 3448	Gx	10 54.7	+54 18	12.1	UMa
NGC 3516	Gx	11 06.8	+72 34	11.7	UMa
NGC 3549	Gx	11 10.9	+53 23	12.1	UMa
NGC 3556 (see M108).					
NGC 3583	Gx	11 14.2	+48 19	11.1	UMa
NGC 3587 (see Owl Nebula).					
NGC 3595	Gx	11 15.4	+47 27	12.1	UMa
NGC 3600	Gx	11 15.9	+41 35	11.7	UMa
NGC 3610	Gx	11 18.4	+58 47	10.8	UMa
NGC 3613	Gx	11 18.6	+58 00	10.9	UMa
NGC 3614	Gx	11 18.3	+45 45	11.6	UMa
NGC 3614A	Gx	11 18.3	+45 43	11.9	UMa
NGC 3619	Gx	11 19.4	+57 46	11.5	UMa
NGC 3631	Gx	11 21.0	+53 10	10.4	UMa
NGC 3642	Gx	11 22.3	+59 05	11.2	UMa
NGC 3669	Gx	11 25.5	+57 43	12.4	UMa
NGC 3675	Gx	11 26.1	+43 35	10.2	UMa
NGC 3683	Gx	11 27.5	+56 53	12.4	UMa
NGC 3683A	Gx	11 29.2	+57 08	11.9	UMa
NGC 3690	Gx	11 28.6	+58 34	11.5	UMa
NGC 3718	Gx	11 32.6	+53 04	10.8	UMa
NGC 3726	Gx	11 33.3	+47 02	10.4	UMa
NGC 3729	Gx	11 33.8	+53 08	11.4	UMa
NGC 3733	Gx	11 35.0	+54 51	12.4	UMa
NGC 3735	Gx	11 36.0	+70 32	11.8	Dra
NGC 3738	Gx	11 35.8	+54 31	11.7	UMa
NGC 3756	Gx	11 36.8	+54 18	11.5	UMa
NGC 3769	Gx	11 37.7	+47 54	11.8	UMa
NGC 3780	Gx	11 39.4	+56 16	11.5	UMa
NGC 3811	Gx	11 41.3	+47 42	12.3	UMa
NGC 3877	Gx	11 46.1	+47 30	11	UMa
NGC 3888	Gx	11 47.6	+55 58	12.1	UMa
NGC 3893	Gx	11 48.7	+48 43	10.5	UMa
NGC 3894	Gx	11 48.9	+59 25	11.6	UMa
NGC 3898	Gx	11 49.3	+56 05	10.7	UMa
NGC 3917	Gx	11 50.8	+51 50	11.8	UMa
NGC 3938	Gx	11 52.8	+44 07	10.4	UMa
NGC 3945	Gx	11 53.2	+60 41	10.8	UMa
NGC 3949	Gx	11 53.7	+47 52	11.1	UMa
NGC 3953	Gx	11 53.8	+52 20	10.1	UMa
NGC 3963	Gx	11 55.0	+58 30	11.9	UMa
NGC 3972	Gx	11 55.8	+55 19	12.3	UMa
NGC 3982	Gx	11 56.5	+55 07	11	UMa
NGC 3992 (see M109).					
NGC 3998	Gx	11 57.9	+55 27	10.7	UMa
NGC 4026	Gx	11 59.4	+50 58	10.8	UMa
NGC 4036	Gx	12 01.5	+61 54	10.7	UMa
NGC 4041	Gx	12 02.2	+62 08	11.3	UMa
NGC 4047	Gx	12 02.9	+48 38	12.2	UMa
NGC 4051	Gx	12 03.2	+44 32	10.2	UMa
NGC 4085	Gx	12 05.4	+50 21	12.4	UMa
NGC 4088	Gx	12 05.6	+50 33	10.6	UMa
NGC 4096	Gx	12 06.0	+47 29	10.8	UMa
NGC 4100	Gx	12 06.1	+49 35	11.2	UMa
NGC 4102	Gx	12 06.4	+52 43	11.2	UMa
NGC 4125	Gx	12 08.1	+65 10	9.5	Dra
NGC 4128	Gx	12 08.5	+68 46	12	Dra
NGC 4144	Gx	12 10.0	+46 27	11.6	UMa
NGC 4157	Gx	12 11.1	+50 29	11.3	UMa
NGC 4194	Gx	12 14.2	+54 32	12.5	UMa
NGC 4217	Gx	12 15.8	+47 06	11.2	CVn
NGC 4220	Gx	12 16.2	+47 53	11.4	CVn
NGC 4236	Gx	12 16.7	+69 28	9.6	Dra
NGC 4248	Gx	12 17.8	+47 25	12.5	CVn
NGC 4256	Gx	12 18.7	+65 54	11.9	Dra
NGC 4258 (see M106).					
NGC 4290	Gx	12 20.8	+58 06	11.8	UMa
NGC 4291	Gx	12 20.3	+75 22	11.5	Dra
NGC 4346	Gx	12 23.5	+47 00	11.1	CVn

Charts 2 – 3

Object	Type	R.A. h m	Dec. ° ′	Mag.	Const.
NGC 4386	Gx	12 24.5	+75 32	11.7	Dra
NGC 4589	Gx	12 37.4	+74 12	10.7	Dra
NGC 4605	Gx	12 40.0	+61 37	10.3	UMa
NGC 4648	Gx	12 41.8	+74 25	12	Dra
NGC 4750	Gx	12 50.1	+72 53	11.2	Dra
NGC 4814	Gx	12 55.4	+58 21	12	UMa
NGC 5204	Gx	13 29.6	+58 25	11.3	UMa
NGC 5308	Gx	13 47.0	+60 58	11.4	UMa
NGC 5322	Gx	13 49.3	+60 11	10.2	UMa
NGC 5376	Gx	13 55.3	+59 31	12.1	UMa
NGC 5389	Gx	13 56.1	+59 45	12.1	UMa
NGC 5422	Gx	14 00.7	+55 10	11.8	UMa
NGC 5430	Gx	14 00.8	+59 20	11.9	UMa
NGC 5457 (see M101).					
NGC 5473	Gx	14 04.7	+54 54	11.4	UMa
NGC 5474	Gx	14 05.0	+53 40	10.8	UMa
NGC 5480	Gx	14 06.4	+50 44	12.1	UMa
NGC 5485	Gx	14 07.2	+55 00	11.4	UMa
NGC 5585	Gx	14 19.8	+56 44	10.7	UMa
NGC 5631	Gx	14 26.6	+56 35	11.5	UMa
NGC 5673	Gx	14 31.5	+49 58	12.1	Boo
NGC 5678	Gx	14 32.1	+57 55	11.3	Dra
NGC 5687	Gx	14 34.9	+54 29	11.8	Boo
NGC 5866	Gx	15 06.5	+55 46	9.9	Dra
NGC 5879	Gx	15 09.8	+57 00	11.6	Dra
NGC 5907	Gx	15 15.9	+56 20	10.3	Dra
NGC 5908	Gx	15 16.7	+55 25	11.8	Dra
NGC 5949	Gx	15 28.0	+64 46	12	Dra
NGC 5963	Gx	15 33.5	+56 34	12.5	Dra
NGC 5965	Gx	15 34.0	+56 41	11.7	Dra
NGC 5982	Gx	15 38.7	+59 21	11.1	Dra
NGC 5985	Gx	15 39.6	+59 20	11.1	Dra
NGC 5987	Gx	15 40.0	+58 05	11.7	Dra
NGC 6015	Gx	15 51.4	+62 19	11.1	Dra
NGC 6068	Gx	15 55.4	+79 00	12.8	UMi
NGC 6127	Gx	16 19.2	+57 59	12	Dra
NGC 6140	Gx	16 21.0	+65 24	11.3	Dra
NGC 6217	Gx	16 32.7	+78 12	11.2	UMi
NGC 6223	Gx	16 43.1	+61 35	11.8	Dra
NGC 6340	Gx	17 10.4	+72 18	11	Dra
NGC 6395	Gx	17 26.5	+71 06	12.3	Dra
NGC 6412	Gx	17 29.6	+75 42	11.8	Dra
NGC 6503	Gx	17 49.5	+70 09	10.2	Dra
NGC 6643	Gx	18 19.8	+74 34	11.1	Dra
NGC 6654	Gx	18 24.1	+73 11	12	Dra
Owl Nebula	PN	11 14.8	+55 01	9.9	UMa
PK164+31.1	PN	07 57.8	+53 25	12.1	Lyn
UGC 4305	Gx	08 19.1	+70 43	10.7	UMa
UGC 4841	Gx	09 14.8	+74 14	12.4	Cam
UGC 4883	Gx	09 18.3	+74 20	13.0p	Cam
UGC 5612	Gx	10 24.1	+70 53	12.7p	UMa
UGC 5720	Gx	10 32.5	+54 24	13.0p	UMa
UGC 6930	Gx	11 57.3	+49 17	12.7p	UMa
UGC 8201	Gx	13 06.4	+67 42	12.9p	Dra
Winnecke 4 (see M40).					

Chart 3

Object	Type	R.A. h m	Dec. ° ′	Mag.	Const.
B148,9	DN	20 49.1	+59 32		Cep
B150	DN	20 50.6	+60 18		Cep
B152	DN	21 14.5	+61 45		Cep
B157	DN	21 33.7	+54 40		Cyg
B160	DN	21 38.0	+56 14		Cep
B161	DN	21 40.3	+57 49		Cep
B162	DN	21 41.1	+56 19		Cep
B163	DN	21 42.2	+56 42		Cep
B164	DN	21 46.5	+51 04		Cyg
B169,70,71	DN	21 58.9	+58 45		Cep
B173,4	DN	22 07.4	+59 20		Cep
B362	DN	21 24.0	+50 10		Cyg
B364	DN	21 33.6	+54 33		Cyg
B365	DN	21 34.9	+56 43		Cep
B367	DN	21 44.4	+57 12		Cep
Blinking Planetary (see NGC 6826).					
Bubble Nebula (see NGC 7635).					
C1 (see NGC 188).					
C2 (see NGC 40).					
C4 (see NGC 7023).					
C6 (see NGC 6543).					
C9 (see Sh2-155).					
C11 (see NGC 7635).					
C12 (see NGC 6946).					
C15 (see NGC 6826).					
Cat's Eye Nebula (see NGC 6543).					
Cave Nebula (see Sh2-155).					
Ced 214	BNer	00 04.7	+67 10		Cep
Cr 463	OC	01 47.7	+71 46	5.7	Cas
Delta Lyrae Cluster (see Steph 1).					
Draco Dwarf	Gx	17 20.2	+57 55	9.9	Dra
Hole in a Cluster (see NGC 6811).					
IC 10	Gx	00 20.4	+59 18	11.3	Cas
IC 334	Gx	03 45.3	+76 38	11.3	Cam
IC 1396	BNe+C	21 39.1	+57 30		Cep
IC 1470	BNe	23 05.2	+60 15		Cep
IC 5217	PN	22 23.9	+50 58	11.3	Lac
K14	OC	00 31.9	+63 10	8.5	Cas
M52	OC	23 24.2	+61 35	6.9	Cas
Magnificent Cluster (see NGC 7789).					
Mrk 50	OC	23 15.3	+60 28	8.5	Cep
NGC 40	PN	00 13.0	+72 31	12.4	Cep
NGC 129	OC	00 29.9	+60 14	6.5	Cas
NGC 188	OC	00 44.4	+85 20	8.1	Cep

Object	Type	R.A. h m	Dec. ° ′	Mag.	Const.
NGC 225	OC	00 43.4	+61 47	7	Cas
NGC 1530	Gx	04 23.5	+75 18	11.4	Cam
NGC 6015	Gx	15 51.4	+62 19	11.1	Dra
NGC 6068	Gx	15 55.4	+79 00	12.8	UMi
NGC 6127	Gx	16 19.2	+57 59	12	Dra
NGC 6140	Gx	16 21.0	+65 24	11.3	Dra
NGC 6217	Gx	16 32.7	+78 12	11.2	UMi
NGC 6223	Gx	16 43.1	+61 35	11.8	Dra
NGC 6340	Gx	17 10.4	+72 18	11	Dra
NGC 6395	Gx	17 26.5	+71 06	12.3	Dra
NGC 6411	Gx	17 35.5	+60 49	11.8	Dra
NGC 6412	Gx	17 29.6	+75 42	11.8	Dra
NGC 6503	Gx	17 49.5	+70 09	10.2	Dra
NGC 6543	PN	17 58.6	+66 38	8.1	Dra
NGC 6643	Gx	18 19.8	+74 34	11.1	Dra
NGC 6654	Gx	18 24.1	+73 11	12	Dra
NGC 6701	Gx	18 43.2	+60 39	12.1	Dra
NGC 6703	Gx	18 47.3	+45 33	11.3	Lyr
NGC 6764	Gx	19 08.3	+50 56	11.8	Cyg
NGC 6811	OC	19 38.2	+46 34	6.8	Cyg
NGC 6824	Gx	19 43.7	+56 07	12.2	Cyg
NGC 6826	PN	19 44.8	+50 31	8.8	Cyg
NGC 6833	PN	19 49.7	+48 58	12.1	Cyg
NGC 6884	PN	20 10.4	+46 28	10.9	Cyg
NGC 6939	OC	20 31.4	+60 38	7.8	Cep
NGC 6946	Gx	20 34.9	+60 09	8.8	Cyg
NGC 6951	Gx	20 37.3	+66 06	10.7	Cep
NGC 7008	PN	21 00.6	+54 33	10.7	Cyg
NGC 7023	BNr+C	21 00.5	+68 10		Cep
NGC 7086	OC	21 30.5	+51 35	8.4	Cyg
NGC 7129	BNr	21 42.5	+66 10		Cep
NGC 7133	BNr	21 44.4	+66 12		Cep
NGC 7139	PN	21 46.1	+63 48	13.3	Cep
NGC 7160	OC	21 53.7	+62 36	6.1	Cep
NGC 7235	OC	22 12.6	+57 17	7.7	Cep
NGC 7261	OC	22 20.4	+58 05	8.4	Cep
NGC 7354	PN	22 40.3	+61 17	12.2	Cep
NGC 7380	C+BNe	22 47.0	+58 06		Cep
NGC 7510	OC	23 11.5	+60 34	7.9	Cep
NGC 7538	BNe	23 13.5	+61 31		Cep
NGC 7635	BNe	23 20.7	+61 12		Cas
NGC 7654 (see M52).					
NGC 7686	OC	23 30.2	+49 08	5.6	And
NGC 7789	OC	23 57.0	+56 44	6.7	Cas
NGC 7790	OC	23 58.4	+61 13	8.5	Cas
NGC 7822	BNe	00 03.6	+68 37		Cep
PK107-2.1	PN	22 56.3	+57 09	14.4	Cep
PK111-2.1	PN	23 26.3	+58 11	11.9	Cas
Sh2-129	BNe	21 11.8	+59 57		Cep
Sh2-132	BNe	22 18.7	+56 08		Cep

Object	Type	R.A. h m	Dec. ° ′	Mag.	Const.
Sh2-155	BNe	22 56.8	+62 37		Cep
Sh2-157	BNe	23 16.1	+60 02		Cas
Steph 1	OC	18 54.2	+36 55	3.8	Lyr
UGC 10822 (see Draco Dwarf).					
UGC 11453	Gx	19 31.1	+54 06	12.9p	Cyg
vdB 1	BNr	00 11.0	+58 46		Cas
vdB 140	BNr	21 17.5	+58 36		Cep
vdB 142	BNr	21 37.1	+57 29		Cep
vdB 143	BNr	21 37.1	+68 12		Cep
vdB 152	BNr	22 13.6	+70 18		Cep

Chart 4

Object	Type	R.A. h m	Dec. ° ′	Mag.	Const.
Andromeda Galaxy	Gx	00 42.7	+41 16	3.4	And
Barbell Nebula (see M76).					
Blue Snowball Nebula (see NGC 7662).					
C5 (see IC 349).					
C17 (see NGC 147).					
C18 (see NGC 185).					
C22 (see NGC 7662).					
C23 (see NGC 891).					
C24 (see NGC 1275).					
C28 (see NGC 752).					
C43 (see NGC 7814).					
California Neb.	BNe	04 00.7	+36 37		Per
Cork Nebula (see M76).					
Cr 21	OC	01 50.2	+27 05	8.2p	Tri
IC 239	Gx	02 36.5	+38 58	11.1	And
IC 284	Gx	03 06.2	+42 22	11.5	Per
IC 348	BNr+C	03 44.6	+32 09		Per
IC 349	BN	03 46.3	+23 56		Tau
IC 351	PN	03 47.5	+35 03	11.9	Per
IC 353	BNr?	03 55.0	+25 29		Tau
IC 1727	Gx	01 47.5	+27 20	11.5	Tri
IC 1995	BN	03 50.3	+25 35		Tau
IC 2003	PN	03 56.4	+33 52	11.5	Per
Little Dumbbell Nebula (see M76).					
M31 (see Andromeda Galaxy).					
M32	Gx	00 42.7	+40 52	8.2	And
M33	Gx	01 33.9	+30 39	5.7	Tri
M34	OC	02 42.0	+42 47	5.2	Per
M45 (see Pleiades).					
M76	PN	01 42.4	+51 34	10.1	Per
M110	Gx	00 40.4	+41 41	8	And
Merope Nebula (see IC 349).					
NGC 14	Gx	00 08.8	+15 49	12.1	Peg
NGC 16	Gx	00 09.1	+27 44	12	Peg
NGC 23	Gx	00 09.9	+25 55	12	Peg
NGC 57	Gx	00 15.5	+17 20	11.6	Psc
NGC 147	Gx	00 33.2	+48 30	9.5	Cas
NGC 185	Gx	00 39.0	+48 20	9.2	Cas

Charts 4 – 5

Object	Type	R.A. h m	Dec. ° ′	Mag.	Const.
NGC 205 (see M110).					
NGC 206	BN	00 40.6	+40 44		And
NGC 214	Gx	00 41.5	+25 30	12.2	And
NGC 221 (see M32).					
NGC 224 (see Andromeda Galaxy).					
NGC 266	Gx	00 49.8	+32 17	11.6	Psc
NGC 278	Gx	00 52.1	+47 33	10.8	Cas
NGC 315	Gx	00 57.8	+30 21	11.2	Psc
NGC 404	Gx	01 09.5	+35 43	10.3	And
NGC 410	Gx	01 11.0	+33 09	11.5	Psc
NGC 507	Gx	01 23.7	+33 15	11.2	Psc
NGC 598 (see M33).					
NGC 650-1 (see M76).					
NGC 672	Gx	01 47.9	+27 26	10.9	Tri
NGC 680	Gx	01 49.8	+21 58	11.9	Ari
NGC 691	Gx	01 50.7	+21 46	11.4	Ari
NGC 697	Gx	01 51.3	+22 22	12	Ari
NGC 744	OC	01 58.4	+55 29	7.9	Per
NGC 750	Gx	01 57.5	+33 12	11.9	Tri
NGC 752	OC	01 57.8	+37 41	5.7	And
NGC 753	Gx	01 57.7	+35 55	12.3	And
NGC 772	Gx	01 59.3	+19 00	10.3	Ari
NGC 777	Gx	02 00.3	+31 26	11.4	Tri
NGC 784	Gx	02 01.3	+28 51	11.7	Tri
NGC 812	Gx	02 06.9	+44 34	11.2	And
NGC 890	Gx	02 22.0	+33 16	11.2	Tri
NGC 891	Gx	02 22.6	+42 21	9.9	And
NGC 925	Gx	02 27.3	+33 35	10.1	Tri
NGC 949	Gx	02 30.8	+37 08	11.8	Tri
NGC 959	Gx	02 32.4	+35 30	12.4	Tri
NGC 972	Gx	02 34.2	+29 19	11.4	Ari
NGC 1003	Gx	02 39.3	+40 52	11.4	Per
NGC 1012	Gx	02 39.3	+30 09	12	Ari
NGC 1023	Gx	02 40.4	+39 04	9.3	Per
NGC 1039 (see M34).					
NGC 1058	Gx	02 43.5	+37 21	11.2	Per
NGC 1060	Gx	02 43.3	+32 25	11.8	Tri
NGC 1122	Gx	02 52.9	+42 12	12.1	Per
NGC 1156	Gx	02 59.7	+25 14	11.7	Ari
NGC 1161	Gx	03 01.2	+44 54	11	Per
NGC 1169	Gx	03 03.6	+46 23	11.2	Per
NGC 1186	Gx	03 05.5	+42 50	11.4	Per
NGC 1245	OC	03 14.7	+47 15	8.4	Per
NGC 1272	Gx	03 19.4	+41 30	11.8	Per
NGC 1275	Gx	03 19.8	+41 31	11.9	Per
NGC 1333	BNr	03 29.3	+31 25		Per
NGC 1342	OC	03 31.6	+37 20	6.7	Per
NGC 1444	OC	03 49.2	+52 40	6.6	Per
NGC 1491	BNe	04 03.4	+51 19		Per
NGC 1499 (see California Nebula).					
NGC 1513	OC	04 10.0	+49 31	8.4	Per
NGC 1514	PN	04 09.2	+30 47	10.9	Tau
NGC 1528	OC	04 15.2	+51 14	6.4	Per
NGC 1545	OC	04 20.9	+50 15	6.2	Per
NGC 1582	OC	04 32.0	+43 51	7.0p	Per
NGC 1624	BNe	04 40.5	+50 27		Per
NGC 7640	Gx	23 22.1	+40 51	11.3	And
NGC 7662	PN	23 25.9	+42 33	8.3	And
NGC 7686	OC	23 30.2	+49 08	5.6	And
NGC 7753	Gx	23 47.1	+29 29	12	Peg
NGC 7798	Gx	23 59.4	+20 45	12.4	Peg
NGC 7814	Gx	00 03.3	+16 09	10.6	Peg
NGC 7817	Gx	00 04.0	+20 45	11.8	Peg
Perseus A (see NGC 1275).					
Pinwheel Galaxy (see M33).					
PK107-13.1	PN	23 23.0	+46 54	13.6	And
PK171-25.1	PN	03 53.5	+19 28	15.1	Tau
Pleiades	C+BNr	03 47.0	+24 07	1.2	Tau
Seven Sisters (see Pleiades).					
Sh2-205	BNe	03 56.1	+53 12		Cam
Tr 2	OC	02 37.3	+55 59	5.9	Per
Triangulum Galaxy (see M33).					
UGC 1281	Gx	01 49.5	+32 36	12.4	Tri
UGC 1886	Gx	02 26.0	+39 28	12.9p	And
UGC 2296	Gx	02 49.2	+18 20	13.0p	Ari
vdB 16	BNr	03 28.3	+29 43		Ari
vdB 24	BNr	03 49.6	+38 59		Per

Chart 5

Object	Type	R.A. h m	Dec. ° ′	Mag.	Const.
B8, 9, 11, 13	DN	04 19.0	+55 03		Cam
B12	DN	04 30.0	+54 17		Cam
B29	DN	05 06.2	+31 44		Aur
B34	DN	05 43.5	+32 39		Aur
Bear Paw Galaxy (see NGC 2537).					
C24 (see NGC 1275).					
C25 (see NGC 2419).					
C31 (see IC 405).					
C39 (see NGC 2392).					
C41 (see Hyades).					
California Neb.	BNe	04 00.7	+36 37		Per
Ced 33	BNr	04 27.1	+26 06		Tau
Ced 34	BNr	04 27.2	+22 57		Tau
Ced 62	BNr	06 07.8	+18 41		Ori
Clown Face Nebula (see NGC 2392).					
Cr 62	OC	05 22.5	+41 00	4.2p	Aur
Cr 89	OC	06 18.0	+23 38	5.7p	Gem
Crab Nebula	SNR	05 34.5	+22 01	8.4:	Tau
Eskimo Nebula (see NGC 2392).					
Flaming Star Nebula (see IC 405).					
Hind's Variable Nebula	BNr	04 21.8	+19 32		Tau

Object	Type	R.A. h m	Dec. ° ′	Mag.	Const.
Hyades	OC	04 27.0	+15 50	0.5	Tau
IC 348	BNr+C	03 44.6	+32 09		Per
IC 351	PN	03 47.5	+35 03	11.9	Per
IC 353	BNr?	03 55.0	+25 29		Tau
IC 405	BNer	05 16.2	+34 16		Aur
IC 410	BNe	05 22.6	+33 31		Aur
IC 417	BNe	05 28.1	+34 26		Aur
IC 443	BNe	06 16.9	+22 47		Gem
IC 444	BNr	06 19.4	+23 16		Gem
IC 1995	BN	03 50.3	+25 35		Tau
IC 2003	PN	03 56.4	+33 52	11.5	Per
IC 2087	BNr	04 40.0	+25 44		Tau
IC 2149	PN	05 56.3	+46 07	10.7	Aur
IC 2157	OC	06 05.0	+24 00	8.4	Gem
IC 2162	BNe	06 12.9	+17 59		Ori
IC 2233	Gx	08 14.0	+45 45	12.6	Lyn
Intergalactic Wanderer (see NGC 2419).					
J900	PN	06 26.0	+17 48	11.7	Gem
Letter Y Cluster (see NGC 1893).					
M1 (see Crab Nebula).					
M35	OC	06 08.9	+24 20	5.1	Gem
M36	OC	05 36.5	+34 08	6	Aur
M37	OC	05 52.4	+32 33	5.6	Aur
M38	OC	05 28.7	+35 50	6.4	Aur
NGC 1245	OC	03 14.7	+47 15	8.4	Per
NGC 1272	Gx	03 19.4	+41 30	11.8	Per
NGC 1275	Gx	03 19.8	+41 31	11.9	Per
NGC 1342	OC	03 31.6	+37 20	6.7	Per
NGC 1444	OC	03 49.4	+52 40	6.6	Per
NGC 1491	BNe	04 03.4	+51 19		Per
NGC 1499 (see California Nebula).					
NGC 1513	OC	04 10.0	+49 31	8.4	Per
NGC 1514	PN	04 09.2	+30 47	10.9	Tau
NGC 1528	OC	04 15.4	+51 14	6.4	Per
NGC 1545	OC	04 20.9	+50 15	6.2	Per
NGC 1554, 5 (see Hind's Variable Nebula).					
NGC 1579	BNr	04 30.2	+35 16		Per
NGC 1582	OC	04 32.0	+43 51	7.0p	Per
NGC 1615	Gx	04 36.0	+19 57	13.6	Tau
NGC 1624	BNe	04 40.5	+50 27		Per
NGC 1647	OC	04 46.0	+19 04	6.4	Tau
NGC 1664	OC	04 51.1	+43 42	7.6	Aur
NGC 1746	OC	05 03.6	+23 49	6.1p	Tau
NGC 1778	OC	05 08.1	+37 03	7.7	Aur
NGC 1807	OC	05 10.7	+16 32	7	Tau
NGC 1817	OC	05 12.1	+16 42	7.7	Tau
NGC 1857	OC	05 20.2	+39 21	7	Aur
NGC 1893	OC	05 22.7	+33 24	7.5	Aur
NGC 1907	OC	05 28.0	+35 19	8.2	Aur
NGC 1912 (see M38).					
NGC 1931	BNer	05 31.4	+34 15		Aur
NGC 1952 (see Crab Nebula).					
NGC 1960 (see M36).					
NGC 2099 (see M37).					
NGC 2129	OC	06 01.0	+23 18	6.7	Gem
NGC 2158	OC	06 07.5	+24 06	8.6	Gem
NGC 2168 (see M35).					
NGC 2174	BNe	06 09.7	+20 30		Ori
NGC 2175	OC	06 09.8	+20 19	6.8	Ori
NGC 2195	BN	06 14.4	+17 39		Ori
NGC 2266	OC	06 43.2	+26 58	9.5p	Gem
NGC 2281	OC	06 49.3	+41 04	5.4	Aur
NGC 2320	Gx	07 05.7	+50 35	11.9	Lyn
NGC 2331	OC	07 07.2	+27 21	8.5p	Gem
NGC 2339	Gx	07 08.3	+18 47	11.8	Gem
NGC 2340	Gx	07 11.2	+50 10	11.7	Lyn
NGC 2344	Gx	07 12.5	+47 10	12	Lyn
NGC 2371, 2	PN	07 25.6	+29 29	11.3	Gem
NGC 2392	PN	07 29.2	+20 55	9.2	Gem
NGC 2415	Gx	07 36.9	+35 15	12.4	Lyn
NGC 2419	GC	07 38.2	+38 53	10.3	Lyn
NGC 2420	OC	07 38.5	+21 34	8.3	Gem
NGC 2493	Gx	08 00.4	+39 50	12	Lyn
NGC 2500	Gx	08 01.9	+50 44	11.6	Lyn
NGC 2532	Gx	08 10.3	+33 57	12.4	Lyn
NGC 2537	Gx	08 13.3	+45 59	11.7	Lyn
NGC 2541	Gx	08 14.7	+49 04	11.8	Lyn
NGC 2543	Gx	08 13.0	+36 15	11.9	Lyn
NGC 2552	Gx	08 19.3	+50 00	12.1	Lyn
NGC 2639	Gx	08 43.6	+50 12	11.7	UMa
Perseus A (see NGC 1275).					
PK158+17.1	PN	06 19.6	+55 37	11.2p	Lyn
PK164+31.1	PN	07 57.8	+53 25	12.1	Lyn
Sh2-205	BNe	03 56.1	+53 12		Cam
Sh2-224	BNe	05 27.3	+42 59		Aur
Sh2-231	BNe	05 39.4	+35 56		Aur
Sh2-235	BNe	05 41.1	+35 52		Aur
Sh2-240 (see Simeis 147).					
Sh2-241	BNer	06 04.1	+30 15		Aur
Sh2-247	BNe	06 08.5	+21 37		Gem
Simeis 147	BNe	05 39.1	+28 00		Tau
Taurus Moving Cluster (see Hyades).					
vdB 24	BNr	03 49.6	+38 59		Per
vdB 29	BNr	04 48.4	+29 47		Tau
vdB 31	BNr	04 55.7	+30 33		Aur

Bear Paw Galaxy (see NGC 2537).
Beehive Cluster (see Praesepe).
C21 (see NGC 4449).
C25 (see NGC 2419).

Chart 6

Object	Type	R.A. h m	Dec. ° ′	Mag.	Const.
C26 (see NGC 4244).					
C40 (see NGC 3626).					
IC 749	Gx	11 58.6	+42 44	12.4	UMa
IC 750	Gx	11 58.9	+42 43	11.9	UMa
IC 2233	Gx	08 14.0	+45 45	12.6	Lyn
Intergalactic Wanderer (see NGC 2419).					
Leo II (see UGC 6253).					
M44 (see Praesepe).					
M97 (see Owl Nebula).					
M106	Gx	12 19.0	+47 18	8.3	CVn
M108	Gx	11 11.5	+55 40	10	UMa
M109	Gx	11 57.6	+53 23	9.8	UMa
M 4-28-110	Gx	11 58.7	+25 03	12.8p	Com
Manger (see Praesepe).					
NGC 2320	Gx	07 05.7	+50 35	11.9	Lyn
NGC 2340	Gx	07 11.2	+50 10	11.7	Lyn
NGC 2344	Gx	07 12.5	+47 10	12	Lyn
NGC 2415	Gx	07 36.9	+35 15	12.4	Lyn
NGC 2419	GC	07 38.2	+38 53	10.3	Lyn
NGC 2493	Gx	08 00.4	+39 50	12	Lyn
NGC 2500	Gx	08 01.9	+50 44	11.6	Lyn
NGC 2532	Gx	08 10.3	+33 57	12.4	Lyn
NGC 2537	Gx	08 13.3	+45 59	11.7	Lyn
NGC 2541	Gx	08 14.7	+49 04	11.8	Lyn
NGC 2543	Gx	08 13.0	+36 15	11.9	Lyn
NGC 2552	Gx	08 19.3	+50 00	12.1	Lyn
NGC 2554	Gx	08 17.9	+23 28	12	Cnc
NGC 2595	Gx	08 27.7	+21 29	12.3	Cnc
NGC 2608	Gx	08 35.3	+28 28	12.3	Cnc
NGC 2632 (see Praesepe).					
NGC 2639	Gx	08 43.6	+50 12	11.7	UMa
NGC 2649	Gx	08 44.1	+34 43	12.3	Lyn
NGC 2672	Gx	08 49.4	+19 05	11.7	Cnc
NGC 2681	Gx	08 53.6	+51 19	10.3	UMa
NGC 2683	Gx	08 52.7	+33 25	9.8	Lyn
NGC 2693	Gx	08 57.0	+51 21	11.9	UMa
NGC 2701	Gx	08 59.1	+53 46	12.3	UMa
NGC 2712	Gx	08 59.5	+44 55	12.1	Lyn
NGC 2749	Gx	09 05.4	+18 19	11.8	Cnc
NGC 2750	Gx	09 05.8	+25 26	11.9	Cnc
NGC 2770	Gx	09 09.6	+33 07	12.2	Lyn
NGC 2776	Gx	09 12.2	+44 57	11.6	Lyn
NGC 2782	Gx	09 14.1	+40 07	11.6	Lyn
NGC 2798	Gx	09 17.4	+42 00	12.3	Lyn
NGC 2832	Gx	09 19.8	+33 45	11.9	Lyn
NGC 2841	Gx	09 22.0	+50 59	9.2	UMa
NGC 2857	Gx	09 24.6	+49 21	12.3	UMa
NGC 2859	Gx	09 24.3	+34 31	10.9	LMi
NGC 2903	Gx	09 32.2	+21 30	9	Leo
NGC 2916	Gx	09 35.0	+21 42	12.1	Leo
NGC 2964	Gx	09 42.9	+31 51	11.3	Leo
NGC 2968	Gx	09 43.2	+31 56	11.7	Leo
NGC 3003	Gx	09 48.5	+33 25	11.9	LMi
NGC 3021	Gx	09 51.0	+33 33	12.1	LMi
NGC 3067	Gx	09 58.4	+32 22	12.1	Leo
NGC 3079	Gx	10 02.0	+55 41	10.9	UMa
NGC 3098	Gx	10 02.3	+24 43	12	Leo
NGC 3158	Gx	10 13.8	+38 46	11.9	LMi
NGC 3162	Gx	10 13.5	+22 44	11.6	Leo
NGC 3177	Gx	10 16.6	+21 07	12.4	Leo
NGC 3184	Gx	10 18.3	+41 25	9.8	UMa
NGC 3185	Gx	10 17.6	+21 41	12.2	Leo
NGC 3190	Gx	10 18.1	+21 50	11.2	Leo
NGC 3193	Gx	10 18.4	+21 54	10.9	Leo
NGC 3198	Gx	10 19.9	+45 33	10.3	UMa
NGC 3226	Gx	10 23.5	+19 54	11.4	Leo
NGC 3227	Gx	10 23.5	+19 52	10.3	Leo
NGC 3239	Gx	10 25.1	+17 10	11.3	Leo
NGC 3245	Gx	10 27.3	+28 30	10.8	LMi
NGC 3254	Gx	10 29.3	+29 29	11.7	LMi
NGC 3264	Gx	10 32.3	+56 05	12	UMa
NGC 3277	Gx	10 32.9	+28 31	11.7	LMi
NGC 3287	Gx	10 34.8	+21 39	12.3	Leo
NGC 3294	Gx	10 36.3	+37 19	11.8	LMi
NGC 3301	Gx	10 36.9	+21 53	11.4	Leo
NGC 3310	Gx	10 38.8	+53 30	10.8	UMa
NGC 3319	Gx	10 39.2	+41 41	11.1	UMa
NGC 3320	Gx	10 39.6	+47 24	12.3	UMa
NGC 3344	Gx	10 43.5	+24 55	9.9	LMi
NGC 3370	Gx	10 47.1	+17 16	11.6	Leo
NGC 3381	Gx	10 48.4	+34 43	11.7	LMi
NGC 3395	Gx	10 49.8	+32 59	12.1	LMi
NGC 3396	Gx	10 49.9	+32 59	12.1	LMi
NGC 3413	Gx	10 51.4	+32 46	12.1	LMi
NGC 3414	Gx	10 51.3	+27 59	11	LMi
NGC 3430	Gx	10 52.2	+32 57	11.6	LMi
NGC 3432	Gx	10 52.5	+36 37	11.2	LMi
NGC 3437	Gx	10 52.6	+22 56	12.1	Leo
NGC 3448	Gx	10 54.7	+54 18	12.1	UMa
NGC 3457	Gx	10 54.8	+17 37	12.6	Leo
NGC 3486	Gx	11 00.4	+28 59	10.5	LMi
NGC 3504	Gx	11 03.2	+27 58	10.9	LMi
NGC 3507	Gx	11 03.4	+18 08	10.9	Leo
NGC 3510	Gx	11 03.7	+28 53	12.2	LMi
NGC 3512	Gx	11 04.0	+28 02	12.3	LMi
NGC 3549	Gx	11 10.9	+53 23	12.1	UMa
NGC 3556 (see M108).					
NGC 3583	Gx	11 14.2	+48 19	11.1	UMa
NGC 3587 (see Owl Nebula).					
NGC 3595	Gx	11 15.4	+47 27	12.1	UMa

Object	Type	R.A. h m	Dec. ° '	Mag.	Const.
NGC 3599	Gx	11 15.5	+18 07	11.9	Leo
NGC 3600	Gx	11 15.9	+41 35	11.7	UMa
NGC 3607	Gx	11 16.9	+18 03	9.9	Leo
NGC 3608	Gx	11 17.0	+18 09	10.8	Leo
NGC 3614	Gx	11 18.3	+45 45	11.6	UMa
NGC 3614A	Gx	11 18.3	+45 43	11.9	UMa
NGC 3626	Gx	11 20.1	+18 21	11	Leo
NGC 3629	Gx	11 20.5	+26 58	12.1	Leo
NGC 3631	Gx	11 21.0	+53 10	10.4	UMa
NGC 3646	Gx	11 21.7	+20 10	11.1	Leo
NGC 3652	Gx	11 22.7	+37 46	12.2	UMa
NGC 3655	Gx	11 22.9	+16 35	11.7	Leo
NGC 3659	Gx	11 23.8	+17 49	12.3	Leo
NGC 3665	Gx	11 24.7	+38 46	10.8	UMa
NGC 3675	Gx	11 26.1	+43 35	10.2	UMa
NGC 3681	Gx	11 26.5	+16 52	11.2	Leo
NGC 3684	Gx	11 27.2	+17 02	11.4	Leo
NGC 3686	Gx	11 27.7	+17 13	11.3	Leo
NGC 3687	Gx	11 28.0	+29 31	12	UMa
NGC 3689	Gx	11 28.2	+25 40	12.3	Leo
NGC 3691	Gx	11 28.1	+16 55	11.8	Leo
NGC 3718	Gx	11 32.6	+53 04	10.8	UMa
NGC 3726	Gx	11 33.3	+47 02	10.4	UMa
NGC 3729	Gx	11 33.8	+53 08	11.4	UMa
NGC 3733	Gx	11 35.0	+54 51	12.4	UMa
NGC 3738	Gx	11 35.8	+54 31	11.7	UMa
NGC 3756	Gx	11 36.8	+54 18	11.5	UMa
NGC 3769	Gx	11 37.7	+47 54	11.8	UMa
NGC 3800	Gx	11 40.2	+15 21	12.7	Leo
NGC 3801	Gx	11 40.3	+17 44	12	Leo
NGC 3811	Gx	11 41.3	+47 42	12.3	UMa
NGC 3813	Gx	11 41.3	+36 33	11.7	UMa
NGC 3842	Gx	11 44.0	+19 57	12	Leo
NGC 3872	Gx	11 45.8	+13 46	11.7	Leo
NGC 3877	Gx	11 46.1	+47 30	11	UMa
NGC 3893	Gx	11 48.7	+48 43	10.5	UMa
NGC 3900	Gx	11 49.1	+27 01	11.3	Leo
NGC 3912	Gx	11 50.1	+26 29	12.4	Leo
NGC 3917	Gx	11 50.8	+51 50	11.8	UMa
NGC 3938	Gx	11 52.8	+44 07	10.4	UMa
NGC 3941	Gx	11 52.9	+36 59	10.3	UMa
NGC 3949	Gx	11 53.7	+47 52	11.1	UMa
NGC 3953	Gx	11 53.8	+52 20	10.1	UMa
NGC 3992 (see M109).					
NGC 3995	Gx	11 57.7	+32 18	12.4	UMa
NGC 4008	Gx	11 58.3	+28 12	12	Leo
NGC 4013	Gx	11 58.5	+43 57	11.2	UMa
NGC 4017	Gx	11 58.8	+27 27	12.2	Com
NGC 4026	Gx	11 59.4	+50 58	10.8	UMa
NGC 4032	Gx	12 00.6	+20 05	12.2	Com
NGC 4047	Gx	12 02.9	+48 38	12.2	UMa
NGC 4051	Gx	12 03.2	+44 32	10.2	UMa
NGC 4062	Gx	12 04.1	+31 54	11.1	UMa
NGC 4085	Gx	12 05.4	+50 21	12.4	UMa
NGC 4088	Gx	12 05.6	+50 33	10.6	UMa
NGC 4096	Gx	12 06.0	+47 29	10.8	UMa
NGC 4100	Gx	12 06.1	+49 35	11.2	UMa
NGC 4102	Gx	12 06.4	+52 43	11.2	UMa
NGC 4111	Gx	12 07.0	+43 04	10.7	CVn
NGC 4136	Gx	12 09.3	+29 56	11	Com
NGC 4138	Gx	12 09.5	+43 41	11.3	CVn
NGC 4143	Gx	12 09.6	+42 32	10.7	CVn
NGC 4144	Gx	12 10.0	+46 27	11.6	UMa
NGC 4145	Gx	12 10.0	+39 53	11.3	CVn
NGC 4150	Gx	12 10.6	+30 24	11.6	Com
NGC 4151	Gx	12 10.5	+39 24	10.8	CVn
NGC 4157	Gx	12 11.1	+50 29	11.3	UMa
NGC 4183	Gx	12 13.3	+43 42	12.3	CVn
NGC 4203	Gx	12 15.1	+33 12	10.9	Com
NGC 4214	Gx	12 15.7	+36 20	9.8	CVn
NGC 4217	Gx	12 15.8	+47 06	11.2	CVn
NGC 4220	Gx	12 16.2	+47 53	11.4	CVn
NGC 4242	Gx	12 17.5	+45 37	10.8	CVn
NGC 4244	Gx	12 17.5	+37 48	10.4	CVn
NGC 4245	Gx	12 17.6	+29 37	11.4	Com
NGC 4248	Gx	12 17.8	+47 25	12.5	CVn
NGC 4258 (see M106).					
NGC 4346	Gx	12 23.5	+47 00	11.1	CVn
NGC 4369	Gx	12 24.6	+39 23	11.7	CVn
NGC 4389	Gx	12 25.6	+45 41	11.7	CVn
NGC 4449	Gx	12 28.2	+44 06	9.6	CVn
NGC 4460	Gx	12 28.8	+44 52	11.3	CVn
NGC 4485	Gx	12 30.5	+41 42	11.9	CVn
NGC 4490	Gx	12 30.6	+41 38	9.8	CVn
NGC 4800	Gx	12 54.6	+46 32	11.5	CVn
Owl Nebula	PN	11 14.8	+55 01	9.9	UMa
PK164+31.1	PN	07 57.8	+53 25	12.1	Lyn
Praesepe	OC	08 40.1	+19 59	3.1	Cnc
UGC 4375	Gx	08 23.2	+22 40	12.8p	Cnc
UGC 5364	Gx	09 59.4	+30 45	13.0p	Leo
UGC 5720	Gx	10 32.5	+54 24	13.0p	UMa
UGC 6253	Gx	11 13.5	+22 09	11.5	Leo
UGC 6887	Gx	11 55.2	+22 42	10.6p	Leo
UGC 6930	Gx	11 57.3	+49 17	12.7p	UMa
UGC 7577	Gx	12 27.7	+43 30	12.9p	CVn

Black Eye Galaxy (see M64).
C21 (see NGC 4449).
C26 (see NGC 4244).
C29 (see NGC 5005).

Chart 7

Object	Type	R.A. h m	Dec. ° '	Mag.	Const.
C32 (see NGC 4631).					
C35 (see NGC 4889).					
C36 (see NGC 4559).					
C38 (see NGC 4565).					
IC 749	Gx	11 58.6	+42 44	12.4	UMa
IC 750	Gx	11 58.9	+42 43	11.9	UMa
IC 983	Gx	14 10.1	+17 44	12.5p	Boo
IC 1029	Gx	14 32.5	+49 54	12.2p	Boo
IC 4182	Gx	13 05.8	+37 36	12.5	CVn
M3	GC	13 42.2	+28 23	6.4	CVn
M51 (see Whirlpool Galaxy).					
M53	GC	13 12.9	+18 10	7.7	Com
M63	Gx	13 15.8	+42 02	8.6	CVn
M64	Gx	12 56.7	+21 41	8.5	Com
M85	Gx	12 25.4	+18 11	9.2	Com
M94	Gx	12 50.9	+41 07	8.1	CVn
M98	Gx	12 13.8	+14 54	10.1	Com
M99	Gx	12 18.8	+14 25	9.8	Com
M100	Gx	12 22.9	+15 49	9.4	Com
M101	Gx	14 03.2	+54 21	7.7	UMa
M102 (same object as M101).					
M106	Gx	12 19.0	+47 18	8.3	CVn
M109	Gx	11 57.6	+53 23	9.8	UMa
M 4-28-110	Gx	11 58.7	+25 03	12.8p	Com
NGC 3583	Gx	11 14.2	+48 19	11.1	UMa
NGC 3595	Gx	11 15.4	+47 27	12.1	UMa
NGC 3614	Gx	11 18.3	+45 45	11.6	UMa
NGC 3614A	Gx	11 18.3	+45 43	11.9	UMa
NGC 3665	Gx	11 24.7	+38 46	10.8	UMa
NGC 3675	Gx	11 26.1	+43 35	10.2	UMa
NGC 3726	Gx	11 33.3	+47 02	10.4	UMa
NGC 3769	Gx	11 37.7	+47 54	11.8	UMa
NGC 3811	Gx	11 41.3	+47 42	12.3	UMa
NGC 3813	Gx	11 41.3	+36 33	11.7	UMa
NGC 3877	Gx	11 46.1	+47 30	11	UMa
NGC 3893	Gx	11 48.7	+48 43	10.5	UMa
NGC 3900	Gx	11 49.1	+27 01	11.3	Leo
NGC 3912	Gx	11 50.1	+26 29	12.4	Leo
NGC 3917	Gx	11 50.8	+51 50	11.8	UMa
NGC 3938	Gx	11 52.8	+44 07	10.4	UMa
NGC 3941	Gx	11 52.9	+36 59	10.3	UMa
NGC 3949	Gx	11 53.7	+47 52	11.1	UMa
NGC 3953	Gx	11 53.8	+52 20	10.1	UMa
NGC 3992 (see M109).					
NGC 3995	Gx	11 57.7	+32 18	12.4	UMa
NGC 4008	Gx	11 58.3	+28 12	12	Leo
NGC 4013	Gx	11 58.5	+43 57	11.2	UMa
NGC 4017	Gx	11 58.8	+27 27	12.2	Com
NGC 4026	Gx	11 59.4	+50 58	10.8	UMa
NGC 4032	Gx	12 00.6	+20 05	12.2	Com
NGC 4047	Gx	12 02.9	+48 38	12.2	UMa
NGC 4051	Gx	12 03.2	+44 32	10.2	UMa
NGC 4062	Gx	12 04.1	+31 54	11.1	UMa
NGC 4064	Gx	12 04.2	+18 27	11.4	Com
NGC 4085	Gx	12 05.4	+50 21	12.4	UMa
NGC 4088	Gx	12 05.6	+50 33	10.6	UMa
NGC 4096	Gx	12 06.0	+47 29	10.8	UMa
NGC 4100	Gx	12 06.1	+49 35	11.2	UMa
NGC 4102	Gx	12 06.4	+52 43	11.2	UMa
NGC 4111	Gx	12 07.0	+43 04	10.7	CVn
NGC 4136	Gx	12 09.3	+29 56	11	Com
NGC 4138	Gx	12 09.5	+43 41	11.3	CVn
NGC 4143	Gx	12 09.6	+42 32	10.7	CVn
NGC 4144	Gx	12 10.0	+46 27	11.6	UMa
NGC 4145	Gx	12 10.0	+39 53	11.3	CVn
NGC 4147	GC	12 10.1	+18 33	10.4	Com
NGC 4150	Gx	12 10.6	+30 24	11.6	Com
NGC 4151	Gx	12 10.5	+39 24	10.8	CVn
NGC 4152	Gx	12 10.6	+16 02	12.2	Com
NGC 4157	Gx	12 11.1	+50 29	11.3	UMa
NGC 4158	Gx	12 11.2	+20 11	12.1	Com
NGC 4162	Gx	12 11.9	+24 07	12.2	Com
NGC 4183	Gx	12 13.3	+43 42	12.3	CVn
NGC 4189	Gx	12 13.8	+13 26	11.7	Com
NGC 4192 (see M98).					
NGC 4194	Gx	12 14.2	+54 32	12.5	UMa
NGC 4203	Gx	12 15.1	+33 12	10.9	Com
NGC 4212	Gx	12 15.7	+13 54	11.2	Com
NGC 4214	Gx	12 15.7	+36 20	9.8	CVn
NGC 4217	Gx	12 15.8	+47 06	11.2	CVn
NGC 4220	Gx	12 16.2	+47 53	11.4	CVn
NGC 4237	Gx	12 17.2	+15 19	11.6	Com
NGC 4242	Gx	12 17.5	+45 37	10.8	CVn
NGC 4244	Gx	12 17.5	+37 48	10.4	CVn
NGC 4245	Gx	12 17.6	+29 37	11.4	Com
NGC 4248	Gx	12 17.8	+47 25	12.5	CVn
NGC 4251	Gx	12 18.1	+28 11	10.7	Com
NGC 4254 (see M99).					
NGC 4258 (see M106).					
NGC 4262	Gx	12 19.5	+14 53	11.6	Com
NGC 4274	Gx	12 19.8	+29 37	10.4	Com
NGC 4278	Gx	12 20.1	+29 17	10.2	Com
NGC 4283	Gx	12 20.3	+29 19	12.1	Com
NGC 4293	Gx	12 21.2	+18 23	10.4	Com
NGC 4298	Gx	12 21.5	+14 36	11.3	Com
NGC 4302	Gx	12 21.7	+14 36	11.6	Com
NGC 4312	Gx	12 22.5	+15 32	11.7	Com
NGC 4314	Gx	12 22.5	+29 54	10.6	Com
NGC 4321 (see M100).					
NGC 4340	Gx	12 23.6	+16 43	11.2	Com

Object	Type	R.A. h m	Dec. ° ′	Mag.	Const.	Object	Type	R.A. h m	Dec. ° ′	Mag.	Const.
NGC 4346	Gx	12 23.5	+47 00	11.1	CVn	NGC 5024 (see M53).					
NGC 4350	Gx	12 24.0	+16 42	11	Com	NGC 5033	Gx	13 13.5	+36 36	10.2	CVn
NGC 4369	Gx	12 24.6	+39 23	11.7	CVn	NGC 5053	GC	13 16.5	+17 42	9	Com
NGC 4377	Gx	12 25.2	+14 46	11.9	Com	NGC 5055 (see M63).					
NGC 4379	Gx	12 25.2	+15 36	11.7	Com	NGC 5112	Gx	13 21.9	+38 44	12.1	CVn
NGC 4382 (see M85).						NGC 5172	Gx	13 29.3	+17 03	11.9	Com
NGC 4383	Gx	12 25.4	+16 28	12.1	Com	NGC 5194 (see Whirlpool Galaxy).					
NGC 4389	Gx	12 25.6	+45 41	11.7	CVn	NGC 5195	Gx	13 30.0	+47 16	9.6	CVn
NGC 4394	Gx	12 25.9	+18 13	10.9	Com	NGC 5198	Gx	13 30.2	+46 40	11.8	CVn
NGC 4395	Gx	12 25.8	+33 33	10.2	CVn	NGC 5272 (see M3).					
NGC 4405	Gx	12 26.1	+16 11	12	Com	NGC 5273	Gx	13 42.1	+35 39	11.6	CVn
NGC 4414	Gx	12 26.5	+31 13	10.1	Com	NGC 5297	Gx	13 46.4	+43 52	11.8	CVn
NGC 4419	Gx	12 26.9	+15 03	11.2	Com	NGC 5313	Gx	13 49.7	+39 59	12	CVn
NGC 4421	Gx	12 27.0	+15 28	11.6	Com	NGC 5320	Gx	13 50.3	+41 22	12.1	CVn
NGC 4448	Gx	12 28.3	+28 37	11.1	Com	NGC 5326	Gx	13 50.8	+39 34	11.9	CVn
NGC 4449	Gx	12 28.2	+44 06	9.6	CVn	NGC 5350	Gx	13 53.4	+40 22	11.3	CVn
NGC 4450	Gx	12 28.5	+17 05	10.1	Com	NGC 5351	Gx	13 53.5	+37 55	12.1	CVn
NGC 4460	Gx	12 28.8	+44 52	11.3	CVn	NGC 5353	Gx	13 53.4	+40 17	11	CVn
NGC 4485	Gx	12 30.5	+41 42	11.9	CVn	NGC 5354	Gx	13 53.4	+40 18	11.4	CVn
NGC 4489	Gx	12 30.9	+16 46	12	Com	NGC 5371	Gx	13 55.7	+40 28	10.6	CVn
NGC 4490	Gx	12 30.6	+41 38	9.8	CVn	NGC 5375	Gx	13 56.9	+29 10	11.5	CVn
NGC 4494	Gx	12 31.4	+25 46	9.8	Com	NGC 5377	Gx	13 56.3	+47 14	11.3	CVn
NGC 4498	Gx	12 31.7	+16 51	12.2	Com	NGC 5383	Gx	13 57.1	+41 51	11.4	CVn
NGC 4525	Gx	12 33.9	+30 17	12.2	Com	NGC 5395	Gx	13 58.6	+37 26	11.4	CVn
NGC 4534	Gx	12 34.1	+35 31	12.3	CVn	NGC 5422	Gx	14 00.7	+55 10	11.8	UMa
NGC 4539	Gx	12 34.6	+18 12	12	Com	NGC 5444	Gx	14 03.4	+35 08	11.8	CVn
NGC 4540	Gx	12 34.8	+15 33	11.7	Com	NGC 5448	Gx	14 02.8	+49 10	11	UMa
NGC 4559	Gx	12 36.0	+27 58	10	Com	NGC 5457 (see M101).					
NGC 4561	Gx	12 36.1	+19 19	12.5	Com	NGC 5466	GC	14 05.5	+28 32	9.2	Boo
NGC 4565	Gx	12 36.3	+25 59	9.6	Com	NGC 5473	Gx	14 04.7	+54 54	11.4	UMa
NGC 4595	Gx	12 39.9	+15 18	12.1	Com	NGC 5474	Gx	14 05.0	+53 40	10.8	UMa
NGC 4618	Gx	12 41.6	+41 09	10.8	CVn	NGC 5480	Gx	14 06.4	+50 44	12.1	UMa
NGC 4625	Gx	12 41.9	+41 16	12.3	CVn	NGC 5485	Gx	14 07.2	+55 00	11.4	UMa
NGC 4627	Gx	12 42.0	+32 34	12.4	CVn	NGC 5523	Gx	14 14.9	+25 19	12.1	Boo
NGC 4631	Gx	12 42.2	+32 33	9.2	CVn	NGC 5529	Gx	14 15.6	+36 14	11.9	Boo
NGC 4651	Gx	12 43.7	+16 24	10.8	Com	NGC 5533	Gx	14 16.1	+35 21	11.8	Boo
NGC 4656	Gx	12 44.0	+32 10	10.5	CVn	NGC 5557	Gx	14 18.4	+36 30	11	Boo
NGC 4725	Gx	12 50.4	+25 30	9.4	Com	NGC 5582	Gx	14 20.7	+39 42	11.6	Boo
NGC 4736 (see M94).						NGC 5585	Gx	14 19.8	+56 44	10.7	UMa
NGC 4747	Gx	12 51.8	+25 46	12.3	Com	NGC 5614	Gx	14 24.1	+34 51	11.7	Boo
NGC 4793	Gx	12 54.7	+28 56	11.6	Com	NGC 5631	Gx	14 26.6	+56 35	11.5	UMa
NGC 4800	Gx	12 54.6	+46 32	11.5	CVn	NGC 5633	Gx	14 27.5	+46 09	12.4	Boo
NGC 4826 (see M64).						NGC 5653	Gx	14 30.2	+31 13	12.2	Boo
NGC 4868	Gx	12 59.2	+37 19	12.2	CVn	NGC 5656	Gx	14 30.4	+35 19	11.8	Boo
NGC 4874	Gx	12 59.6	+27 58	11.7	Com	NGC 5660	Gx	14 29.8	+49 37	11.9	Boo
NGC 4889	Gx	13 00.1	+27 59	11.5	Com	NGC 5673	Gx	14 31.5	+49 58	12.1	Boo
NGC 4914	Gx	13 00.7	+37 19	11.6	CVn	NGC 5676	Gx	14 32.8	+49 27	11.2	Boo
NGC 5005	Gx	13 10.9	+37 03	9.8	CVn	NGC 5687	Gx	14 34.9	+54 29	11.8	Boo
NGC 5012	Gx	13 11.6	+22 55	12.2	Com	NGC 5689	Gx	14 35.5	+48 45	11.9	Boo
NGC 5023	Gx	13 12.2	+44 02	12.3	CVn	NGC 5866	Gx	15 06.5	+55 46	9.9	Dra

Charts 7 – 8

Object	Type	R.A. h m	Dec. ° ′	Mag.	Const.
NGC 5899	Gx	15 15.1	+42 03	11.7	Boo
NGC 5908	Gx	15 16.7	+55 25	11.8	Dra
NGC 5962	Gx	15 36.5	+16 37	11.3	Ser
NGC 5984	Gx	15 42.9	+14 14	12.5	Ser
NGC 5996	Gx	15 47.0	+17 53	12.8	Ser
NGC 6012	Gx	15 54.2	+14 36	12	Ser
NGC 6058	PN	16 04.4	+40 41	12.9	Her
NGC 6166	Gx	16 28.6	+39 33	11.8	Her
NGC 6229	GC	16 47.0	+47 32	9.4	Her
Sunflower Galaxy (see M63).					
UGC 6887	Gx	11 55.2	+22 42	10.6p	Leo
UGC 6930	Gx	11 57.3	+49 17	12.7p	UMa
UGC 7577	Gx	12 27.7	+43 30	12.9p	CVn
UGC 8320	Gx	13 14.5	+45 55	12.7p	CVn
Whirlpool Gal.	Gx	13 29.9	+47 12	8.1	CVn

Chart 8

Object	Type	R.A. h m	Dec. ° ′	Mag.	Const.
B145	DN	20 02.8	+37 40		Cyg
B146	DN	20 03.5	+36 02		Cyg
B343	DN	20 13.5	+40 16		Cyg
B346	DN	20 26.7	+43 45		Cyg
B350	DN	20 49.1	+45 53		Cyg
Bas 6	OC	20 06.8	+38 21	7.7	Cyg
Be 86	OC	20 20.4	+38 42	7.9	Cyg
Bi 2	OC	20 09.2	+35 29	6.3	Cyg
Blinking Planetary (see NGC 6826).					
Brocchi's Cluster	OC	19 25.4	+20 11	3.6	Vul
C15 (see NGC 6826).					
C27 (see NGC 6888).					
C37 (see NGC 6885).					
Campbell's Star (see PK64+5.1).					
Ced 174	BNe	20 02.8	+36 58		Cyg
Coathanger (see Brocchi's Cluster).					
Cr 399 (see Brocchi's Cluster).					
Cr 419	OC	20 18.1	+40 43	5.4p	Cyg
Crescent Nebula (see NGC 6888).					
Cygnus Nebula (see IC 1318).					
Delta Lyrae Cluster (see Steph 1).					
Donut Nebula (see Ring Nebula).					
Dumbbell Neb.	PN	19 59.6	+22 43	7.3:	Vul
Footprint Nebula (see M1-92).					
H20	OC	19 53.1	+18 20	7.7	Sge
Hercules Cluster (see M13).					
Hole in a Cluster (see NGC 6811).					
Hourglass Nebula (see Dumbbell Nebula).					
IC 1311	BNe+C	20 10.8	+41 11		Cyg
IC 1318	BNe	20 21.0	+39 54		Cyg
IC 4954, 5	BNr	20 04.8	+29 15		Vul
IC 4996	OC	20 16.5	+37 38	7.3	Cyg
IC 5067	BN	20 47.8	+44 22		Cyg
IC 5076	BNr	20 55.9	+47 25		Cyg

Object	Type	R.A. h m	Dec. ° ′	Mag.	Const.
LDN 889	DN	20 24.8	+40 10		Cyg
M1-92	BNr	19 36.3	+29 33		Cyg
M13	GC	16 41.7	+36 28	5.9	Her
M27 (see Dumbbell Nebula).					
M29	OC	20 23.9	+38 32	6.6	Cyg
M56	GC	19 16.6	+30 11	8.2	Lyr
M57 (see Ring Nebula).					
M71	GC	19 53.8	+18 47	8.3	Sge
M92	GC	17 17.1	+43 08	6.5	Her
NGC 6058	PN	16 04.4	+40 41	12.9	Her
NGC 6166	Gx	16 28.6	+39 33	11.8	Her
NGC 6181	Gx	16 32.3	+19 50	11.9	Her
NGC 6205 (see M13).					
NGC 6207	Gx	16 43.1	+36 50	11.6	Her
NGC 6210	PN	16 44.5	+23 49	8.8	Her
NGC 6229	GC	16 47.0	+47 32	9.4	Her
NGC 6239	Gx	16 50.1	+42 44	12.4	Her
NGC 6341 (see M92).					
NGC 6482	Gx	17 51.8	+23 04	11.4	Her
NGC 6487	Gx	17 52.7	+29 50	11.9	Her
NGC 6500	Gx	17 56.0	+18 20	12.2	Her
NGC 6548	Gx	18 06.0	+18 35	11.7	Her
NGC 6555	Gx	18 07.8	+17 36	12.4	Her
NGC 6632	Gx	18 25.1	+27 32	12.1	Her
NGC 6674	Gx	18 38.6	+25 23	12.2	Her
NGC 6703	Gx	18 47.3	+45 33	11.3	Lyr
NGC 6720 (see Ring Nebula).					
NGC 6764	Gx	19 08.3	+50 56	11.8	Cyg
NGC 6779 (see M56).					
NGC 6811	OC	19 38.2	+46 34	6.8	Cyg
NGC 6813	BNe	19 40.4	+27 18		Vul
NGC 6819	OC	19 41.3	+40 11	7.3	Cyg
NGC 6820	BNe	19 43.1	+23 17		Vul
NGC 6823	OC	19 43.1	+23 18	7.1	Vul
NGC 6826	PN	19 44.8	+50 31	8.8	Cyg
NGC 6830	OC	19 51.0	+23 04	7.9	Vul
NGC 6833	PN	19 49.7	+48 58	12.1	Cyg
NGC 6834	OC	19 52.2	+29 25	7.8	Cyg
NGC 6838 (see M71).					
NGC 6842	PN	19 55.0	+29 17	13.1	Vul
NGC 6853 (see Dumbbell Nebula).					
NGC 6866	OC	20 03.7	+44 00	7.6	Cyg
NGC 6871	OC	20 05.9	+35 47	5.2	Cyg
NGC 6882	OC	20 11.7	+26 33	8.1:	Vul
NGC 6883	OC	20 11.3	+35 51	8.0p	Cyg
NGC 6884	PN	20 10.4	+46 28	10.9	Cyg
NGC 6885	OC	20 12.0	+26 29	8.1	Vul
NGC 6888	BNe	20 12.0	+38 21		Cyg
NGC 6910	OC	20 23.1	+40 47	7.4	Cyg
NGC 6913 (see M29).					

Object	Type	R.A. h m	Dec. ° ′	Mag.	Const.
NGC 6914	BNr	20 24.7	+42 29		Cyg
Pelican Nebula	BN	20 47.8	+44 22		Cyg
PK47+42.1	PN	16 27.5	+27 54	13	Her
PK51+9.1	PN	18 49.7	+20 51	11.4	Her
PK52-2.2	PN	19 39.2	+15 57	11.8	Aql
PK64+5.1	PN	19 34.8	+30 31	11.3	Cyg
Ring Nebula	PN	18 53.6	+33 02	8.8	Lyr
Sh2-82	BNer	19 30.3	+18 16		Sge
Sh2-84	BNe	19 49.0	+18 24		Sge
Sh2-88	BNe	19 46.0	+25 20		Vul
Sh2-90	BNr	19 49.3	+26 52		Vul
Sh2-101	BNe	20 00.0	+35 17		Cyg
Sh2-104	BNe	20 17.8	+36 44		Cyg
Sh2-108	BNe	20 19.1	+39 21		Cyg
Sh2-112	BNe	20 33.9	+45 39		Cyg
Sh2-115	BNe	20 34.5	+46 52		Cyg
Smoke Ring Nebula (see Ring Nebula).					
St 1	OC	19 35.8	+25 13	5.3:	Vul
Steph 1	OC	18 54.2	+36 55	3.8	Lyr
UGC 11453	Gx	19 31.1	+54 06	12.9p	Cyg
vdB 126	BNr	19 27.1	+22 43		Vul
vdB 128	BNr	20 04.6	+32 15		Cyg
vdB 133	BNr	20 30.7	+36 56		Cyg

Chart 9

Object	Type	R.A. h m	Dec. ° ′	Mag.	Const.
B145	DN	20 02.8	+37 40		Cyg
B146	DN	20 03.5	+36 02		Cyg
B157	DN	21 33.7	+54 40		Cyg
B160	DN	21 38.0	+56 14		Cep
B162	DN	21 41.1	+56 19		Cep
B163	DN	21 42.2	+56 42		Cep
B164	DN	21 46.5	+51 04		Cyg
B168	DN	21 53.2	+47 12		Cyg
B343	DN	20 13.5	+40 16		Cyg
B346	DN	20 26.7	+43 45		Cyg
B350	DN	20 49.1	+45 53		Cyg
B352	DN	20 57.1	+45 54		Cyg
B361	DN	21 12.9	+47 22		Cyg
B362	DN	21 24.0	+50 10		Cyg
B364	DN	21 33.6	+54 33		Cyg
B365	DN	21 34.9	+56 43		Cep
Bas 6	OC	20 06.8	+38 21	7.7	Cyg
Be 86	OC	20 20.4	+38 42	7.9	Cyg
Bi 2	OC	20 09.2	+35 29	6.3	Cyg
Blinking Planetary (see NGC 6826).					
Blue Flash Nebula (see NGC 6905).					
Blue Snowball Nebula (see NGC 7662).					
C15 (see NGC 6826).					
C16 (see NGC 7243).					
C17 (see NGC 147).					
C18 (see NGC 185).					
C19 (see IC 5146).					
C20 (see North America Nebula).					
C22 (see NGC 7662).					
C27 (see NGC 6888).					
C30 (see NGC 7331).					
C33 (see NGC 6992 and NGC 6995).					
C34 (see NGC 6960).					
C37 (see NGC 6885).					
C42 (see NGC 7006).					
Ced 174	BNe	20 02.8	+36 58		Cyg
Cirrus Nebula (see Veil Nebula).					
Cocoon Nebula (see IC 5146).					
Cr 419	OC	20 18.1	+40 43	5.4p	Cyg
Crescent Nebula (see NGC 6888).					
Cygnus Nebula (see IC 1318).					
Dumbbell Neb.	PN	19 59.6	+22 43	7.3:	Vul
Egg Nebula (see PK80-6.1).					
Filamentary Nebula (see NGC 6960).					
Hole in a Cluster (see NGC 6811).					
Hourglass Nebula (see Dumbbell Nebula).					
IC 1311	BNe+C	20 10.8	+41 11		Cyg
IC 1318	BNe	20 21.0	+39 54		Cyg
IC 1340	BNe	20 56.2	+31 04		Cyg
IC 1369	OC	21 12.1	+47 44	6.8	Cyg
IC 1392	Gx	21 35.5	+35 24	12.9p	Cyg
IC 1396	BNe+C	21 39.1	+57 30		Cep
IC 4954,5	BNr	20 04.8	+29 15		Vul
IC 4996	OC	20 16.5	+37 38	7.3	Cyg
IC 4997	PN	20 20.2	+16 44	10.5	Sge
IC 5067	BN	20 47.8	+44 22		Cyg
IC 5068	BNe	20 50.8	+42 31		Cyg
IC 5070	BNe	20 50.8	+44 21		Cyg
IC 5076	BNr	20 55.9	+47 25		Cyg
IC 5117	PN	21 32.5	+44 35	11.5	Cyg
IC 5146	BNe+C	21 53.4	+47 16		Cyg
IC 5217	PN	22 23.9	+50 58	11.3	Lac
LDN 889	DN	20 24.8	+40 10		Cyg
LDN 935	DN	20 56.8	+43 52		Cyg
M27 (see Dumbbell Nebula).					
M29	OC	20 23.9	+38 32	6.6	Cyg
M39	OC	21 32.2	+48 26	4.6	Cyg
M110	Gx	00 40.4	+41 41	8	And
Network Nebula (see NGC 6992).					
NGC 16	Gx	00 09.1	+27 44	12	Peg
NGC 23	Gx	00 09.9	+25 55	12	Peg
NGC 147	Gx	00 33.2	+48 30	9.5	Cas
NGC 185	Gx	00 39.0	+48 20	9.2	Cas
NGC 205 (see M110).					
NGC 278	Gx	00 52.1	+47 33	10.8	Cas
NGC 6764	Gx	19 08.3	+50 56	11.8	Cyg

Charts 9 – 10

Object	Type	R.A. h m	Dec. ° ′	Mag.	Const.
NGC 6811	OC	19 38.2	+46 34	6.8	Cyg
NGC 6819	OC	19 41.3	+40 11	7.3	Cyg
NGC 6826	PN	19 44.8	+50 31	8.8	Cyg
NGC 6833	PN	19 49.7	+48 58	12.1	Cyg
NGC 6834	OC	19 52.2	+29 25	7.8	Cyg
NGC 6842	PN	19 55.0	+29 17	13.1	Vul
NGC 6853 (see Dumbbell Nebula).					
NGC 6866	OC	20 03.7	+44 00	7.6	Cyg
NGC 6871	OC	20 05.9	+35 47	5.2	Cyg
NGC 6879	PN	20 10.5	+16 55	12.5	Sge
NGC 6882	OC	20 11.7	+26 33	8.1:	Vul
NGC 6883	OC	20 11.3	+35 51	8.0p	Cyg
NGC 6884	PN	20 10.4	+46 28	10.9	Cyg
NGC 6885	OC	20 12.0	+26 29	8.1	Vul
NGC 6886	PN	20 12.7	+19 59	11.4	Sge
NGC 6888	BNe	20 12.0	+38 21		Cyg
NGC 6905	PN	20 22.4	+20 06	11.1	Del
NGC 6910	OC	20 23.1	+40 47	7.4	Cyg
NGC 6913 (see M29).					
NGC 6914	BNr	20 24.7	+42 29		Cyg
NGC 6940	OC	20 34.6	+28 18	6.3	Vul
NGC 6960	BNe	20 45.7	+30 43		Cyg
NGC 6992	BNe	20 56.4	+31 43		Cyg
NGC 6995	BNe	20 57.1	+31 13		Cyg
NGC 7000 (see North America Nebula).					
NGC 7006	GC	21 01.5	+16 11	10.6	Del
NGC 7008	PN	21 00.6	+54 33	10.7	Cyg
NGC 7013	Gx	21 03.6	+29 54	11.3	Cyg
NGC 7026	PN	21 06.3	+47 51	10.9	Cyg
NGC 7027	PN	21 07.1	+42 14	8.5	Cyg
NGC 7039	OC	21 11.2	+45 39	7.6	Cyg
NGC 7048	PN	21 14.2	+46 16	12.1	Cyg
NGC 7062	OC	21 23.2	+46 23	8.3	Cyg
NGC 7063	OC	21 24.4	+36 30	7	Cyg
NGC 7082	OC	21 29.4	+47 05	7.2	Cyg
NGC 7086	OC	21 30.5	+51 35	8.4	Cyg
NGC 7092 (see M39).					
NGC 7177	Gx	22 00.7	+17 44	11.2	Peg
NGC 7209	OC	22 05.2	+46 30	6.7	Lac
NGC 7217	Gx	22 07.9	+31 22	10.1	Peg
NGC 7243	OC	22 15.3	+49 53	6.4	Lac
NGC 7292	Gx	22 28.4	+30 18	12.5	Peg
NGC 7331	Gx	22 37.1	+34 25	9.5	Peg
NGC 7332	Gx	22 37.4	+23 48	11.1	Peg
NGC 7448	Gx	23 00.1	+15 59	11.7	Peg
NGC 7454	Gx	23 01.1	+16 23	11.8	Peg
NGC 7457	Gx	23 01.0	+30 09	11.2	Peg
NGC 7497	Gx	23 09.1	+18 11	12.2	Peg
NGC 7625	Gx	23 20.5	+17 14	12.1	Peg
NGC 7640	Gx	23 22.1	+40 51	11.3	And

Object	Type	R.A. h m	Dec. ° ′	Mag.	Const.
NGC 7662	PN	23 25.9	+42 33	8.3	And
NGC 7678	Gx	23 28.5	+22 25	11.8	Peg
NGC 7686	OC	23 30.2	+49 08	5.6	And
NGC 7741	Gx	23 43.9	+26 05	11.3	Peg
NGC 7753	Gx	23 47.1	+29 29	12	Peg
NGC 7769	Gx	23 51.1	+20 09	12	Peg
NGC 7798	Gx	23 59.4	+20 45	12.4	Peg
NGC 7817	Gx	00 04.0	+20 45	11.8	Peg
North America Nebula	BNe	20 58.8	+44 20		Cyg
Pelican Nebula	BN	20 47.8	+44 22		Cyg
PK72-17.1	PN	21 16.8	+24 10	>12.0p	Vul
PK80-6.1	PN	21 02.3	+36 42	12.0p	Cyg
PK86-8.1	PN	21 33.1	+39 38	12	Cyg
PK107-13.1	PN	23 23.0	+46 54	13.6	And
Sh2-90	BNr	19 49.3	+26 52		Vul
Sh2-101	BNe	20 00.0	+35 17		Cyg
Sh2-104	BNe	20 17.8	+36 44		Cyg
Sh2-108	BNe	20 19.1	+39 21		Cyg
Sh2-112	BNe	20 33.9	+45 39		Cyg
Sh2-115	BNe	20 34.5	+46 52		Cyg
Sh2-132	BNe	22 18.7	+56 08		Cep
Stephan's Quintet	—	22 36.0	+33 58	13.1	Peg
UGC 11920	Gx	22 08.5	+48 26	12.9p	Lac
UGC 12613	Gx	23 28.6	+14 45	12.8p	Peg
vdB 128	BNr	20 04.6	+32 15		Cyg
vdB 133	BNr	20 30.7	+36 56		Cyg
vdB 145	BNr	21 43.7	+48 55		Cyg
Veil Nebula	BNe	20 50.0	+31 00		Cyg

Chart 10

C43 (see NGC 7814).
C51 (see IC 1613).
C56 (see NGC 246).

Object	Type	R.A. h m	Dec. ° ′	Mag.	Const.
IC 1613	Gx	01 04.9	+02 08	9.2	Cet
IC 1870	Gx	02 57.9	−02 21	12.6p	Eri
II Zw 5	Gx	02 41.3	+04 13	11.8p	Cet
M74	Gx	01 36.7	+15 47	9.2	Psc
M77	Gx	02 42.7	−00 01	8.8	Cet
M-1-3-85	Gx	01 05.1	−06 13	12.4p	Cet
M-2-9-36	Gx	03 22.9	−11 12	13.0p	Eri
M-3-1-15	Gx	00 01.9	−15 27	11.1p	Cet
M-3-10-42	Gx	03 44.0	−14 22	13.0p	Eri
M-3-10-45	Gx	03 46.6	−16 33	13.0p	Eri
NGC 14	Gx	00 08.8	+15 49	12.1	Peg
NGC 50	Gx	00 14.7	−07 21	11.6	Cet
NGC 57	Gx	00 15.5	+17 20	11.6	Psc
NGC 63	Gx	00 17.8	+11 27	11.6	Psc
NGC 128	Gx	00 29.3	+02 52	11.8	Psc
NGC 151	Gx	00 34.0	−09 42	11.6	Cet
NGC 157	Gx	00 34.8	−08 24	10.4	Cet

Chart 10

Object	Type	R.A. h m	Dec. ° ′	Mag.	Const.
NGC 175	Gx	00 37.4	−19 56	12.1	Cet
NGC 198	Gx	00 39.4	+02 48	13.2	Psc
NGC 210	Gx	00 40.6	−13 52	10.9	Cet
NGC 245	Gx	00 46.1	−01 43	12.2	Cet
NGC 246	PN	00 47.1	−11 52	10.9	Cet
NGC 255	Gx	00 47.8	−11 28	11.9	Cet
NGC 271	Gx	00 50.7	−01 55	12	Cet
NGC 274	Gx	00 51.0	−07 03	11.8	Cet
NGC 275	Gx	00 51.1	−07 04	12.5	Cet
NGC 309	Gx	00 56.7	−09 55	11.9	Cet
NGC 337	Gx	00 59.8	−07 35	11.6	Cet
NGC 337A	Gx	01 01.6	−07 35	12.2	Cet
NGC 428	Gx	01 12.9	+00 59	11.5	Cet
NGC 448	Gx	01 15.3	−01 38	12.1	Cet
NGC 450	Gx	01 15.5	−00 52	11.5	Cet
NGC 467	Gx	01 19.2	+03 18	11.8	Psc
NGC 470	Gx	01 19.8	+03 25	11.8	Psc
NGC 474	Gx	01 20.1	+03 25	11.5	Psc
NGC 488	Gx	01 21.8	+05 15	10.3	Psc
NGC 493	Gx	01 22.1	+00 57	12.3	Cet
NGC 514	Gx	01 24.1	+12 55	11.7	Psc
NGC 520	Gx	01 24.6	+03 48	11.4	Psc
NGC 521	Gx	01 24.6	+01 44	11.7	Cet
NGC 524	Gx	01 24.8	+09 32	10.2	Psc
NGC 530	Gx	01 24.7	−01 35	13	Cet
NGC 533	Gx	01 25.5	+01 46	11.4	Cet
NGC 584	Gx	01 31.3	−06 52	10.5	Cet
NGC 596	Gx	01 32.9	−07 02	10.9	Cet
NGC 600	Gx	01 33.1	−07 19	12.4	Cet
NGC 615	Gx	01 35.1	−07 20	11.6	Cet
NGC 628 (see M74).					
NGC 636	Gx	01 39.1	−07 31	11.5	Cet
NGC 660	Gx	01 43.0	+13 39	11.2	Psc
NGC 676	Gx	01 49.0	+05 54	9.6	Psc
NGC 681	Gx	01 49.2	−10 26	12	Cet
NGC 701	Gx	01 51.1	−09 42	12.2	Cet
NGC 718	Gx	01 53.2	+04 12	11.7	Psc
NGC 720	Gx	01 53.0	−13 44	10.2	Cet
NGC 731	Gx	01 54.9	−09 01	12.1	Cet
NGC 741	Gx	01 56.3	+05 38	11.2	Psc
NGC 772	Gx	01 59.3	+19 00	10.3	Ari
NGC 779	Gx	01 59.7	−05 58	11.2	Cet
NGC 788	Gx	02 01.1	−06 49	12.1	Cet
NGC 803	Gx	02 03.8	+16 02	12.6	Ari
NGC 821	Gx	02 08.4	+11 00	10.7	Ari
NGC 835	Gx	02 09.4	−10 08	12.1	Cet
NGC 864	Gx	02 15.5	+06 00	10.8	Cet
NGC 873	Gx	02 16.5	−11 21	12.4	Cet
NGC 877	Gx	02 18.0	+14 33	11.9	Ari
NGC 887	Gx	02 19.5	−16 04	12	Cet
NGC 895	Gx	02 21.6	−05 31	11.7	Cet
NGC 936	Gx	02 27.6	−01 09	10.2	Cet
NGC 941	Gx	02 28.5	−01 09	12.4	Cet
NGC 945	Gx	02 28.6	−10 32	12.1	Cet
NGC 955	Gx	02 30.6	−01 06	12	Cet
NGC 958	Gx	02 30.7	−02 56	12.2	Cet
NGC 988	Gx	02 35.5	−09 22	11	Cet
NGC 991	Gx	02 35.5	−07 09	11.7	Cet
NGC 1015	Gx	02 38.2	−01 19	12.1	Cet
NGC 1016	Gx	02 38.3	+02 07	11.6	Cet
NGC 1022	Gx	02 38.5	−06 41	11.3	Cet
NGC 1032	Gx	02 39.4	+01 06	11.6	Cet
NGC 1035	Gx	02 39.5	−08 08	12.2	Cet
NGC 1042	Gx	02 40.4	−08 26	11	Cet
NGC 1045	Gx	02 40.5	−11 17	12.1	Cet
NGC 1052	Gx	02 41.1	−08 15	10.5	Cet
NGC 1055	Gx	02 41.7	+00 27	10.6	Cet
NGC 1068 (see M77).					
NGC 1070	Gx	02 43.4	+04 58	11.9	Cet
NGC 1073	Gx	02 43.7	+01 23	11	Cet
NGC 1084	Gx	02 46.0	−07 35	10.7	Eri
NGC 1087	Gx	02 46.4	−00 30	10.9	Cet
NGC 1090	Gx	02 46.6	−00 15	11.8	Cet
NGC 1134	Gx	02 53.7	+13 01	12.1	Ari
NGC 1140	Gx	02 54.6	−10 02	12.5	Eri
NGC 1172	Gx	03 01.6	−14 50	11.9	Eri
NGC 1179	Gx	03 02.6	−18 54	11.9	Eri
NGC 1199	Gx	03 03.6	−15 37	11.3	Eri
NGC 1200	Gx	03 03.9	−12 00	12.7	Eri
NGC 1209	Gx	03 06.1	−15 37	11.4	Eri
NGC 1241	Gx	03 11.2	−08 55	12	Eri
NGC 1253	Gx	03 14.2	−02 49	11.7	Eri
NGC 1300	Gx	03 19.7	−19 25	10.4	Eri
NGC 1305	Gx	03 21.4	−02 19	13.3	Eri
NGC 1309	Gx	03 22.1	−15 24	11.5	Eri
NGC 1337	Gx	03 28.1	−08 23	11.9	Eri
NGC 1338	Gx	03 28.9	−12 09	12.7	Eri
NGC 1357	Gx	03 33.3	−13 40	11.5	Eri
NGC 1359	Gx	03 33.8	−19 29	12.1	Eri
NGC 1376	Gx	03 37.1	−05 03	12.1	Eri
NGC 1393	Gx	03 38.6	−18 26	12	Eri
NGC 1400	Gx	03 39.5	−18 41	11	Eri
NGC 1407	Gx	03 40.2	−18 35	9.7	Eri
NGC 1417	Gx	03 42.0	−04 42	12.1	Eri
NGC 1421	Gx	03 42.5	−13 29	11.4	Eri
NGC 1440	Gx	03 45.1	−18 16	11.5	Eri
NGC 1452	Gx	03 45.4	−18 38	12.1	Eri
NGC 1453	Gx	03 46.5	−03 58	11.5	Eri
NGC 1461	Gx	03 48.5	−16 24	11.8	Eri
NGC 7814	Gx	00 03.3	+16 09	10.6	Peg

Charts 10 – 11

Object	Type	R.A. h m	Dec. ° ′	Mag.	Const.
PK171-25.1	PN	03 53.5	+19 28	15.1	Tau
UGC 2296	Gx	02 49.2	+18 20	13.0p	Ari

Chart 11

Object	Type	R.A. h m	Dec. ° ′	Mag.	Const.
B30-32	DN	05 29.8	+12 32		Ori
B33 (see Horsehead Nebula).					
B35	DN	05 45.5	+09 03		Ori
B225	DN	05 29.8	+12 32		Ori
Barnard's Loop	BNe	05 20.0	−04 00		Ori
C41 (see Hyades).					
C46 (see Hubble's Variable Nebula).					
C49 (see Rosette Nebula).					
C50 (see NGC 2244).					
Ced 59	BNer	05 45.3	+09 04		Ori
Ced 62	BNr	06 07.8	+18 41		Ori
Christmas Tree Cluster (see NGC 2264).					
Cone Nebula (see NGC 2264).					
Cr 69	OC	05 35.1	+09 56	2.8p	Ori
Cr 91	OC	06 21.7	+02 22	6.4p	Mon
Cr 92	OC	06 22.9	+05 07	8.6p	Mon
Cr 96	OC	06 30.3	+02 52	7.3	Mon
Cr 97	OC	06 31.3	+05 55	5.4p	Mon
Cr 106	OC	06 37.1	+05 57	4.6p	Mon
Cr 107	OC	06 37.7	+04 44	5.1	Mon
Cr 111	OC	06 38.7	+06 54	7.0p	Mon
Do 25	OC	06 45.1	+00 18	7.6	Mon
E 556-15	Gx	06 21.1	−20 03	12.7p	CMa
Great Orion Nebula	BNer	05 35.4	−05 27	4	Ori
Hind's Variable Nebula	BNr	04 21.8	+19 32		Tau
Horsehead Nebula	DN	05 40.9	−02 28		Ori
Hubble's Variable Nebula	BNer	06 39.2	+08 44		Mon
Hyades	OC	04 27.0	+15 50	0.5	Tau
IC 382	Gx	04 37.9	−09 31	12.2	Eri
IC 418	PN	05 27.5	−12 42	9.3	Lep
IC 423	BNr	05 33.4	−00 37		Ori
IC 426	BNr	05 36.8	−00 15		Ori
IC 430	BNr	05 38.5	−07 05		Ori
IC 431	BNr	05 40.3	−01 27		Ori
IC 432	BNr	05 40.9	−01 29		Ori
IC 434	BNe	05 41.0	−02 24		Ori
IC 435	BNr	05 43.0	−02 19		Ori
IC 438	Gx	05 53.0	−17 53	12.8p	Lep
IC 446	BNr	06 31.0	+10 27		Mon
IC 448	BNr	06 32.7	+07 19		Mon
IC 2118	BNr	05 06.9	−07 13		Eri
IC 2162	BNe	06 12.9	+17 59		Ori
IC 2165	PN	06 21.7	−12 59	10.5	CMa

Object	Type	R.A. h m	Dec. ° ′	Mag.	Const.
IC 2169	BNr	06 31.2	+09 54		Mon
J320	PN	05 05.6	+10 42	12	Ori
J900	PN	06 26.0	+17 48	11.7	Gem
LDN 1616	DN	05 06.5	−03 30		Ori
Lower's Nebula (see Sh2-261).					
M42 (see Great Orion Nebula).					
M43	BNer	05 35.6	−05 16	9	Ori
M78	BNr	05 46.7	+00 03	8	Ori
M-2-9-36	Gx	03 22.9	−11 12	13.0p	Eri
M-2-15-11	Gx	05 50.9	−14 47	11.7p	Lep
M-3-10-42	Gx	03 44.0	−14 22	13.0p	Eri
M-3-10-45	Gx	03 46.6	−16 33	13.0p	Eri
NGC 1172	Gx	03 01.6	−14 50	11.9	Eri
NGC 1179	Gx	03 02.6	−18 54	11.9	Eri
NGC 1199	Gx	03 03.6	−15 37	11.3	Eri
NGC 1200	Gx	03 03.9	−12 00	12.7	Eri
NGC 1209	Gx	03 06.1	−15 37	11.4	Eri
NGC 1241	Gx	03 11.2	−08 55	12	Eri
NGC 1253	Gx	03 14.2	−02 49	11.7	Eri
NGC 1300	Gx	03 19.7	−19 25	10.4	Eri
NGC 1305	Gx	03 21.4	−02 19	13.3	Eri
NGC 1309	Gx	03 22.1	−15 24	11.5	Eri
NGC 1337	Gx	03 28.1	−08 23	11.9	Eri
NGC 1338	Gx	03 28.9	−12 09	12.7	Eri
NGC 1357	Gx	03 33.3	−13 40	11.5	Eri
NGC 1359	Gx	03 33.8	−19 29	12.1	Eri
NGC 1376	Gx	03 37.1	−05 03	12.1	Eri
NGC 1393	Gx	03 38.6	−18 26	12	Eri
NGC 1400	Gx	03 39.5	−18 41	11	Eri
NGC 1407	Gx	03 40.2	−18 35	9.7	Eri
NGC 1417	Gx	03 42.0	−04 42	12.1	Eri
NGC 1421	Gx	03 42.5	−13 29	11.4	Eri
NGC 1440	Gx	03 45.1	−18 16	11.5	Eri
NGC 1452	Gx	03 45.4	−18 38	12.1	Eri
NGC 1453	Gx	03 46.5	−03 58	11.5	Eri
NGC 1461	Gx	03 48.5	−16 24	11.8	Eri
NGC 1507	Gx	04 04.5	−02 11	12.3	Eri
NGC 1535	PN	04 14.3	−12 44	9.6p	Eri
NGC 1554, 5 (see Hind's Variable Nebula).					
NGC 1575	Gx	04 26.3	−10 06	12.2	Eri
NGC 1587	Gx	04 30.7	+00 40	11.7	Tau
NGC 1589	Gx	04 30.8	+00 52	11.8	Tau
NGC 1600	Gx	04 31.7	−05 05	10.9	Eri
NGC 1615	Gx	04 36.0	+19 57	13.6	Tau
NGC 1620	Gx	04 36.6	−00 09	12.3	Eri
NGC 1637	Gx	04 41.5	−02 51	10.8	Eri
NGC 1638	Gx	04 41.6	−01 49	12	Eri
NGC 1647	OC	04 46.0	+19 04	6.4	Tau
NGC 1653	Gx	04 45.8	−02 23	12	Eri
NGC 1662	OC	04 48.5	+10 56	6.4	Ori

Object	Type	R.A. h m	Dec. ° ′	Mag.	Const.
NGC 1667	Gx	04 48.6	−06 19	12.1	Eri
NGC 1684	Gx	04 52.5	−03 06	11.7	Ori
NGC 1691	Gx	04 54.6	+03 16	12	Ori
NGC 1700	Gx	04 56.9	−04 52	11.2	Eri
NGC 1723	Gx	04 59.4	−10 59	11.7	Eri
NGC 1726	Gx	04 59.7	−07 45	11.7	Eri
NGC 1730	Gx	04 59.5	−15 49	12.3	Lep
NGC 1779	Gx	05 05.3	−09 09	12.1	Eri
NGC 1784	Gx	05 05.5	−11 52	11.7	Lep
NGC 1788	BNr	05 06.9	−03 21		Ori
NGC 1807	OC	05 10.7	+16 32	7	Tau
NGC 1817	OC	05 12.1	+16 42	7.7	Tau
NGC 1832	Gx	05 12.1	−15 41	11.3	Lep
NGC 1888	Gx	05 22.6	−11 30	11.9	Lep
NGC 1954	Gx	05 32.8	−14 04	11.8	Lep
NGC 1973	BNer	05 35.1	−04 44		Ori
NGC 1975	BNer	05 35.4	−04 41		Ori
NGC 1976 (see Great Orion Nebula).					
NGC 1977	BNer	05 35.5	−04 52		Ori
NGC 1981	OC	05 35.2	−04 26	4.6	Ori
NGC 1982 (see M43).					
NGC 1999	BNer	05 36.5	−06 42		Ori
NGC 2022	PN	05 42.1	+09 05	11.9	Ori
NGC 2023	BNer	05 41.6	−02 14		Ori
NGC 2024	BNe	05 41.9	−01 51		Ori
NGC 2064	BNr	05 46.3	00 00		Ori
NGC 2067	BNr	05 46.5	+00 06		Ori
NGC 2068 (see M78).					
NGC 2071	BNr	05 47.2	+00 18		Ori
NGC 2089	Gx	05 47.9	−17 36	11.9	Lep
NGC 2149	BNr	06 03.5	−09 44		Mon
NGC 2169	OC	06 08.4	+13 57	5.9	Ori
NGC 2170	BNr	06 07.5	−06 24		Mon
NGC 2182	BNr	06 09.5	−06 20		Mon
NGC 2183, 85	BNr	06 11.0	−06 13		Mon
NGC 2194	OC	06 13.8	+12 48	8.5	Ori
NGC 2195	BN	06 14.4	+17 39		Ori
NGC 2204	OC	06 15.7	−18 39	8.6	CMa
NGC 2215	OC	06 21.0	−07 17	8.4	Mon
NGC 2232	OC	06 28.0	−04 54	3.9	Mon
NGC 2236	OC	06 29.7	+06 50	8.5	Mon
NGC 2237-39, 46 (see Rosette Nebula).					
NGC 2244	OC	06 32.0	+04 55	4.8	Mon
NGC 2245	BNr	06 32.7	+10 10		Mon
NGC 2247	BNr	06 33.2	+10 20		Mon
NGC 2251	OC	06 34.7	+08 22	7.3	Mon
NGC 2252	OC	06 35.0	+05 23	7.7p	Mon
NGC 2261 (see Hubble's Variable Nebula).					
NGC 2264	C+BNe	06 41.1	+09 53	3.9	Mon
NGC 2282	BNr	06 46.9	+01 19		Mon

Object	Type	R.A. h m	Dec. ° ′	Mag.	Const.
NGC 2283	Gx	06 45.9	−18 13	12.2	CMa
NGC 2286	OC	06 47.6	−03 10	7.5	Mon
NGC 2301	OC	06 51.8	+00 28	6	Mon
NGC 2316	BNer	06 59.7	−07 46		Mon
Orion Nebula (see Great Orion Nebula).					
PK171-25.1	PN	03 53.5	+19 28	15.1	Tau
PK198-6.1	PN	06 02.4	+09 39	12	Ori
PK215-30.1	PN	05 03.2	−15 36	13.2p	Lep
Rosette Neb.	BNe	06 32.3	+05 03		Mon
Sh2-261	BNe	06 08.9	+15 49		Ori
Sh2-276 (see Barnard's Loop).					
Sh2-282	BNe	06 38.0	+01 31		Mon
Tank Trap Nebula (see NGC 2024).					
Taurus Moving Cluster (see Hyades).					
Trapezium	C+BNer	05 35.3	−05 23	4.7	Ori
vdB 26	BNr	04 13.6	+10 13		Tau
vdB 37	BNr	05 18.2	+13 24		Ori
vdB 38	BNr	05 21.7	+08 27		Ori
vdB 49	BNr	05 39.2	+04 10		Ori
Witch's Head Nebula (see IC 2118).					

Chart 12

Object	Type	R.A. h m	Dec. ° ′	Mag.	Const.
Bas 11A	OC	07 17.1	−13 58	8.2	CMa
Beehive Cluster (see Praesepe).					
Butterfly Wing Nebula (see NGC 2346).					
C46 (see Hubble's Variable Nebula).					
C48 (see NGC 2775).					
C49 (see Rosette Nebula).					
C50 (see NGC 2244).					
C54 (see NGC 2506).					
C58 (see NGC 2360).					
Ced 62	BNr	06 07.8	+18 41		Ori
Ced 90	BNer	07 05.2	−12 20		CMa
Christmas Tree Cluster (see NGC 2264).					
Cone Nebula (see NGC 2264).					
Cr 91	OC	06 21.7	+02 22	6.4p	Mon
Cr 92	OC	06 22.9	+05 07	8.6p	Mon
Cr 96	OC	06 30.3	+02 52	7.3	Mon
Cr 97	OC	06 31.3	+05 55	5.4p	Mon
Cr 106	OC	06 37.1	+05 57	4.6p	Mon
Cr 107	OC	06 37.7	+04 44	5.1	Mon
Cr 111	OC	06 38.7	+06 54	7.0p	Mon
Do 25	OC	06 45.1	+00 18	7.6	Mon
E 556-15	Gx	06 21.1	−20 03	12.7p	CMa
E 563-31	Gx	08 52.3	−17 45	13.0p	Hya
Gum 1	BNer	07 04.3	−10 28		Mon
Hubble's Variable Nebula	BNer	06 39.2	+08 44		Mon
IC 446	BNr	06 31.0	+10 27		Mon
IC 448	BNr	06 32.7	+07 19		Mon
IC 466	BNer	07 08.6	−04 19		Mon

Chart 12

Object	Type	R.A. h m	Dec. ° ′	Mag.	Const.
IC 2162	BNe	06 12.9	+17 59		Ori
IC 2165	PN	06 21.7	−12 59	10.5	CMa
IC 2169	BNr	06 31.2	+09 54		Mon
IC 2177	BNe	07 05.1	−10 42		Mon
IC 2367	Gx	08 24.2	−18 47	12.5p	Pup
IC 2482	Gx	09 27.0	−12 07	12.5p	Hya
J900	PN	06 26.0	+17 48	11.7	Gem
LBN 1036	BNe	07 16.0	−10 40		Mon
Lower's Nebula (see Sh2-261).					
M44 (see Praesepe).					
M46	OC	07 41.8	−14 49	6.1	Pup
M47	OC	07 36.6	−14 30	4.4	Pup
M48	OC	08 13.8	−05 48	5.8	Hya
M50	OC	07 02.8	−08 23	5.9	Mon
M67	OC	08 51.4	+11 49	6.9	Cnc
M-1-23-19	Gx	09 06.6	−07 14	12.9p	Hya
M-1-24-1	Gx	09 10.9	−08 53	11.9p	Hya
M-2-25-6	Gx	09 36.5	−11 20	12.6p	Hya
M-3-25-4	Gx	09 33.3	−16 46	12.6p	Hya
Manger (see Praesepe).					
Medusa Nebula (see PK205+14.1).					
Mel 71	OC	07 37.5	−12 04	7.1	Pup
Mrk 1236	Gx	09 49.9	+00 37	13.0p	Sex
NGC 2149	BNr	06 03.5	−09 44		Mon
NGC 2169	OC	06 08.4	+13 57	5.9	Ori
NGC 2170	BNr	06 07.5	−06 24		Mon
NGC 2182	BNr	06 09.5	−06 20		Mon
NGC 2183, 85	BNr	06 11.0	−06 13		Mon
NGC 2194	OC	06 13.8	+12 48	8.5	Ori
NGC 2195	BN	06 14.4	+17 39		Ori
NGC 2215	OC	06 21.0	−07 17	8.4	Mon
NGC 2232	OC	06 28.0	−04 54	3.9	Mon
NGC 2236	OC	06 29.7	+06 50	8.5	Mon
NGC 2237-39, 46 (see Rosette Nebula).					
NGC 2244	OC	06 32.0	+04 55	4.8	Mon
NGC 2245	BNr	06 32.7	+10 10		Mon
NGC 2247	BNr	06 33.2	+10 20		Mon
NGC 2251	OC	06 34.7	+08 22	7.3	Mon
NGC 2252	OC	06 35.0	+05 23	7.7p	Mon
NGC 2261 (see Hubble's Variable Nebula).					
NGC 2264	C+BNe	06 41.1	+09 53	3.9	Mon
NGC 2282	BNr	06 46.9	+01 19		Mon
NGC 2283	Gx	06 45.9	−18 13	12.2	CMa
NGC 2286	OC	06 47.6	−03 10	7.5	Mon
NGC 2301	OC	06 51.8	+00 28	6	Mon
NGC 2316	BNer	06 59.7	−07 46		Mon
NGC 2323 (see M50).					
NGC 2324	OC	07 04.2	+01 03	8.4	Mon
NGC 2335	OC	07 06.6	−10 05	7.2	Mon
NGC 2339	Gx	07 08.3	+18 47	11.8	Gem
NGC 2343	OC	07 08.3	−10 39	6.7	Mon
NGC 2345	OC	07 08.3	−13 10	7.7	CMa
NGC 2346	PN	07 09.4	−00 48	11.6	Mon
NGC 2353	OC	07 14.6	−10 18	7.1	Mon
NGC 2359	BNe	07 18.6	−13 12		CMa
NGC 2360	OC	07 17.8	−15 37	7.2	CMa
NGC 2374	OC	07 24.0	−13 16	8	CMa
NGC 2395	OC	07 27.1	+13 35	8	Gem
NGC 2396	OC	07 28.1	−11 44	7.4p	Pup
NGC 2414	OC	07 33.3	−15 27	7.9	Pup
NGC 2422 (see M47).					
NGC 2423	OC	07 37.1	−13 52	6.7	Pup
NGC 2437 (see M46).					
NGC 2438	PN	07 41.8	−14 44	11	Pup
NGC 2440	PN	07 41.9	−18 12	9.4	Pup
NGC 2485	Gx	07 56.8	+07 29	12.2	CMi
NGC 2506	OC	08 00.2	−10 47	7.6	Mon
NGC 2513	Gx	08 02.4	+09 25	11.6	Cnc
NGC 2517	Gx	08 02.8	−12 19	11.8	Pup
NGC 2525	Gx	08 05.6	−11 26	11.6	Pup
NGC 2539	OC	08 10.7	−12 50	6.5	Pup
NGC 2548 (see M48).					
NGC 2610	PN	08 33.4	−16 09	12.8	Hya
NGC 2612	Gx	08 33.8	−13 10	12.7	Hya
NGC 2632 (see Praesepe).					
NGC 2648	Gx	08 42.7	+14 17	11.8	Cnc
NGC 2665	Gx	08 46.0	−19 18	12.7	Hya
NGC 2672	Gx	08 49.4	+19 05	11.7	Cnc
NGC 2682 (see M67).					
NGC 2695	Gx	08 54.5	−03 04	11.9	Hya
NGC 2698	Gx	08 55.6	−03 11	12.6	Hya
NGC 2708	Gx	08 56.1	−03 22	12	Hya
NGC 2713	Gx	08 57.3	+02 55	11.8	Hya
NGC 2716	Gx	08 57.6	+03 05	11.8	Hya
NGC 2721	Gx	08 58.9	−04 54	11.7	Hya
NGC 2749	Gx	09 05.4	+18 19	11.8	Cnc
NGC 2763	Gx	09 06.8	−15 30	12	Hya
NGC 2775	Gx	09 10.3	+07 02	10.1	Cnc
NGC 2781	Gx	09 11.5	−14 49	11.6	Hya
NGC 2811	Gx	09 16.2	−16 19	11.3	Hya
NGC 2817	Gx	09 17.2	−04 45	12.6	Hya
NGC 2848	Gx	09 20.2	−16 32	11.8	Hya
NGC 2855	Gx	09 21.5	−11 55	11.7	Hya
NGC 2872	Gx	09 25.7	+11 26	11.9	Leo
NGC 2889	Gx	09 27.2	−11 39	11.7	Hya
NGC 2906	Gx	09 32.1	+08 27	12.7	Leo
NGC 2907	Gx	09 31.6	−16 44	11.7	Hya
NGC 2911	Gx	09 33.8	+10 09	11.5	Leo
NGC 2947	Gx	09 36.1	−12 26	12.4	Hya
NGC 2962	Gx	09 40.9	+05 10	11.9	Hya

Object	Type	R.A. h m	Dec. ° ′	Mag.	Const.
NGC 2967	Gx	09 42.1	+00 20	11.6	Sex
NGC 2974	Gx	09 42.6	−03 42	10.9	Sex
NGC 2993	Gx	09 45.8	−14 22	12.6	Hya
NGC 3020	Gx	09 50.1	+12 49	11.9	Leo
NGC 3023	Gx	09 49.9	+00 37	12.2	Sex
NGC 3041	Gx	09 53.1	+16 41	11.5	Leo
NGC 3044	Gx	09 53.7	+01 35	11.9	Sex
NGC 3052	Gx	09 54.5	−18 38	12.2	Hya
NGC 3055	Gx	09 55.3	+04 16	12.1	Sex
Opened Box Cluster (see NGC 2360).					
PK198-6.1	PN	06 02.4	+09 39	12	Ori
PK205+14.1	PN	07 29.0	+13 15	10.3	Gem
PK217+14.1	PN	07 51.7	+03 00	13.5	CMi
PK219+31.1	PN	08 54.2	+08 55	12	Cnc
PK238+34.1	PN	09 39.1	−02 48	12.4	Hya
Praesepe	OC	08 40.1	+19 59	3.1	Cnc
Rosette Neb.	BNe	06 32.3	+05 03		Mon
Sh2-261	BNe	06 08.9	+15 49		Ori
Sh2-282	BNe	06 38.0	+01 31		Mon
Sh2-294	BNe	07 16.6	−09 26		Mon
Sh2-301	BNe	07 09.8	−18 29		CMa
Sh2-302	BNe	07 31.6	−16 58		Pup
Sh2-307	BNe	07 35.5	−18 46		Pup
Thor's Helmet (see NGC 2359).					
UGC 5373	Gx	10 00.0	+05 20	11.8p	Sex
vdB 97	BNr	07 32.6	−16 54		Pup

Chart 13 Antennae (see NGC 4038 and NGC 4039).
C40 (see NGC 3626).
C48 (see NGC 2775).
C52 (see NGC 4697).
C53 (see NGC 3115).
C59 (see NGC 3242).
C60 (see NGC 4038).
C61 (see NGC 4039).
CBS Eye Nebula (see NGC 3242).
Eyes (see NGC 4435 and NGC 4438).
Ghost of Jupiter Nebula (see NGC 3242).

Object	Type	R.A. h m	Dec. ° ′	Mag.	Const.
IC 600	Gx	10 17.2	−03 30	13.0p	Sex
IC 630	Gx	10 38.5	−07 10	12.9p	Sex
IC 651	Gx	10 51.0	−02 09	12.7p	Sex
IC 2482	Gx	09 27.0	−12 07	12.5p	Hya
III Zw 66	Gx	12 27.2	+14 07	12.3p	Com
Leo I	Gx	10 08.5	+12 18	9.8	Leo
Leo's Triplet	Gx	11 20.0	+13 20	9	Com
Lost Galaxy (see NGC 4526).					
M49	Gx	12 29.8	+08 00	8.4	Vir
M58	Gx	12 37.7	+11 49	9.8	Vir
M59	Gx	12 42.0	+11 39	9.8	Vir
M60	Gx	12 43.7	+11 33	8.8	Vir
M61	Gx	12 21.9	+04 28	9.7	Vir
M65	Gx	11 18.9	+13 05	9.3	Leo
M66	Gx	11 20.2	+12 59	9	Leo
M84	Gx	12 25.1	+12 53	9.3	Vir
M85	Gx	12 25.4	+18 11	9.2	Com
M86	Gx	12 26.2	+12 57	9.2	Vir
M87	Gx	12 30.8	+12 24	8.6	Vir
M88	Gx	12 32.0	+14 25	9.5	Com
M89	Gx	12 35.7	+12 33	9.8	Vir
M90	Gx	12 36.8	+13 10	9.5	Vir
M91	Gx	12 35.4	+14 30	10.2	Com
M95	Gx	10 44.0	+11 42	9.7	Leo
M96	Gx	10 46.8	+11 49	9.2	Leo
M98	Gx	12 13.8	+14 54	10.1	Com
M99	Gx	12 18.8	+14 25	9.8	Com
M100	Gx	12 22.9	+15 49	9.4	Com
M104 (see Sombrero Galaxy).					
M105	Gx	10 47.8	+12 35	9.3	Leo
M-0-27-5	Gx	10 23.9	−03 11	13.0p	Sex
M-1-23-19	Gx	09 06.6	−07 14	12.9p	Hya
M-1-24-1	Gx	09 10.9	−08 53	11.9p	Hya
M-1-26-30	Gx	10 11.0	−04 43	11.7p	Sex
M-1-29-15	Gx	11 22.7	−07 41	13.0p	Crt
M-1-32-19	Gx	12 30.3	−08 24	12.5p	Vir
M-1-32-28	Gx	12 35.6	−07 53	12.6p	Vir
M-1-33-1	Gx	12 44.1	−05 41	12.9p	Vir
M-1-33-3	Gx	12 45.7	−06 04	13.0p	Vir
M-1-33-27	Gx	12 51.2	−06 34	12.4p	Vir
M-2-25-6	Gx	09 36.5	−11 20	12.6p	Hya
M-2-30-14	Gx	11 40.6	−10 05	12.6p	Crt
M-2-33-15	Gx	12 49.4	−10 07	11.6	Vir
M-2-33-17	Gx	12 50.1	−14 44	12.9p	Crv
M-3-25-4	Gx	09 33.3	−16 46	12.6p	Hya
Mrk 1236	Gx	09 49.9	+00 37	13.0p	Sex
NGC 2749	Gx	09 05.4	+18 19	11.8	Cnc
NGC 2763	Gx	09 06.8	−15 30	12	Hya
NGC 2775	Gx	09 10.3	+07 02	10.1	Cnc
NGC 2781	Gx	09 11.5	−14 49	11.6	Hya
NGC 2811	Gx	09 16.2	−16 19	11.3	Hya
NGC 2817	Gx	09 17.2	−04 45	12.6	Hya
NGC 2848	Gx	09 20.2	−16 32	11.8	Hya
NGC 2855	Gx	09 21.5	−11 55	11.7	Hya
NGC 2872	Gx	09 25.7	+11 26	11.9	Leo
NGC 2889	Gx	09 27.2	−11 39	11.7	Hya
NGC 2906	Gx	09 32.1	+08 27	12.7	Leo
NGC 2907	Gx	09 31.6	−16 44	11.7	Hya
NGC 2911	Gx	09 33.8	+10 09	11.5	Leo
NGC 2947	Gx	09 36.1	−12 26	12.4	Hya
NGC 2962	Gx	09 40.9	+05 10	11.9	Hya
NGC 2967	Gx	09 42.1	+00 20	11.6	Sex

Chart 13

Object	Type	R.A. h　m	Dec. °　′	Mag.	Const.	Object	Type	R.A. h　m	Dec. °　′	Mag.	Const.
NGC 2974	Gx	09 42.6	−03 42	10.9	Sex	NGC 3630	Gx	11 20.3	+02 58	11.9	Leo
NGC 2993	Gx	09 45.8	−14 22	12.6	Hya	NGC 3640	Gx	11 21.1	+03 14	10.4	Leo
NGC 3020	Gx	09 50.1	+12 49	11.9	Leo	NGC 3655	Gx	11 22.9	+16 35	11.7	Leo
NGC 3023	Gx	09 49.9	+00 37	12.2	Sex	NGC 3659	Gx	11 23.8	+17 49	12.3	Leo
NGC 3041	Gx	09 53.1	+16 41	11.5	Leo	NGC 3660	Gx	11 23.5	−08 39	13.2	Crt
NGC 3044	Gx	09 53.7	+01 35	11.9	Sex	NGC 3663	Gx	11 24.0	−12 18	12.5	Crt
NGC 3052	Gx	09 54.5	−18 38	12.2	Hya	NGC 3666	Gx	11 24.4	+11 21	12	Leo
NGC 3055	Gx	09 55.3	+04 16	12.1	Sex	NGC 3672	Gx	11 25.0	−09 48	11.4	Crt
NGC 3091	Gx	10 00.2	−19 38	11	Hya	NGC 3681	Gx	11 26.5	+16 52	11.2	Leo
NGC 3115	Gx	10 05.2	−07 43	8.9	Sex	NGC 3684	Gx	11 27.2	+17 02	11.4	Leo
NGC 3124	Gx	10 06.7	−19 13	12	Hya	NGC 3686	Gx	11 27.7	+17 13	11.3	Leo
NGC 3145	Gx	10 10.2	−12 26	11.7	Hya	NGC 3691	Gx	11 28.1	+16 55	11.8	Leo
NGC 3166	Gx	10 13.7	+03 26	10.4	Sex	NGC 3692	Gx	11 28.4	+09 24	12.1	Leo
NGC 3169	Gx	10 14.2	+03 28	10.2	Sex	NGC 3705	Gx	11 30.1	+09 17	11.1	Leo
NGC 3200	Gx	10 18.6	−17 59	12.2	Hya	NGC 3732	Gx	11 34.2	−09 51	12.5	Crt
NGC 3226	Gx	10 23.5	+19 54	11.4	Leo	NGC 3800	Gx	11 40.2	+15 21	12.7	Leo
NGC 3227	Gx	10 23.5	+19 52	10.3	Leo	NGC 3801	Gx	11 40.3	+17 44	12	Leo
NGC 3239	Gx	10 25.1	+17 10	11.3	Leo	NGC 3810	Gx	11 41.0	+11 28	10.8	Leo
NGC 3242	PN	10 24.8	−18 38	7.8	Hya	NGC 3818	Gx	11 42.0	−06 09	11.7	Vir
NGC 3338	Gx	10 42.1	+13 45	11.1	Leo	NGC 3836	Gx	11 43.5	−16 48	12.7	Crt
NGC 3346	Gx	10 43.6	+14 52	11.7	Leo	NGC 3842	Gx	11 44.0	+19 57	12	Leo
NGC 3351 (see M95).						NGC 3865	Gx	11 44.9	−09 14	12	Crt
NGC 3367	Gx	10 46.6	+13 45	11.5	Leo	NGC 3872	Gx	11 45.8	+13 46	11.7	Leo
NGC 3368 (see M96).						NGC 3887	Gx	11 47.1	−16 51	10.6	Crt
NGC 3370	Gx	10 47.1	+17 16	11.6	Leo	NGC 3892	Gx	11 48.0	−10 58	11.5	Crt
NGC 3377	Gx	10 47.7	+13 59	10.4	Leo	NGC 3957	Gx	11 54.0	−19 34	12	Crt
NGC 3379 (see M105).						NGC 3962	Gx	11 54.7	−13 58	10.7	Crt
NGC 3384	Gx	10 48.3	+12 38	9.9	Leo	NGC 3968	Gx	11 55.5	+11 58	11.8	Leo
NGC 3389	Gx	10 48.5	+12 32	11.9	Leo	NGC 3976	Gx	11 55.9	+06 45	11.5	Vir
NGC 3411	Gx	10 50.4	−12 51	11.9	Hya	NGC 3981	Gx	11 56.1	−19 54	11	Crt
NGC 3412	Gx	10 50.9	+13 25	10.5	Leo	NGC 4024	Gx	11 58.5	−18 21	11.9	Crv
NGC 3423	Gx	10 51.2	+05 51	11.1	Sex	NGC 4027	Gx	11 59.5	−19 16	11.2	Crv
NGC 3433	Gx	10 52.1	+10 09	11.6	Leo	NGC 4030	Gx	12 00.4	−01 06	10.6	Vir
NGC 3457	Gx	10 54.8	+17 37	12.6	Leo	NGC 4033	Gx	12 00.6	−17 51	11.8	Crv
NGC 3485	Gx	11 00.0	+14 51	11.8	Leo	NGC 4037	Gx	12 01.4	+13 24	11.9	Com
NGC 3489	Gx	11 00.3	+13 54	10.3	Leo	NGC 4038	Gx	12 01.9	−18 52	10.5	Crv
NGC 3495	Gx	11 01.3	+03 38	11.8	Leo	NGC 4039	Gx	12 01.9	−18 53	10.3	Crv
NGC 3507	Gx	11 03.4	+18 08	10.9	Leo	NGC 4045	Gx	12 02.7	+01 59	12	Vir
NGC 3521	Gx	11 05.8	−00 02	9	Leo	NGC 4064	Gx	12 04.2	+18 27	11.4	Com
NGC 3528	Gx	11 07.3	−19 28	11.8	Crt	NGC 4067	Gx	12 04.2	+10 51	12.5	Vir
NGC 3593	Gx	11 14.6	+12 49	10.4	Leo	NGC 4073	Gx	12 04.4	+01 54	11.4	Vir
NGC 3596	Gx	11 15.1	+14 47	11.2	Leo	NGC 4094	Gx	12 05.9	−14 32	11.8	Crv
NGC 3599	Gx	11 15.5	+18 07	11.9	Leo	NGC 4116	Gx	12 07.6	+02 42	12	Vir
NGC 3607	Gx	11 16.9	+18 03	9.9	Leo	NGC 4123	Gx	12 08.2	+02 53	11.4	Vir
NGC 3608	Gx	11 17.0	+18 09	10.8	Leo	NGC 4124	Gx	12 08.2	+10 23	11.3	Vir
NGC 3611	Gx	11 17.5	+04 33	12.2	Leo	NGC 4147	GC	12 10.1	+18 33	10.4	Com
NGC 3623 (see M65).						NGC 4152	Gx	12 10.6	+16 02	12.2	Com
NGC 3626	Gx	11 20.1	+18 21	11	Leo	NGC 4168	Gx	12 12.3	+13 12	11.2	Vir
NGC 3627 (see M66).						NGC 4177	Gx	12 12.7	−14 01	13.2	Crv
NGC 3628	Gx	11 20.3	+13 35	9.5	Leo	NGC 4178	Gx	12 12.8	+10 52	11.4	Vir

Object	Type	R.A. h m	Dec. ° ′	Mag.	Const.
NGC 4179	Gx	12 12.9	+01 18	11	Vir
NGC 4189	Gx	12 13.8	+13 26	11.7	Com
NGC 4192 (see M98).					
NGC 4193	Gx	12 13.9	+13 10	12.3	Vir
NGC 4206	Gx	12 15.3	+13 02	12.2	Vir
NGC 4212	Gx	12 15.7	+13 54	11.2	Com
NGC 4215	Gx	12 15.9	+06 24	12.1	Vir
NGC 4216	Gx	12 15.9	+13 09	10	Vir
NGC 4224	Gx	12 16.6	+07 28	11.8	Vir
NGC 4233	Gx	12 17.1	+07 37	11.9	Vir
NGC 4235	Gx	12 17.1	+07 12	11.6	Vir
NGC 4237	Gx	12 17.2	+15 19	11.6	Com
NGC 4240	Gx	12 17.4	−09 57	12.7	Vir
NGC 4241	Gx	12 17.4	+06 41	11.9	Vir
NGC 4254 (see M99).					
NGC 4260	Gx	12 19.4	+06 06	11.8	Vir
NGC 4261	Gx	12 19.4	+05 50	10.4	Vir
NGC 4262	Gx	12 19.5	+14 53	11.6	Com
NGC 4267	Gx	12 19.8	+12 48	10.9	Vir
NGC 4270	Gx	12 19.8	+05 28	12.2	Vir
NGC 4273	Gx	12 19.9	+05 21	11.9	Vir
NGC 4281	Gx	12 20.4	+05 23	11.3	Vir
NGC 4293	Gx	12 21.2	+18 23	10.4	Com
NGC 4294	Gx	12 21.3	+11 31	12.1	Vir
NGC 4298	Gx	12 21.5	+14 36	11.3	Com
NGC 4299	Gx	12 21.7	+11 30	12.5	Vir
NGC 4302	Gx	12 21.7	+14 36	11.6	Com
NGC 4303 (see M61).					
NGC 4307	Gx	12 22.1	+09 03	12	Vir
NGC 4312	Gx	12 22.5	+15 32	11.7	Com
NGC 4313	Gx	12 22.6	+11 48	11.6	Vir
NGC 4321 (see M100).					
NGC 4324	Gx	12 23.1	+05 15	11.6	Vir
NGC 4339	Gx	12 23.6	+06 05	11.3	Vir
NGC 4340	Gx	12 23.6	+16 43	11.2	Com
NGC 4343	Gx	12 23.6	+06 57	12.1	Vir
NGC 4350	Gx	12 24.0	+16 42	11	Com
NGC 4351	Gx	12 24.0	+12 12	12.6	Vir
NGC 4361	PN	12 24.5	−18 48	10.9	Crv
NGC 4365	Gx	12 24.5	+07 19	9.6	Vir
NGC 4371	Gx	12 24.9	+11 42	10.8	Vir
NGC 4374 (see M84).					
NGC 4377	Gx	12 25.2	+14 46	11.9	Com
NGC 4378	Gx	12 25.3	+04 56	11.7	Vir
NGC 4379	Gx	12 25.2	+15 36	11.7	Com
NGC 4380	Gx	12 25.4	+10 01	11.7	Vir
NGC 4382 (see M85).					
NGC 4383	Gx	12 25.4	+16 28	12.1	Com
NGC 4387	Gx	12 25.7	+12 49	12.1	Vir
NGC 4388	Gx	12 25.8	+12 40	11	Vir
NGC 4394	Gx	12 25.9	+18 13	10.9	Com
NGC 4402	Gx	12 26.1	+13 07	11.8	Vir
NGC 4405	Gx	12 26.1	+16 11	12	Com
NGC 4406 (see M86).					
NGC 4411B	Gx	12 26.8	+08 53	12.3	Vir
NGC 4413	Gx	12 26.5	+12 37	12.2	Vir
NGC 4417	Gx	12 26.8	+09 35	11.1	Vir
NGC 4419	Gx	12 26.9	+15 03	11.2	Com
NGC 4420	Gx	12 27.0	+02 30	12.1	Vir
NGC 4421	Gx	12 27.0	+15 28	11.6	Com
NGC 4424	Gx	12 27.2	+09 25	11.7	Vir
NGC 4425	Gx	12 27.2	+12 44	11.8	Vir
NGC 4429	Gx	12 27.4	+11 06	10	Vir
NGC 4430	Gx	12 27.4	+06 16	12	Vir
NGC 4434	Gx	12 27.6	+08 09	12.2	Vir
NGC 4435	Gx	12 27.7	+13 05	10.8	Vir
NGC 4438	Gx	12 27.8	+13 01	10.2	Vir
NGC 4440	Gx	12 27.9	+12 18	11.7	Vir
NGC 4442	Gx	12 28.1	+09 48	10.4	Vir
NGC 4450	Gx	12 28.5	+17 05	10.1	Com
NGC 4452	Gx	12 28.7	+11 45	12	Vir
NGC 4454	Gx	12 28.9	−01 56	11.9	Vir
NGC 4457	Gx	12 29.0	+03 34	10.9	Vir
NGC 4458	Gx	12 29.0	+13 15	12.1	Vir
NGC 4459	Gx	12 29.0	+13 59	10.4	Com
NGC 4461	Gx	12 29.0	+13 11	11.2	Vir
NGC 4469	Gx	12 29.5	+08 45	11.2	Vir
NGC 4470	Gx	12 29.6	+07 49	12.1	Vir
NGC 4472 (see M49).					
NGC 4473	Gx	12 29.8	+13 26	10.2	Com
NGC 4474	Gx	12 29.9	+14 04	11.5	Com
NGC 4477	Gx	12 30.0	+13 38	10.4	Com
NGC 4478	Gx	12 30.3	+12 20	11.4	Vir
NGC 4480	Gx	12 30.4	+04 15	12.4	Vir
NGC 4486 (see M87).					
NGC 4487	Gx	12 31.1	−08 03	10.9	Vir
NGC 4488	Gx	12 30.9	+08 22	12.2	Vir
NGC 4489	Gx	12 30.9	+16 46	12	Com
NGC 4496A	Gx	12 31.7	+03 56	11.4	Vir
NGC 4496B	Gx	12 31.7	+03 56	13.5	Vir
NGC 4498	Gx	12 31.7	+16 51	12.2	Com
NGC 4501 (see M88).					
NGC 4503	Gx	12 32.1	+11 11	11.1	Vir
NGC 4504	Gx	12 32.3	−07 34	11.2	Vir
NGC 4517	Gx	12 32.8	+00 07	10.4	Vir
NGC 4517A	Gx	12 32.5	+00 23	12.5	Vir
NGC 4519	Gx	12 33.5	+08 39	11.8	Vir
NGC 4522	Gx	12 33.7	+09 10	12.3	Vir
NGC 4526	Gx	12 34.0	+07 42	9.7	Vir
NGC 4527	Gx	12 34.1	+02 39	10.5	Vir

Charts 13 – 14

Object	Type	R.A. h m	Dec. ° ′	Mag.	Const.
NGC 4528	Gx	12 34.1	+11 19	12.1	Vir
NGC 4531	Gx	12 34.3	+13 05	11.4	Vir
NGC 4532	Gx	12 34.3	+06 28	11.9	Vir
NGC 4535	Gx	12 34.3	+08 12	10	Vir
NGC 4536	Gx	12 34.4	+02 11	10.6	Vir
NGC 4539	Gx	12 34.6	+18 12	12	Com
NGC 4540	Gx	12 34.8	+15 33	11.7	Com
NGC 4546	Gx	12 35.5	–03 48	10.3	Vir
NGC 4548 (see M91).					
NGC 4550	Gx	12 35.5	+12 13	11.7	Vir
NGC 4551	Gx	12 35.6	+12 16	12	Vir
NGC 4552 (see M89).					
NGC 4561	Gx	12 36.1	+19 19	12.5	Com
NGC 4564	Gx	12 36.5	+11 26	11.1	Vir
NGC 4567	Gx	12 36.5	+11 16	11.3	Vir
NGC 4568	Gx	12 36.6	+11 14	10.8	Vir
NGC 4569 (see M90).					
NGC 4570	Gx	12 36.9	+07 15	10.9	Vir
NGC 4571	Gx	12 36.9	+14 13	11.3	Com
NGC 4578	Gx	12 37.5	+09 33	11.5	Vir
NGC 4579 (see M58).					
NGC 4580	Gx	12 37.8	+05 22	11.8	Vir
NGC 4586	Gx	12 38.5	+04 19	11.7	Vir
NGC 4592	Gx	12 39.3	–00 32	11.7	Vir
NGC 4593	Gx	12 39.7	–05 21	10.9	Vir
NGC 4594 (see Sombrero Galaxy).					
NGC 4595	Gx	12 39.9	+15 18	12.1	Com
NGC 4596	Gx	12 39.9	+10 11	10.4	Vir
NGC 4597	Gx	12 40.2	–05 48	12.1	Vir
NGC 4602	Gx	12 40.6	–05 08	11.5	Vir
NGC 4606	Gx	12 41.0	+11 55	11.8	Vir
NGC 4608	Gx	12 41.2	+10 09	11	Vir
NGC 4612	Gx	12 41.5	+07 19	10.9	Vir
NGC 4621 (see M59).					
NGC 4632	Gx	12 42.5	–00 05	11.7	Vir
NGC 4636	Gx	12 42.8	+02 41	9.5	Vir
NGC 4638	Gx	12 42.8	+11 27	11.2	Vir
NGC 4639	Gx	12 42.9	+13 15	11.5	Vir
NGC 4643	Gx	12 43.3	+01 59	10.8	Vir
NGC 4647	Gx	12 43.5	+11 35	11.3	Vir
NGC 4649 (see M60).					
NGC 4651	Gx	12 43.7	+16 24	10.8	Com
NGC 4653	Gx	12 43.8	–00 34	12.2	Vir
NGC 4654	Gx	12 43.9	+13 08	10.6	Vir
NGC 4658	Gx	12 44.6	–10 05	12.5	Vir
NGC 4660	Gx	12 44.5	+11 11	11.2	Vir
NGC 4665	Gx	12 45.1	+03 03	10.5	Vir
NGC 4666	Gx	12 45.1	–00 28	10.7	Vir
NGC 4684	Gx	12 47.3	–02 44	11.4	Vir
NGC 4689	Gx	12 47.8	+13 46	10.9	Com

Object	Type	R.A. h m	Dec. ° ′	Mag.	Const.
NGC 4691	Gx	12 48.2	–03 20	11.1	Vir
NGC 4694	Gx	12 48.3	+10 59	11.4	Vir
NGC 4697	Gx	12 48.6	–05 48	9.2	Vir
NGC 4698	Gx	12 48.4	+08 29	10.6	Vir
NGC 4699	Gx	12 49.0	–08 40	9.5	Vir
NGC 4700	Gx	12 49.1	–11 25	11.9	Vir
NGC 4701	Gx	12 49.2	+03 23	12.4	Vir
NGC 4710	Gx	12 49.7	+15 10	11	Com
NGC 4713	Gx	12 50.0	+05 19	11.7	Vir
NGC 4727	Gx	12 51.0	–14 20	13	Crv
NGC 4731	Gx	12 51.0	–06 23	11.5	Vir
NGC 4733	Gx	12 51.1	+10 55	11.8	Vir
NGC 4739	Gx	12 51.6	–08 25	12.6	Vir
NGC 4742	Gx	12 51.8	–10 27	11.3	Vir
NGC 4753	Gx	12 52.4	–01 12	9.9	Vir
NGC 4754	Gx	12 52.3	+11 19	10.6	Vir
NGC 4760	Gx	12 53.1	–10 30	11.4	Vir
NGC 4762	Gx	12 52.9	+11 14	10.3	Vir
NGC 4771	Gx	12 53.4	+01 16	12.3	Vir
NGC 4772	Gx	12 53.5	+02 10	11	Vir
NGC 4775	Gx	12 53.8	–06 37	11.1	Vir
NGC 4781	Gx	12 54.4	–10 32	11.1	Vir
NGC 4782	Gx	12 54.6	–12 34	11.7	Crv
NGC 4783	Gx	12 54.6	–12 33	11.5	Crv
NGC 4786	Gx	12 54.5	–06 51	11.7	Vir
NGC 4790	Gx	12 54.9	–10 15	12.1	Vir
NGC 4802	Gx	12 55.8	–12 03	11.8	Crv
NGC 4808	Gx	12 55.8	+04 18	11.7	Vir
NGC 4818	Gx	12 56.8	–08 32	11.1	Vir
NGC 4825	Gx	12 57.2	–13 40	11.7	Vir
NGC 4845	Gx	12 58.0	+01 35	11.2	Vir
NGC 4856	Gx	12 59.4	–15 03	10.5	Vir
NGC 4866	Gx	12 59.5	+14 10	11.2	Vir
PK238+34.1	PN	09 39.1	–02 48	12.4	Hya
Ring Tail (see NGC 4038 and NGC 4039).					
Siamese Twins (see NGC 4567 and NGC 4568).					
Sombrero Galaxy	Gx	12 40.0	–11 37	8.3	Vir
Spindle Galaxy (see NGC 3115).					
UGC 5373	Gx	10 00.0	+05 20	11.8p	Sex
UGC 5470 (see Leo I).					
Virgo A (see M87).					

Object	Type	R.A. h m	Dec. ° ′	Mag.	Const.
Antennae (see NGC 4038 and NGC 4039).					
C45 (see NGC 5248).					
C52 (see NGC 4697).					
C60 (see NGC 4038).					
C61 (see NGC 4039).					
Eyes (see NGC 4435 and NGC 4438).					
IC 983	Gx	14 10.1	+17 44	12.5p	Boo
IC 1055	Gx	14 47.4	–13 43	12.7p	Lib

Chart 14

Chart 14

Object	Type	R.A. h m	Dec. ° ′	Mag.	Const.
III Zw 66	Gx	12 27.2	+14 07	12.3p	Com
LDN 134	DN	15 53.6	−04 39		Lib
Lost Galaxy (see NGC 4526).					
M5	GC	15 18.6	+02 05	5.8	Ser
M49	Gx	12 29.8	+08 00	8.4	Vir
M53	GC	13 12.9	+18 10	7.7	Com
M58	Gx	12 37.7	+11 49	9.8	Vir
M59	Gx	12 42.0	+11 39	9.8	Vir
M60	Gx	12 43.7	+11 33	8.8	Vir
M61	Gx	12 21.9	+04 28	9.7	Vir
M84	Gx	12 25.1	+12 53	9.3	Vir
M85	Gx	12 25.4	+18 11	9.2	Com
M86	Gx	12 26.2	+12 57	9.2	Vir
M87	Gx	12 30.8	+12 24	8.6	Vir
M88	Gx	12 32.0	+14 25	9.5	Com
M89	Gx	12 35.7	+12 33	9.8	Vir
M90	Gx	12 36.8	+13 10	9.5	Vir
M91	Gx	12 35.4	+14 30	10.2	Com
M98	Gx	12 13.8	+14 54	10.1	Com
M99	Gx	12 18.8	+14 25	9.8	Com
M100	Gx	12 22.9	+15 49	9.4	Com
M104 (see Sombrero Galaxy).					
M-1-32-19	Gx	12 30.3	−08 24	12.5p	Vir
M-1-32-28	Gx	12 35.6	−07 53	12.6p	Vir
M-1-33-1	Gx	12 44.1	−05 41	12.9p	Vir
M-1-33-3	Gx	12 45.7	−06 04	13.0p	Vir
M-1-33-27	Gx	12 51.2	−06 34	12.4p	Vir
M-2-33-15	Gx	12 49.4	−10 07	11.6	Vir
M-2-33-17	Gx	12 50.1	−14 44	12.9p	Crv
M-2-34-6	Gx	13 09.8	−10 20	11.0p	Vir
M-2-34-54	Gx	13 27.9	−13 25	13.0p	Vir
M-2-35-10	Gx	13 38.2	−09 48	12.8p	Vir
M-3-34-14	Gx	13 12.6	−17 32	12.8p	Vir
NGC 4030	Gx	12 00.4	−01 06	10.6	Vir
NGC 4033	Gx	12 00.6	−17 51	11.8	Crv
NGC 4037	Gx	12 01.4	+13 24	11.9	Com
NGC 4038	Gx	12 01.9	−18 52	10.5	Crv
NGC 4039	Gx	12 01.9	−18 53	10.3	Crv
NGC 4045	Gx	12 02.7	+01 59	12	Vir
NGC 4064	Gx	12 04.2	+18 27	11.4	Com
NGC 4067	Gx	12 04.2	+10 51	12.5	Vir
NGC 4073	Gx	12 04.4	+01 54	11.4	Vir
NGC 4094	Gx	12 05.9	−14 32	11.8	Crv
NGC 4116	Gx	12 07.6	+02 42	12	Vir
NGC 4123	Gx	12 08.2	+02 53	11.4	Vir
NGC 4124	Gx	12 08.2	+10 23	11.3	Vir
NGC 4147	GC	12 10.1	+18 33	10.4	Com
NGC 4152	Gx	12 10.6	+16 02	12.2	Com
NGC 4168	Gx	12 12.3	+13 12	11.2	Vir
NGC 4177	Gx	12 12.7	−14 01	13.2	Crv

Object	Type	R.A. h m	Dec. ° ′	Mag.	Const.
NGC 4178	Gx	12 12.8	+10 52	11.4	Vir
NGC 4179	Gx	12 12.9	+01 18	11	Vir
NGC 4189	Gx	12 13.8	+13 26	11.7	Com
NGC 4192 (see M98).					
NGC 4193	Gx	12 13.9	+13 10	12.3	Vir
NGC 4206	Gx	12 15.3	+13 02	12.2	Vir
NGC 4212	Gx	12 15.7	+13 54	11.2	Com
NGC 4215	Gx	12 15.9	+06 24	12.1	Vir
NGC 4216	Gx	12 15.9	+13 09	10	Vir
NGC 4224	Gx	12 16.6	+07 28	11.8	Vir
NGC 4233	Gx	12 17.1	+07 37	11.9	Vir
NGC 4235	Gx	12 17.1	+07 12	11.6	Vir
NGC 4237	Gx	12 17.2	+15 19	11.6	Com
NGC 4240	Gx	12 17.4	−09 57	12.7	Vir
NGC 4241	Gx	12 17.4	+06 41	11.9	Vir
NGC 4254 (see M99).					
NGC 4260	Gx	12 19.4	+06 06	11.8	Vir
NGC 4261	Gx	12 19.4	+05 50	10.4	Vir
NGC 4262	Gx	12 19.5	+14 53	11.6	Com
NGC 4267	Gx	12 19.8	+12 48	10.9	Vir
NGC 4270	Gx	12 19.8	+05 28	12.2	Vir
NGC 4273	Gx	12 19.9	+05 21	11.9	Vir
NGC 4281	Gx	12 20.4	+05 23	11.3	Vir
NGC 4293	Gx	12 21.2	+18 23	10.4	Com
NGC 4294	Gx	12 21.3	+11 31	12.1	Vir
NGC 4298	Gx	12 21.5	+14 36	11.3	Com
NGC 4299	Gx	12 21.7	+11 30	12.5	Vir
NGC 4302	Gx	12 21.7	+14 36	11.6	Com
NGC 4303 (see M61).					
NGC 4307	Gx	12 22.1	+09 03	12	Vir
NGC 4312	Gx	12 22.5	+15 32	11.7	Com
NGC 4313	Gx	12 22.6	+11 48	11.6	Vir
NGC 4321 (see M100).					
NGC 4324	Gx	12 23.1	+05 15	11.6	Vir
NGC 4339	Gx	12 23.6	+06 05	11.3	Vir
NGC 4340	Gx	12 23.6	+16 43	11.2	Com
NGC 4343	Gx	12 23.6	+06 57	12.1	Vir
NGC 4350	Gx	12 24.0	+16 42	11	Com
NGC 4351	Gx	12 24.0	+12 12	12.6	Vir
NGC 4361	PN	12 24.5	−18 48	10.9	Crv
NGC 4365	Gx	12 24.5	+07 19	9.6	Vir
NGC 4371	Gx	12 24.9	+11 42	10.8	Vir
NGC 4374 (see M84).					
NGC 4377	Gx	12 25.2	+14 46	11.9	Com
NGC 4378	Gx	12 25.3	+04 56	11.7	Vir
NGC 4379	Gx	12 25.2	+15 36	11.7	Com
NGC 4380	Gx	12 25.4	+10 01	11.7	Vir
NGC 4382 (see M85).					
NGC 4383	Gx	12 25.4	+16 28	12.1	Com
NGC 4387	Gx	12 25.7	+12 49	12.1	Vir

Chart 14

Object	Type	R.A. h m	Dec. ° ′	Mag.	Const.
NGC 4388	Gx	12 25.8	+12 40	11	Vir
NGC 4394	Gx	12 25.9	+18 13	10.9	Com
NGC 4402	Gx	12 26.1	+13 07	11.8	Vir
NGC 4405	Gx	12 26.1	+16 11	12	Com
NGC 4406 (see M86).					
NGC 4411B	Gx	12 26.8	+08 53	12.3	Vir
NGC 4413	Gx	12 26.5	+12 37	12.2	Vir
NGC 4417	Gx	12 26.8	+09 35	11.1	Vir
NGC 4419	Gx	12 26.9	+15 03	11.2	Com
NGC 4420	Gx	12 27.0	+02 30	12.1	Vir
NGC 4421	Gx	12 27.0	+15 28	11.6	Com
NGC 4424	Gx	12 27.2	+09 25	11.7	Vir
NGC 4425	Gx	12 27.2	+12 44	11.8	Vir
NGC 4429	Gx	12 27.4	+11 06	10	Vir
NGC 4430	Gx	12 27.4	+06 16	12	Vir
NGC 4434	Gx	12 27.6	+08 09	12.2	Vir
NGC 4435	Gx	12 27.7	+13 05	10.8	Vir
NGC 4438	Gx	12 27.8	+13 01	10.2	Vir
NGC 4440	Gx	12 27.9	+12 18	11.7	Vir
NGC 4442	Gx	12 28.1	+09 48	10.4	Vir
NGC 4450	Gx	12 28.5	+17 05	10.1	Com
NGC 4452	Gx	12 28.7	+11 45	12	Vir
NGC 4454	Gx	12 28.9	−01 56	11.9	Vir
NGC 4457	Gx	12 29.0	+03 34	10.9	Vir
NGC 4458	Gx	12 29.0	+13 15	12.1	Vir
NGC 4459	Gx	12 29.0	+13 59	10.4	Com
NGC 4461	Gx	12 29.0	+13 11	11.2	Vir
NGC 4469	Gx	12 29.5	+08 45	11.2	Vir
NGC 4470	Gx	12 29.6	+07 49	12.1	Vir
NGC 4472 (see M49).					
NGC 4473	Gx	12 29.8	+13 26	10.2	Com
NGC 4474	Gx	12 29.9	+14 04	11.5	Com
NGC 4477	Gx	12 30.0	+13 38	10.4	Com
NGC 4478	Gx	12 30.3	+12 20	11.4	Vir
NGC 4480	Gx	12 30.4	+04 15	12.4	Vir
NGC 4486 (see M87).					
NGC 4487	Gx	12 31.1	−08 03	10.9	Vir
NGC 4488	Gx	12 30.9	+08 22	12.2	Vir
NGC 4489	Gx	12 30.9	+16 46	12	Com
NGC 4496A	Gx	12 31.7	+03 56	11.4	Vir
NGC 4496B	Gx	12 31.7	+03 56	13.5	Vir
NGC 4498	Gx	12 31.7	+16 51	12.2	Com
NGC 4501 (see M88).					
NGC 4503	Gx	12 32.1	+11 11	11.1	Vir
NGC 4504	Gx	12 32.3	−07 34	11.2	Vir
NGC 4517	Gx	12 32.8	+00 07	10.4	Vir
NGC 4517A	Gx	12 32.5	+00 23	12.5	Vir
NGC 4519	Gx	12 33.5	+08 39	11.8	Vir
NGC 4522	Gx	12 33.7	+09 10	12.3	Vir
NGC 4526	Gx	12 34.0	+07 42	9.7	Vir
NGC 4527	Gx	12 34.1	+02 39	10.5	Vir
NGC 4528	Gx	12 34.1	+11 19	12.1	Vir
NGC 4531	Gx	12 34.3	+13 05	11.4	Vir
NGC 4532	Gx	12 34.3	+06 28	11.9	Vir
NGC 4535	Gx	12 34.3	+08 12	10	Vir
NGC 4536	Gx	12 34.4	+02 11	10.6	Vir
NGC 4539	Gx	12 34.6	+18 12	12	Com
NGC 4540	Gx	12 34.8	+15 33	11.7	Com
NGC 4546	Gx	12 35.5	−03 48	10.3	Vir
NGC 4548 (see M91).					
NGC 4550	Gx	12 35.5	+12 13	11.7	Vir
NGC 4551	Gx	12 35.6	+12 16	12	Vir
NGC 4552 (see M89).					
NGC 4561	Gx	12 36.1	+19 19	12.5	Com
NGC 4564	Gx	12 36.5	+11 26	11.1	Vir
NGC 4567	Gx	12 36.5	+11 16	11.3	Vir
NGC 4568	Gx	12 36.6	+11 14	10.8	Vir
NGC 4569 (see M90).					
NGC 4570	Gx	12 36.9	+07 15	10.9	Vir
NGC 4571	Gx	12 36.9	+14 13	11.3	Com
NGC 4578	Gx	12 37.5	+09 33	11.5	Vir
NGC 4579 (see M58).					
NGC 4580	Gx	12 37.8	+05 22	11.8	Vir
NGC 4586	Gx	12 38.5	+04 19	11.7	Vir
NGC 4592	Gx	12 39.3	−00 32	11.7	Vir
NGC 4593	Gx	12 39.7	−05 21	10.9	Vir
NGC 4594 (see Sombrero Galaxy).					
NGC 4595	Gx	12 39.9	+15 18	12.1	Com
NGC 4596	Gx	12 39.9	+10 11	10.4	Vir
NGC 4597	Gx	12 40.2	−05 48	12.1	Vir
NGC 4602	Gx	12 40.6	−05 08	11.5	Vir
NGC 4606	Gx	12 41.0	+11 55	11.8	Vir
NGC 4608	Gx	12 41.2	+10 09	11	Vir
NGC 4612	Gx	12 41.5	+07 19	10.9	Vir
NGC 4621 (see M59).					
NGC 4632	Gx	12 42.5	−00 05	11.7	Vir
NGC 4636	Gx	12 42.8	+02 41	9.5	Vir
NGC 4638	Gx	12 42.8	+11 27	11.2	Vir
NGC 4639	Gx	12 42.9	+13 15	11.5	Vir
NGC 4643	Gx	12 43.3	+01 59	10.8	Vir
NGC 4647	Gx	12 43.5	+11 35	11.3	Vir
NGC 4649 (see M60).					
NGC 4651	Gx	12 43.7	+16 24	10.8	Com
NGC 4653	Gx	12 43.8	−00 34	12.2	Vir
NGC 4654	Gx	12 43.9	+13 08	10.6	Vir
NGC 4658	Gx	12 44.6	−10 05	12.5	Vir
NGC 4660	Gx	12 44.5	+11 11	11.2	Vir
NGC 4665	Gx	12 45.1	+03 03	10.5	Vir
NGC 4666	Gx	12 45.1	−00 28	10.7	Vir
NGC 4684	Gx	12 47.3	−02 44	11.4	Vir

Object	Type	R.A. h m	Dec. ° ′	Mag.	Const.
NGC 4689	Gx	12 47.8	+13 46	10.9	Com
NGC 4691	Gx	12 48.2	−03 20	11.1	Vir
NGC 4694	Gx	12 48.3	+10 59	11.4	Vir
NGC 4697	Gx	12 48.6	−05 48	9.2	Vir
NGC 4698	Gx	12 48.4	+08 29	10.6	Vir
NGC 4699	Gx	12 49.0	−08 40	9.5	Vir
NGC 4700	Gx	12 49.1	−11 25	11.9	Vir
NGC 4701	Gx	12 49.2	+03 23	12.4	Vir
NGC 4710	Gx	12 49.7	+15 10	11	Com
NGC 4713	Gx	12 50.0	+05 19	11.7	Vir
NGC 4727	Gx	12 51.0	−14 20	13	Crv
NGC 4731	Gx	12 51.0	−06 23	11.5	Vir
NGC 4733	Gx	12 51.1	+10 55	11.8	Vir
NGC 4739	Gx	12 51.6	−08 25	12.6	Vir
NGC 4742	Gx	12 51.8	−10 27	11.3	Vir
NGC 4753	Gx	12 52.4	−01 12	9.9	Vir
NGC 4754	Gx	12 52.3	+11 19	10.6	Vir
NGC 4760	Gx	12 53.1	−10 30	11.4	Vir
NGC 4762	Gx	12 52.9	+11 14	10.3	Vir
NGC 4771	Gx	12 53.4	+01 16	12.3	Vir
NGC 4772	Gx	12 53.5	+02 10	11	Vir
NGC 4775	Gx	12 53.8	−06 37	11.1	Vir
NGC 4781	Gx	12 54.4	−10 32	11.1	Vir
NGC 4782	Gx	12 54.6	−12 34	11.7	Crv
NGC 4783	Gx	12 54.6	−12 33	11.5	Crv
NGC 4786	Gx	12 54.5	−06 51	11.7	Vir
NGC 4790	Gx	12 54.9	−10 15	12.1	Vir
NGC 4802	Gx	12 55.8	−12 03	11.8	Crv
NGC 4808	Gx	12 55.8	+04 18	11.7	Vir
NGC 4818	Gx	12 56.8	−08 32	11.1	Vir
NGC 4825	Gx	12 57.2	−13 40	11.7	Vir
NGC 4845	Gx	12 58.0	+01 35	11.2	Vir
NGC 4856	Gx	12 59.4	−15 03	10.5	Vir
NGC 4866	Gx	12 59.5	+14 10	11.2	Vir
NGC 4877	Gx	13 00.4	−15 17	12.4	Vir
NGC 4880	Gx	13 00.2	+12 29	11.4	Vir
NGC 4899	Gx	13 00.9	−13 57	11.9	Vir
NGC 4900	Gx	13 00.7	+02 30	11.4	Vir
NGC 4902	Gx	13 01.0	−14 31	10.9	Vir
NGC 4904	Gx	13 01.0	−00 02	12	Vir
NGC 4915	Gx	13 01.3	−04 34	12.1	Vir
NGC 4928	Gx	13 03.0	−08 05	12.5	Vir
NGC 4933A	Gx	13 04.0	−11 30	11.7	Vir
NGC 4939	Gx	13 04.2	−10 20	11.3	Vir
NGC 4941	Gx	13 04.2	−05 33	11.1	Vir
NGC 4948	Gx	13 04.9	−07 57	14.4	Vir
NGC 4951	Gx	13 05.1	−06 30	11.9	Vir
NGC 4958	Gx	13 05.8	−08 01	10.7	Vir
NGC 4981	Gx	13 08.8	−06 47	11.3	Vir
NGC 4984	Gx	13 09.0	−15 31	11.3	Vir
NGC 4995	Gx	13 09.7	−07 50	11.1	Vir
NGC 4999	Gx	13 09.6	+01 40	11.8	Vir
NGC 5015	Gx	13 12.4	−04 20	12.1	Vir
NGC 5018	Gx	13 13.0	−19 31	10.7	Vir
NGC 5020	Gx	13 12.7	+12 36	11.7	Vir
NGC 5024 (see M53).					
NGC 5028	Gx	13 13.8	−13 03	12.7	Vir
NGC 5037	Gx	13 15.0	−16 35	12.2	Vir
NGC 5044	Gx	13 15.4	−16 23	10.8	Vir
NGC 5053	GC	13 16.5	+17 42	9	Com
NGC 5054	Gx	13 17.0	−16 38	10.9	Vir
NGC 5073	Gx	13 19.3	−14 51	12.3	Vir
NGC 5077	Gx	13 19.5	−12 39	11.3	Vir
NGC 5079	Gx	13 19.6	−12 42	13	Vir
NGC 5105	Gx	13 21.8	−13 12	11.8	Vir
NGC 5129	Gx	13 24.2	+13 59	12.1	Vir
NGC 5147	Gx	13 26.3	+02 06	11.8	Vir
NGC 5170	Gx	13 29.8	−17 58	11.3	Vir
NGC 5172	Gx	13 29.3	+17 03	11.9	Com
NGC 5230	Gx	13 35.5	+13 40	12.1	Vir
NGC 5247	Gx	13 38.1	−17 53	10.1	Vir
NGC 5248	Gx	13 37.5	+08 53	10.3	Boo
NGC 5254	Gx	13 39.6	−11 30	12.3	Vir
NGC 5300	Gx	13 48.3	+03 57	11.4	Vir
NGC 5324	Gx	13 52.1	−06 03	11.7	Vir
NGC 5334	Gx	13 52.9	−01 07	11.3	Vir
NGC 5363	Gx	13 56.1	+05 15	10.1	Vir
NGC 5364	Gx	13 56.2	+05 01	10.5	Vir
NGC 5426	Gx	14 03.4	−06 04	12.1	Vir
NGC 5427	Gx	14 03.4	−06 02	11.4	Vir
NGC 5468	Gx	14 06.6	−05 27	12.5	Vir
NGC 5493	Gx	14 11.5	−05 03	11.4	Vir
NGC 5496	Gx	14 11.6	−01 09	12.1	Vir
NGC 5506	Gx	14 13.2	−03 12	11.9	Vir
NGC 5532	Gx	14 16.9	+10 48	11.9	Boo
NGC 5534	Gx	14 17.7	−07 25	12.3	Vir
NGC 5566	Gx	14 20.3	+03 56	10.6	Vir
NGC 5576	Gx	14 21.1	+03 16	11	Vir
NGC 5584	Gx	14 22.4	−00 23	11.4	Vir
NGC 5595	Gx	14 24.2	−16 43	12	Lib
NGC 5597	Gx	14 24.5	−16 46	12	Lib
NGC 5600	Gx	14 23.8	+14 38	12.1	Boo
NGC 5605	Gx	14 25.1	−13 10	12.3	Lib
NGC 5634	GC	14 29.6	−05 59	9.5	Vir
NGC 5638	Gx	14 29.7	+03 14	11.2	Vir
NGC 5645	Gx	14 30.7	+07 17	12.5	Vir
NGC 5665	Gx	14 32.4	+08 05	12	Boo
NGC 5668	Gx	14 33.4	+04 27	11.5	Vir
NGC 5669	Gx	14 32.7	+09 53	11.3	Boo
NGC 5690	Gx	14 37.7	+02 17	11.8	Vir

Charts 14 – 15

Object	Type	R.A. h m	Dec. ° ′	Mag.	Const.
NGC 5691	Gx	14 37.9	−00 24	12.3	Vir
NGC 5701	Gx	14 39.2	+05 22	10.9	Vir
NGC 5713	Gx	14 40.2	−00 17	11.2	Vir
NGC 5716	Gx	14 41.1	−17 29	12.9	Lib
NGC 5728	Gx	14 42.4	−17 15	11.5	Lib
NGC 5729	Gx	14 42.1	−09 01	12.6	Lib
NGC 5740	Gx	14 44.4	+01 41	11.9	Vir
NGC 5746	Gx	14 44.9	+01 57	10.3	Vir
NGC 5750	Gx	14 46.2	−00 13	11.6	Vir
NGC 5774	Gx	14 53.7	+03 35	12.1	Vir
NGC 5775	Gx	14 54.0	+03 33	11.4	Vir
NGC 5791	Gx	14 58.8	−19 16	11.6	Lib
NGC 5792	Gx	14 58.4	−01 05	11.2	Lib
NGC 5796	Gx	14 59.4	−16 37	11.6	Lib
NGC 5806	Gx	15 00.0	+01 53	11.7	Vir
NGC 5812	Gx	15 00.9	−07 27	11.2	Lib
NGC 5813	Gx	15 01.2	+01 42	10.5	Vir
NGC 5831	Gx	15 04.1	+01 13	11.5	Vir
NGC 5838	Gx	15 05.4	+02 06	10.9	Vir
NGC 5846	Gx	15 06.5	+01 36	10	Vir
NGC 5850	Gx	15 07.1	+01 33	10.8	Vir
NGC 5854	Gx	15 07.8	+02 34	11.9	Vir
NGC 5861	Gx	15 09.3	−11 19	11.6	Lib
NGC 5864	Gx	15 09.6	+03 03	11.8	Vir
NGC 5869	Gx	15 09.8	+00 28	12.9p	Vir
NGC 5878	Gx	15 13.8	−14 16	11.5	Lib
NGC 5885	Gx	15 15.1	−10 05	11.8	Lib
NGC 5892	Gx	15 13.8	−15 28	11.7	Lib
NGC 5904 (see M5).					
NGC 5915	Gx	15 21.6	−13 06	12.3	Lib
NGC 5916	Gx	15 21.6	−13 10	13.4	Lib
NGC 5921	Gx	15 21.9	+05 04	10.8	Ser
NGC 5937	Gx	15 30.8	−02 50	12.3	Ser
NGC 5957	Gx	15 35.4	+12 03	11.7	Ser
NGC 5962	Gx	15 36.5	+16 37	11.3	Ser
NGC 5970	Gx	15 38.5	+12 11	11.5	Ser
NGC 5984	Gx	15 42.9	+14 14	12.5	Ser
NGC 5990	Gx	15 46.3	+02 25	12.4	Ser
NGC 5996	Gx	15 47.0	+17 53	12.8	Ser
NGC 6010	Gx	15 54.3	+00 33	12.6	Ser
NGC 6012	Gx	15 54.2	+14 36	12	Ser
PK318+41.1	PN	13 40.6	−19 53	11.8	Vir
Ring Tail (see NGC 4038 and NGC 4039).					
Siamese Twins (see NGC 4567 and NGC 4568).					
Sombrero Galaxy	Gx	12 40.0	−11 37	8.3	Vir
Virgo A (see M87).					
Chart 15 B40	DN	16 14.7	−18 59		Sco
B41	DN	16 22.5	−19 40		Sco
B43	DN	16 31.0	−19 20		Oph

Object	Type	R.A. h m	Dec. ° ′	Mag.	Const.
B64	DN	17 17.2	−18 32		Oph
B79	DN	17 39.5	−19 47		Oph
B83a	DN	17 45.3	−20 00		Sgr
B84	DN	17 46.5	−20 11		Sgr
B84a	DN	17 57.5	−17 40		Sgr
B92	DN	18 15.5	−18 11		Sgr
B93	DN	18 16.9	−18 04		Sgr
B95	DN	18 25.6	−11 45		Sct
B97	DN	18 29.1	−09 56		Sct
B100, 1	DN	18 32.7	−09 08		Sct
B103	DN	18 39.2	−06 37		Sct
B104	DN	18 47.3	−04 32		Sct
B108	DN	18 49.6	−06 19		Sct
B110	DN	18 50.2	−04 46		Sct
B111	DN	18 51.0	−05 00		Sct
B112	DN	18 51.2	−06 40		Sct
B113	DN	18 51.4	−04 19		Sct
B114-7	DN	18 53.2	−07 06		Sct
B118	DN	18 53.9	−07 27		Sct
B259	DN	17 22.0	−19 19		Oph
B270	DN	17 32.0	−19 40		Oph
B276	DN	17 39.5	−19 47		Oph
B312	DN	18 30.9	−15 08		Sct
B314	DN	18 37.7	−09 37		Sct
B320	DN	18 53.0	−05 50		Sct
Box Nebula (see NGC 6309).					
Cr 350	OC	17 48.1	+01 18	6.1p	Oph
Eagle Nebula	BNe+C	18 18.8	−13 47	6	Ser
IC 1276	GC	18 10.7	−07 12	10.3:	Ser
IC 1283, 84	BNer	18 17.8	−19 40		Sgr
IC 1287	BNr	18 31.3	−10 50		Sct
IC 4592	BNr	16 12.0	−19 28		Sco
IC 4593	PN	16 12.2	+12 04	10.7	Her
IC 4603	BNr	16 25.6	−24 28		Oph
IC 4665	OC	17 46.3	+05 43	4.2	Oph
IC 4703	BNe	18 18.6	−13 58		Sgr
IC 4706	BN	18 19.6	−16 01		Sgr
IC 4725 (see M25).					
IC 4756	OC	18 39.0	+05 27	4.6	Ser
LDN 134	DN	15 53.6	−04 39		Lib
LDN 557	DN	18 38.6	−01 47		Ser
LDN 582	DN	18 52.6	−01 56		Aql
LDN 617	DN	18 57.5	+01 04		Aql
M5	GC	15 18.6	+02 05	5.8	Ser
M9	GC	17 19.2	−18 31	7.9:	Oph
M10	GC	16 57.1	−04 06	6.6	Oph
M11	OC	18 51.1	−06 16	5.8	Sct
M12	GC	16 47.2	−01 57	6.6	Oph
M14	GC	17 37.6	−03 15	7.6	Oph
M16 (see Eagle Nebula).					

Object	Type	R.A. h m	Dec. ° ′	Mag.	Const.
M17 (see Omega Nebula).					
M18	OC	18 19.9	−17 08	6.9	Sgr
M23	OC	17 56.8	−19 01	5.5	Sgr
M24	—	18 16.9	−18 29	4.5:	Sgr
M25	OC	18 31.6	−19 15	4.6	Sgr
M26	OC	18 45.2	−09 24	8	Sct
M107	GC	16 32.5	−13 03	8.1	Oph
M-2-41-1	Gx	16 17.3	−11 44	13.0p	Sco
Milky Way Star Cloud or Patch (see M24).					
NGC 5806	Gx	15 00.0	+01 53	11.7	Vir
NGC 5812	Gx	15 00.9	−07 27	11.2	Lib
NGC 5813	Gx	15 01.2	+01 42	10.5	Vir
NGC 5831	Gx	15 04.1	+01 13	11.5	Vir
NGC 5838	Gx	15 05.4	+02 06	10.9	Vir
NGC 5846	Gx	15 06.5	+01 36	10	Vir
NGC 5850	Gx	15 07.1	+01 33	10.8	Vir
NGC 5854	Gx	15 07.8	+02 34	11.9	Vir
NGC 5861	Gx	15 09.3	−11 19	11.6	Lib
NGC 5864	Gx	15 09.6	+03 03	11.8	Vir
NGC 5869	Gx	15 09.8	+00 28	12.9p	Vir
NGC 5878	Gx	15 13.8	−14 16	11.5	Lib
NGC 5885	Gx	15 15.1	−10 05	11.8	Lib
NGC 5892	Gx	15 13.8	−15 28	11.7	Lib
NGC 5904 (see M5).					
NGC 5915	Gx	15 21.6	−13 06	12.3	Lib
NGC 5916	Gx	15 21.6	−13 10	13.4	Lib
NGC 5921	Gx	15 21.9	+05 04	10.8	Ser
NGC 5937	Gx	15 30.8	−02 50	12.3	Ser
NGC 5957	Gx	15 35.4	+12 03	11.7	Ser
NGC 5962	Gx	15 36.5	+16 37	11.3	Ser
NGC 5970	Gx	15 38.5	+12 11	11.5	Ser
NGC 5984	Gx	15 42.9	+14 14	12.5	Ser
NGC 5990	Gx	15 46.3	+02 25	12.4	Ser
NGC 5996	Gx	15 47.0	+17 53	12.8	Ser
NGC 6010	Gx	15 54.3	+00 33	12.6	Ser
NGC 6012	Gx	15 54.2	+14 36	12	Ser
NGC 6070	Gx	16 10.0	+00 43	11.8	Ser
NGC 6106	Gx	16 18.8	+07 25	12.2	Her
NGC 6118	Gx	16 21.8	−02 17	11.7	Ser
NGC 6171 (see M107).					
NGC 6181	Gx	16 32.3	+19 50	11.9	Her
NGC 6218 (see M12).					
NGC 6254 (see M10).					
NGC 6309	PN	17 14.1	−12 55	11.5	Oph
NGC 6333 (see M9).					
NGC 6342	GC	17 21.2	−19 35	9.5	Oph
NGC 6356	GC	17 23.6	−17 49	8.2	Oph
NGC 6366	GC	17 27.7	−05 05	9.5:	Oph
NGC 6384	Gx	17 32.4	+07 04	10.4	Oph
NGC 6389	Gx	17 32.7	+16 24	12.1	Her

Object	Type	R.A. h m	Dec. ° ′	Mag.	Const.
NGC 6402 (see M14).					
NGC 6426	GC	17 44.9	+03 10	10.9	Oph
NGC 6439	PN	17 48.3	−16 28	12.6	Sgr
NGC 6445	PN	17 49.3	−20 01	11.2	Sgr
NGC 6494 (see M23).					
NGC 6500	Gx	17 56.0	+18 20	12.2	Her
NGC 6517	GC	18 01.9	−08 57	10.1	Oph
NGC 6535	GC	18 03.9	−00 18	9.3	Ser
NGC 6537	PN	18 05.2	−19 51	11.6	Sgr
NGC 6539	GC	18 04.8	−07 35	8.9	Ser
NGC 6548	Gx	18 06.0	+18 35	11.7	Her
NGC 6555	Gx	18 07.8	+17 36	12.4	Her
NGC 6567	PN	18 13.8	−19 05	11	Sgr
NGC 6572	PN	18 12.1	+06 51	9.1	Oph
NGC 6574	Gx	18 11.8	+14 59	12	Her
NGC 6589	BNr	18 16.3	−19 48		Sgr
NGC 6590	BNr	18 17.0	−19 53		Sgr
NGC 6595	OC	18 17.0	−19 53	7.0p	Sgr
NGC 6604	OC	18 18.1	−12 14	6.5	Ser
NGC 6611 (see Eagle Nebula).					
NGC 6613 (see M18).					
NGC 6618 (see Omega Nebula).					
NGC 6633	OC	18 27.7	+06 34	4.6	Oph
NGC 6645	OC	18 32.6	−16 54	8.5p	Sgr
NGC 6664	OC	18 36.7	−08 13	7.8	Sct
NGC 6694 (see M26).					
NGC 6705 (see M11).					
NGC 6709	OC	18 51.5	+10 21	6.7	Aql
NGC 6712	GC	18 53.1	−08 42	8.1	Sct
NGC 6716	OC	18 54.6	−19 53	6.9	Sgr
Omega Neb.	BNe+C	18 20.8	−16 11	7	Sgr
Pal 8	GC	18 41.5	−19 50	10.9	Sgr
PK7+7.1	PN	17 35.2	−18 34	13.5	Oph
PK18+20.1	PN	17 12.9	−03 16	13.4	Oph
PK27-3.2	PN	18 54.0	−06 26	13.4p	Sct
PK38+12.1	PN	18 17.6	+10 09	13.6	Oph
Sh2-46	BNe	18 06.1	−14 10		Ser
Sh2-53	BNe	18 25.2	−13 13		Sct
Sh2-55	BNe	18 32.2	−11 46		Sct
Sh2-64	BNe	18 31.6	−01 55		Ser
Star Queen Nebula (see Eagle Nebula).					
Swan Nebula (see Omega Nebula).					
vdB 111	BNr	17 19.0	+06 05		Oph
vdB 123	BNr	18 30.5	+01 11		Ser
White-Eyed Pea (see IC 4593).					
Wild Duck Cluster (see M11).					
B92	DN	18 15.5	−18 11		Sgr
B93	DN	18 16.9	−18 04		Sgr
B95	DN	18 25.6	−11 45		Sct

Chart 16

Chart 16

Object	Type	R.A. h m	Dec. ° ′	Mag.	Const.
B97	DN	18 29.1	−09 56		Sct
B100, 1	DN	18 32.7	−09 08		Sct
B103	DN	18 39.2	−06 37		Sct
B104	DN	18 47.3	−04 32		Sct
B108	DN	18 49.6	−06 19		Sct
B110	DN	18 50.2	−04 46		Sct
B111	DN	18 51.0	−05 00		Sct
B112	DN	18 51.2	−06 40		Sct
B113	DN	18 51.4	−04 19		Sct
B114-7	DN	18 53.2	−07 06		Sct
B118	DN	18 53.9	−07 27		Sct
B127, 29, 30	DN	19 02.0	−05 26		Aql
B132	DN	19 04.1	−04 28		Aql
B133	DN	19 06.1	−06 50		Aql
B134	DN	19 06.9	−06 14		Aql
B135	DN	19 07.7	−03 55		Aql
B136	DN	19 08.8	−04 02		Aql
B139	DN	19 18.1	−01 28		Aql
B142	DN	19 40.7	+10 30		Aql
B143	DN	19 41.5	+11 00		Aql
B312	DN	18 30.9	−15 08		Sct
B314	DN	18 37.7	−09 37		Sct
B320	DN	18 53.0	−05 50		Sct
B336	DN	19 36.8	+12 20		Aql
B337, 34	DN	19 36.0	+12 25		Aql
Barnard's Galaxy (see NGC 6822).					
Brocchi's Cluster	OC	19 25.4	+20 11	3.6	Vul
C42 (see NGC 7006).					
C47 (see NGC 6934).					
C55 (see Saturn Nebula).					
C57 (see NGC 6822).					
Coathanger (see Brocchi's Cluster).					
Cr 399 (see Brocchi's Cluster).					
Cr 401	OC	19 38.4	+00 20	7.0p	Aql
Eagle Nebula	BNe+C	18 18.8	−13 47	6	Ser
H20	OC	19 53.1	+18 20	7.7	Sge
IC 1276	GC	18 10.7	−07 12	10.3:	Ser
IC 1283, 84	BNer	18 17.8	−19 40		Sgr
IC 1287	BNr	18 31.3	−10 50		Sct
IC 4703	BNe	18 18.6	−13 58		Sgr
IC 4706	BN	18 19.6	−16 01		Sgr
IC 4725 (see M25).					
IC 4756	OC	18 39.0	+05 27	4.6	Ser
IC 4846	PN	19 16.5	−09 03	11.9	Aql
IC 4997	PN	20 20.2	+16 44	10.5	Sge
LDN 557	DN	18 38.6	−01 47		Ser
LDN 582	DN	18 52.6	−01 56		Aql
LDN 617	DN	18 57.5	+01 04		Aql
LDN 663	DN	19 36.9	+07 34		Aql
LDN 673	DN	19 20.9	+11 16		Aql

Object	Type	R.A. h m	Dec. ° ′	Mag.	Const.
LDN 684	DN	19 21.8	+12 26		Aql
Little Gem Nebula (see NGC 6818).					
M2	GC	21 33.5	−00 49	6.5	Aqr
M11	OC	18 51.1	−06 16	5.8	Sct
M15	GC	21 30.0	+12 10	6.4	Peg
M16 (see Eagle Nebula).					
M17 (see Omega Nebula).					
M18	OC	18 19.9	−17 08	6.9	Sgr
M24	—	18 16.9	−18 29	4.5:	Sgr
M25	OC	18 31.6	−19 15	4.6	Sgr
M26	OC	18 45.2	−09 24	8	Sct
M71	GC	19 53.8	+18 47	8.3	Sge
M72	GC	20 53.5	−12 32	9.4	Aqr
M73	—	20 59.0	−12 38	9.3	Aqr
Milky Way Star Cloud or Patch (see M24).					
NGC 6517	GC	18 01.9	−08 57	10.1	Oph
NGC 6535	GC	18 03.9	−00 18	9.3	Ser
NGC 6537	PN	18 05.2	−19 51	11.6	Sgr
NGC 6539	GC	18 04.8	−07 35	8.9	Ser
NGC 6548	Gx	18 06.0	+18 35	11.7	Her
NGC 6555	Gx	18 07.8	+17 36	12.4	Her
NGC 6567	PN	18 13.8	−19 05	11	Sgr
NGC 6572	PN	18 12.1	+06 51	9.1	Oph
NGC 6574	Gx	18 11.8	+14 59	12	Her
NGC 6589	BNr	18 16.3	−19 48		Sgr
NGC 6590	BNr	18 17.0	−19 53		Sgr
NGC 6595	OC	18 17.0	−19 53	7.0p	Sgr
NGC 6604	OC	18 18.1	−12 14	6.5	Ser
NGC 6611 (see Eagle Nebula).					
NGC 6613 (see M18).					
NGC 6618 (see Omega Nebula).					
NGC 6633	OC	18 27.7	+06 34	4.6	Oph
NGC 6645	OC	18 32.6	−16 54	8.5p	Sgr
NGC 6664	OC	18 36.7	−08 13	7.8	Sct
NGC 6694 (see M26).					
NGC 6705 (see M11).					
NGC 6709	OC	18 51.5	+10 21	6.7	Aql
NGC 6712	GC	18 53.1	−08 42	8.1	Sct
NGC 6716	OC	18 54.6	−19 53	6.9	Sgr
NGC 6738	OC	19 01.4	+11 36	8.3p	Aql
NGC 6741	PN	19 02.6	−00 27	11.4	Aql
NGC 6751	PN	19 05.9	−06 00	11.9	Aql
NGC 6755	OC	19 07.8	+04 14	7.5	Aql
NGC 6760	GC	19 11.2	+01 02	9	Aql
NGC 6778	PN	19 18.4	−01 36	12.3	Aql
NGC 6781	PN	19 18.4	+06 33	11.4	Aql
NGC 6790	PN	19 23.0	+01 31	10.5	Aql
NGC 6803	PN	19 31.3	+10 03	11.4	Aql
NGC 6804	PN	19 31.6	+09 13	12	Aql
NGC 6807	PN	19 34.6	+05 41	12	Aql

Charts 16 – 17

Object	Type	R.A. h m	Dec. ° ′	Mag.	Const.
NGC 6814	Gx	19 42.7	−10 19	11.2	Aql
NGC 6818	PN	19 44.0	−14 09	9.3	Sgr
NGC 6822	Gx	19 45.0	−14 48	8.8	Sgr
NGC 6838 (see M71).					
NGC 6879	PN	20 10.5	+16 55	12.5	Sge
NGC 6886	PN	20 12.7	+19 59	11.4	Sge
NGC 6891	PN	20 15.2	+12 42	10.5	Del
NGC 6903	Gx	20 23.7	−19 19	11.9	Cap
NGC 6934	GC	20 34.2	+07 24	8.9	Del
NGC 6941	Gx	20 36.4	−04 37	12.7	Aql
NGC 6962	Gx	20 47.3	+00 19	12.1	Aqr
NGC 6981 (see M72).					
NGC 6994 (see M73).					
NGC 7006	GC	21 01.5	+16 11	10.6	Del
NGC 7009 (see Saturn Nebula).					
NGC 7042	Gx	21 13.8	+13 35	12	Peg
NGC 7078 (see M15).					
NGC 7089 (see M2).					
Omega Neb.	BNe+C	18 20.8	−16 11	7	Sgr
Pal 8	GC	18 41.5	−19 50	10.9	Sgr
Pal 11	GC	19 45.2	−08 00	9.8:	Aql
Phantom Streak (see NGC 6741).					
PK26-11.1	PN	19 18.3	−11 06	14	Aql
PK27-3.2	PN	18 54.0	−06 26	13.4p	Sct
PK38+12.1	PN	18 17.6	+10 09	13.6	Oph
PK45-2.1	PN	19 24.4	+09 54	12.7	Aql
PK52-2.2	PN	19 39.2	+15 57	11.8	Aql
Saturn Nebula	PN	21 04.2	−11 22	8.3p	Aqr
Sh2-46	BNe	18 06.1	−14 10		Ser
Sh2-53	BNe	18 25.2	−13 13		Sct
Sh2-55	BNe	18 32.2	−11 46		Sct
Sh2-64	BNe	18 31.6	−01 55		Ser
Sh2-82	BNer	19 30.3	+18 16		Sge
Sh2-84	BNe	19 49.0	+18 24		Sge
Star Queen Nebula (see Eagle Nebula).					
Swan Nebula (see Omega Nebula).					
vdB 123	BNr	18 30.5	+01 11		Ser
Wild Duck Cluster (see M11).					

Chart 17

Object	Type	R.A. h m	Dec. ° ′	Mag.	Const.
C42 (see NGC 7006).					
C43 (see NGC 7814).					
C44 (see NGC 7479).					
C55 (see Saturn Nebula).					
C56 (see NGC 246).					
Ced 211	BNe	23 43.8	−15 17		Aqr
IC 1447	Gx	22 30.0	−05 07	12.9p	Aqr
M2	GC	21 33.5	−00 49	6.5	Aqr
M15	GC	21 30.0	+12 10	6.4	Peg
M-1-57-18	Gx	22 38.9	−05 51	12.9p	Aqr
M-3-1-15	Gx	00 01.9	−15 27	11.1p	Cet
NGC 14	Gx	00 08.8	+15 49	12.1	Peg
NGC 50	Gx	00 14.7	−07 21	11.6	Cet
NGC 57	Gx	00 15.5	+17 20	11.6	Psc
NGC 63	Gx	00 17.8	+11 27	11.6	Psc
NGC 128	Gx	00 29.3	+02 52	11.8	Psc
NGC 151	Gx	00 34.0	−09 42	11.6	Cet
NGC 157	Gx	00 34.8	−08 24	10.4	Cet
NGC 175	Gx	00 37.4	−19 56	12.1	Cet
NGC 198	Gx	00 39.4	+02 48	13.2	Psc
NGC 210	Gx	00 40.6	−13 52	10.9	Cet
NGC 245	Gx	00 46.1	−01 43	12.2	Cet
NGC 246	PN	00 47.1	−11 52	10.9	Cet
NGC 255	Gx	00 47.8	−11 28	11.9	Cet
NGC 271	Gx	00 50.7	−01 55	12	Cet
NGC 274	Gx	00 51.0	−07 03	11.8	Cet
NGC 275	Gx	00 51.1	−07 04	12.5	Cet
NGC 309	Gx	00 56.7	−09 55	11.9	Cet
NGC 337	Gx	00 59.8	−07 35	11.6	Cet
NGC 7006	GC	21 01.5	+16 11	10.6	Del
NGC 7009 (see Saturn Nebula).					
NGC 7042	Gx	21 13.8	+13 35	12	Peg
NGC 7078 (see M15).					
NGC 7089 (see M2).					
NGC 7171	Gx	22 01.0	−13 16	12.2	Aqr
NGC 7177	Gx	22 00.7	+17 44	11.2	Peg
NGC 7183	Gx	22 02.4	−18 55	11.9	Aqr
NGC 7218	Gx	22 10.2	−16 40	12	Aqr
NGC 7280	Gx	22 26.5	+16 09	12.1	Peg
NGC 7309	Gx	22 34.4	−10 21	12.5	Aqr
NGC 7371	Gx	22 46.1	−11 00	11.5	Aqr
NGC 7448	Gx	23 00.1	+15 59	11.7	Peg
NGC 7454	Gx	23 01.1	+16 23	11.8	Peg
NGC 7469	Gx	23 03.3	+08 52	12.3	Peg
NGC 7479	Gx	23 05.0	+12 19	10.8	Peg
NGC 7497	Gx	23 09.1	+18 11	12.2	Peg
NGC 7541	Gx	23 14.7	+04 32	11.7	Psc
NGC 7556	Gx	23 15.7	−02 23	11.7	Psc
NGC 7562	Gx	23 16.0	+06 41	11.6	Psc
NGC 7585	Gx	23 18.0	−04 39	11.4	Aqr
NGC 7600	Gx	23 18.9	−07 35	11.9	Aqr
NGC 7606	Gx	23 19.1	−08 29	10.8	Aqr
NGC 7619	Gx	23 20.2	+08 12	11.1	Peg
NGC 7625	Gx	23 20.5	+17 14	12.1	Peg
NGC 7626	Gx	23 20.7	+08 13	11.1	Peg
NGC 7714	Gx	23 36.2	+02 09	12.5	Psc
NGC 7716	Gx	23 36.5	+00 18	12.1	Psc
NGC 7721	Gx	23 38.8	−06 31	11.6	Aqr
NGC 7723	Gx	23 39.0	−12 58	11.2	Aqr
NGC 7727	Gx	23 39.9	−12 18	10.6	Aqr
NGC 7742	Gx	23 44.3	+10 46	11.6	Peg

Charts 17 – 18

Object	Type	R.A. h m	Dec. ° ′	Mag.	Const.
NGC 7743	Gx	23 44.4	+09 56	11.5	Peg
NGC 7785	Gx	23 55.3	+05 55	11.6	Psc
NGC 7814	Gx	00 03.3	+16 09	10.6	Peg
Saturn Nebula	PN	21 04.2	−11 22	8.3p	Aqr
UGC 12613	Gx	23 28.6	+14 45	12.8p	Peg

Chart 18

Object	Type	R.A. h m	Dec. ° ′	Mag.	Const.
Blanco 1	OC	00 04.3	−29 56	4.5	Scl
C62 (see NGC 247).					
C65 (see NGC 253).					
C67 (see NGC 1097).					
C70 (see NGC 300).					
C72 (see NGC 55).					
C87 (see NGC 1261).					
E 151-36A	Gx	01 14.3	−55 24	13.0p	Phe
E 154-23	Gx	02 56.9	−54 34	12.2	Hor
E 240-10	Gx	23 37.7	−47 30	12.6p	Phe
E 245-5	Gx	01 45.1	−43 36	12.7p	Phe
E 249-31B	Gx	03 55.8	−42 22	12.3p	Eri
E 300-14	Gx	03 09.6	−41 02	13.0p	Eri
E 351-30 (see Sculptor Dwarf Galaxy).					
E 356-4 (see Fornax Dwarf Galaxy).					
E 358-63	Gx	03 46.3	−34 57	12.6p	For
E 479-4	Gx	02 26.4	−24 18	12.9p	For
E 548-81	Gx	03 42.1	−21 15	12.9p	Eri
Fornax A (see NGC 1316).					
Fornax Dwarf Galaxy	Gx	02 40.0	−34 27	9.0p	For
Grus Quartet (see NGC 7552, NGC 7582, NGC 7590, and NGC 7599).					
IC 335	Gx	03 35.5	−34 27	12.9p	For
IC 1558	Gx	00 35.8	−25 22	12.8p	Scl
IC 1625	Gx	01 07.7	−46 55	13.0p	Phe
IC 1633	Gx	01 09.9	−45 56	12.5p	Phe
IC 1788	Gx	02 15.8	−31 12	12.4	For
IC 1933	Gx	03 25.7	−52 47	12.8	Hor
IC 1953	Gx	03 33.7	−21 29	11.7	Eri
IC 1954	Gx	03 31.5	−51 54	11.4	Hor
IC 1970	Gx	03 36.5	−43 57	12.9p	Hor
IC 1993	Gx	03 47.1	−33 42	12.5p	For
IC 2006	Gx	03 54.5	−35 58	11.4	Eri
IC 2035	Gx	04 09.0	−45 31	11.5	Hor
IC 5325	Gx	23 28.7	−41 20	11.1	Phe
IC 5328	Gx	23 33.3	−45 01	11.2	Phe
IC 5332	Gx	23 34.5	−36 06	10.3	Scl
M-3-10-42	Gx	03 44.0	−14 22	13.0p	Eri
M-3-10-45	Gx	03 46.6	−16 33	13.0p	Eri
M-6-51-14	Gx	23 37.1	−37 43	12.9p	Scl
NGC 24	Gx	00 09.9	−24 58	11.3	Scl
NGC 45	Gx	00 14.1	−23 11	10.8	Cet
NGC 55	Gx	00 15.1	−39 13	8.1	Scl

Object	Type	R.A. h m	Dec. ° ′	Mag.	Const.
NGC 134	Gx	00 30.4	−33 15	10.4	Scl
NGC 150	Gx	00 34.3	−27 48	11.3	Scl
NGC 175	Gx	00 37.4	−19 56	12.1	Cet
NGC 247	Gx	00 47.1	−20 46	9.2	Cet
NGC 253	Gx	00 47.6	−25 17	7.6	Scl
NGC 254	Gx	00 47.5	−31 25	11.6	Scl
NGC 288	GC	00 52.8	−26 35	8.1:	Scl
NGC 289	Gx	00 52.7	−31 12	10.6	Scl
NGC 300	Gx	00 54.9	−37 41	8.1	Scl
NGC 439	Gx	01 13.8	−31 45	11.4	Scl
NGC 578	Gx	01 30.5	−22 40	11	Cet
NGC 613	Gx	01 34.3	−29 25	10	Scl
NGC 625	Gx	01 35.1	−41 26	11	Phe
NGC 685	Gx	01 47.7	−52 46	11.3	Eri
NGC 692	Gx	01 48.7	−48 39	13.0p	Phe
NGC 897	Gx	02 21.1	−33 43	12.2	For
NGC 908	Gx	02 23.1	−21 14	10.4	Cet
NGC 922	Gx	02 25.1	−24 47	12	For
NGC 986	Gx	02 33.6	−39 03	10.9	For
NGC 1049	GC	02 39.7	−34 17	12.9	For
NGC 1079	Gx	02 43.7	−29 00	11.3	For
NGC 1097	Gx	02 46.3	−30 16	9.2	For
NGC 1179	Gx	03 02.6	−18 54	11.9	Eri
NGC 1187	Gx	03 02.6	−22 52	10.7	Eri
NGC 1201	Gx	03 04.1	−26 04	10.8	For
NGC 1232	Gx	03 09.8	−20 35	10	Eri
NGC 1249	Gx	03 10.0	−53 20	11.5	Hor
NGC 1255	Gx	03 13.5	−25 44	11	For
NGC 1261	GC	03 12.3	−55 13	8.3	Hor
NGC 1288	Gx	03 17.2	−32 35	12	For
NGC 1291	Gx	03 17.3	−41 06	8.5	Eri
NGC 1292	Gx	03 18.3	−27 37	11.9	For
NGC 1300	Gx	03 19.7	−19 25	10.4	Eri
NGC 1302	Gx	03 19.8	−26 04	10.9	For
NGC 1309	Gx	03 22.1	−15 24	11.5	Eri
NGC 1310	Gx	03 21.1	−37 06	12.3	For
NGC 1316	Gx	03 22.7	−37 12	8.2	For
NGC 1317	Gx	03 22.7	−37 06	10.8	For
NGC 1325	Gx	03 24.4	−21 33	11.5	Eri
NGC 1326	Gx	03 23.9	−36 28	10.9	For
NGC 1332	Gx	03 26.3	−21 20	10.5	Eri
NGC 1339	Gx	03 28.1	−32 17	11.6	For
NGC 1341	Gx	03 28.0	−37 09	12.1	For
NGC 1344	Gx	03 28.3	−31 04	10.2	For
NGC 1350	Gx	03 31.1	−33 38	10.3	For
NGC 1351	Gx	03 30.6	−34 51	11.3	For
NGC 1353	Gx	03 32.1	−20 49	11.4	Eri
NGC 1359	Gx	03 33.8	−19 29	12.1	Eri
NGC 1365	Gx	03 33.6	−36 08	9.3	For
NGC 1366	Gx	03 33.9	−31 12	11.9	For

Object	Type	R.A. h m	Dec. ° ′	Mag.	Const.
NGC 1371	Gx	03 35.0	−24 56	10.6	For
NGC 1374	Gx	03 35.3	−35 14	11	For
NGC 1379	Gx	03 36.1	−35 26	11	For
NGC 1380	Gx	03 36.4	−34 59	10	For
NGC 1381	Gx	03 36.5	−35 18	11.5	For
NGC 1385	Gx	03 37.5	−24 30	10.7	For
NGC 1386	Gx	03 36.8	−36 00	11.2	Eri
NGC 1387	Gx	03 37.0	−35 30	10.8	For
NGC 1389	Gx	03 37.2	−35 45	11.4	Eri
NGC 1393	Gx	03 38.6	−18 26	12	Eri
NGC 1395	Gx	03 38.5	−23 02	9.7	Eri
NGC 1398	Gx	03 38.9	−26 20	9.5	For
NGC 1399	Gx	03 38.5	−35 27	8.8	For
NGC 1400	Gx	03 39.5	−18 41	11	Eri
NGC 1404	Gx	03 38.9	−35 36	9.7	Eri
NGC 1406	Gx	03 39.4	−31 19	11.6	For
NGC 1407	Gx	03 40.2	−18 35	9.7	Eri
NGC 1411	Gx	03 38.7	−44 06	11.1	Hor
NGC 1415	Gx	03 40.9	−22 34	11.5	Eri
NGC 1425	Gx	03 42.2	−29 54	10.8	For
NGC 1426	Gx	03 42.8	−22 07	11.2	Eri
NGC 1427	Gx	03 42.3	−35 24	10.9	For
NGC 1433	Gx	03 42.0	−47 13	10	Hor
NGC 1437	Gx	03 43.6	−35 51	11.7	Eri
NGC 1439	Gx	03 44.8	−21 55	11.2	Eri
NGC 1440	Gx	03 45.1	−18 16	11.5	Eri
NGC 1448	Gx	03 44.5	−44 39	10.8	Hor
NGC 1452	Gx	03 45.4	−18 38	12.1	Eri
NGC 1461	Gx	03 48.5	−16 24	11.8	Eri
NGC 1487	Gx	03 55.8	−42 22	11.4	Eri
NGC 1493	Gx	03 57.5	−46 13	11.2	Hor
NGC 1494	Gx	03 57.7	−48 54	11.6	Hor
NGC 1512	Gx	04 03.9	−43 21	10.2	Hor
NGC 1527	Gx	04 08.4	−47 54	10.7	Hor
NGC 1531	Gx	04 12.0	−32 51	12.1	Eri
NGC 1532	Gx	04 12.1	−32 52	9.9	Eri
NGC 1537	Gx	04 13.7	−31 39	10.6	Eri
NGC 7531	Gx	23 14.8	−43 36	11.2	Gru
NGC 7552	Gx	23 16.2	−42 35	10.4	Gru
NGC 7582	Gx	23 18.4	−42 22	10.1	Gru
NGC 7590	Gx	23 18.9	−42 14	11.3	Gru
NGC 7599	Gx	23 19.3	−42 15	11.1	Gru
NGC 7632	Gx	23 22.0	−42 29	12.1	Gru
NGC 7690	Gx	23 33.0	−51 42	12.1	Phe
NGC 7713	Gx	23 36.2	−37 56	11.1	Scl
NGC 7744	Gx	23 45.0	−42 55	11.5	Phe
NGC 7755	Gx	23 47.9	−30 31	11.4	Scl
NGC 7764	Gx	23 50.9	−40 44	12.5	Phe
NGC 7793	Gx	23 57.8	−32 35	9.3	Scl

Object	Type	R.A. h m	Dec. ° ′	Mag.	Const.	
Sculptor Dwarf Galaxy	Gx	01 00.2	−33 43	8.8	Scl	
Sculptor Galaxy (see NGC 253).						
Zeta Sculptoris Cluster (see Blanco 1).						
Be 135	DN	07 19.0	−44 35		Pup	Chart 19
C58 (see NGC 2360).						
C64 (see NGC 2362).						
C71 (see NGC 2477).						
C73 (see NGC 1851).						
Cr 121	OC	06 54.1	−24 11	2.6	CMa	
Cr 132	OC	07 13.6	−30 50	3.6	CMa	
Cr 135	OC	07 17.0	−36 50	2.1	Pup	
Cr 140	OC	07 23.9	−32 00	3.5	CMa	
Cr 185	OC	08 22.5	−36 10	7.8	Pup	
E 208-21	Gx	07 33.9	−50 27	12.2p	Pup	
E 209-9	Gx	07 58.2	−49 51	12.4p	Pup	
E 249-31B	Gx	03 55.8	−42 22	12.3p	Eri	
E 311-12	Gx	07 47.6	−41 27	12.9p	Pup	
E 358-63	Gx	03 46.3	−34 57	12.6p	For	
E 362-11	Gx	05 16.7	−37 06	13.0p	Col	
E 428-11	Gx	07 15.5	−29 21	12.9p	CMa	
E 490-37	Gx	06 44.4	−26 07	13.0p	CMa	
E 492-2	Gx	07 11.7	−26 42	13.0p	CMa	
E 494-26	Gx	08 06.2	−27 31	12.5p	Pup	
E 550-24	Gx	04 21.2	−21 51	12.7p	Eri	
E 556-15	Gx	06 21.1	−20 03	12.7p	CMa	
IC 335	Gx	03 35.5	−34 27	12.9p	For	
IC 438	Gx	05 53.0	−17 53	12.8p	Lep	
IC 456	Gx	07 00.3	−30 10	13.0p	CMa	
IC 1954	Gx	03 31.5	−51 54	11.4	Hor	
IC 1970	Gx	03 36.5	−43 57	12.9p	Hor	
IC 1993	Gx	03 47.1	−33 42	12.5p	For	
IC 2006	Gx	03 54.5	−35 58	11.4	Eri	
IC 2035	Gx	04 09.0	−45 31	11.5	Hor	
IC 2158	Gx	06 05.3	−27 51	12.9p	Col	
IC 2163	Gx	06 16.5	−21 23	12.4p	CMa	
IC 2395	OC	08 41.1	−48 12	4.6	Vel	
M41	OC	06 46.0	−20 44	4.5	CMa	
M46	OC	07 41.8	−14 49	6.1	Pup	
M47	OC	07 36.6	−14 30	4.4	Pup	
M79	GC	05 24.5	−24 33	8	Lep	
M93	OC	07 44.6	−23 52	6.2:	Pup	
Mel 66	OC	07 26.3	−47 44	7.8	Pup	
Mrk 18	OC	09 00.6	−48 59	7.8	Vel	
NGC 1341	Gx	03 28.0	−37 09	12.1	For	
NGC 1365	Gx	03 33.6	−36 08	9.3	For	
NGC 1374	Gx	03 35.3	−35 14	11	For	
NGC 1379	Gx	03 36.1	−35 26	11	For	
NGC 1380	Gx	03 36.4	−34 59	10	For	

Chart 19

Object	Type	R.A. h m	Dec. ° ′	Mag.	Const.
NGC 1381	Gx	03 36.5	−35 18	11.5	For
NGC 1386	Gx	03 36.8	−36 00	11.2	Eri
NGC 1387	Gx	03 37.0	−35 30	10.8	For
NGC 1389	Gx	03 37.2	−35 45	11.4	Eri
NGC 1399	Gx	03 38.5	−35 27	8.8	For
NGC 1404	Gx	03 38.9	−35 36	9.7	Eri
NGC 1406	Gx	03 39.4	−31 19	11.6	For
NGC 1411	Gx	03 38.7	−44 06	11.1	Hor
NGC 1425	Gx	03 42.2	−29 54	10.8	For
NGC 1427	Gx	03 42.3	−35 24	10.9	For
NGC 1433	Gx	03 42.0	−47 13	10	Hor
NGC 1437	Gx	03 43.6	−35 51	11.7	Eri
NGC 1448	Gx	03 44.5	−44 39	10.8	Hor
NGC 1487	Gx	03 55.8	−42 22	11.4	Eri
NGC 1493	Gx	03 57.5	−46 13	11.2	Hor
NGC 1494	Gx	03 57.7	−48 54	11.6	Hor
NGC 1512	Gx	04 03.9	−43 21	10.2	Hor
NGC 1515	Gx	04 04.0	−54 06	11.2	Dor
NGC 1518	Gx	04 06.8	−21 11	11.7	Eri
NGC 1521	Gx	04 08.3	−21 03	11.3	Eri
NGC 1527	Gx	04 08.4	−47 54	10.7	Hor
NGC 1531	Gx	04 12.0	−32 51	12.1	Eri
NGC 1532	Gx	04 12.1	−32 52	9.9	Eri
NGC 1537	Gx	04 13.7	−31 39	10.6	Eri
NGC 1566	Gx	04 20.0	−54 56	9.4	Dor
NGC 1596	Gx	04 27.6	−55 02	11	Dor
NGC 1617	Gx	04 31.7	−54 36	10.5	Dor
NGC 1640	Gx	04 42.2	−20 26	11.7	Eri
NGC 1679	Gx	04 49.9	−31 58	11.5	Cae
NGC 1705	Gx	04 54.2	−53 22	11.8	Pic
NGC 1744	Gx	05 00.0	−26 02	11.3	Lep
NGC 1792	Gx	05 05.3	−37 59	9.9	Col
NGC 1800	Gx	05 06.4	−31 57	12.6	Col
NGC 1808	Gx	05 07.7	−37 31	9.9	Col
NGC 1851	GC	05 14.1	−40 03	7.1	Col
NGC 1904 (see M79).					
NGC 1964	Gx	05 33.4	−21 57	10.7	Lep
NGC 1979	Gx	05 34.0	−23 19	11.8	Lep
NGC 2089	Gx	05 47.9	−17 36	11.9	Lep
NGC 2090	Gx	05 47.0	−34 15	11	Col
NGC 2139	Gx	06 01.1	−23 41	11.4	Lep
NGC 2188	Gx	06 10.2	−34 06	11.6	Col
NGC 2196	Gx	06 12.2	−21 48	11.1	Lep
NGC 2204	OC	06 15.7	−18 39	8.6	CMa
NGC 2206	Gx	06 16.0	−26 46	12.2	CMa
NGC 2207	Gx	06 16.4	−21 22	10.8	CMa
NGC 2217	Gx	06 21.7	−27 14	10.2	CMa
NGC 2223	Gx	06 24.6	−22 50	11.8	CMa
NGC 2263	Gx	06 38.5	−24 51	11.9	CMa
NGC 2272	Gx	06 42.7	−27 28	11.7	CMa
NGC 2280	Gx	06 44.8	−27 38	10.5	CMa
NGC 2283	Gx	06 45.9	−18 13	12.2	CMa
NGC 2287 (see M41).					
NGC 2292	Gx	06 47.7	−26 45	10.8	CMa
NGC 2293	Gx	06 47.7	−26 45	12.2p	CMa
NGC 2298	GC	06 49.0	−36 00	9.3	Pup
NGC 2310	Gx	06 53.9	−40 52	11.7	Pup
NGC 2325	Gx	07 02.7	−28 42	11.2	CMa
NGC 2354	OC	07 14.3	−25 44	6.5	CMa
NGC 2360	OC	07 17.8	−15 37	7.2	CMa
NGC 2362	OC	07 18.8	−24 57	4.1	CMa
NGC 2367	OC	07 20.1	−21 56	7.9	CMa
NGC 2380	Gx	07 23.9	−27 32	11.5	CMa
NGC 2383	OC	07 24.8	−20 56	8.4:	CMa
NGC 2384	OC	07 25.1	−21 02	7.4:	CMa
NGC 2414	OC	07 33.3	−15 27	7.9	Pup
NGC 2421	OC	07 36.3	−20 37	8.3	Pup
NGC 2422 (see M47).					
NGC 2427	Gx	07 36.5	−47 38	11.5	Pup
NGC 2437 (see M46).					
NGC 2438	PN	07 41.8	−14 44	11	Pup
NGC 2439	OC	07 40.8	−31 39	6.9	Pup
NGC 2440	PN	07 41.9	−18 12	9.4	Pup
NGC 2447 (see M93).					
NGC 2451	OC	07 45.4	−37 58	2.8	Pup
NGC 2452	PN	07 47.4	−27 20	12	Pup
NGC 2453	OC	07 47.8	−27 14	8.3	Pup
NGC 2467	C+BNe	07 52.6	−26 23	7.1p	Pup
NGC 2477	OC	07 52.3	−38 33	5.8	Pup
NGC 2482	OC	07 54.9	−24 18	7.3	Pup
NGC 2483	OC	07 55.9	−27 56	7.6	Pup
NGC 2489	OC	07 56.2	−30 04	7.9	Pup
NGC 2527	OC	08 05.3	−28 10	6.5	Pup
NGC 2533	OC	08 07.0	−29 54	7.6	Pup
NGC 2546	OC	08 12.4	−37 38	6.3	Pup
NGC 2547	OC	08 10.7	−49 16	4.7	Vel
NGC 2567	OC	08 18.6	−30 38	7.4	Pup
NGC 2579	OC	08 21.1	−36 11	7.5	Pup
NGC 2626	BNer	08 35.6	−40 40		Vel
NGC 2659	OC	08 42.6	−44 57	8.6:	Vel
NGC 2670	OC	08 45.5	−48 47	7.8	Vel
Opened Box Cluster (see NGC 2360).					
Pi 4	OC	08 34.5	−44 16	5.9	Vel
Pi 6	OC	08 39.3	−46 13	7.2	Vel
PK264-8.1	PN	08 11.5	−48 43	12.4	Vel
Ru 32	OC	07 45.0	−25 31	8.4	Pup
Ru 44	OC	07 59.0	−28 35	7.2	Pup
Ru 55	OC	08 12.3	−32 36	7.8	Pup
Sh2-301	BNe	07 09.8	−18 29		CMa
Sh2-302	BNe	07 31.6	−16 58		Pup

Object	Type	R.A. h m	Dec. ° ′	Mag.	Const.
Sh2-307	BNe	07 35.5	−18 46		Pup
Starfish Cluster (see NGC 2482).					
Tr 7	OC	07 27.3	−23 58	7.9	Pup
Tuft in the Tail of the Dog (see Cr 140).					
vdB 96	BNr	07 19.6	−23 58		CMa
vdB 97	BNr	07 32.6	−16 54		Pup
vdB 98	BNr	07 36.4	−25 20		Pup
Vela SNR	BN	08 35.0	−44 00		Vel
Wa 6	OC	08 40.4	−46 09	8.4	Vel

Chart 20

Object	Type	R.A. h m	Dec. ° ′	Mag.	Const.
Antennae (see NGC 4038 and NGC 4039).					
Be 135	DN	07 19.0	−44 35		Pup
C59 (see NGC 3242).					
C60 (see NGC 4038).					
C61 (see NGC 4039).					
C71 (see NGC 2477).					
C74 (see NGC 3132).					
C79 (see NGC 3201).					
C85 (see IC 2391).					
CBS Eye Nebula (see NGC 3242).					
Cr 185	OC	08 22.5	−36 10	7.8	Pup
Cr 197	OC	08 44.7	−41 22	6.7	Vel
E 208-21	Gx	07 33.9	−50 27	12.2p	Pup
E 209-9	Gx	07 58.2	−49 51	12.4p	Pup
E 213-11	Gx	10 16.9	−48 53	12.0p	Vel
E 263-48	Gx	10 31.2	−46 15	12.6p	Vel
E 265-7	Gx	11 07.8	−46 31	12.4p	Cen
E 266-15	Gx	11 40.9	−44 29	13.0p	Cen
E 311-12	Gx	07 47.6	−41 27	12.9p	Pup
E 320-26	Gx	11 49.8	−38 47	12.8p	Cen
E 321-25	Gx	12 21.7	−39 46	12.8p	Cen
E 371-16	Gx	08 47.1	−33 46	13.0p	Pyx
E 373-5	Gx	09 30.9	−35 41	12.9p	Ant
E 373-8	Gx	09 33.3	−33 02	12.7p	Ant
E 380-1	Gx	12 14.7	−35 31	12.9p	Hya
E 380-6	Gx	12 15.6	−35 38	12.6p	Hya
E 436-27	Gx	10 28.9	−31 37	12.6p	Ant
E 494-26	Gx	08 06.2	−27 31	12.5p	Pup
E 495-21	Gx	08 36.3	−26 25	12.5p	Pyx
E 499-23	Gx	09 56.4	−26 06	12.8p	Hya
E 501-51	Gx	10 37.5	−26 19	12.9p	Hya
E 562-23	Gx	08 36.6	−20 28	13.0p	Pyx
E 563-17	Gx	08 44.5	−20 21	13.0p	Pyx
E 563-31	Gx	08 52.3	−17 45	13.0p	Hya
Eight Burst Nebula (see NGC 3132).					
Ghost of Jupiter Nebula (see NGC 3242).					
Gum 15	BNe	08 44.6	−41 17		Vel
Gum 17	BNe	08 50.5	−42 07		Vel
Gum 23	BNe	08 59.7	−47 27		Vel
Gum 25	BNe	09 02.4	−48 42		Vel

Object	Type	R.A. h m	Dec. ° ′	Mag.	Const.
IC 764	Gx	12 10.2	−29 44	12.3	Hya
IC 2311	Gx	08 18.8	−25 22	12.5p	Pup
IC 2367	Gx	08 24.2	−18 47	12.5p	Pup
IC 2391	OC	08 40.2	−53 04	2.5	Vel
IC 2395	OC	08 41.1	−48 12	4.6	Vel
IC 2469	Gx	09 23.0	−32 27	11.3p	Pyx
IC 2511	Gx	09 49.4	−32 51	13.0p	Ant
IC 2522	Gx	09 55.2	−33 08	11.9	Ant
IC 2531	Gx	09 59.9	−29 37	12.9p	Ant
IC 2533	Gx	10 00.5	−31 15	12.9p	Ant
IC 2537	Gx	10 03.9	−27 34	12.2	Ant
IC 2552	Gx	10 10.8	−34 51	13.0p	Ant
IC 2560	Gx	10 16.3	−33 34	12.6p	Ant
IC 2597	Gx	10 37.8	−27 05	12.9p	Hya
IC 2627	Gx	11 09.9	−23 44	12	Crt
IC 2995	Gx	12 05.8	−27 56	12.4	Hya
IC 3253	Gx	12 23.8	−34 37	12.1	Cen
IC 3370	Gx	12 27.6	−39 20	11	Cen
IC 3896	Gx	12 56.7	−50 21	11.5	Cen
IC 3896A	Gx	12 55.5	−50 04	13.0p	Cen
M-6-20-4	Gx	08 57.5	−39 16	12.9p	Vel
Mel 66	OC	07 26.3	−47 44	7.8	Pup
Mrk 18	OC	09 00.6	−48 59	7.8	Vel
NGC 2427	Gx	07 36.5	−47 38	11.5	Pup
NGC 2439	OC	07 40.8	−31 39	6.9	Pup
NGC 2451	OC	07 45.4	−37 58	2.8	Pup
NGC 2452	PN	07 47.4	−27 20	12	Pup
NGC 2453	OC	07 47.8	−27 14	8.3	Pup
NGC 2467	C+BNe	07 52.6	−26 23	7.1p	Pup
NGC 2477	OC	07 52.3	−38 33	5.8	Pup
NGC 2482	OC	07 54.9	−24 18	7.3	Pup
NGC 2483	OC	07 55.9	−27 56	7.6	Pup
NGC 2489	OC	07 56.2	−30 04	7.9	Pup
NGC 2527	OC	08 05.3	−28 10	6.5	Pup
NGC 2533	OC	08 07.0	−29 54	7.6	Pup
NGC 2546	OC	08 12.4	−37 38	6.3	Pup
NGC 2547	OC	08 10.7	−49 16	4.7	Vel
NGC 2559	Gx	08 17.1	−27 27	10.9	Pup
NGC 2566	Gx	08 18.8	−25 30	11	Pup
NGC 2567	OC	08 18.6	−30 38	7.4	Pup
NGC 2571	OC	08 18.9	−29 44	7	Pup
NGC 2579	OC	08 21.1	−36 11	7.5	Pup
NGC 2610	PN	08 33.4	−16 09	12.8	Hya
NGC 2613	Gx	08 33.4	−22 58	10.5	Pyx
NGC 2626	BNer	08 35.6	−40 40		Vel
NGC 2627	OC	08 37.3	−29 57	8.4p	Pyx
NGC 2640	Gx	08 37.4	−55 07	12.7p	Car
NGC 2659	OC	08 42.6	−44 57	8.6:	Vel
NGC 2663	Gx	08 45.1	−33 48	10.9	Pyx
NGC 2665	Gx	08 46.0	−19 18	12.7	Hya

Chart 20

Object	Type	R.A. h m	Dec. ° ′	Mag.	Const.	Object	Type	R.A. h m	Dec. ° ′	Mag.	Const.
NGC 2669	OC	08 46.3	−52 55	6.1	Vel	NGC 3275	Gx	10 30.9	−36 44	11.6	Ant
NGC 2670	OC	08 45.5	−48 47	7.8	Vel	NGC 3281	Gx	10 31.9	−34 51	11.7	Ant
NGC 2784	Gx	09 12.3	−24 10	10	Hya	NGC 3285	Gx	10 33.6	−27 27	12	Hya
NGC 2792	PN	09 12.4	−42 26	11.7	Vel	NGC 3309	Gx	10 36.6	−27 31	11	Hya
NGC 2815	Gx	09 16.3	−23 38	11.8	Hya	NGC 3311	Gx	10 36.7	−27 32	10.9	Hya
NGC 2818	OC	09 16.0	−36 37	8.2	Pyx	NGC 3312	Gx	10 37.0	−27 34	11.8	Hya
NGC 2818A	PN	09 16.0	−36 38	11.6	Pyx	NGC 3313	Gx	10 37.4	−25 19	11.6	Hya
NGC 2835	Gx	09 17.9	−22 21	10.4	Hya	NGC 3318	Gx	10 37.3	−41 38	12	Vel
NGC 2845	Gx	09 18.6	−38 01	12.7	Vel	NGC 3330	OC	10 38.6	−54 09	7.4	Vel
NGC 2848	Gx	09 20.2	−16 32	11.8	Hya	NGC 3336	Gx	10 40.3	−27 47	12.3	Hya
NGC 2865	Gx	09 23.5	−23 10	11.4	Hya	NGC 3347	Gx	10 42.8	−36 21	11.4	Ant
NGC 2910	OC	09 30.4	−52 54	7.2	Vel	NGC 3358	Gx	10 43.6	−36 25	11.5	Ant
NGC 2921	Gx	09 34.5	−20 55	12	Hya	NGC 3366	Gx	10 35.1	−43 42	12	Vel
NGC 2925	OC	09 33.7	−53 26	8.3p	Vel	NGC 3390	Gx	10 48.1	−31 32	12.4	Hya
NGC 2935	Gx	09 36.7	−21 08	11.1	Hya	NGC 3449	Gx	10 52.9	−32 56	12.1	Ant
NGC 2983	Gx	09 43.7	−20 29	11.8	Hya	NGC 3450	Gx	10 48.1	−20 51	11.9	Hya
NGC 2986	Gx	09 44.3	−21 17	10.6	Hya	NGC 3464	Gx	10 54.7	−21 04	12.5	Hya
NGC 2997	Gx	09 45.7	−31 11	9.3	Ant	NGC 3511	Gx	11 03.4	−23 05	11	Crt
NGC 3001	Gx	09 46.3	−30 26	11.4	Ant	NGC 3513	Gx	11 03.8	−23 15	11.5	Crt
NGC 3038	Gx	09 51.3	−32 45	11.7	Ant	NGC 3528	Gx	11 07.3	−19 28	11.8	Crt
NGC 3051	Gx	09 54.0	−27 17	11.8	Ant	NGC 3557	Gx	11 10.0	−37 32	10.5	Cen
NGC 3052	Gx	09 54.5	−18 38	12.2	Hya	NGC 3568	Gx	11 10.8	−37 27	12.3	Cen
NGC 3054	Gx	09 54.5	−25 42	11.5	Hya	NGC 3585	Gx	11 13.3	−26 45	9.7	Hya
NGC 3056	Gx	09 54.6	−28 18	11.6	Ant	NGC 3621	Gx	11 18.3	−32 49	8.9	Hya
NGC 3078	Gx	09 58.4	−26 56	11	Hya	NGC 3673	Gx	11 25.2	−26 44	11.5	Hya
NGC 3081	Gx	09 59.5	−22 50	12	Hya	NGC 3680	OC	11 25.7	−43 15	7.6	Cen
NGC 3087	Gx	09 59.1	−34 13	11.7	Ant	NGC 3706	Gx	11 29.7	−36 24	11.2	Cen
NGC 3091	Gx	10 00.2	−19 38	11	Hya	NGC 3717	Gx	11 31.5	−30 18	11.4	Hya
NGC 3095	Gx	10 00.1	−31 33	11.8	Ant	NGC 3742	Gx	11 35.6	−37 57	12	Cen
NGC 3100	Gx	10 00.7	−31 40	11.2	Ant	NGC 3783	Gx	11 39.0	−37 44	11.6	Cen
NGC 3108	Gx	10 02.5	−31 41	11.5	Ant	NGC 3836	Gx	11 43.5	−16 48	12.7	Crt
NGC 3109	Gx	10 03.1	−26 10	9.8	Hya	NGC 3885	Gx	11 46.8	−27 55	11.8	Hya
NGC 3124	Gx	10 06.7	−19 13	12	Hya	NGC 3887	Gx	11 47.1	−16 51	10.6	Crt
NGC 3132	PN	10 07.0	−40 26	9.7	Vel	NGC 3904	Gx	11 49.2	−29 17	10.8	Hya
NGC 3137	Gx	10 09.1	−29 04	11.5	Ant	NGC 3923	Gx	11 51.0	−28 48	9.6	Hya
NGC 3175	Gx	10 14.7	−28 52	11.3	Ant	NGC 3936	Gx	11 52.3	−26 54	11.8	Hya
NGC 3200	Gx	10 18.6	−17 59	12.2	Hya	NGC 3955	Gx	11 54.0	−23 10	11.3	Crt
NGC 3201	GC	10 17.6	−46 25	6.9	Vel	NGC 3956	Gx	11 54.0	−20 34	12.2	Crt
NGC 3223	Gx	10 21.6	−34 15	11	Ant	NGC 3957	Gx	11 54.0	−19 34	12	Crt
NGC 3228	OC	10 21.8	−51 43	6	Vel	NGC 3962	Gx	11 54.7	−13 58	10.7	Crt
NGC 3241	Gx	10 24.3	−32 29	12.1	Ant	NGC 3981	Gx	11 56.1	−19 54	11	Crt
NGC 3242	PN	10 24.8	−18 38	7.8	Hya	NGC 4024	Gx	11 58.5	−18 21	11.9	Crv
NGC 3244	Gx	10 25.5	−39 50	12.4	Ant	NGC 4027	Gx	11 59.5	−19 16	11.2	Crv
NGC 3250	Gx	10 26.5	−39 57	11.3	Ant	NGC 4033	Gx	12 00.6	−17 51	11.8	Crv
NGC 3256	Gx	10 27.9	−43 54	10.8	Vel	NGC 4038	Gx	12 01.9	−18 52	10.5	Crv
NGC 3258	Gx	10 28.9	−35 36	11.5	Ant	NGC 4039	Gx	12 01.9	−18 53	10.3	Crv
NGC 3261	Gx	10 29.0	−44 39	11.2	Vel	NGC 4105	Gx	12 06.7	−29 46	10.4	Hya
NGC 3263	Gx	10 29.2	−44 06	10.8	Vel	NGC 4106	Gx	12 06.7	−29 46	10.6	Hya
NGC 3268	Gx	10 30.0	−35 20	11.6	Ant	NGC 4112	Gx	12 07.2	−40 12	12	Cen
NGC 3271	Gx	10 30.4	−35 22	11.7	Ant	NGC 4219	Gx	12 16.5	−43 19	11.9	Cen

Object	Type	R.A. h m	Dec. ° '	Mag.	Const.
NGC 4304	Gx	12 22.2	−33 29	11.7	Hya
NGC 4373	Gx	12 25.3	−39 46	10.6	Cen
NGC 4373A	Gx	12 25.6	−39 19	12.1	Cen
NGC 4444	Gx	12 28.6	−43 16	12.3	Cen
NGC 4507	Gx	12 35.6	−39 55	12.1	Cen
Omicron Velorum Cluster (see IC 2391).					
Pi 4	OC	08 34.5	−44 16	5.9	Vel
Pi 6	OC	08 39.3	−46 13	7.2	Vel
Pi 16	OC	09 51.1	−53 11	8	Vel
PK264-8.1	PN	08 11.5	−48 43	12.4	Vel
Ring Tail (see NGC 4038 and NGC 4039).					
Ru 44	OC	07 59.0	−28 35	7.2	Pup
Ru 55	OC	08 12.3	−32 36	7.8	Pup
Ru 82	OC	09 45.6	−53 59	8.1	Vel
Ru 106	GC	12 38.7	−51 09	10.9:	Cen
SL 4	DN	08 53.6	−42 13		Vel
Starfish Cluster (see NGC 2482).					
Tr 10	OC	08 47.8	−42 29	4.6	Vel
Vela SNR	BN	08 35.0	−44 00		Vel
Wa 6	OC	08 40.4	−46 09	8.4	Vel

Chart 21

Object	Type	R.A. h m	Dec. ° '	Mag.	Const.
Antennae (see NGC 4038 and NGC 4039).					
B228	DN	15 45.5	−34 24		Lup
Be 146	DN	13 57.6	−40 00		Cen
Be 149	DN	16 09.4	−39 08		Sco
C60 (see NGC 4038).					
C61 (see NGC 4039).					
C66 (see NGC 5694).					
C75 (see NGC 6124).					
C77 (see NGC 5128).					
C80 (see Omega Centauri).					
C82 (see NGC 6193).					
C83 (see NGC 4945).					
C84 (see NGC 5286).					
C88 (see NGC 5823).					
Centaurus A (see NGC 5128).					
E 219-21	Gx	13 02.3	−50 20	12.7p	Cen
E 219-41	Gx	13 14.0	−49 29	12.9p	Cen
E 221-6	Gx	13 50.4	−48 23	13.0p	Cen
E 221-10	Gx	13 51.0	−49 03	13.0p	Cen
E 221-26	Gx	14 08.4	−47 58	12.0p	Cen
E 221-32	Gx	14 12.2	−49 23	12.7p	Cen
E 221-34A	Gx	14 16.1	−48 08	12.6p	Cen
E 265-7	Gx	11 07.8	−46 31	12.4p	Cen
E 266-15	Gx	11 40.9	−44 29	13.0p	Cen
E 269-57	Gx	13 10.1	−46 26	12.5p	Cen
E 269-85	Gx	13 20.0	−47 17	12.7p	Cen
E 270-17	Gx	13 34.7	−45 32	11.8p	Cen
E 271-10	Gx	14 00.8	−45 25	12.1	Cen
E 273-14	Gx	14 58.4	−47 42	12.9p	Lup
E 274-1	Gx	15 14.2	−46 49	11.7p	Lup
E 320-26	Gx	11 49.8	−38 47	12.8p	Cen
E 321-25	Gx	12 21.7	−39 46	12.8p	Cen
E 323-34	Gx	12 53.4	−41 12	12.9p	Cen
E 324-24	Gx	13 27.6	−41 29	12.9p	Cen
E 380-1	Gx	12 14.7	−35 31	12.9p	Hya
E 380-6	Gx	12 15.6	−35 38	12.6p	Hya
E 383-76	Gx	13 47.5	−32 52	13.0p	Cen
E 383-87	Gx	13 49.3	−36 04	11.4p	Cen
E 384-2	Gx	13 51.3	−33 49	12.6p	Cen
E 385-30	Gx	14 29.3	−33 27	13.0p	Cen
E 442-26	Gx	12 52.2	−29 51	12.6p	Hya
E 443-24	Gx	13 01.0	−32 26	12.9p	Cen
E 445-2	Gx	13 39.4	−30 47	12.8p	Cen
E 507-25	Gx	12 51.5	−26 27	12.6p	Hya
E 507-45	Gx	12 55.6	−26 49	13.0p	Hya
Ho 18	OC	14 50.7	−52 15	8	Lup
Ho 22	OC	16 46.7	−47 06	6.7	Ara
IC 764	Gx	12 10.2	−29 44	12.3	Hya
IC 2995	Gx	12 05.8	−27 56	12.4	Hya
IC 3253	Gx	12 23.8	−34 37	12.1	Cen
IC 3370	Gx	12 27.6	−39 20	11	Cen
IC 3896	Gx	12 56.7	−50 21	11.5	Cen
IC 3896A	Gx	12 55.5	−50 04	13.0p	Cen
IC 4214	Gx	13 17.7	−32 06	12.3p	Cen
IC 4296	Gx	13 36.6	−33 58	10.5	Cen
IC 4329	Gx	13 49.1	−30 18	11	Cen
IC 4351	Gx	13 57.9	−29 19	11.8	Hya
IC 4367	Gx	14 05.6	−39 12	13.0p	Cen
IC 4386	Gx	14 15.1	−43 58	13.0p	Cen
IC 4402	Gx	14 21.2	−46 18	12.1p	Lup
IC 4406	PN	14 22.4	−44 09	10.2	Lup
IC 4444	Gx	14 31.7	−43 25	11.4	Lup
IC 4538	Gx	15 21.2	−23 39	12.9p	Lib
M68	GC	12 39.5	−26 45	8.2	Hya
M83	Gx	13 37.0	−29 52	7.6	Hya
M-3-34-14	Gx	13 12.6	−17 32	12.8p	Vir
NGC 3680	OC	11 25.7	−43 15	7.6	Cen
NGC 3706	Gx	11 29.7	−36 24	11.2	Cen
NGC 3742	Gx	11 35.6	−37 57	12	Cen
NGC 3783	Gx	11 39.0	−37 44	11.6	Cen
NGC 3885	Gx	11 46.8	−27 55	11.8	Hya
NGC 3904	Gx	11 49.2	−29 17	10.8	Hya
NGC 3923	Gx	11 51.0	−28 48	9.6	Hya
NGC 3936	Gx	11 52.3	−26 54	11.8	Hya
NGC 3955	Gx	11 54.0	−23 10	11.3	Crt
NGC 4027	Gx	11 59.5	−19 16	11.2	Crv
NGC 4033	Gx	12 00.6	−17 51	11.8	Crv
NGC 4038	Gx	12 01.9	−18 52	10.5	Crv
NGC 4039	Gx	12 01.9	−18 53	10.3	Crv

Chart 21

Object	Type	R.A. h m	Dec. ° ′	Mag.	Const.
NGC 4094	Gx	12 05.9	−14 32	11.8	Crv
NGC 4105	Gx	12 06.7	−29 46	10.4	Hya
NGC 4106	Gx	12 06.7	−29 46	10.6	Hya
NGC 4112	Gx	12 07.2	−40 12	12	Cen
NGC 4177	Gx	12 12.7	−14 01	13.2	Crv
NGC 4219	Gx	12 16.5	−43 19	11.9	Cen
NGC 4304	Gx	12 22.2	−33 29	11.7	Hya
NGC 4361	PN	12 24.5	−18 48	10.9	Crv
NGC 4373	Gx	12 25.3	−39 46	10.6	Cen
NGC 4373A	Gx	12 25.6	−39 19	12.1	Cen
NGC 4444	Gx	12 28.6	−43 16	12.3	Cen
NGC 4462	Gx	12 29.4	−23 10	11.9	Crv
NGC 4507	Gx	12 35.6	−39 55	12.1	Cen
NGC 4590 (see M68).					
NGC 4645	Gx	12 44.2	−41 45	11.8	Cen
NGC 4650	Gx	12 44.3	−40 44	11.8	Cen
NGC 4696	Gx	12 48.8	−41 19	10.2	Cen
NGC 4709	Gx	12 50.1	−41 23	11.5	Cen
NGC 4751	Gx	12 52.9	−42 40	12	Cen
NGC 4767	Gx	12 53.9	−39 43	11.6	Cen
NGC 4835	Gx	12 58.1	−46 16	11.5	Cen
NGC 4930	Gx	13 04.1	−41 25	11.7	Cen
NGC 4936	Gx	13 04.3	−30 31	10.8	Cen
NGC 4945	Gx	13 05.4	−49 28	8.8	Cen
NGC 4947	Gx	13 05.3	−35 20	11.9	Cen
NGC 4965	Gx	13 07.2	−28 14	12.1	Hya
NGC 4976	Gx	13 08.6	−49 30	10.1	Cen
NGC 5011	Gx	13 12.9	−43 06	11.1	Cen
NGC 5018	Gx	13 13.0	−19 31	10.7	Vir
NGC 5026	Gx	13 14.2	−42 58	11.5	Cen
NGC 5037	Gx	13 15.0	−16 35	12.2	Vir
NGC 5042	Gx	13 15.5	−23 59	11.8	Hya
NGC 5054	Gx	13 17.0	−16 38	10.9	Vir
NGC 5061	Gx	13 18.1	−26 50	10.2	Hya
NGC 5064	Gx	13 19.0	−47 55	12	Cen
NGC 5068	Gx	13 18.9	−21 02	9.6	Vir
NGC 5078	Gx	13 19.8	−27 25	10.6	Hya
NGC 5084	Gx	13 20.3	−21 50	10.4	Vir
NGC 5087	Gx	13 20.4	−20 37	11	Vir
NGC 5090	Gx	13 21.2	−43 42	11.6	Cen
NGC 5101	Gx	13 21.8	−27 26	10.4	Hya
NGC 5102	Gx	13 22.0	−36 38	8.8	Cen
NGC 5121	Gx	13 24.8	−37 41	11.6	Cen
NGC 5128	Gx	13 25.5	−43 01	6.7	Cen
NGC 5134	Gx	13 25.3	−21 08	11.2	Vir
NGC 5135	Gx	13 25.7	−29 50	11.8	Hya
NGC 5139 (see Omega Centauri).					
NGC 5140	Gx	13 26.3	−33 52	11.8	Cen
NGC 5153	Gx	13 27.9	−29 37	12.3	Hya
NGC 5156	Gx	13 28.7	−48 55	11.7	Cen
NGC 5161	Gx	13 29.2	−33 10	11.4	Cen
NGC 5170	Gx	13 29.8	−17 58	11.3	Vir
NGC 5188	Gx	13 31.5	−34 48	11.8	Cen
NGC 5193	Gx	13 31.9	−33 14	11.7	Cen
NGC 5206	Gx	13 33.7	−48 09	10.6	Cen
NGC 5236 (see M83).					
NGC 5247	Gx	13 38.1	−17 53	10.1	Vir
NGC 5253	Gx	13 39.9	−31 39	10.2	Cen
NGC 5264	Gx	13 41.6	−29 55	12.1	Hya
NGC 5266	Gx	13 43.0	−48 10	10.9	Cen
NGC 5266A	Gx	13 40.6	−48 21	12.1	Cen
NGC 5286	GC	13 46.5	−51 22	7.4	Cen
NGC 5292	Gx	13 47.7	−30 56	11.9	Cen
NGC 5307	PN	13 51.1	−51 12	11.2	Cen
NGC 5328	Gx	13 52.9	−28 29	11.7	Hya
NGC 5333	Gx	13 54.4	−48 31	12.8p	Cen
NGC 5357	Gx	13 56.0	−30 20	12.1	Cen
NGC 5365	Gx	13 57.8	−43 56	11.2	Cen
NGC 5367	BNr	13 57.7	−39 59		Cen
NGC 5408	Gx	14 03.4	−41 23	12.1	Cen
NGC 5419	Gx	14 03.6	−33 59	10.8	Cen
NGC 5460	OC	14 07.6	−48 19	5.6	Cen
NGC 5483	Gx	14 10.4	−43 20	11.2	Cen
NGC 5488	Gx	14 08.1	−33 19	11.9	Cen
NGC 5516	Gx	14 15.9	−48 07	12	Cen
NGC 5530	Gx	14 18.5	−43 23	11.1	Lup
NGC 5556	Gx	14 20.6	−29 15	11.8	Hya
NGC 5643	Gx	14 32.7	−44 10	10.4	Lup
NGC 5662	OC	14 35.2	−56 33	5.5	Cen
NGC 5670	Gx	14 35.6	−45 58	13.0p	Lup
NGC 5688	Gx	14 39.6	−45 01	11.9	Lup
NGC 5694	GC	14 39.6	−26 32	10.2	Hya
NGC 5716	Gx	14 41.1	−17 29	12.9	Lib
NGC 5728	Gx	14 42.4	−17 15	11.5	Lib
NGC 5786	Gx	14 58.9	−42 01	11.2	Cen
NGC 5791	Gx	14 58.8	−19 16	11.6	Lib
NGC 5796	Gx	14 59.4	−16 37	11.6	Lib
NGC 5822	OC	15 05.2	−54 21	6.5p	Lup
NGC 5823	OC	15 05.7	−55 36	7.9	Cir
NGC 5824	GC	15 04.0	−33 04	9.1	Lup
NGC 5843	Gx	15 07.5	−36 20	12.3	Lup
NGC 5873	PN	15 12.9	−38 07	11	Lup
NGC 5882	PN	15 16.8	−45 39	9.4	Lup
NGC 5892	Gx	15 13.8	−15 28	11.7	Lib
NGC 5897	GC	15 17.4	−21 01	8.4	Lib
NGC 5898	Gx	15 18.2	−24 06	11.4	Lib
NGC 5903	Gx	15 18.6	−24 04	11.1	Lib
NGC 5925	OC	15 27.7	−54 31	8.4p	Nor
NGC 5927	GC	15 28.0	−50 40	8	Lup
NGC 5946	GC	15 35.5	−50 40	8.4	Nor

Charts 21 – 22

Object	Type	R.A. h m	Dec. ° ′	Mag.	Const.
NGC 5986	GC	15 46.1	–37 47	7.6	Lup
NGC 6124	OC	16 25.6	–40 40	5.8:	Sco
NGC 6134	OC	16 27.7	–49 09	7.2	Nor
NGC 6139	GC	16 27.7	–38 51	9.1	Sco
NGC 6153	PN	16 31.5	–40 15	10.9	Sco
NGC 6167	OC	16 34.4	–49 36	6.7	Nor
NGC 6169	OC	16 34.1	–44 03	6.6p	Nor
NGC 6178	OC	16 35.7	–45 38	7.2	Sco
NGC 6188	BNer	16 40.5	–48 47		Ara
NGC 6192	OC	16 40.3	–43 22	8.5p	Sco
NGC 6193	OC	16 41.3	–48 46	5.2	Ara
NGC 6200	OC	16 44.2	–47 29	7.4	Ara
NGC 6204	OC	16 46.5	–47 01	8.2	Ara
Omega Centauri	GC	13 26.8	–47 29	3.9	Cen
PK303+40.1	PN	12 53.6	–22 52	12.7	Hya
PK318+41.1	PN	13 40.6	–19 53	11.8	Vir
PK329+2.1	PN	15 51.7	–51 31	12.6	Nor
Ring Tail (see NGC 4038 and NGC 4039).					
Ru 106	GC	12 38.7	–51 09	10.9:	Cen
Sh2-1	BNer	15 58.9	–26 09		Sco
SL 7	DN	16 01.8	–41 52		Lup
SL 8	DN	16 14.2	–44 04		Nor
SL 11	DN	15 57.0	–37 48		Lup
SL 15	DN	16 46.6	–44 30		Sco

Chart 22

Object	Type	R.A. h m	Dec. ° ′	Mag.	Const.
B40	DN	16 14.7	–18 59		Sco
B41	DN	16 22.5	–19 40		Sco
B42	DN	16 25.0	–23 30		Oph
B43	DN	16 31.0	–19 20		Oph
B44	DN	16 40.0	–24 20		Oph
B45	DN	16 38.0	–22 30		Oph
B46	DN	16 57.3	–22 40		Oph
B47	DN	17 01.0	–22 40		Oph
B48	DN	17 01.0	–40 47		Sco
B50	DN	17 03.0	–34 26		Sco
B51	DN	17 04.0	–22 15		Oph
B53	DN	17 06.1	–33 15		Sco
B55	DN	17 07.5	–32 00		Sco
B57	DN	17 08.3	–22 50		Oph
B58	DN	17 11.2	–40 25		Sco
B60	DN	17 11.8	–22 27		Oph
B61	DN	17 15.2	–20 21		Oph
B62	DN	17 16.2	–20 53		Oph
B63	DN	17 16.0	–21 23		Oph
B64	DN	17 17.2	–18 32		Oph
B65, 6, 7 (see Pipe Nebula).					
B67a	DN	17 22.5	–21 53		Oph
B68	DN	17 22.6	–23 44		Oph
B69	DN	17 22.9	–23 53		Oph
B70	DN	17 23.5	–23 58		Oph

Object	Type	R.A. h m	Dec. ° ′	Mag.	Const.
B72 (see Snake Nebula).					
B74	DN	17 25.2	–24 12		Oph
B77 (see Pipe Nebula).					
B78 (see Pipe Nebula).					
B79	DN	17 39.5	–19 47		Oph
B83a	DN	17 45.3	–20 00		Sgr
B84	DN	17 46.5	–20 11		Sgr
B84a	DN	17 57.5	–17 40		Sgr
B86	DN	18 02.7	–27 50		Sgr
B87	DN	18 04.3	–32 30		Sgr
B90	DN	18 10.2	–28 19		Sgr
B91	DN	18 10.0	–23 39		Sgr
B92	DN	18 15.5	–18 11		Sgr
B93	DN	18 16.9	–18 04		Sgr
B228	DN	15 45.5	–34 24		Lup
B231	DN	16 37.5	–35 12		Sco
B233	DN	16 44.1	–35 21		Sco
B238	DN	16 52.5	–23 05		Oph
B244	DN	17 10.1	–28 24		Oph
B246	DN	17 11.8	–22 27		Oph
B252	DN	17 15.2	–32 13		Sco
B256	DN	17 12.2	–28 51		Oph
B257	DN	17 22.0	–35 35		Sco
B259	DN	17 22.0	–19 19		Oph
B261	DN	17 25.3	–23 00		Oph
B262	DN	17 26.0	–22 28		Oph
B263	DN	17 26.3	–42 38		Sco
B268	DN	17 31.0	–21 00		Oph
B270	DN	17 32.0	–19 40		Oph
B276	DN	17 39.5	–19 47		Oph
B283	DN	17 51.3	–33 53		Sco
B287	DN	17 54.4	–35 12		Sco
B303	DN	18 09.2	–24 07		Sgr
Barnard's Galaxy (see NGC 6822).					
Be 149	DN	16 09.4	–39 08		Sco
Be 157	DN	19 02.9	–37 08		CrA
Bowl of Pipe Nebula (see Pipe Nebula).					
Bug Nebula	PN	17 13.7	–37 06	9.6	Sco
Butterfly Cluster (see M6).					
C57 (see NGC 6822).					
C68 (see NGC 6729).					
C69 (see Bug Nebula).					
C75 (see NGC 6124).					
C76 (see NGC 6231).					
C78 (see NGC 6541).					
C81 (see NGC 6352).					
C82 (see NGC 6193).					
C86 (see NGC 6397).					
Cr 338	OC	17 38.2	–37 34	8.0p	Sco

Chart 22

Object	Type	R.A. h m	Dec. ° ′	Mag.	Const.
Cr 367	OC	18 09.6	−23 59	6.4p	Sgr
Cr 394	OC	18 52.5	−20 23	6.3p	Sgr
E 183-30	Gx	18 56.9	−54 33	11.8	Tel
E 274-1	Gx	15 14.2	−46 49	11.7p	Lup
Ho 22	OC	16 46.7	−47 06	6.7	Ara
IC 1274	BNe	18 09.5	−23 44		Sgr
IC 1275	BNe	18 10.0	−23 50		Sgr
IC 1283,84	BNer	18 17.8	−19 40		Sgr
IC 1297	PN	19 17.4	−39 37	10.7	CrA
IC 4592	BNr	16 12.0	−19 28		Sco
IC 4601	BNr	16 20.0	−20 02		Sco
IC 4603	BNr	16 25.6	−24 28		Oph
IC 4604	BNr	16 25.6	−23 26		Oph
IC 4605	BNr	16 30.2	−25 06		Sco
IC 4606	BN	16 31.6	−26 03		Sco
IC 4628	BN	16 57.0	−40 20		Sco
IC 4634	PN	17 01.6	−21 50	10.9	Oph
IC 4637	PN	17 05.2	−40 53	12.5:	Sco
IC 4642	PN	17 11.8	−55 24	12.4	Ara
IC 4651	OC	17 25.0	−49 57	6.9	Ara
IC 4663	PN	17 45.5	−44 54	12.3	Sco
IC 4679	Gx	18 11.4	−56 15	12.8p	Tel
IC 4684	BNr	18 09.1	−23 25		Sgr
IC 4685	BNr	18 09.3	−23 59		Sgr
IC 4699	PN	18 18.5	−45 59	13.3:	Tel
IC 4725 (see M25).					
IC 4732	PN	18 33.9	−22 39	12.1	Sgr
IC 4776	PN	18 45.9	−33 21	10.4	Sgr
IC 4797	Gx	18 56.5	−54 18	11.3	Tel
IC 4808	Gx	19 01.1	−45 19	12.9p	CrA
IC 4812	BNr	19 01.1	−37 04		CrA
IC 4837	Gx	19 15.2	−54 40	12.4	Tel
IC 4837A	Gx	19 15.3	−54 08	12.5p	Tel
IC 4889	Gx	19 45.3	−54 21	11.3	Tel
IC 4931	Gx	20 00.8	−38 34	12.9p	Sgr
IC 4946	Gx	20 24.0	−44 00	12.6p	Sgr
IC 4991	Gx	20 18.4	−41 03	12.8p	Sgr
IC 5013	Gx	20 28.6	−36 02	12.6p	Mic
Lagoon Nebula	BNe	18 03.8	−24 23	5.8:	Sgr
LDN 1710	DN	17 20.7	−31 57		Sco
LDN 1773 (see Pipe Nebula).					
Little Gem Nebula (see NGC 6818).					
Little Ghost Nebula (see NGC 6369).					
M4	GC	16 23.6	−26 32	5.9	Sco
M6	OC	17 40.1	−32 13	4.2	Sco
M7	OC	17 53.9	−34 49	3.3	Sco
M8 (see Lagoon Nebula).					
M9	GC	17 19.2	−18 31	7.9:	Oph
M18	OC	18 19.9	−17 08	6.9	Sgr
M19	GC	17 02.6	−26 16	7.2	Oph
M20 (see Trifid Nebula).					
M21	OC	18 04.6	−22 30	5.9	Sgr
M22	GC	18 36.4	−23 54	5.1	Sgr
M23	OC	17 56.8	−19 01	5.5	Sgr
M24	—	18 16.9	−18 29	4.5:	Sgr
M25	OC	18 31.6	−19 15	4.6	Sgr
M28	GC	18 24.5	−24 52	6.8	Sgr
M54	GC	18 55.1	−30 29	7.7	Sgr
M55	GC	19 40.0	−30 58	7	Sgr
M62	GC	17 01.2	−30 07	6.6	Oph
M69	GC	18 31.4	−32 21	7.7	Sgr
M70	GC	18 43.2	−32 18	8.1	Sgr
M75	GC	20 06.1	−21 55	8.6	Sgr
M80	GC	16 17.0	−22 59	7.2	Sco
Milky Way Star Cloud or Patch (see M24).					
NGC 5882	PN	15 16.8	−45 39	9.4	Lup
NGC 5927	GC	15 28.0	−50 40	8	Lup
NGC 5946	GC	15 35.5	−50 40	8.4	Nor
NGC 5986	GC	15 46.1	−37 47	7.6	Lup
NGC 6031	OC	16 07.6	−54 04	8.5	Nor
NGC 6067	OC	16 13.2	−54 13	5.6	Nor
NGC 6093 (see M80).					
NGC 6121 (see M4).					
NGC 6124	OC	16 25.6	−40 40	5.8:	Sco
NGC 6134	OC	16 27.7	−49 09	7.2	Nor
NGC 6139	GC	16 27.7	−38 51	9.1	Sco
NGC 6144	GC	16 27.2	−26 02	9	Sco
NGC 6152	OC	16 32.7	−52 37	8.1p	Nor
NGC 6153	PN	16 31.5	−40 15	10.9	Sco
NGC 6167	OC	16 34.4	−49 36	6.7	Nor
NGC 6169	OC	16 34.1	−44 03	6.6p	Nor
NGC 6178	OC	16 35.7	−45 38	7.2	Sco
NGC 6188	BNer	16 40.5	−48 47		Ara
NGC 6192	OC	16 40.3	−43 22	8.5p	Sco
NGC 6193	OC	16 41.3	−48 46	5.2	Ara
NGC 6200	OC	16 44.2	−47 29	7.4	Ara
NGC 6204	OC	16 46.5	−47 01	8.2	Ara
NGC 6208	OC	16 49.5	−53 49	7.2	Ara
NGC 6231	OC	16 54.0	−41 48	2.6	Sco
NGC 6235	GC	16 53.4	−22 11	8.9	Oph
NGC 6242	OC	16 55.6	−39 30	6.4	Sco
NGC 6249	OC	16 57.6	−44 47	8.2	Sco
NGC 6250	OC	16 58.0	−45 48	5.9	Ara
NGC 6259	OC	17 00.7	−44 40	8	Sco
NGC 6266 (see M62).					
NGC 6273 (see M19).					
NGC 6281	OC	17 04.8	−37 54	5.4	Sco
NGC 6284	GC	17 04.5	−24 46	8.9	Oph
NGC 6287	GC	17 05.2	−22 42	9.3	Oph
NGC 6293	GC	17 10.2	−26 35	8.3	Oph

Object	Type	R.A. h m	Dec. ° ′	Mag.	Const.
NGC 6302 (*see* Bug Nebula).					
NGC 6304	GC	17 14.5	−29 28	8.3	Oph
NGC 6316	GC	17 16.6	−28 08	8.1	Oph
NGC 6322	OC	17 18.5	−42 57	6	Sco
NGC 6325	GC	17 18.0	−23 46	10.2	Oph
NGC 6326	PN	17 20.8	−51 45	11	Ara
NGC 6333 (*see* M9).					
NGC 6334	BNe	17 20.5	−35 43		Sco
NGC 6342	GC	17 21.2	−19 35	9.5	Oph
NGC 6352	GC	17 25.5	−48 25	7.8	Ara
NGC 6355	GC	17 24.0	−26 21	8.6	Oph
NGC 6356	GC	17 23.6	−17 49	8.2	Oph
NGC 6357	BNe	17 24.6	−34 10		Sco
NGC 6369	PN	17 29.3	−23 46	11.4	Oph
NGC 6383	OC	17 34.8	−32 34	5.5	Sco
NGC 6388	GC	17 36.3	−44 44	6.8	Sco
NGC 6396	OC	17 38.1	−35 00	8.5	Sco
NGC 6397	GC	17 40.7	−53 40	5.3	Ara
NGC 6401	GC	17 38.6	−23 55	7.4	Oph
NGC 6405 (*see* M6).					
NGC 6416	OC	17 44.4	−32 21	5.7	Sco
NGC 6425	OC	17 46.9	−31 32	7.2	Sco
NGC 6440	GC	17 48.9	−20 22	9.3	Sgr
NGC 6441	GC	17 50.2	−37 03	7.2	Sco
NGC 6445	PN	17 49.3	−20 01	11.2	Sgr
NGC 6451	OC	17 50.7	−30 13	8.2p	Sco
NGC 6453	GC	17 50.9	−34 36	10.2	Sco
NGC 6469	OC	17 52.9	−22 21	8.2p	Sgr
NGC 6475 (*see* M7).					
NGC 6494 (*see* M23).					
NGC 6496	GC	17 59.0	−44 16	8.6	Sco
NGC 6514 (*see* Trifid Nebula).					
NGC 6520	OC	18 03.4	−27 54	7.6p	Sgr
NGC 6522	GC	18 03.6	−30 02	9.9	Sgr
NGC 6523 (*see* Lagoon Nebula).					
NGC 6526	BNe	18 02.6	−23 35		Sgr
NGC 6528	GC	18 04.8	−30 03	9.6	Sgr
NGC 6530	OC	18 04.8	−24 20	4.6	Sgr
NGC 6531 (*see* M21).					
NGC 6537	PN	18 05.2	−19 51	11.6	Sgr
NGC 6541	GC	18 08.0	−43 42	6.3	CrA
NGC 6544	GC	18 07.4	−25 00	7.5:	Sgr
NGC 6546	OC	18 07.2	−23 20	8	Sgr
NGC 6553	GC	18 09.3	−25 54	8.3	Sgr
NGC 6558	GC	18 10.3	−31 46	8.6	Sgr
NGC 6559	BNe	18 10.0	−24 06		Sgr
NGC 6563	PN	18 12.0	−33 52	11	Sgr
NGC 6565	PN	18 11.9	−28 11	13.2	Sgr
NGC 6567	PN	18 13.8	−19 05	11	Sgr
NGC 6568	OC	18 12.8	−21 36	8.6p	Sgr
NGC 6569	GC	18 13.7	−31 50	8.4	Sgr
NGC 6578	PN	18 16.3	−20 27	12.9	Sgr
NGC 6584	GC	18 18.6	−52 13	7.9	Tel
NGC 6589	BNr	18 16.3	−19 48		Sgr
NGC 6590	BNr	18 17.0	−19 53		Sgr
NGC 6595	OC	18 17.0	−19 53	7.0p	Sgr
NGC 6613 (*see* M18).					
NGC 6624	GC	18 23.7	−30 22	7.6	Sgr
NGC 6626 (*see* M28).					
NGC 6629	PN	18 25.7	−23 12	11.3	Sgr
NGC 6637 (*see* M69).					
NGC 6638	GC	18 30.9	−25 30	9.2	Sgr
NGC 6642	GC	18 31.9	−23 28	8.9	Sgr
NGC 6644	PN	18 32.6	−25 08	10.7	Sgr
NGC 6645	OC	18 32.6	−16 54	8.5p	Sgr
NGC 6652	GC	18 35.8	−32 59	8.5	Sgr
NGC 6656 (*see* M22).					
NGC 6681 (*see* M70).					
NGC 6715 (*see* M54).					
NGC 6716	OC	18 54.6	−19 53	6.9	Sgr
NGC 6717	GC	18 55.1	−22 42	8.4	Sgr
NGC 6723	GC	18 59.6	−36 38	6.8	Sgr
NGC 6726, 7	BNr	19 01.7	−36 53		CrA
NGC 6729	BNer	19 01.9	−36 57		CrA
NGC 6754	Gx	19 11.4	−50 39	12.1	Tel
NGC 6788	Gx	19 26.8	−54 57	12.8p	Tel
NGC 6809 (*see* M55).					
NGC 6818	PN	19 44.0	−14 09	9.3	Sgr
NGC 6822	Gx	19 45.0	−14 48	8.8	Sgr
NGC 6849	Gx	20 06.3	−40 12	11.9	Sgr
NGC 6851	Gx	20 03.6	−48 17	11.8	Tel
NGC 6861	Gx	20 07.3	−48 22	11.1	Tel
NGC 6864 (*see* M75).					
NGC 6868	Gx	20 09.9	−48 23	10.6	Tel
NGC 6875	Gx	20 13.2	−46 10	11.9	Tel
NGC 6887	Gx	20 17.3	−52 48	12	Tel
NGC 6890	Gx	20 18.3	−44 48	12.1	Sgr
NGC 6893	Gx	20 20.8	−48 14	11.7	Tel
NGC 6902	Gx	20 24.5	−43 39	11	Sgr
NGC 6909	Gx	20 27.6	−47 02	11.8	Tel
Pal 8	GC	18 41.5	−19 50	10.9	Sgr
Parrot's Head (*see* B87).					
Pipe Nebula	DN	17 30.0	−26 00		Oph
PK1-6.1	PN	18 15.4	−30 32	14.0p	Sgr
PK1-6.2	PN	18 16.2	−30 52	11.8	Sgr
PK2-9.1	PN	18 29.2	−31 30	11.9	Sgr
PK3-6.1	PN	18 17.7	−29 08	13.0p	Sgr
PK3-14.1	PN	18 55.6	−32 16	12.8	Sgr
PK3-17.1	PN	19 05.6	−33 12	12.5	Sgr
PK4-11.1	PN	18 39.4	−30 41	13.3	Sgr

Charts 22 – 23

Object	Type	R.A. h m	Dec. ° '	Mag.	Const.
PK7+7.1	PN	17 35.2	−18 34	13.5	Oph
PK7-6.2	PN	18 28.0	−26 07	14	Sgr
PK329+2.1	PN	15 51.7	−51 31	12.6	Nor
PK342-14.1	PN	18 07.3	−51 02	11.9p	Ara
PK352-7.1	PN	18 00.2	−38 50	11.4p	CrA
PK353+8.1	PN	16 55.8	−29 50	12.8p	Oph
PK356-4.1	PN	17 54.6	−34 22	12.2	Sco
PK359-0.1	PN	17 47.9	−30 00	11.8	Sgr
R126	BNe	17 17.0	−36 20		Sco
Rho Ophiuchi Nebula (see IC 4604).					
S Nebula (see Snake Nebula).					
Sh2-1	BNer	15 58.9	−26 09		Sco
Sh2-3	BNe	17 12.3	−38 29		Sco
Sh2-9	BNer	16 21.1	−25 35		Sco
Sh2-13	BNe	17 29.1	−31 33		Sco
Sh2-16	BNe	17 46.6	−29 18		Sgr
Sh2-35	BNe	18 15.9	−20 15		Sgr
SL 7	DN	16 01.8	−41 52		Lup
SL 8	DN	16 14.2	−44 04		Nor
SL 11	DN	15 57.0	−37 48		Lup
SL 15	DN	16 46.6	−44 30		Sco
SL 17	DN	16 53.0	−43 35		Sco
SL 18	DN	16 44.8	−40 23		Sco
SL 26	DN	17 34.2	−40 25		Sco
SL 28	DN	17 35.3	−39 14		Sco
SL 42	DN	19 10.3	−37 08		CrA
Snake Nebula	DN	17 23.0	−23 32		Oph
Stem of Pipe Nebula (see Pipe Nebula).					
Table of Scorpius (see NGC 6231).					
Tom Thumb Cluster (see NGC 6451).					
Tr 24	OC	16 57.0	−40 40	8.6p	Sco
Tr 27	OC	17 36.2	−33 29	6.7	Sco
Tr 28	OC	17 36.8	−32 29	7.7	Sco
Tr 29	OC	17 41.6	−40 06	7.5p	Sco
Trifid Nebula	BNer+C	18 02.6	−23 02	6.3	Sgr
vdB 107	BNr	16 29.2	−26 27		Sco
vdBH 81	BNr	17 04.0	−51 05		Ara

Chart 23

Object	Type	R.A. h m	Dec. ° '	Mag.	Const.
Blanco 1	OC	00 04.3	−29 56	4.5	Scl
C63 (see Helix Nebula).					
C72 (see NGC 55).					
Ced 211	BNe	23 43.8	−15 17		Aqr
E 186-62	Gx	20 34.0	−52 59	12.9p	Ind
E 235-55	Gx	21 05.9	−48 12	12.7p	Ind
E 240-10	Gx	23 37.7	−47 30	12.6p	Phe
E 342-27	Gx	21 16.9	−42 16	12.9p	Mic
E 342-50	Gx	21 28.3	−37 52	12.9p	Gru
E 404-12	Gx	21 57.1	−34 35	12.8p	PsA
E 462-15	Gx	20 23.2	−27 43	12.9p	Sgr
F 591	Gx	21 48.3	−54 59	12.8p	Ind

Object	Type	R.A. h m	Dec. ° '	Mag.	Const.
Grus Quartet (see NGC 7552, NGC 7582, NGC 7590, and NGC 7599).					
Helix Nebula	PN	22 29.6	−20 48	7.3	Aqr
IC 1438	Gx	22 16.5	−21 26	12.7p	Aqr
IC 1459	Gx	22 57.2	−36 28	10	Gru
IC 4931	Gx	20 00.8	−38 34	12.9p	Sgr
IC 4946	Gx	20 24.0	−44 00	12.6p	Sgr
IC 4991	Gx	20 18.4	−41 03	12.8p	Sgr
IC 5013	Gx	20 28.6	−36 02	12.6p	Mic
IC 5020	Gx	20 30.6	−33 29	12.1	Mic
IC 5105	Gx	21 24.4	−40 32	11.4	Mic
IC 5148	PN	21 59.5	−39 23	11:	Gru
IC 5152	Gx	22 02.7	−51 18	10.6	Ind
IC 5156	Gx	22 03.2	−33 50	12.1	PsA
IC 5179	Gx	22 16.1	−36 51	12.3p	Gru
IC 5181	Gx	22 13.4	−46 01	11.4	Gru
IC 5186	Gx	22 18.8	−36 48	12.6p	Gru
IC 5201	Gx	22 21.0	−46 02	11.1	Gru
IC 5240	Gx	22 41.9	−44 46	11.4	Gru
IC 5267	Gx	22 57.2	−43 24	10.3	Gru
IC 5270	Gx	22 57.9	−35 51	12.9p	PsA
IC 5271	Gx	22 58.0	−33 45	11.7	PsA
IC 5273	Gx	22 59.4	−37 42	11.1	Gru
IC 5325	Gx	23 28.7	−41 20	11.1	Phe
IC 5328	Gx	23 33.3	−45 01	11.2	Phe
IC 5332	Gx	23 34.5	−36 06	10.3	Scl
M30	GC	21 40.4	−23 11	7.5	Cap
M55	GC	19 40.0	−30 58	7	Sgr
M75	GC	20 06.1	−21 55	8.6	Sgr
M-6-51-14	Gx	23 37.1	−37 43	12.9p	Scl
NGC 24	Gx	00 09.9	−24 58	11.3	Scl
NGC 55	Gx	00 15.1	−39 13	8.1	Scl
NGC 6754	Gx	19 11.4	−50 39	12.1	Tel
NGC 6809 (see M55).					
NGC 6849	Gx	20 06.3	−40 12	11.9	Sgr
NGC 6851	Gx	20 03.6	−48 17	11.8	Tel
NGC 6861	Gx	20 07.3	−48 22	11.1	Tel
NGC 6864 (see M75).					
NGC 6868	Gx	20 09.9	−48 23	10.6	Tel
NGC 6875	Gx	20 13.2	−46 10	11.9	Tel
NGC 6887	Gx	20 17.3	−52 48	12	Tel
NGC 6890	Gx	20 18.3	−44 48	12.1	Sgr
NGC 6893	Gx	20 20.8	−48 14	11.7	Tel
NGC 6902	Gx	20 24.5	−43 39	11	Sgr
NGC 6903	Gx	20 23.7	−19 19	11.9	Cap
NGC 6907	Gx	20 25.1	−24 48	11.1	Cap
NGC 6909	Gx	20 27.6	−47 02	11.8	Tel
NGC 6923	Gx	20 31.6	−30 50	12	Mic
NGC 6925	Gx	20 34.3	−31 59	11.3	Mic
NGC 6935	Gx	20 38.3	−52 07	11.9	Ind

Object	Type	R.A. h m	Dec. ° ′	Mag.	Const.
NGC 6942	Gx	20 40.6	−54 18	11.9	Ind
NGC 6958	Gx	20 48.7	−38 00	11.3	Mic
NGC 7007	Gx	21 05.5	−52 33	11.9	Ind
NGC 7029	Gx	21 11.9	−49 17	11.5	Ind
NGC 7038	Gx	21 15.1	−47 13	12	Ind
NGC 7041	Gx	21 16.5	−48 22	11.2	Ind
NGC 7049	Gx	21 19.0	−48 34	10.3	Ind
NGC 7064	Gx	21 29.0	−52 46	12.2	Ind
NGC 7070	Gx	21 30.4	−43 05	12	Gru
NGC 7079	Gx	21 32.6	−44 04	11.6	Gru
NGC 7090	Gx	21 36.5	−54 33	10.7	Ind
NGC 7097	Gx	21 40.2	−42 32	11.6	Gru
NGC 7099 (*see* M30).					
NGC 7130	Gx	21 48.3	−34 57	12	PsA
NGC 7135	Gx	21 49.7	−34 53	11.3	PsA
NGC 7141	Gx	21 52.3	−55 34	11.9	Ind
NGC 7144	Gx	21 52.7	−48 15	10.9	Gru
NGC 7145	Gx	21 53.3	−47 53	11.1	Gru
NGC 7154	Gx	21 55.4	−34 49	12.4	PsA
NGC 7166	Gx	22 00.5	−43 23	11.6	Gru
NGC 7167	Gx	22 00.5	−24 38	12.5	Aqr
NGC 7168	Gx	22 02.1	−51 45	11.9	Ind
NGC 7172	Gx	22 02.0	−31 52	11.8	PsA
NGC 7173	Gx	22 02.1	−31 58	11.1	PsA
NGC 7176	Gx	22 02.1	−31 59	11.1	PsA
NGC 7183	Gx	22 02.4	−18 55	11.9	Aqr
NGC 7184	Gx	22 02.6	−20 49	11.2	Aqr
NGC 7196	Gx	22 05.9	−50 07	11.4	Ind
NGC 7213	Gx	22 09.3	−47 10	10	Gru
NGC 7221	Gx	22 11.2	−30 34	12.2	PsA
NGC 7232	Gx	22 15.6	−45 51	11.6	Gru
NGC 7252	Gx	22 20.7	−24 41	11.4	Aqr
NGC 7267	Gx	22 24.4	−33 42	12.1	PsA
NGC 7285	Gx	22 28.6	−24 50	11.9	Aqr
NGC 7293 (*see* Helix Nebula).					
NGC 7307	Gx	22 33.9	−40 56	12.2	Gru
NGC 7314	Gx	22 35.8	−26 03	10.9	PsA
NGC 7361	Gx	22 42.3	−30 03	12.2	PsA
NGC 7377	Gx	22 47.8	−22 19	10.4	Aqr
NGC 7392	Gx	22 51.8	−20 36	11.8	Aqr
NGC 7410	Gx	22 55.0	−39 40	10.5	Gru
NGC 7412	Gx	22 55.8	−42 38	11.1	Gru
NGC 7418	Gx	22 56.6	−37 02	11	Gru
NGC 7421	Gx	22 56.9	−37 21	11.9	Gru
NGC 7424	Gx	22 57.3	−41 04	10.2	Gru
NGC 7456	Gx	23 02.2	−39 34	11.6	Gru
NGC 7462	Gx	23 02.8	−40 50	11.3	Gru
NGC 7484	Gx	23 07.1	−36 16	11.8	Scl
NGC 7496	Gx	23 09.8	−43 26	11.1	Gru
NGC 7507	Gx	23 12.1	−28 32	10.6	Scl

Object	Type	R.A. h m	Dec. ° ′	Mag.	Const.
NGC 7513	Gx	23 13.2	−28 22	11.9	Scl
NGC 7531	Gx	23 14.8	−43 36	11.2	Gru
NGC 7552	Gx	23 16.2	−42 35	10.4	Gru
NGC 7582	Gx	23 18.4	−42 22	10.1	Gru
NGC 7590	Gx	23 18.9	−42 14	11.3	Gru
NGC 7599	Gx	23 19.3	−42 15	11.1	Gru
NGC 7632	Gx	23 22.0	−42 29	12.1	Gru
NGC 7689	Gx	23 33.3	−54 06	11.4	Phe
NGC 7690	Gx	23 33.0	−51 42	12.1	Phe
NGC 7713	Gx	23 36.2	−37 56	11.1	Scl
NGC 7744	Gx	23 45.0	−42 55	11.5	Phe
NGC 7755	Gx	23 47.9	−30 31	11.4	Scl
NGC 7764	Gx	23 50.9	−40 44	12.5	Phe
NGC 7793	Gx	23 57.8	−32 35	9.3	Scl
Zeta Sculptoris Cluster (*see* Blanco 1).					

Chart 24

Object	Type	R.A. h m	Dec. ° ′	Mag.	Const.
30 Doradus (*see* Tarantula Nebula).					
47 Tucanae	GC	00 24.1	−72 05	4	Tuc
C73 (*see* NGC 1851).					
C87 (*see* NGC 1261).					
C96 (*see* NGC 2516).					
C103 (*see* Tarantula Nebula).					
C104 (*see* NGC 362).					
C106 (*see* 47 Tucanae).					
E 15-8	Gx	04 07.2	−82 17	12.8p	Men
E 18-2	Gx	08 19.2	−78 42	12.9p	Cha
E 27-1	Gx	21 52.5	−81 32	12.1p	Oct
E 27-8	Gx	22 23.0	−80 00	12.8p	Oct
E 54-21	Gx	03 49.8	−71 38	12.7p	Hyi
E 115-21	Gx	02 37.8	−61 20	13.0p	Hor
E 116-12	Gx	03 13.1	−57 21	12.9p	Hor
E 121-6	Gx	06 07.5	−61 48	10.7p	Pic
E 121-26	Gx	06 21.6	−59 44	12.5p	Pic
E 151-36A	Gx	01 14.3	−55 24	13.0p	Phe
E 154-23	Gx	02 56.9	−54 34	12.2	Hor
E 208-21	Gx	07 33.9	−50 27	12.2p	Pup
E 209-9	Gx	07 58.2	−49 51	12.4p	Pup
E 362-11	Gx	05 16.7	−37 06	13.0p	Col
IC 1660	BN	01 12.6	−71 45		Tuc
IC 1933	Gx	03 25.7	−52 47	12.8	Hor
IC 1954	Gx	03 31.5	−51 54	11.4	Hor
IC 2035	Gx	04 09.0	−45 31	11.5	Hor
IC 2051	Gx	03 52.0	−83 50	12.1p	Men
IC 2056	Gx	04 16.4	−60 12	11.6	Ret
IC 5250	Gx	22 47.3	−65 03	12.1p	Tuc
IC 5250A	Gx	22 47.4	−65 03	12.2p	Tuc
Large Magellanic Cloud	Gx	05 23.6	−69 45	0.1	Dor
LH 120	BNe	05 44.0	−67 50		Dor
Mel 227	OC	20 18.0	−79 09	5.3p	Oct

Object	Type	R.A. h m	Dec. ° ′	Mag.	Const.
NGC 104 (see 47 Tucanae).					
NGC 121	GC	00 26.8	−71 32	10.6	Tuc
NGC 248	BNe	00 45.4	−73 23		Tuc
NGC 249	BNe	00 45.4	−73 05		Tuc
NGC 261	BN	00 46.5	−73 07		Tuc
NGC 294	BN	00 52.1	−73 21		Tuc
NGC 346	BNe+C	00 59.1	−72 11		Tuc
NGC 362	GC	01 03.2	−70 51	6.8	Tuc
NGC 371	BN+C	01 03.3	−72 05	9.3	Tuc
NGC 406	Gx	01 07.4	−69 53	12.3	Tuc
NGC 416	GC	01 08.0	−72 21	11	Tuc
NGC 419	GC	01 08.3	−72 53	10	Tuc
NGC 434	Gx	01 12.2	−58 15	12.1	Tuc
NGC 460	BN	01 14.8	−72 18		Tuc
NGC 685	Gx	01 47.7	−52 46	11.3	Eri
NGC 782	Gx	01 57.6	−57 47	11.8	Eri
NGC 1249	Gx	03 10.0	−53 20	11.5	Hor
NGC 1261	GC	03 12.3	−55 13	8.3	Hor
NGC 1313	Gx	03 18.3	−66 30	8.9	Ret
NGC 1433	Gx	03 42.0	−47 13	10	Hor
NGC 1493	Gx	03 57.5	−46 13	11.2	Hor
NGC 1494	Gx	03 57.7	−48 54	11.6	Hor
NGC 1511	Gx	03 59.6	−67 38	11.1	Hyi
NGC 1515	Gx	04 04.0	−54 06	11.2	Dor
NGC 1527	Gx	04 08.4	−47 54	10.7	Hor
NGC 1533	Gx	04 09.9	−56 07	10.8	Dor
NGC 1543	Gx	04 12.7	−57 44	9.7	Ret
NGC 1546	Gx	04 14.6	−56 04	11.3	Dor
NGC 1549	Gx	04 15.8	−55 36	9.5	Dor
NGC 1553	Gx	04 16.2	−55 47	9.1	Dor
NGC 1559	Gx	04 17.6	−62 47	10.4	Ret
NGC 1566	Gx	04 20.0	−54 56	9.4	Dor
NGC 1574	Gx	04 22.0	−56 58	10.2	Ret
NGC 1596	Gx	04 27.6	−55 02	11	Dor
NGC 1617	Gx	04 31.7	−54 36	10.5	Dor
NGC 1672	Gx	04 45.7	−59 15	9.8	Dor
NGC 1688	Gx	04 48.4	−59 48	12	Dor
NGC 1703	Gx	04 52.9	−59 45	11.6	Dor
NGC 1705	Gx	04 54.2	−53 22	11.8	Pic
NGC 1763	BNe	04 56.8	−66 24		Dor
NGC 1786	GC	04 59.1	−67 45	10.1	Dor
NGC 1792	Gx	05 05.3	−37 59	9.9	Col
NGC 1796	Gx	05 02.7	−61 08	12.3	Dor
NGC 1808	Gx	05 07.7	−37 31	9.9	Col
NGC 1824	Gx	05 06.9	−59 43	12.6	Dor
NGC 1835	GC	05 05.2	−69 24	9.8	Dor
NGC 1851	GC	05 14.1	−40 03	7.1	Col
NGC 1892	Gx	05 17.1	−64 58	12	Dor
NGC 1929, 34-6	BNe	05 22.0	−67 58		Dor
NGC 1947	Gx	05 26.8	−63 46	10.8	Dor

Object	Type	R.A. h m	Dec. ° ′	Mag.	Const.
NGC 1966	BNe	05 26.8	−68 49		Dor
NGC 1968	BN	05 27.2	−67 26		Dor
NGC 1974	BN	05 27.9	−67 24		Dor
NGC 1978	GC	05 28.6	−66 14	9.9	Dor
NGC 2014	BN	05 32.2	−67 40	8.0p	Dor
NGC 2018	BNe	05 30.6	−71 04		Men
NGC 2048	BNe	05 35.2	−69 46		Dor
NGC 2070 (see Tarantula Nebula).					
NGC 2074	BNe	05 39.6	−69 27		Dor
NGC 2077-80	BNe	05 40.4	−69 38		Dor
NGC 2081	BNe	05 39.6	−69 27		Dor
NGC 2082	Gx	05 41.9	−64 18	12	Dor
NGC 2187A	Gx	06 03.7	−69 36	12.9p	Dor
NGC 2210	GC	06 11.5	−69 08	10.2	Dor
NGC 2305	Gx	06 48.6	−64 16	12.7p	Vol
NGC 2307	Gx	06 48.9	−64 20	13.0p	Vol
NGC 2397	Gx	07 21.3	−69 00	12	Vol
NGC 2417	Gx	07 30.2	−62 15	12.4	Car
NGC 2434	Gx	07 34.9	−69 17	11.5	Vol
NGC 2442	Gx	07 36.4	−69 32	11.4p	Vol
NGC 2516	OC	07 58.3	−60 52	3.8	Car
NGC 7098	Gx	21 44.3	−75 07	11.4	Oct
NGC 7329	Gx	22 40.4	−66 29	11.8	Tuc
NGC 7796	Gx	23 59.0	−55 27	11.5	Phe
Small Magellanic Cloud	Gx	00 52.7	−72 50	2.3	Tuc
Tarantula Neb.	BNe	05 38.6	−69 05	8.3	Dor

Chart 25

30 Doradus (see Tarantula Nebula).					
Bas 18	OC	13 28.3	−62 22	8.2	Cen
Be 145	DN	14 48.6	−65 15		Cir
Blue Planetary Nebula (see NGC 3918).					
C77 (see NGC 5128).					
C80 (see Omega Centauri).					
C83 (see NGC 4945).					
C84 (see NGC 5286).					
C85 (see IC 2391).					
C88 (see NGC 5823).					
C89 (see NGC 6087).					
C90 (see NGC 2867).					
C91 (see NGC 3532).					
C92 (see Eta Carinae Nebula).					
C94 (see Jewel Box Cluster).					
C95 (see NGC 6025).					
C96 (see NGC 2516).					
C97 (see NGC 3766).					
C98 (see NGC 4609).					
C99 (see Coal Sack).					
C100 (see IC 2944, 48).					
C102 (see Southern Pleiades).					

Object	Type	R.A. h m	Dec. ° ′	Mag.	Const.
C103 (see Tarantula Nebula).					
C105 (see NGC 4833).					
C108 (see NGC 4372).					
C109 (see NGC 3195).					
Ced 122	BNe	13 25.4	−64 01		Cen
Centaurus A (see NGC 5128).					
Centaurus Cluster (see IC 2944, 48).					
Circinus Galaxy (see E 97-13).					
Coal Sack	DN	12 53.0	−63 00		Cru
Cr 228	OC	10 43.0	−60 01	4.4	Car
Cr 236	OC	10 57.0	−61 02	7.7p	Car
Cr 240	OC	11 11.2	−60 17	3.9	Car
Cr 272	OC	13 30.6	−61 16	7.7	Cen
Cr 292	OC	15 50.7	−57 40	7.9p	Nor
Cr 299	OC	16 18.4	−55 07	6.9p	Nor
E 15-8	Gx	04 07.2	−82 17	12.8p	Men
E 18-2	Gx	08 19.2	−78 42	12.9p	Cha
E 19-3	Gx	10 38.0	−81 06	13.0p	Cha
E 37-10	Gx	10 04.3	−75 29	13.0p	Car
E 60-19	Gx	08 57.5	−69 04	12.7p	Vol
E 91-3	Gx	09 13.5	−63 38	12.9p	Car
E 97-13	Gx	14 13.2	−65 20	10.1	Cir
E 137-10	Gx	16 15.8	−60 48	12.4p	TrA
E 137-12	Gx	16 16.3	−61 18	13.0p	TrA
E 219-21	Gx	13 02.3	−50 20	12.7p	Cen
E 219-41	Gx	13 14.0	−49 29	12.9p	Cen
E 221-6	Gx	13 50.4	−48 23	13.0p	Cen
E 221-10	Gx	13 51.0	−49 03	13.0p	Cen
E 221-26	Gx	14 08.4	−47 58	12.0p	Cen
E 221-32	Gx	14 12.2	−49 23	12.7p	Cen
E 221-34A	Gx	14 16.1	−48 08	12.6p	Cen
E 269-57	Gx	13 10.1	−46 26	12.5p	Cen
E 269-85	Gx	13 20.0	−47 17	12.7p	Cen
E 270-17	Gx	13 34.7	−45 32	11.8p	Cen
E 271-10	Gx	14 00.8	−45 25	12.1	Cen
E 273-14	Gx	14 58.4	−47 42	12.9p	Lup
E 323-34	Gx	12 53.4	−41 12	12.9p	Cen
E 324-24	Gx	13 27.6	−41 29	12.9p	Cen
Eta Carinae Nebula	BNe	10 43.8	−59 52		Car
Gum 32	BNe	10 46.3	−58 39		Car
Gum 39	BNe	11 28.9	−62 41		Cen
Gum 41	BNer	11 30.4	−63 50		Cen
H5	OC	12 27.3	−60 46	7.1	Cru
Ho 16	OC	13 29.3	−61 12	8.4	Cen
Ho 17	OC	14 33.7	−61 23	8.3	Cen
Ho 18	OC	14 50.7	−52 15	8	Lup
IC 2051	Gx	03 52.0	−83 50	12.1p	Men
IC 2391	OC	08 40.2	−53 04	2.5	Vel
IC 2395	OC	08 41.1	−48 12	4.6	Vel

Object	Type	R.A. h m	Dec. ° ′	Mag.	Const.
IC 2448	PN	09 07.1	−69 57	10.4	Car
IC 2488	OC	09 27.6	−56 59	7.4p	Vel
IC 2501	PN	09 38.8	−60 06	10.4	Car
IC 2553	PN	10 09.4	−62 37	10.3	Car
IC 2554	Gx	10 08.9	−67 02	12.5p	Car
IC 2581	OC	10 27.4	−57 38	4.3	Car
IC 2602 (see Southern Pleiades).					
IC 2714	OC	11 17.5	−62 46	8.2p	Car
IC 2872	BNe	11 29.0	−62 57		Cen
IC 2944, 48	BNe+C	11 36.6	−63 02		Cen
IC 2966	BNr	11 50.4	−64 54		Mus
IC 3896	Gx	12 56.7	−50 21	11.5	Cen
IC 3896A	Gx	12 55.5	−50 04	13.0p	Cen
IC 4191	PN	13 08.8	−67 39	10.6	Mus
IC 4402	Gx	14 21.2	−46 18	12.1p	Lup
IC 4499	GC	15 00.3	−82 13	10.1	Aps
IC 4585	Gx	16 00.3	−66 19	13.0p	TrA
Jewel Box Cluster	OC	12 53.6	−60 20	4.2	Cru
Keyhole Nebula (see Eta Carinae Nebula).					
Large Magellanic Cloud	Gx	05 23.6	−69 45	0.1	Dor
Ly 2	OC	14 24.5	−61 20	6.4	Cen
Mel 101	OC	10 42.1	−65 06	8.0:	Car
Mel 105	OC	11 19.5	−63 30	8.5	Car
NGC 2018	BNe	05 30.6	−71 04		Men
NGC 2048	BNe	05 35.2	−69 46		Dor
NGC 2070 (see Tarantula Nebula).					
NGC 2074	BNe	05 39.6	−69 27		Dor
NGC 2077-80	BNe	05 40.4	−69 38		Dor
NGC 2081	BNe	05 39.6	−69 27		Dor
NGC 2187A	Gx	06 03.7	−69 36	12.9p	Dor
NGC 2210	GC	06 11.5	−69 08	10.2	Dor
NGC 2305	Gx	06 48.6	−64 16	12.7p	Vol
NGC 2307	Gx	06 48.9	−64 20	13.0p	Vol
NGC 2397	Gx	07 21.3	−69 00	12	Vol
NGC 2417	Gx	07 30.2	−62 15	12.4	Car
NGC 2434	Gx	07 34.9	−69 17	11.5	Vol
NGC 2442	Gx	07 36.4	−69 32	11.4p	Vol
NGC 2516	OC	07 58.3	−60 52	3.8	Car
NGC 2547	OC	08 10.7	−49 16	4.7	Vel
NGC 2640	Gx	08 37.4	−55 07	12.7p	Car
NGC 2669	OC	08 46.3	−52 55	6.1	Vel
NGC 2670	OC	08 45.5	−48 47	7.8	Vel
NGC 2808	GC	09 12.1	−64 52	6.2	Car
NGC 2822	Gx	09 13.8	−69 39	10.7	Car
NGC 2836	Gx	09 13.7	−69 20	11.8	Car
NGC 2867	PN	09 21.4	−58 19	9.7	Car
NGC 2887	Gx	09 23.4	−63 49	12.8p	Car
NGC 2910	OC	09 30.4	−52 54	7.2	Vel
NGC 2925	OC	09 33.7	−53 26	8.3p	Vel

Chart 25

Object	Type	R.A. h m	Dec. ° ′	Mag.	Const.
NGC 3059	Gx	09 50.1	−73 55	10.8	Car
NGC 3114	OC	10 02.7	−60 07	4.2	Car
NGC 3136	Gx	10 05.8	−67 23	10.6	Car
NGC 3136B	Gx	10 10.2	−67 00	12.3	Car
NGC 3149	Gx	10 03.8	−80 25	12.6p	Cha
NGC 3195	PN	10 09.5	−80 52	11.6	Cha
NGC 3199	BNe	10 17.1	−57 55		Car
NGC 3211	PN	10 17.8	−62 40	10.7	Car
NGC 3228	OC	10 21.8	−51 43	6	Vel
NGC 3247	C+BNe	10 25.9	−57 56	7.6	Car
NGC 3293	C+BNer	10 35.8	−58 14	4.7	Car
NGC 3324	BNer+C	10 37.3	−58 38		Car
NGC 3330	OC	10 38.6	−54 09	7.4	Vel
NGC 3372 (see Eta Carinae Nebula).					
NGC 3496	OC	10 59.8	−60 20	8.2	Car
NGC 3503	BNer	11 01.3	−59 51		Car
NGC 3532	OC	11 06.4	−58 40	3	Car
NGC 3572	OC	11 10.4	−60 14	6.6	Car
NGC 3576, 79	BNe	11 12.0	−61 12		Car
NGC 3590	OC	11 12.9	−60 47	8.2	Car
NGC 3603	BN	11 15.0	−61 12		Car
NGC 3699	PN	11 28.0	−59 57	11.3	Cen
NGC 3766	OC	11 36.1	−61 37	5.3	Cen
NGC 3882	Gx	11 46.1	−56 23	12.5	Cen
NGC 3918	PN	11 50.3	−57 11	8.1	Cen
NGC 3960	OC	11 50.9	−55 42	8.3	Cen
NGC 4103	OC	12 06.7	−61 15	7.4p	Cru
NGC 4219	Gx	12 16.5	−43 19	11.9	Cen
NGC 4349	OC	12 24.5	−61 54	7.4	Cru
NGC 4372	GC	12 25.8	−72 40	7.2	Mus
NGC 4439	OC	12 28.4	−60 06	8.4	Cru
NGC 4444	Gx	12 28.6	−43 16	12.3	Cen
NGC 4463	OC	12 30.0	−64 48	7.2	Mus
NGC 4507	Gx	12 35.6	−39 55	12.1	Cen
NGC 4609	OC	12 42.3	−62 58	6.9	Cru
NGC 4645	Gx	12 44.2	−41 45	11.8	Cen
NGC 4650	Gx	12 44.3	−40 44	11.8	Cen
NGC 4696	Gx	12 48.8	−41 19	10.2	Cen
NGC 4709	Gx	12 50.1	−41 23	11.5	Cen
NGC 4751	Gx	12 52.9	−42 40	12	Cen
NGC 4755 (see Jewel Box Cluster).					
NGC 4767	Gx	12 53.9	−39 43	11.6	Cen
NGC 4815	OC	12 58.0	−64 57	8.6	Mus
NGC 4833	GC	12 59.6	−70 53	8.4	Mus
NGC 4835	Gx	12 58.1	−46 16	11.5	Cen
NGC 4930	Gx	13 04.1	−41 25	11.7	Cen
NGC 4945	Gx	13 05.4	−49 28	8.8	Cen
NGC 4947	Gx	13 05.3	−35 20	11.9	Cen
NGC 4976	Gx	13 08.6	−49 30	10.1	Cen
NGC 5011	Gx	13 12.9	−43 06	11.1	Cen
NGC 5026	Gx	13 14.2	−42 58	11.5	Cen
NGC 5064	Gx	13 19.0	−47 55	12	Cen
NGC 5090	Gx	13 21.2	−43 42	11.6	Cen
NGC 5102	Gx	13 22.0	−36 38	8.8	Cen
NGC 5121	Gx	13 24.8	−37 41	11.6	Cen
NGC 5128	Gx	13 25.5	−43 01	6.7	Cen
NGC 5138	OC	13 27.3	−59 01	7.6	Cen
NGC 5139 (see Omega Centauri).					
NGC 5156	Gx	13 28.7	−48 55	11.7	Cen
NGC 5189	PN	13 33.6	−65 59	9.9	Mus
NGC 5206	Gx	13 33.7	−48 09	10.6	Cen
NGC 5266	Gx	13 43.0	−48 10	10.9	Cen
NGC 5266A	Gx	13 40.6	−48 21	12.1	Cen
NGC 5281	OC	13 46.6	−62 54	5.9	Cen
NGC 5286	GC	13 46.5	−51 22	7.4	Cen
NGC 5307	PN	13 51.1	−51 12	11.2	Cen
NGC 5315	PN	13 54.0	−66 31	9.8	Cir
NGC 5316	OC	13 53.9	−61 52	6	Cen
NGC 5333	Gx	13 54.4	−48 31	12.8p	Cen
NGC 5365	Gx	13 57.8	−43 56	11.2	Cen
NGC 5460	OC	14 07.6	−48 19	5.6	Cen
NGC 5516	Gx	14 15.9	−48 07	12	Cen
NGC 5606	OC	14 27.8	−59 38	7.7	Cen
NGC 5612	Gx	14 34.0	−78 23	12.1	Aps
NGC 5617	OC	14 29.8	−60 43	6.3	Cen
NGC 5662	OC	14 35.2	−56 33	5.5	Cen
NGC 5670	Gx	14 35.6	−45 58	13.0p	Lup
NGC 5822	OC	15 05.2	−54 21	6.5p	Lup
NGC 5823	OC	15 05.7	−55 36	7.9	Cir
NGC 5833	Gx	15 11.9	−72 52	12	Aps
NGC 5925	OC	15 27.7	−54 31	8.4p	Nor
NGC 5927	GC	15 28.0	−50 40	8	Lup
NGC 5946	GC	15 35.5	−50 40	8.4	Nor
NGC 5967	Gx	15 48.3	−75 40	12	Aps
NGC 5979	PN	15 47.7	−61 13	11.5	TrA
NGC 6025	OC	16 03.7	−60 30	5.1	TrA
NGC 6031	OC	16 07.6	−54 04	8.5	Nor
NGC 6067	OC	16 13.2	−54 13	5.6	Nor
NGC 6087	OC	16 18.9	−57 54	5.4	Nor
Omega Centauri	GC	13 26.8	−47 29	3.9	Cen
Omicron Velorum Cluster (see IC 2391).					
Pi 16	OC	09 51.1	−53 11	8	Vel
Pi 20	OC	15 15.4	−59 04	7.8	Cir
PK264-8.1	PN	08 11.5	−48 43	12.4	Vel
PK307-4.1	PN	13 39.6	−67 23	12.9	Mus
PK322-2.1	PN	15 34.3	−59 09	12.1	Nor
PK329+2.1	PN	15 51.7	−51 31	12.6	Nor
PK329-2.2	PN	16 14.5	−54 57	11.9	Nor
R47	BNe	10 05.0	−58 55		Car
R50	BNe	10 27.0	−57 10		Car

Object	Type	R.A. h m	Dec. ° ′	Mag.	Const.
R58	BNe	11 06.0	−65 35		Car
Ru 82	OC	09 45.6	−53 59	8.1	Vel
Ru 92	OC	10 53.9	−61 44	8.6	Car
Ru 93	OC	11 04.4	−61 22	7.7	Car
Ru 98	OC	11 58.0	−64 29	7	Cru
Ru 106	GC	12 38.7	−51 09	10.9:	Cen
Ru 108	OC	13 32.2	−58 29	7.5	Cen
Running Chicken Nebula (see IC 2944, 48).					
Southern Pleiades	OC	10 43.2	−64 24	1.9	Car
St 13	OC	11 13.1	−58 55	7	Car
St 14	OC	11 44.0	−62 30	6.3	Cen
Tarantula Neb.	BNe	05 38.6	−69 05	8.3	Dor
Tr 15	OC	10 44.8	−59 22	7	Car
Tr 17	OC	10 56.2	−59 13	8.4	Car
Tr 18	OC	11 11.4	−60 40	6.9	Car
Tr 21	OC	13 32.2	−62 47	7.7	Cen
Tr 22	OC	14 31.2	−61 10	7.9	Cen
vdBH 63	BNr	14 49.4	−65 14		Cir
vdBH 65a	BNr	15 01.1	−63 17		Cir

Chart 26

Object	Type	R.A. h m	Dec. ° ′	Mag.	Const.
Be 145	DN	14 48.6	−65 15		Cir
C82 (see NGC 6193).					
C86 (see NGC 6397).					
C89 (see NGC 6087).					
C93 (see NGC 6752).					
C95 (see NGC 6025).					
C101 (see NGC 6744).					
C105 (see NGC 4833).					
C107 (see NGC 6101).					
C108 (see NGC 4372).					
Cr 292	OC	15 50.7	−57 40	7.9p	Nor
Cr 299	OC	16 18.4	−55 07	6.9p	Nor
E 27-1	Gx	21 52.5	−81 32	12.1p	Oct
E 27-8	Gx	22 23.0	−80 00	12.8p	Oct
E 69-14	Gx	16 52.3	−69 08	12.4p	TrA
E 101-14	Gx	16 48.7	−62 36	13.0p	Ara
E 107-4	Gx	21 03.5	−67 11	12.9p	Pav
E 137-10	Gx	16 15.8	−60 48	12.4p	TrA
E 137-12	Gx	16 16.3	−61 18	13.0p	TrA
E 137-18	Gx	16 21.0	−60 29	12.1p	TrA
E 137-34	Gx	16 35.2	−58 05	12.1p	Nor
E 137-38	Gx	16 40.9	−60 24	12.6p	Ara
E 138-5	Gx	16 53.9	−58 47	12.8p	Ara
E 138-10	Gx	16 59.1	−60 13	11.6p	Ara
E 138-29	Gx	17 29.2	−62 27	12.7p	Ara
E 183-30	Gx	18 56.9	−54 33	11.8	Tel
E 185-54	Gx	20 03.5	−55 57	12.0p	Tel
E 186-62	Gx	20 34.0	−52 59	12.9p	Ind
E 235-55	Gx	21 05.9	−48 12	12.7p	Ind
E 342-27	Gx	21 16.9	−42 16	12.9p	Mic
E 342-50	Gx	21 28.3	−37 52	12.9p	Gru
F 591	Gx	21 48.3	−54 59	12.8p	Ind
IC 4499	GC	15 00.3	−82 13	10.1	Aps
IC 4585	Gx	16 00.3	−66 19	13.0p	TrA
IC 4595	Gx	16 20.7	−70 09	12.7p	TrA
IC 4618	Gx	16 57.9	−77 00	12.7p	Aps
IC 4633	Gx	17 13.8	−77 32	12.4p	Aps
IC 4642	PN	17 11.8	−55 24	12.4	Ara
IC 4646	Gx	17 23.9	−60 00	12.6p	Ara
IC 4651	OC	17 25.0	−49 57	6.9	Ara
IC 4654	Gx	17 37.1	−74 23	13.0p	Aps
IC 4662	Gx	17 47.1	−64 38	11.3	Pav
IC 4679	Gx	18 11.4	−56 15	12.8p	Tel
IC 4704	Gx	18 27.9	−71 37	13.0p	Pav
IC 4710	Gx	18 28.6	−66 59	12	Pav
IC 4712	Gx	18 31.1	−71 42	13.0p	Pav
IC 4721	Gx	18 34.4	−58 30	11.9	Pav
IC 4742	Gx	18 41.9	−63 52	13.0p	Pav
IC 4765	Gx	18 47.3	−63 20	12.4p	Pav
IC 4785	Gx	18 52.9	−59 15	13.0p	Pav
IC 4797	Gx	18 56.5	−54 18	11.3	Tel
IC 4831	Gx	19 14.7	−62 16	12.7p	Pav
IC 4837	Gx	19 15.2	−54 40	12.4	Tel
IC 4837A	Gx	19 15.3	−54 08	12.5p	Tel
IC 4845	Gx	19 20.4	−60 23	12.4p	Pav
IC 4889	Gx	19 45.3	−54 21	11.3	Tel
IC 4901	Gx	19 54.4	−58 43	12.1p	Pav
IC 4933	Gx	20 03.5	−54 59	13.0p	Tel
IC 4946	Gx	20 24.0	−44 00	12.6p	Sgr
IC 5052	Gx	20 52.1	−69 12	11	Pav
IC 5063	Gx	20 52.0	−57 04	11.9	Ind
IC 5092	Gx	21 16.2	−64 28	13.0p	Pav
IC 5105	Gx	21 24.4	−40 32	11.4	Mic
IC 5152	Gx	22 02.7	−51 18	10.6	Ind
IC 5181	Gx	22 13.4	−46 01	11.4	Gru
IC 5201	Gx	22 21.0	−46 02	11.1	Gru
IC 5250	Gx	22 47.3	−65 03	12.1p	Tuc
IC 5250A	Gx	22 47.4	−65 03	12.2p	Tuc
Mel 227	OC	20 18.0	−79 09	5.3p	Oct
NGC 4372	GC	12 25.8	−72 40	7.2	Mus
NGC 4833	GC	12 59.6	−70 53	8.4	Mus
NGC 5612	Gx	14 34.0	−78 23	12.1	Aps
NGC 5833	Gx	15 11.9	−72 52	12	Aps
NGC 5967	Gx	15 48.3	−75 40	12	Aps
NGC 6025	OC	16 03.7	−60 30	5.1	TrA
NGC 6031	OC	16 07.6	−54 04	8.5	Nor
NGC 6067	OC	16 13.2	−54 13	5.6	Nor
NGC 6087	OC	16 18.9	−57 54	5.4	Nor
NGC 6101	GC	16 25.8	−72 12	9.2	Aps

Charts 26 – A2

Object	Type	R.A. h m	Dec. ° ′	Mag.	Const.
NGC 6134	OC	16 27.7	−49 09	7.2	Nor
NGC 6152	OC	16 32.7	−52 37	8.1p	Nor
NGC 6156	Gx	16 34.9	−60 37	11.5	TrA
NGC 6167	OC	16 34.4	−49 36	6.7	Nor
NGC 6188	BNer	16 40.5	−48 47		Ara
NGC 6193	OC	16 41.3	−48 46	5.2	Ara
NGC 6208	OC	16 49.5	−53 49	7.2	Ara
NGC 6215	Gx	16 51.1	−59 00	10.9	Ara
NGC 6221	Gx	16 52.8	−59 13	10.1	Ara
NGC 6300	Gx	17 17.0	−62 49	10.1	Ara
NGC 6326	PN	17 20.8	−51 45	11	Ara
NGC 6362	GC	17 31.9	−67 03	8.1	Ara
NGC 6392	Gx	17 43.5	−69 47	12.5p	Aps
NGC 6397	GC	17 40.7	−53 40	5.3	Ara
NGC 6438	Gx	18 22.3	−85 24	12.4	Oct
NGC 6438A	Gx	18 22.7	−85 24	12.1	Oct
NGC 6483	Gx	17 59.5	−63 40	12.2	Pav
NGC 6492	Gx	18 02.8	−66 26	12.3p	Pav
NGC 6584	GC	18 18.6	−52 13	7.9	Tel
NGC 6673	Gx	18 45.1	−62 18	11.7	Pav
NGC 6684	Gx	18 49.0	−65 10	10.4	Pav
NGC 6699	Gx	18 52.0	−57 19	12.1	Pav
NGC 6730	Gx	19 07.6	−68 55	13.0p	Pav
NGC 6744	Gx	19 09.8	−63 51	8.6	Pav
NGC 6752	GC	19 10.9	−59 59	5.3	Pav
NGC 6753	Gx	19 11.4	−57 03	11.1	Pav
NGC 6754	Gx	19 11.4	−50 39	12.1	Tel
NGC 6758	Gx	19 13.9	−56 19	11.5	Tel
NGC 6769	Gx	19 18.4	−60 30	11.5	Pav
NGC 6770	Gx	19 18.6	−60 30	11.6	Pav
NGC 6776	Gx	19 25.3	−63 52	12.2	Pav
NGC 6782	Gx	19 24.0	−59 55	11.8	Pav
NGC 6788	Gx	19 26.8	−54 57	12.8p	Tel
NGC 6810	Gx	19 43.6	−58 39	11.4	Pav
NGC 6851	Gx	20 03.6	−48 17	11.8	Tel
NGC 6861	Gx	20 07.3	−48 22	11.1	Tel
NGC 6868	Gx	20 09.9	−48 23	10.6	Tel
NGC 6872	Gx	20 16.9	−70 46	11.2	Pav
NGC 6875	Gx	20 13.2	−46 10	11.9	Tel
NGC 6876	Gx	20 18.3	−70 51	10.8	Pav
NGC 6887	Gx	20 17.3	−52 48	12	Tel
NGC 6890	Gx	20 18.3	−44 48	12.1	Sgr
NGC 6893	Gx	20 20.8	−48 14	11.7	Tel
NGC 6902	Gx	20 24.5	−43 39	11	Sgr
NGC 6909	Gx	20 27.6	−47 02	11.8	Tel
NGC 6935	Gx	20 38.3	−52 07	11.9	Ind
NGC 6942	Gx	20 40.6	−54 18	11.9	Ind
NGC 6943	Gx	20 44.6	−68 45	11.2	Pav
NGC 6958	Gx	20 48.7	−38 00	11.3	Mic
NGC 7007	Gx	21 05.5	−52 33	11.9	Ind
NGC 7020	Gx	21 11.3	−64 02	11.8	Pav
NGC 7029	Gx	21 11.9	−49 17	11.5	Ind
NGC 7038	Gx	21 15.1	−47 13	12	Ind
NGC 7041	Gx	21 16.5	−48 22	11.2	Ind
NGC 7049	Gx	21 19.0	−48 34	10.3	Ind
NGC 7059	Gx	21 27.4	−60 01	11.9	Pav
NGC 7064	Gx	21 29.0	−52 46	12.2	Ind
NGC 7070	Gx	21 30.4	−43 05	12	Gru
NGC 7079	Gx	21 32.6	−44 04	11.6	Gru
NGC 7083	Gx	21 35.8	−63 54	11.2	Ind
NGC 7090	Gx	21 36.5	−54 33	10.7	Ind
NGC 7096	Gx	21 41.3	−63 55	12	Ind
NGC 7097	Gx	21 40.2	−42 32	11.6	Gru
NGC 7098	Gx	21 44.3	−75 07	11.4	Oct
NGC 7125	Gx	21 49.3	−60 43	12	Ind
NGC 7126	Gx	21 49.3	−60 37	12.2	Ind
NGC 7141	Gx	21 52.3	−55 34	11.9	Ind
NGC 7144	Gx	21 52.7	−48 15	10.9	Gru
NGC 7145	Gx	21 53.3	−47 53	11.1	Gru
NGC 7166	Gx	22 00.5	−43 23	11.6	Gru
NGC 7168	Gx	22 02.1	−51 45	11.9	Ind
NGC 7192	Gx	22 06.8	−64 19	11.2	Ind
NGC 7196	Gx	22 05.9	−50 07	11.4	Ind
NGC 7205	Gx	22 08.6	−57 27	11.1	Tuc
NGC 7213	Gx	22 09.3	−47 10	10	Gru
NGC 7232	Gx	22 15.6	−45 51	11.6	Gru
NGC 7329	Gx	22 40.4	−66 29	11.8	Tuc
NGC 7689	Gx	23 33.3	−54 06	11.4	Phe
NGC 7690	Gx	23 33.0	−51 42	12.1	Phe
NGC 7796	Gx	23 59.0	−55 27	11.5	Phe
PK322-2.1	PN	15 34.3	−59 09	12.1	Nor
PK329-2.2	PN	16 14.5	−54 57	11.9	Nor
PK342-14.1	PN	18 07.3	−51 02	11.9p	Ara
vdBH 63	BNr	14 49.4	−65 14		Cir
vdBH 65a	BNr	15 01.1	−63 17		Cir
vdBH 81	BNr	17 04.0	−51 05		Ara

Chart A1

Object	Type	R.A. h m	Dec. ° ′	Mag.	Const.
Cr 350	OC	17 48.1	+01 18	6.1p	Oph
IC 4665	OC	17 46.3	+05 43	4.2	Oph
NGC 6426	GC	17 44.9	+03 10	10.9	Oph

Chart A2

Object	Type	R.A. h m	Dec. ° ′	Mag.	Const.
C5 (see IC 349).					
IC 349	BN	03 46.3	+23 56		Tau
IC 353	BNr?	03 55.0	+25 29		Tau
IC 1995	BN	03 50.3	+25 35		Tau
M45 (see Pleiades).					
Merope Nebula (see IC 349).					
Pleiades	C+BNr	03 47.0	+24 07	1.2	Tau
Seven Sisters (see Pleiades).					

	Object	Type	R.A. h m	Dec. ° ′	Mag.	Const.	Object	Type	R.A. h m	Dec. ° ′	Mag.	Const.
Chart A3	Ho 17	OC	14 33.7	−61 23	8.3	Cen	NGC 4206	Gx	12 15.3	+13 02	12.2	Vir
	Ly 2	OC	14 24.5	−61 20	6.4	Cen	NGC 4212	Gx	12 15.7	+13 54	11.2	Com
	NGC 5316	OC	13 53.9	−61 52	6	Cen	NGC 4215	Gx	12 15.9	+06 24	12.1	Vir
	NGC 5606	OC	14 27.8	−59 38	7.7	Cen	NGC 4216	Gx	12 15.9	+13 09	10	Vir
	NGC 5617	OC	14 29.8	−60 43	6.3	Cen	NGC 4224	Gx	12 16.6	+07 28	11.8	Vir
	Tr 22	OC	14 31.2	−61 10	7.9	Cen	NGC 4233	Gx	12 17.1	+07 37	11.9	Vir
							NGC 4235	Gx	12 17.1	+07 12	11.6	Vir
Chart A4	C1 (see NGC 188).						NGC 4237	Gx	12 17.2	+15 19	11.6	Com
	NGC 188	OC	00 44.4	+85 20	8.1	Cep	NGC 4241	Gx	12 17.4	+06 41	11.9	Vir
	NGC 2268	Gx	07 14.3	+84 23	11.5	Cam	NGC 4254 (see M99).					
	NGC 2276	Gx	07 27.2	+85 45	11.4	Cep	NGC 4260	Gx	12 19.4	+06 06	11.8	Vir
	NGC 2300	Gx	07 32.3	+85 43	11	Cep	NGC 4261	Gx	12 19.4	+05 50	10.4	Vir
							NGC 4262	Gx	12 19.5	+14 53	11.6	Com
Chart A5	IC 2051	Gx	03 52.0	−83 50	12.1p	Men	NGC 4267	Gx	12 19.8	+12 48	10.9	Vir
	NGC 6438	Gx	18 22.3	−85 24	12.4	Oct	NGC 4270	Gx	12 19.8	+05 28	12.2	Vir
	NGC 6438A	Gx	18 22.7	−85 24	12.1	Oct	NGC 4273	Gx	12 19.9	+05 21	11.9	Vir
							NGC 4281	Gx	12 20.4	+05 23	11.3	Vir
Chart B1	Eyes (see NGC 4435 and NGC 4438).						NGC 4293	Gx	12 21.2	+18 23	10.4	Com
	III Zw 66	Gx	12 27.2	+14 07	12.3p	Com	NGC 4294	Gx	12 21.3	+11 31	12.1	Vir
	Lost Galaxy (see NGC 4526).						NGC 4298	Gx	12 21.5	+14 36	11.3	Com
	M49	Gx	12 29.8	+08 00	8.4	Vir	NGC 4299	Gx	12 21.7	+11 30	12.5	Vir
	M58	Gx	12 37.7	+11 49	9.8	Vir	NGC 4302	Gx	12 21.7	+14 36	11.6	Com
	M59	Gx	12 42.0	+11 39	9.8	Vir	NGC 4303 (see M61).					
	M60	Gx	12 43.7	+11 33	8.8	Vir	NGC 4307	Gx	12 22.1	+09 03	12	Vir
	M61	Gx	12 21.9	+04 28	9.7	Vir	NGC 4312	Gx	12 22.5	+15 32	11.7	Com
	M84	Gx	12 25.1	+12 53	9.3	Vir	NGC 4313	Gx	12 22.6	+11 48	11.6	Vir
	M85	Gx	12 25.4	+18 11	9.2	Com	NGC 4321 (see M100).					
	M86	Gx	12 26.2	+12 57	9.2	Vir	NGC 4324	Gx	12 23.1	+05 15	11.6	Vir
	M87	Gx	12 30.8	+12 24	8.6	Vir	NGC 4339	Gx	12 23.6	+06 05	11.3	Vir
	M88	Gx	12 32.0	+14 25	9.5	Com	NGC 4340	Gx	12 23.6	+16 43	11.2	Com
	M89	Gx	12 35.7	+12 33	9.8	Vir	NGC 4343	Gx	12 23.6	+06 57	12.1	Vir
	M90	Gx	12 36.8	+13 10	9.5	Vir	NGC 4350	Gx	12 24.0	+16 42	11	Com
	M91	Gx	12 35.4	+14 30	10.2	Com	NGC 4351	Gx	12 24.0	+12 12	12.6	Vir
	M98	Gx	12 13.8	+14 54	10.1	Com	NGC 4365	Gx	12 24.5	+07 19	9.6	Vir
	M99	Gx	12 18.8	+14 25	9.8	Com	NGC 4371	Gx	12 24.9	+11 42	10.8	Vir
	M100	Gx	12 22.9	+15 49	9.4	Com	NGC 4374 (see M84).					
	NGC 4032	Gx	12 00.6	+20 05	12.2	Com	NGC 4377	Gx	12 25.2	+14 46	11.9	Com
	NGC 4037	Gx	12 01.4	+13 24	11.9	Com	NGC 4378	Gx	12 25.3	+04 56	11.7	Vir
	NGC 4064	Gx	12 04.2	+18 27	11.4	Com	NGC 4379	Gx	12 25.2	+15 36	11.7	Com
	NGC 4067	Gx	12 04.2	+10 51	12.5	Vir	NGC 4380	Gx	12 25.4	+10 01	11.7	Vir
	NGC 4123	Gx	12 08.2	+02 53	11.4	Vir	NGC 4382 (see M85).					
	NGC 4124	Gx	12 08.2	+10 23	11.3	Vir	NGC 4383	Gx	12 25.4	+16 28	12.1	Com
	NGC 4147	GC	12 10.1	+18 33	10.4	Com	NGC 4387	Gx	12 25.7	+12 49	12.1	Vir
	NGC 4152	Gx	12 10.6	+16 02	12.2	Com	NGC 4388	Gx	12 25.8	+12 40	11	Vir
	NGC 4158	Gx	12 11.2	+20 11	12.1	Com	NGC 4394	Gx	12 25.9	+18 13	10.9	Com
	NGC 4168	Gx	12 12.3	+13 12	11.2	Vir	NGC 4402	Gx	12 26.1	+13 07	11.8	Vir
	NGC 4178	Gx	12 12.8	+10 52	11.4	Vir	NGC 4405	Gx	12 26.1	+16 11	12	Com
	NGC 4189	Gx	12 13.8	+13 26	11.7	Com	NGC 4406 (see M86).					
	NGC 4192 (see M98).						NGC 4411B	Gx	12 26.8	+08 53	12.3	Vir
	NGC 4193	Gx	12 13.9	+13 10	12.3	Vir	NGC 4413	Gx	12 26.5	+12 37	12.2	Vir

Object	Type	R.A. h m	Dec. ° '	Mag.	Const.
NGC 4417	Gx	12 26.8	+09 35	11.1	Vir
NGC 4419	Gx	12 26.9	+15 03	11.2	Com
NGC 4421	Gx	12 27.0	+15 28	11.6	Com
NGC 4424	Gx	12 27.2	+09 25	11.7	Vir
NGC 4425	Gx	12 27.2	+12 44	11.8	Vir
NGC 4429	Gx	12 27.4	+11 06	10	Vir
NGC 4430	Gx	12 27.4	+06 16	12	Vir
NGC 4434	Gx	12 27.6	+08 09	12.2	Vir
NGC 4435	Gx	12 27.7	+13 05	10.8	Vir
NGC 4438	Gx	12 27.8	+13 01	10.2	Vir
NGC 4440	Gx	12 27.9	+12 18	11.7	Vir
NGC 4442	Gx	12 28.1	+09 48	10.4	Vir
NGC 4450	Gx	12 28.5	+17 05	10.1	Com
NGC 4452	Gx	12 28.7	+11 45	12	Vir
NGC 4457	Gx	12 29.0	+03 34	10.9	Vir
NGC 4458	Gx	12 29.0	+13 15	12.1	Vir
NGC 4459	Gx	12 29.0	+13 59	10.4	Com
NGC 4461	Gx	12 29.0	+13 11	11.2	Vir
NGC 4469	Gx	12 29.5	+08 45	11.2	Vir
NGC 4470	Gx	12 29.6	+07 49	12.1	Vir
NGC 4472 (see M49).					
NGC 4473	Gx	12 29.8	+13 26	10.2	Com
NGC 4474	Gx	12 29.9	+14 04	11.5	Com
NGC 4477	Gx	12 30.0	+13 38	10.4	Com
NGC 4478	Gx	12 30.3	+12 20	11.4	Vir
NGC 4480	Gx	12 30.4	+04 15	12.4	Vir
NGC 4486 (see M87).					
NGC 4488	Gx	12 30.9	+08 22	12.2	Vir
NGC 4489	Gx	12 30.9	+16 46	12	Com
NGC 4496A	Gx	12 31.7	+03 56	11.4	Vir
NGC 4496B	Gx	12 31.7	+03 56	13.5	Vir
NGC 4498	Gx	12 31.7	+16 51	12.2	Com
NGC 4501 (see M88).					
NGC 4503	Gx	12 32.1	+11 11	11.1	Vir
NGC 4519	Gx	12 33.5	+08 39	11.8	Vir
NGC 4522	Gx	12 33.7	+09 10	12.3	Vir
NGC 4526	Gx	12 34.0	+07 42	9.7	Vir
NGC 4528	Gx	12 34.1	+11 19	12.1	Vir
NGC 4531	Gx	12 34.3	+13 05	11.4	Vir
NGC 4532	Gx	12 34.3	+06 28	11.9	Vir
NGC 4535	Gx	12 34.3	+08 12	10	Vir
NGC 4539	Gx	12 34.6	+18 12	12	Com
NGC 4540	Gx	12 34.8	+15 33	11.7	Com
NGC 4548 (see M91).					
NGC 4550	Gx	12 35.5	+12 13	11.7	Vir
NGC 4551	Gx	12 35.6	+12 16	12	Vir
NGC 4552 (see M89).					
NGC 4561	Gx	12 36.1	+19 19	12.5	Com
NGC 4564	Gx	12 36.5	+11 26	11.1	Vir
NGC 4567	Gx	12 36.5	+11 16	11.3	Vir
NGC 4568	Gx	12 36.6	+11 14	10.8	Vir
NGC 4569 (see M90).					
NGC 4570	Gx	12 36.9	+07 15	10.9	Vir
NGC 4571	Gx	12 36.9	+14 13	11.3	Com
NGC 4578	Gx	12 37.5	+09 33	11.5	Vir
NGC 4579 (see M58).					
NGC 4580	Gx	12 37.8	+05 22	11.8	Vir
NGC 4586	Gx	12 38.5	+04 19	11.7	Vir
NGC 4595	Gx	12 39.9	+15 18	12.1	Com
NGC 4596	Gx	12 39.9	+10 11	10.4	Vir
NGC 4606	Gx	12 41.0	+11 55	11.8	Vir
NGC 4608	Gx	12 41.2	+10 09	11	Vir
NGC 4612	Gx	12 41.5	+07 19	10.9	Vir
NGC 4621 (see M59).					
NGC 4638	Gx	12 42.8	+11 27	11.2	Vir
NGC 4639	Gx	12 42.9	+13 15	11.5	Vir
NGC 4647	Gx	12 43.5	+11 35	11.3	Vir
NGC 4649 (see M60).					
NGC 4651	Gx	12 43.7	+16 24	10.8	Com
NGC 4654	Gx	12 43.9	+13 08	10.6	Vir
NGC 4660	Gx	12 44.5	+11 11	11.2	Vir
NGC 4665	Gx	12 45.1	+03 03	10.5	Vir
NGC 4689	Gx	12 47.8	+13 46	10.9	Com
NGC 4694	Gx	12 48.3	+10 59	11.4	Vir
NGC 4698	Gx	12 48.4	+08 29	10.6	Vir
NGC 4701	Gx	12 49.2	+03 23	12.4	Vir
NGC 4710	Gx	12 49.7	+15 10	11	Com
NGC 4713	Gx	12 50.0	+05 19	11.7	Vir
NGC 4733	Gx	12 51.1	+10 55	11.8	Vir
NGC 4754	Gx	12 52.3	+11 19	10.6	Vir
NGC 4762	Gx	12 52.9	+11 14	10.3	Vir
NGC 4808	Gx	12 55.8	+04 18	11.7	Vir
Siamese Twins (see NGC 4567 and NGC 4568).					
Virgo A (see M87).					

Chart B2

Object	Type	R.A. h m	Dec. ° '	Mag.	Const.
B33 (see Horsehead Nebula).					
Great Orion Nebula	BNer	05 35.4	−05 27	4	Ori
Horsehead Nebula	DN	05 40.9	−02 28		Ori
IC 423	BNr	05 33.4	−00 37		Ori
IC 426	BNr	05 36.8	−00 15		Ori
IC 430	BNr	05 38.5	−07 05		Ori
IC 431	BNr	05 40.3	−01 27		Ori
IC 432	BNr	05 40.9	−01 29		Ori
IC 434	BNe	05 41.0	−02 24		Ori
IC 435	BNr	05 43.0	−02 19		Ori
M42 (see Great Orion Nebula).					
M43	BNer	05 35.6	−05 16	9	Ori
M78	BNr	05 46.7	+00 03	8	Ori

Object	Type	R.A. h m	Dec. ° ′	Mag.	Const.
NGC 1973	BNer	05 35.1	−04 44		Ori
NGC 1975	BNer	05 35.4	−04 41		Ori
NGC 1976 (see Great Orion Nebula).					
NGC 1977	BNer	05 35.5	−04 52		Ori
NGC 1981	OC	05 35.2	−04 26	4.6	Ori
NGC 1982 (see M43).					
NGC 1999	BNer	05 36.5	−06 42		Ori
NGC 2023	BNer	05 41.6	−02 14		Ori

Object	Type	R.A. h m	Dec. ° ′	Mag.	Const.
NGC 2024	BNe	05 41.9	−01 51		Ori
NGC 2064	BNr	05 46.3	00 00		Ori
NGC 2067	BNr	05 46.5	+00 06		Ori
NGC 2068 (see M78).					
NGC 2071	BNr	05 47.2	+00 18		Ori
Orion Nebula (see Great Orion Nebula).					
Tank Trap Nebula (see NGC 2024).					
Trapezium	C+BNer	05 35.3	−05 23	4.7	Ori